T0292639

Studies in Computational Intelligence

Volume 639

Series editor

Janusz Kacprzyk, Polish Academy of Sciences, Warsaw, Poland
e-mail: kacprzyk@ibspan.waw.pl

About this Series

The series "Studies in Computational Intelligence" (SCI) publishes new developments and advances in the various areas of computational intelligence—quickly and with a high quality. The intent is to cover the theory, applications, and design methods of computational intelligence, as embedded in the fields of engineering, computer science, physics and life sciences, as well as the methodologies behind them. The series contains monographs, lecture notes and edited volumes in computational intelligence spanning the areas of neural networks, connectionist systems, genetic algorithms, evolutionary computation, artificial intelligence, cellular automata, self-organizing systems, soft computing, fuzzy systems, and hybrid intelligent systems. Of particular value to both the contributors and the readership are the short publication timeframe and the worldwide distribution, which enable both wide and rapid dissemination of research output.

More information about this series at http://www.springer.com/series/7092

Witold Pedrycz · Shyi-Ming Chen
Editors

Sentiment Analysis and Ontology Engineering

An Environment of Computational Intelligence

 Springer

Editors
Witold Pedrycz
Department of Electrical and Computer
 Engineering
University of Alberta
Edmonton, AB
Canada

Shyi-Ming Chen
Department of Computer Science
 and Information Engineering
National Taiwan University of Science
 and Technology
Taipei
Taiwan

ISSN 1860-949X   ISSN 1860-9503   (electronic)
Studies in Computational Intelligence
ISBN 978-3-319-30317-8   ISBN 978-3-319-30319-2   (eBook)
DOI 10.1007/978-3-319-30319-2

Library of Congress Control Number: 2016932511

This Springer imprint is published by Springer Nature
The registered company is Springer International Publishing AG Switzerland

Preface

Sentiment analysis arises as one of the important approaches to realize social analytics. Sentiment analysis and prediction along with ontology design, analysis, and usage are essential to numerous pursuits in knowledge engineering and a variety application areas embracing business, health sector, Internet, social networking, politics, and economy. We have been witnessing a spectrum of new concepts, methods, and innovative applications. Capturing the sentiments and emotional states present in textual information embraces a wide range of web-oriented activities, such as those detecting the sentiments associated to the product reviews, developing marketing programs enhancing customer services, identifying new opportunities, and presenting financial market prediction. The rapid growth of user-generated data (e.g., consumer reviews) associated with various social media sites has triggered the development of social analytics tools useful to automatically extract, analyze, and summarize user-generated contents. In the era of Web 2.0, a sheer volume of user-contributed data is profoundly visible. A representative example of so-called sentic computing is described as a paradigm, which combines opinion mining and sentiment analysis, by exploiting artificial intelligence and semantic web techniques to recognize, interpret, and process opinions and sentiments in natural language. Feature-based summarization systems are developed to extract explicit product features and sentiments at sentence level.

Currently, many studies have been focused on the construction of fuzzy ontologies originating from different sources. For instance, one can refer to some standard entity-relationship (ER) models that have shown a widespread usage. We may also allude to some approaches that have been proposed for modeling fuzzy information in ER and extended entity-relationship (EER) models. Building fuzzy ontologies from fuzzy EER models arose as one of long-term interesting pursuits. Then the constructed fuzzy ontologies could be represented in the fuzzy OWL language.

Evidently, human centricity is positioned at the forefront of the conceptual, methodological, and algorithmic developments in sentiment analysis and ontologies. Human knowledge is formalized through constructs of ontology. People interact with systems and anticipate that any communication is helpful and becomes realized in a friendly and highly supportive environment. Consumer comments are subjective in nature and inherently loaded with emotions. They might be also contradictory and inconsistent. Such comments present on the web call for advanced processing tools so that we can transform them into coherent and meaningful views at products and services. Subsequently, the constructed views help organizations draw sound conclusions and enhance their business practices.

Computational Intelligence (CI) forming an advanced framework of concepts and algorithms that are visibly human-centric plays here a crucial role. From this perspective, CI is positioned in a unique manner to deliver a methodological and computational setting to sentiment analysis and ontology-driven constructs and processing. As a matter of fact, some of its roles with this regard have already been vividly demonstrated; however, there are a number of innovative avenues still worth pursuing.

The ultimate objectives of this edited volume is to provide the reader with a fully updated, in-depth treatise on the emerging principles, conceptual underpinnings, algorithms, and practices of CI in the realization of concepts and implementation of models of sentiment analysis and ontology-oriented engineering.

This volume involves studies devoted to the key issues of sentiment analysis, sentiment models, and ontology engineering. The book is structured into three main parts. The first one, composed of three chapters, offers a comprehensive and prudently structured exposure to the fundamentals of sentiment analysis and natural language processing stressing different points of view. The second part of the book consists of studies devoted to the concepts, methodologies, and algorithmic developments elaborating on fuzzy linguistic aggregation to emotion analysis, carrying out interpretability of computational sentiment models, emotion classification, and sentiment-oriented information retrieval, a methodology of adaptive dynamics in knowledge acquisition. The chapters forming the third part of the volume build a plethora of applications showing how sentiment analysis and ontologies become successfully applied to investment strategies, customer experience management, disaster relief, monitoring in social media, customer review rating prediction, and ontology learning.

An overall concise characterization of the objectives of the edited volume can be offered through several focal points:

- Systematic exposure of the concepts, design methodology, and detailed algorithms. The structure of the volume mostly adheres to the top-down strategy starting with the concepts and motivation and then proceeding with the detailed design that materializes in specific algorithms as well as case studies and applications.
- Individual chapters with clearly delineated agenda and well-defined focus and additional reading material available via carefully selected references.
- A wealth of carefully structured illustrative material. The volume includes a series of brief illustrative numeric experiments, detailed schemes, and more advanced problems.

- Self-containment. The intent is to make a material self-contained and, if necessary, augment some parts with a step-by-step explanation of more advanced concepts supported by a significant amount of illustrative numeric material and some application scenarios. Several introductory chapters fully satisfy this objective.

Given the theme of this undertaking, this book is aimed at a broad audience of researchers and practitioners. Owing to the nature of the material being covered and a way it has been organized, the volume will appeal to the well-established communities including those active in various disciplines in which sentiment analysis and ontologies, their analysis and optimization are of genuine relevance. Those involved in intelligent systems, data analysis, Internet engineering, CI, and knowledge-based systems will benefit from the exposure to the subject matter.

Considering the way in which the edited volume has been structured, this book may serve as a highly useful reference material for graduate students and senior undergraduate students.

We would like to take this opportunity to express our sincere thanks to the authors for presenting results of their innovative research and sharing their expertise in the area. We have received a large number of high quality submissions; it is needless to say that we had to make a number of difficult decisions to choose a limited number of papers to be included in the volume. The reviewers deserve our thanks for their constructive and timely input. We very much appreciate the continuous support and encouragement from the Editor-in-Chief, Prof. Janusz Kacprzyk whose leadership and dedication makes this book series a highly visible vehicle to disseminate the most recent, highly relevant, and far-fetching publications in the domain of computational intelligence and Internet engineering and its various applications.

We hope that this volume will be of genuine interest to a broad audience of readers and the research reported here will help continue with further vibrant progress in research, education, and help foster numerous practical endeavors.

Witold Pedrycz
Shyi-Ming Chen

Contents

Fundamentals of Sentiment Analysis and Its Applications

Mohsen Farhadloo and Erik Rolland

Abstract The problem of identifying people's opinions expressed in written language is a relatively new and very active field of research. Having access to huge amount of data due to the ubiquity of Internet, has enabled researchers in different fields—such as natural language processing, machine learning and data mining, text mining, management and marketing and even psychology—to conduct research in order to discover people's opinions and sentiments from the publicly available data sources. Sentiment analysis and opinion mining are typically done at various level of abstraction: document, sentence and aspect. Recently researchers are also investigating concept-level sentiment analysis, which is a form of aspect-level sentiment analysis in which aspects can be multi terms. Also recently research has started addressing sentiment analysis and opinion mining by using, modifying and extending topic modeling techniques. Topic models are probabilistic techniques for discovering the main themes existing in a collection of unstructured documents. In this book chapter we aim at addressing recent approaches to sentiment analysis, and explain this in the context of wider use. We start the chapter with a brief contextual introduction to the problem of sentiment analysis and opinion mining and extend our introduction with some of its applications in different domains. The main challenges in sentiment analysis and opinion mining are discussed, and different existing approaches to address these challenges are explained. Recent directions with respect to applying sentiment analysis and opinion mining are discussed. We will review these studies towards the end of this chapter, and conclude the chapter with new opportunities for research.

Keywords Sentiment Analysis · Opinion mining · Customer satisfaction · Probabilistic approaches

M. Farhadloo (✉) · E. Rolland
School of Engineering University of California, Merced, USA
e-mail: mfarhadloo@ucmerced.edu

E. Rolland
e-mail: erolland@ucmerced.edu

© Springer International Publishing Switzerland 2016
W. Pedrycz and S.-M. Chen (eds.), *Sentiment Analysis and Ontology Engineering*,
Studies in Computational Intelligence 639, DOI 10.1007/978-3-319-30319-2_1

1

1 Introduction

With the increasing popularity of analytics and data science, computational intelligence methods are proving to be important competitive tools in many industries.
For example, in business analytics data is mined for patterns that would help better
understand customers, and improve sales and marketing. Computational intelligence
methods allow for probabilistic methods to be used in finding patterns in data. These
methods typically work on low-level data, and are not guided by absolute knowledge
as is the case with general AI methods [3]. In addition, a huge amount of the data
that warrants analysis is now generated in written form. For example, users could
leave written comments about a product or service in online sites like Yelp and Tri-
pAdviser. The written text is subject to interpretation, and representing the data in
an absolute syntax (such as a binary system) is difficult. However, computational
intelligence methods allows for such fuzziness, and may be the most appropriate
methods for finding patterns in such data.

Sentiment analysis brings together various research areas such as natural language
processing, data mining and text mining, and is fast becoming of major importance
to organizations as they strive to integrate computational intelligence methods into
their operations, and attempts to shed more light on, and improve, their products
and services [32]. In sentiment analysis, or opinion mining, (SAOM), the goal is to
discover people's opinions expressed in written language (text). Sentiment in term
means "what one feels about something", "personal experience, one's own feeling",
"an attitude toward something" or "an opinion".

Opinions are central to almost all human activities and are key influencers of our
behaviors. Our beliefs and perceptions of reality, and the choices we make, are, to
a considerable degree, conditioned upon how others see and evaluate the world. For
this reason, when we need to make a decision we often seek out the opinions of others.
This is not only true for individuals but also true for organizations [38]. Traditionally, closed-form customer satisfaction questionnaires would be used to determine
the significant components, or aspects, of overall customer satisfaction [57]. However, the development and execution of questionnaires are expensive or may not be
available. In some cases, public agencies are even prohibited by law from collecting
satisfaction questionnaires from customers. In cases such as this, the only alternative
may be to analyze publicly available free-form text comments, for example from
sources such as Trip Advisor.

Organizations are increasingly focused on understanding how their value-creating
activities are perceived by their customers. Customers' opinions drive the organization's image and the demand for their products or services. For non-profits and government organizations, better serving the customers' needs help generate political
and taxpayer support, while in for-profit organizations improved customer understanding drives the organizational ability to generate revenues and compete in the
marketplace. Thus, understanding customers' perception is a key to organizational
success.

Customers' perceptions of products or services are often established through surveys, focus groups, observation, and other fairly labor intensive and expensive methods. With the advent of the Internet, survey tools have become more readily available (and cheaper to use), but obtaining accurate and relevant data from customers surveys is still a challenge. At the same time, the Internet has bred a whole industry of product/service review services, such as Yelp, that provides newfound capabilities and an ocean of information regarding customer's opinions. Examples include sites such as Trip Advisor for travel, Yelp and Urban Spoon for restaurants, Patagonia for outdoor clothing and gear, Lands' End for clothing and Epinions for product reviews have become commonplace. These sites allow customers to both read and provide reviews of the product or service. The customer reviews, which are often openly available on the web, contain a wealth of information usable by management, competitors, investors, and other stakeholders to discern customer concerns driving overall customer satisfaction for a particular service or product (for simplicity, we use the terms overall satisfaction and overall rating interchangeably). Usually the number of online reviews about an object is of a very large scale, such as hundreds of thousands, and the number is consistently growing as more and more people keep contributing online. Thus, we are now faced with a relatively new challenge of extracting customer opinions and sentiments from (often) unstructured data in the form of text comments.

An ideal method is one which would enable us to automatically identify the dimensions that drive customer satisfaction without *a priori* knowledge about those dimensions and their impact on overall customer satisfaction. The benefits of such a system would be that it would be able to discover the components and their impact on overall satisfaction without human intervention. Such a system enables analyzing large amounts of data which otherwise could not be analyzed by humans.

Both individuals and organizations can take advantage of sentiment analysis and opinion mining. When an individual wants to buy a product or decides whether or not use a service, he/she has access to a large number of user reviews, but reading and analyzing all of them could be a lengthy and perhaps frustrating process. Also when an organization or a manager seeks to elicit the public opinion about their products, market theirs products, identify new opportunities, predict sales trends, or manage its reputation, it needs to deal with an overwhelming number of available customers comments. With sentiment analysis techniques, we can automatically analyze a large amount of available data, and extract opinions that may help both customers and organization to achieve their goals. This is one of the reasons why sentiment analysis has been spread in popularity from computer science to management and social sciences. Sentiment analysis has also applications in review-oriented search engines, review summarization, and for fixing the errors in users ratings (such as for cases where users have clearly accidentally selected an incorrect rating when their review otherwise indicates a different evaluation) [43]. In public policy there are typically huge amounts of data (newspapers or other) documents containing opinions regarding an issue. For instance, during the recent financial crisis is Greece, there were vast amounts of documented opinions for or against Greece taking a new Eurozone loan.

One might establish a pretty good predictor for the Greek referendum by analyzing Greek media over the months leading up to the crisis. If sentiment analysis can be done, search engine could search for opinions and find the answers of these type of questions, as typically attempted with social media analytics.

Sentiment analysis can be used as a complement to other systems such as recommendation systems, information extraction and question answering systems [43]. The performance of recommendation systems may improve by not recommending items that receive a lot of negative feedback [51]. Information extraction systems may benefit from opinion mining systems by discarding certain types of information found in subjective sentences. In question answering systems different types of questions (opinion-oriented and definitional questions) may acquire different types of treatments. This is another field that can leverage from opinion mining.

This chapter provides an introduction to the problem of sentiment analysis. The focus of this chapter is on highlighting the various challenges researchers encounter in aspect-level sentiment analysis and some of the recent methods to address them. Since recently some researchers have started addressing sentiment analysis and opinion mining by using, modifying and extending topic modeling techniques, in this chapter a comprehensive survey about these methods will be given. Topic models are probabilistic techniques for discovering the main themes existing in a collection of unstructured documents. There are previous books and surveys on sentiment analysis [16, 37]. In this book chapter we aim at addressing recent approaches to sentiment analysis, and explain these in the context of wider use.

The remainder of this chapter is structured as follows: in Sect. 2 we first define the problem of sentiment analysis and then in Sect. 3 we will describe various challenges and approaches to them explained in the literature.

2 Problem Definition

In this section we will define the problem of SAOM and in particular the problem of aspect-level sentiment analysis. It is useful to first define some of the terms and concepts that are related to the problem. Here we have borrowed some of the definitions from the [37].

Contributor: A Contributor is the person or organization who is expressing his/her/its opinions in written language or text. Contributor may also called opinion holder or opinion source.
Object: An object is an entity which can be a product, service, person, event, organization, or topic [37]. It may be associated with a set of components and attributes that in sentiment analysis literature these components and attributes are called aspects.
Review: A Review is a contributor-generated text that contains the opinions of the contributor about some aspects of the object. A review may also be called opinionated document.

Overall rating: This is a user reported overall satisfaction with the object for example on a Likert scale from 1 to 5.

Opinion: An opinion on an aspect is a positive, neutral or negative view, attitude, emotion or appraisal on that aspect from a contributor.

Aspect: An aspect is an important attribute of the object with respect to overall costumer satisfaction that the contributor has commented on in their review.

An opinionated review may be generated by a number of contributors or contains opinions from a number of sources. Also it should be noticed that the review may be a direct review about a single object or a comparative review in that compares 2 or more objects to each other. In general a review d_i is a collection of sentences $d_i = \{s_{i1}, s_{i2}, \ldots, s_{im}\}$ that hold opinions from different contributers about different aspects of different objects. Liu [37] has defined a direct opinion as a quintuple of $(o_j, f_{jk}, oo_{ijkl}, h_i, t_l)$ where oo_{ijkl} is the opinion orientation of contributer h_i about the kth aspect f_{jk} of object o_j. Although this model is not a comprehensive model that contains all information about all cases, it is a sufficient model for practical applications. It is worth mentioning that most of the literature have considered the orientation of the opinion as a binary variable: either positive or negative. There are some works that include the neutral class as well. In general the opinion orientation could be considered in different scales rather than just binary or ternary.

The objective of SAOM can be written as:

Given a collection of reviews $D = \{d_1, d_2, \ldots, d_D\}$ all about an object, discover all **aspects** and corresponding **sentiments** expressed in that collection.

This objective can be reached at different levels. If the focus is on each document and the sentiment orientation of the whole document is found, this is called **document-level** sentiment analysis. Although a document may convey an overall positive or negative sentiment, it is quite likely that not all sentences in a document are all positive or negative. For example if the reviews of interest are movie reviews, a particular contributor may give an overall positive rating to the movie, however they may not like all aspects of that movie and in their review they mention what aspects they like and what they do not like about that movie. When the focus of SAOM is on sentences and a sentiment orientation for sentences of the review is found, it is called **sentence-level** sentiment analysis. In a finer analysis, one may be interested in knowing what particular aspects of the object the contributor is commenting on and also whether they like them or not. This level os SAOM is called **aspect-level** SAOM.

The visualization and reporting of the SAOM results are essential issues that need particular attention. There are different ways to visualize the results of the analysis. One way could be making a list of discovered aspects and linking to each of them the corresponding positive and negative reviews. This could be augmented with a rating for each aspect. This way people can easily find out their peers' opinions about different aspects and specifically pick what particular reviews they want to read. It is likely that some contributors give a positive/negative score to an object just because they are (not) happy with a particular aspect of the object. Other users may/not care about those particular aspects as much as others. Having an aspect-centric review report makes it possible to be more focused on the specific aspects that you care and seek others' opinion regarding them. If one wants to get an idea about popular or unpopular aspects of an object, creating a frequency list of different discovered aspects could be beneficial. One of the benefits of SAOM is that it empowers individuals and organizations to track the people's opinions over time. How things change over time has critical values for individuals and organizations. For example, Blackberry was a market leader in the mobile phone market for years, but lost out to the market pressure due to lacking features in their mobile phones. Once the importance of the missing aspects of a Blackberry phone, such as the lack of ability to text, outweighed the importance of secure email, their market-share diminished severely.

3 Challenges

In order to illustrate the need for fuzziness in the computational intelligence approaches to extracting patterns from user comments, we summarize below some major challenges associated with SAOM.

3.1 Synonymy and Polysemy

Users in different contexts or with different needs, knowledge or linguistic habits will describe the same information using different terms (synonymy) [11, 13]. Furnas et al. [20] showed that people generate the same keyword to describe well-known objects only 20 percent of the time. Because searchers and authors often use different words, relevant materials are missed. People also use the same word to refer to different things (polysemy) [13]. Words like saturn, jaguar, or chip have several different meanings. In different contexts or when used by different people the same term takes on varying referential significance [11]. In general synonymy and polysemy are caused due to the possible variability in word usage. Synonymy and polysemy issues pose difficult challenges on sentiment analysis and opinion mining problem.

Several approaches have been proposed in order for addressing the synonymy and polysemy challenges. *Stemming* which can be seen as normalizing some kind of surface-level variability is a popular technique. The purpose of stemming is to bring variant forms of a word together (to their morphological roots). A stemming algorithm for English reduces the words "stemmer", "stemming", "stemming" and "stems" to the root word "stem". Stemming algorithms have been studied in computer science since 1960s [39, 46, 59]. Stemming may sometimes help information retrieval but, it does not address cases where related words are not morphologically related (e.g., physician and doctor) [13].

Controlled vocabulary is another approach that has been shown effective in dealing with the issues caused by variability in word usage [2, 35, 50]. However since in controlled vocabulary approach it is a requirement that terms be restricted to a pre-determined list of words, it is not applicable in SAOM (it is not feasible to restrict contributors to limit their vocabulary to a pre-determined one). In review websites the contributors can usually express their opinions in any way that they prefer and it is not feasible to enforce a particular set of words.

Latent Semantic Analysis (LSA) [11] is an other approach to work around synonymy challenge. With advent of large scale collections of text data, statistical techniques are being used more and more to discover the relationship among terms and documents. LSA simultaneously models the relationship among documents based on their constituent words and relationship among words based on their occurrence in the documents. LSA can be seen as a linear dimensionality reduction technique based on singular value decomposition of *term-document* matrix. By reducing the dimensions and using fewer dimensions than the number of unique words, LSA induces similarities among words. LSA constructs a *semantic space* wherein terms and documents that are closely associated are place near one another [11]. It is worth noting that while LSA method deals nicely with the synonymy problem, it offers only a partial solution to polysemy problem. Since the meaning of a word can be conditioned by other words of the in the document, LSA provides some help regarding polysemy problem. However the failure arises from every single word having only a single representative point in the semantic space of LSA not more than one representatives.

3.2 Sarcasm

Identifying sarcasm is a very hard task for humans and it is even harder for machines. The ability to reliably identify sarcasm in text can improve the performance of many natural language processing tasks particularly SAOM. Sarcasm is a form of expression where the literal meaning is opposite to the intended. "The restaurant was great in that it will make all future meals seem more delicious," is an example of a sarcastic sentence, in which, although there is technically no negative term in the language, it is intended to convey a negative sentiment. This example clearly shows some difficulties in dealing with sarcastic phrases. Dealing with a sarcastic situation requires a good understanding of the context, the culture of the situation, the topic, the people and also the language involved in the sarcastic statement. Having access to all of these pieces of information is a difficult task in itself, but trying to make use of them is especially challenging for a machine. Although the phenomenon of sarcasm has extensively been studied in fields such as psychology, cognitive science and linguistics [22, 23, 53], very few attempts have been done on analyzing it computationally. Lack of a dataset with reliably labeled instances of sarcasm and not-sarcasm is one of the reason that computational analysis of sarcasm is very young. Recently [18] has generated a corpus consisting of Amazon product reviews that can be used for understanding sarcasm on two levels: document and text utterance level. The corpus contains a pair of sarcastic and non-sarcastic reviews written about the same object. Filatova [18] has used quality control algorithms to quantify the quality of their collection. There are some works [18, 24, 41] that have tried to automatically detect sarcasm in small text phrases such as Twitter data. However the drawback of working on short phrases in order to detect sarcasm is the ignorance of the fact that context plays an important role in sarcastic situations. Indeed [24] has shown that lexical features are not sufficient for sarcasm detection and pragmatic and contextual information are necessary in order to enhance the performance. Analyzing sarcasm is an area that is in need of more thorough research in SAOM research community.

3.3   Compound Sentences

A compound sentence has two independent clauses or sentences. Two independent clauses can be joined by a coordinating conjunction (such as "and", "or", "but" and "for") or a semicolon. Dealing with compound sentences makes the problem of SAOM difficult. For instance a sentences like "The kids enjoyed the beach but we did not," or "Despite a pleasant experience I cant support the many reviews that it was a great restaurant," are challenging for sentiment analysis. Dealing with compound sentences is still largely an open area of research in SAOM.

3.4   Unstructured Data

The reviews that the contributors have written for each object are the input which are in the plain text format. A challenge in the problem is the transformation of the unstructured input data which is available in the form of written reviews into a semi-structured data. Semi-structured data is data that is neither raw data, nor conformal with the formal structure of data models associated with relational databases or other forms of data tables, but nonetheless contains tags or other markers to separate semantic elements.

3.5   Aspect Identification

A body of works in the literature have addressed the problem of aspect-level SAOM in two stages: first, aspect identification and second, sentiment identification. The goal of the aspect identification is to discover the particular aspects of the object that contributors are expressing their opinions about. There are 2 categories of works in the literature addressing aspect identification:

1. **Automatic extraction**: There is no prior knowledge about the aspects an the aspects are automatically extracted from the given reviews [4, 15, 21, 29, 40]. This category can be divided into *supervised* and *unsupervised* subcategories.
2. **(Semi) Manual extraction**: Either a subset of aspects or the whole set of desired aspects are known a priori. In cases where a subset of aspects are known, the subset is being used as a seed set and expanded in some ways [55].

Of the early works in the aspect level sentiment analysis, is the research by Hu and Liu [29]. Hu and Liu propose to first find the frequent aspects using association mining, and then using them, they extract the infrequent aspects. Any sentence that contains one of the frequent aspects is being analyzed to find out the infrequent aspects.

In [21] authors aim at presenting an unsupervised aspect identification algorithm that employs clustering over sentences with each cluster representing an aspect. They finally proposed applying an empirical weighting scheme to the list of terms, which are sorted according to their frequency of occurrence.

Clustering over sentences has been used in [15] in order to find similar sentences which are most likely about similar aspects. In [15] instead of representing the sentences using the commonly Bag-Of-Word (BOW) method, they propose to use Bag-Of-Nouns that makes clustering more effective.

Blair-Goldensohn et al. [4] propose a sentiment summarization system for local services. In their system the aspects are divided into two types: dynamic aspects (string-based frequent nouns/noun phrases) which are extracted similar to the method of [29] and static aspects (generic and coarse-grained ones), which are extracted by designing classifier for each one using hand-labeled sentences.

PLSA [28] is the probabilistic version of LSA that has evolved from it. PLSA is a model for a collection of documents in which each document is modeled as a mixture of topics. Lu et al. [40] have taken advantage of PLSA and the structure of the sentences for aspect identification. Each document is represented as a bag of phrases and each phrase is a pair of head and modifier terms $<h_i, m_i>$. Each aspect is also modeled as a distribution over *head terms*. In the unstructured version a document is considered as a collection of head terms (the modifiers are ignored). In this model the log likelihood of all documents can be written as:

$$log p(D|\Lambda) = \sum_{d=1}^{D} \sum_{w_h \in V_h} \{c(w_h, d) log \sum_{k=1}^{K} [\pi_{d,k} p(w_h | \Theta_k)]\} \qquad (1)$$

where V_h is the set of all head terms in the vocabulary, $\pi_{d,k}$ is the proportion of topic k in document d, $c(w_h, d)$ is the number of times head term w_h has been occurred in document d, Θ_k is the kth aspect-topic and Λ is the set of all model parameters. In the structured version (*structured* PLSA) since a single modifier term can modify different head terms, each modifier term is being modeled as a mixture of K topics. Each modifier term w_m is represented as a set of head terms that it modifies and again each aspect is a distribution over head terms (Θ_k). The modifier can be regarded as a sample of the following mixture model:

$$p_{d(w_m)}(w_h) = \sum_{k=1}^{K} [\pi_{d(w_m),k} p(w_h)|\Theta_k)] \tag{2}$$

and the log likelihood of the collection of modifiers V_m is:

$$logp(V_m|\Lambda) = \sum_{w_m \in V_m} \sum_{w_h \in V_h} \{c(w_h, d(w_m))log \sum_{k=1}^{K} [\pi_{d(w_m),k} p(w_h|\Theta_k)]\} \tag{3}$$

In both equations of (1) and (3) Λ is the model parameters set (including Θ_k) and is estimated using Expectation-Maximization (EM) algorithm. The details of the EM algorithms can be found in [40]. It is worth noting that the structured PLSA estimated the parameters based on the co-occurrence of head terms at the level of modifiers, not at the level of documents. Since the reviews are usually short reviews and have few phrases, the structured PLSA is typically more informative.

3.6   Sentiment Identification

The sentiment identification is the process of discovering the opinion orientation of the text fragments of interest. The sentiment orientation can be expressed in different scales. Rating scales are used widely online in an attempt to provide indications of consumer opinions of products. Many sites like *TripAdvisor.com, Amazon.com, Epinions.com* use the 5-star overall rating scale. Users can cast their vote about movies in a $1-10$ rating scale on *IMDb* and an on a 5-star scale on *rottentomatos.com*.

In SAOM dealing with negative sentiments is particularly difficult. As Leo Tolstoy said in his book *Anna Karenina* "All happy families resemble one another, each unhappy family is unhappy in its own way", in SAOM all happy sentiments are pretty much alike, each unhappy sentiment is uniquely expressed. In some cases, people choose to cover their criticisms with qualifiers to be polite (even on the Internet), and more broadly, people just tend to be more creative in how they choose to describe the things that they do not like. All of these make the automatic solution of the problem challenging. Automatic techniques make use of machine learning algorithms to address the problem and these algorithms aim at finding patterns from data and try to learn from training examples. When there are not enough training examples or there are training examples with very limited variety, one can imagine how hard the pattern extraction would be.

Most of the work in the literature have treated sentiment in a binary scale of negative and positive [29, 44], some have added a third class of sentiment (the neutral sentiment) into their consideration [15]. As in some actual cases that each review comes with an overall rating in a 5-star scale, the number of sentiment orientations can vary. [33] have considered 5-level sentiment categories in their experiments.

There are 2 types of fragments in a textual document: *objective fragments* are those text fragments that express factual data and *subjective fragments* that express personal feelings. Although both subjective and objective fragments may carry an opinion about an object, it is more likely for a subjective sentence to be opinionated. The process of sentiment identification can be written as the following steps:

1. Extract all the opinionated fragments.
2. Identify the sentiment of each opinionated fragments.
3. Infer the overall sentiment from the opinion of individual fragments.

3.6.1 Extract Opinionated Fragments

Subjectivity classification is the task of determining whether a fragment of text is subjective or objective. If a text fragment is being categorized as subjective then it is very likely that it is an opinionated piece of text and then its opinion orientation needs to be identified. There are works on subjectivity determination [19, 27], that can be used in this regard.

It is sensible to conjecture that opinionated words and phrases are dominant sentiment indicators. Since adjective, adverbs and also some nouns and verbs are strong sentiment indicators, Part-Of-Speech (POS) tags and syntactic structure of the sentences are also beneficial in extracting opinionated fragments. POS tagging is the process of marking up a word in a text (corpus) as corresponding to a particular grammatical category such as nouns, adjective, adverb, verb or etc. categories. The POS tagging algorithms are based on terms definitions as well as the context i.e. the term's relationship with adjacent and related words and phrases.

In aspect-level SAOM the sentiment orientation of all text fragments that contains one of the extracted aspects has to be identified. So if in the first stage the aspects have been identified they can be used in order for extracting opinionated fragments.

3.6.2 Identification of Sentiment Using Unsupervised Techniques

Once the opinionated fragment has been extracted in the next step the sentiment orientation of it should be identified. Turney [52] has determined the sentiment polarity of a word by computing its similarity to 2 seed terms: "Excellent" for positive and "Poor" for negative polarity. The similarity between an unknown word and each of the seed terms is computed by counting the number of occurrences and co-occurrences of them using the results of a web search engine. Depending upon which one of the seed terms the unknown term is more similar to, the sentiment orientation is determined.

3.6.3   Identification of Sentiment Using Supervised Techniques

Sentiment identification can easily been formulated as a classification problem. In the literature of SAOM various classification techniques have been employed from machine learning to identify sentiments. Pang et al. [44] have experimented a variety of features with both naive Bayes classifiers and also Support Vector Machines (SVM) to classify the whole input document (movie reviews) as either positive or negative. The best results came from the unigrams in a presence-base frequency model run through SVMs. For sentiment classification of sentences [21] implement a naive Bayes classifier using expectation maximization and bootstrapping from a small set of labeled data to a large set of unlabeled data. Feature engineering, finding the most suitable set of features for sentiment identification, is a critical issue in designing classifiers. Different features have been used in SAOM such as *terms and their frequencies, POS tags, opinion words and phrases, syntactic structures and negations*. In [15] a new feature set called *score representation* was introduced for classification of sentiments. Score representation is a low dimensional representation and is computed using the scores of the terms in the vocabulary that are learned from the data. Score representation has been used for designing 3-class SVM classifiers and outperforms the classifiers that are based on BOW representation.

Another feature engineering approach based on a hierarchical deep learning framework that simultaneously discovers aspects and their corresponding sentiments has been proposed in [34]. In their framework classifiers based on deep learning (recursive neural networks) have been designed that jointly predict the aspect and sentiment labels of the input phrase. The main idea is to learn a vector (or a matrix or a tensor) representation for words using hierarchical deep learning that can explain the aspect-sentiment labels at the phrase level. In this framework each node in the parse tree of the given phrase is represented by a vector (or a vector and a matrix) and the model is defined by leveraging the syntactic structure of the phrase encoded in the parse tree. Figure 1 shows an example parse tree in which each node is associated by a d-dimensional vector.

Fig. 1 An example parse tree. Each node is associated with e vector and the vector representation of the nodes are computed *bottom-up*

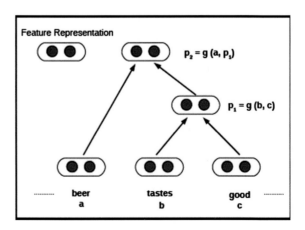

The vector representation of the nodes are calculated bottom-up as:

$$p_1 = f\left(W\begin{bmatrix} b \\ c \end{bmatrix}\right), \qquad p_2 = f\left(W\begin{bmatrix} a \\ p_1 \end{bmatrix}\right) \tag{4}$$

where p_1 and p_2 are the parent vectors and b and c are the leaf nodes and $f = tanh$ is a standard element-wise non-linear function. The goal is to learn the combination matrix $W \in R^{d \times 2d}$ and also the compositional feature representation of the words (the associated vectors and matrices). In this model the class label of each given phrase is predicted using the vector representation of its root node of the parse tree as:

$$y_i = softmax(W_s p_i^{root})$$

where $W_s \in R^{C \times d}$ is the classification matrix and needs to be estimated. The true label of the given phrase in the training set is $t_i \in R^{C \times 1}$ that has an entry 1 at the correct label and a 0 at other indices. Since the aspects and sentiments meant to be captured jointly the labels are supposed to be provided as aspect-sentiment pairs. For instance (Taste, Positive) correspond to one class label. If the whole set of model's parameters is denoted by Θ then Θ is estimated in a way that the softmax function of the vector representation at the root level of the parse tree $y_i \in R^{C \times 1}$ matches the class label of the ith text snippet as closely as possible. This can be achieved by minimizing the cross entropy error between y_i and t_i. The error function to be minimized is:

$$E(\Theta) = \sum_i \sum_j t_{i,j} log y_{i,j} + \lambda \|\Theta\|^2 \tag{5}$$

The non-convex objective function in Eq. (5) is minimized using Adagrad optimization [12] procedures. The estimation involves computation of the sub-gradients of the $E(\Theta)$ and the forward calculation of the vectors and matrices and back-propagation of the softmax error at the root node.

3.6.4   Identification of Sentiment Using Lexicons

Many work in the literature have identified the sentiment orientation of the fragment of interest using some opinion lexicons [1, 14, 26, 29, 49, 58]. Lexicon based approaches discover the sentiment orientation of an opinionated phrase of interest by looking-up it from their existing lexicons. The lexicon generation process usually start with a seed list of opinion words that their opinion orientations are known a priori and then using different approaches the seed list is expanded and more opinion words are added to it. There exist 2 main strategies to expand the initial seed list of opinion words: *dictionary-based* and *corpus-based* strategies. Dictionary-based methods make use of online dictionaries like *WordNet* [8, 14, 17] and some additional information in them (like their glosses) in order to expand the initial seed set. These approaches take advantage of synonym and antonym relationships among words and also some machine learning techniques to generate better list. The generated lexicon based on online dictionaries is a context independent opinion lexicon. In order to generate domain specific lexicon that can capture the opinion words in a particular domain, corpus-based strategies have been proposed. In this strategy the initial seed list of opinion words is expanded using some syntactic rules and co-occurrence patterns. In this approach rules are designed for connective terms such as *And, Or, But, Neither-or, Neither-nor* and leveraging them the seed set is expanded. Using the particular corpus of interest domain specific lexicons can be generated.

For sentiment identification, [29] follow a lexicon based approach, which involves identifying the closest adjectives to the nouns in the under process sentence and look-up those adjectives or their synonyms in their lists of positive and negative adjectives. Blair-Goldensohn et al. [4] first compute a single sentiment score for each term in their lexicon. These scores are computed starting with an initial seed with arbitrary scores, and then propagation of these scores using a modified version of standard label propagation algorithms over a graph [62]. A shortcoming of this approach is that it is an iterative algorithm and it is not clear when to stop propagating the scores. Using the computed scores and considering the neighbors of each sentence, they design maximum entropy classifiers for positive and negative classes. Concept-based approaches use Web ontologies or semantic networks to accomplish semantic text analysis [9]. In this approach the document is represented with bag-of-concepts instead of bag-of-words. In concept-based approach for identifying the sentiment of each document a dictionary (SenticNet) which contains the affective labels of concepts is used [45]. However it is not clear how to identify the aspects using the bag-of-concepts representation.

3.7 Topic Model Based Approaches

Fundamental characteristics of human word usage is that people use a wide variety of words to describe the same object or concept [13]. As it was mentioned in Sect. 3.1 dealing with synonymy and polysemy in natural language is a challenging issue in many text mining tasks. This illustrates the need for the probabilistic features of computational intelligence to address SAOM. In topic modeling, each topic is defined as a probability distribution over terms (terms of considered vocabulary list). One of the outputs of topic modeling algorithms is a probability distribution over terms for each topic that determines which words have higher probability in the context of a particular topic. It is reasonable to expect that in each context there are certain terms that have higher probability of occurrence than other terms. Topic modeling techniques by definition are able to capture this characteristic of natural language.

Topic models are probabilistic techniques based on hierarchical Bayesian networks for discovering the main themes existing in a collection of unstructured documents. Most of the existing works in the literature solve the problem of SAOM in 2 stages: first aspect identification and second sentiment identification [4, 15, 29].

One of the advantages of methods that are based upon topic modeling is that they are able to find aspects and sentiments simultaneously. Also these algorithms do not require labeled training data and they find the topics from the analysis of the original texts.

Latent Dirichlet Allocation (LDA) was proposed in order to find short descriptions of the members of a collection. LDA enables efficient processing of large collections while preserving the essential statistical relationships that are useful for basic tasks such as classification, novelty detection, summarization and similarity and relevance judgments [6]. LDA assumes that each document has been generated from a mixture of topics. Therefore for each document to be generated, the proportion of each topic in that document should be known and also for each word of a document to which topic that word belongs, should be known. The generative process of LDA can be summarized as [5]:

(1) **For each document**:

    (a) Randomly choose a distribution over topics.

(2) **For each word in the document**:

    (a) Randomly choose a topic from the distribution over topics in step 1.
    (b) Randomly choose a word from the corresponding distribution over the vocabulary (topic).

It is worth mentioning that LDA is based on the assumption that there exist a hidden structure that has generated the given collection of documents. Given the collection of documents (observation) the goal is to find the topics ($\{\boldsymbol{\beta}_1, \boldsymbol{\beta}_2, \ldots, \boldsymbol{\beta}_K\}$), distribution over topics for each document ($\boldsymbol{\theta}_d$) and topic assignment ($z_{d,n}$) for each word $w_{d,n}$ in each document d. $\{\boldsymbol{\beta}, \boldsymbol{\theta}, \mathbf{Z}\}$ construct the hidden structure behind the observed data and that is why the model is called *Latent* Dirichlet Allocation.

A neat way to describe the topic models is by using graphical models that provide a graphical language for describing families of probability distributions. The graphical model of LDA is depicted in Fig. 2.

In every graphical model a number of statistical dependencies are encoded that define that particular model. For instance in Fig. 2, the probability of nth word in document d ($w_{d,n}$) depends on the topic assignment $z_{d,n}$ of that word and all topics $\{\boldsymbol{\beta}_{1:K}\}$. Topic assignment determines which topic to be used to get the probability of $w_{d,n}$. Also the topic assignments of the words in document d depend on distribution of topics $\boldsymbol{\theta}_d$ in that document. The joint distribution of the hidden and observed variable of the graphical model in Fig. 2 can be written as:

$$p(\boldsymbol{\beta}_{1:K}, \boldsymbol{\theta}_{1:D}, \mathbf{Z}_{1:D}, \mathbf{W}_{1:D}) = \prod_{k=1}^{K} p(\boldsymbol{\beta}_k) \prod_{d=1}^{D} p(\boldsymbol{\theta}_d) \prod_{n=1}^{N} p(z_{d,n} | \boldsymbol{\theta}_d) p(w_{d,n} | z_{d,n}, \boldsymbol{\beta}_{1:K}))$$

(6)

Given a collection of documents $D = \{d_1, d_2, \ldots, d_D\}$ the goal of training is to learn the hidden structure $\{\boldsymbol{\beta}, \boldsymbol{\theta}, \mathbf{Z}\}$. In Bayesian framework, the particular computational problem in order to use the model is the *inference problem* in which computing

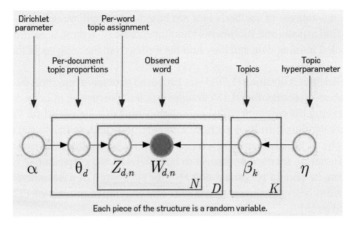

Fig. 2 Latent Dirichlet Allocation graphical model. The *shaded* node is an observed random variable and the *unshaded* ones are hidden random variables

the posterior distribution of the hidden variables given the observed variables is of interest. The exact computation of the posterior probability is intractable and approximation methods have been considered.

There are 2 general approximation methods in topic modeling algorithms:

(1) **Sampling-based algorithms**: These algorithms are Markov Chain Monte Carlo (MCMC) techniques which provide a principled way to approximate an integral (expected value). Monte Carlo techniques are algorithms that aim at obtaining a desired value by performing simulations involving probabilistic choices [47]. Sampling-based algorithms like *Gibbs sampling* try to approximate the posterior distribution by drawing samples from it without explicitly computing the posterior distribution. In Gibbs sampling the hidden structure defines a *vector* that represents the *state* of the system. The basic idea in Gibbs sampling is instead of sampling a multivariate random vector all at once, each dimension is being sampled using the samples of the other dimensions. For an introduction to Gibbs sampling you can refer to [47] and see [48] for a good description of Gibbs sampling for LDA.

(2) **Variational algorithms**: These algorithms approximate the posterior probability by a parameterized family of distribution over hidden variables and using optimization techniques find the set of parameters that makes the approximated distribution closest to the exact posterior. We can put variational methods in other words: these algorithms are based on Jensen's inequality and try to obtain an adjustable lower bound on the log likelihood. Essentially one considers a family of lower bounds, indexed by a set of variational parameters. Th variational parameters are chose by an optimization procedure that attempts to find the tightest possible lower bound [6].

The LDA model is the simplest topic model that provides a powerful tool for discovering the hidden structure in a large collection of documents. Since its introduction LDA has been extended and modified in many ways. One of the areas that topic models can very well fit is the area of SAOM. In particular in aspect-level sentiment analysis, people are talking about different aspects of an object. Different contributors may use different terms in order to point to the same aspect of an object. For example in reviews about a camera, you may see people commenting about *image quality* and people commenting about *resolution* of the camera and most likely they are speaking about the same aspect. Therefore when people are talking about a particular aspect, a number of terms are most likely to be exploited. In topic modeling terminology, this means that each aspect has a distribution over terms or should be considered as a topic in our model. Also when contributors want to give a particular rating to an aspect in their review, a specific range of words have higher probability and for another level of rating a different set of words have higher probability mass than the rest of the vocabulary. This means that each rating (sentiment) like each aspect should be considered as a topic in order to extend topic modeling algorithms for SAOM.

Just simply increasing the number of topics (to count for both aspect-topics and rating-topics) in LDA is not sufficient to apply it to SAOM. Researchers have started modifying LDA in many ways in order to adapt it to address the problem of aspect-level sentiment analysis.

One of the assumptions of LDA is the Bag-of-Word (BOW) assumption. In LDA it is assumed that the order of the terms in the document can be neglected and the document can be represented as a collection of words. Since this assumption is not realistic, researchers have tried to address this shortcoming. To go beyond BOW [54] has suggested bigram topic model. In bigram-topic model [54] has integrated bigram-based and topic-based approaches to document modeling. In this model each topic instead of being a single distribution over words, has multiple distributions over words depending on the previous word. If the size of the vocabulary list is N then each topic is characterized by N distributions specific to that topic. It can easily be realized that the parameter space in this approach will expand rapidly. The LDA collocation (LDA-Col) model [25] is another attempt to go beyond BOW assumption of LDA. In LDA-Col a new hidden variable is introduced for each word in each document that determines whether or not the word should be drawn from a unigram model or should be drawn from a bigram model considering its previous word. Topical N-grams (TNG) model [56] is very similar to LDA-Col and the only difference is that in TNG it is possible to decide whether to form a bigram for the same two consecutive word tokens depending on their nearby context. The TNG model automatically determines to form an n-gram (and further assign a topic) or not, based on its surrounding context. Examples of topics found bt TNG are more interpretable than its LDA counterpart [56].

Fig. 3 Topical N-grams
graphical model. The *shaded*
node is an observed random
variable and the *unshaded*
ones are hidden random
variables

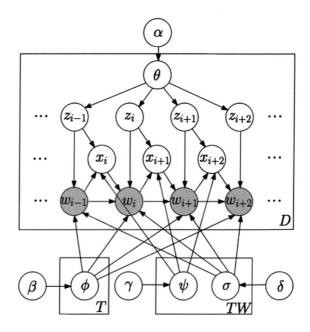

In bigram topic model there is no extra hidden variable to decide whether or not to form a bigram or a unigram (all the terms are drawn from bigram models), in LDA-Col the binary random variable x_i for each word w_i determines unigram or bigram status however the value of x_i does not depends on the previous topic z_{i-1}. The graphical model of LDS-Col is exactly as Fig. 3 except that the links from z_i to x_{i+1} are not there.

In another attempt in the direction of relaxing BOW assumption of LDA in that it ignores the order of the words, [33] has introduced the notion of aspects and senti- ments coherency and has also tried to leverage from the existing syntactic structure in natural sentences. In that model (CFACTS) they have considered each document as a collection of *windows* and all the words inside a window have the same aspect or sentiment topic (coherency). Therefore in spite of LDA that each word of a document has a topic assignment, each window has a topic or sentiment assignment variable and all the words in that window have the same assignment. In the CFACTS model each word could be either aspect word, sentiment word or background word, so they have considered a new hidden variable c_i for each word that determines to which category that word belongs. The syntactic dependencies among words have been captured through c_{i-1} and c_i in the model of [33].

Since the sentiments and aspects for adjacent windows may still be dependent for a specific review document in many ways, in the CFACTS model in Fig. 4, the multinomial variable $\psi_{d,x}$ for each window x of a review d has been introduced. $\psi_{d,x}$ can take on 3 different values that captures the following scenarios:

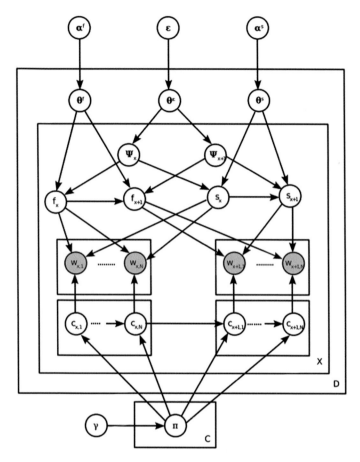

Fig. 4 Coherent Facet and Sentiments graphical model. The *shaded* node is an observed random variable and the *unshaded* ones are hidden random variables. This model introduces two new variables: $\psi_{d,x}$ which determines the dependency between windows and c_i that specifies to which category each word belongs: aspects, sentiments or background category

$\phi_{d,x} = 0$: indicates that both aspect and sentiment of current window is the same as those of the previous window.

$\phi_{d,x} = 1$: indicates that the aspect topic of the current window is independent from the previous window but the sentiment topic of the current window is the same as previous one.

$\phi_{d,x} = 2$: indicates that both aspect and sentiment topics of the current window are independent form the previous window.

On the same line as coherency idea the model of [7] assumes all words in a single sentence are generated from a single topic and applies LDA on each sentence to extract topics. The authors of [31] further extend the idea to discover sentiments related to each aspect. In this model, each review has a distribution over sentiments

Fig. 5 Phrase Latent
Dirichlet Allocation
graphical model. The *shaded*
node is an observed random
variable and the *unshaded*
ones are hidden random
variables

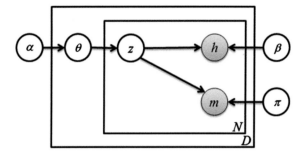

and each sentiment has a distribution over aspects. To generate each sentence of a
review, first a sentiment is drawn from the distribution over sentiments of that review,
and then an aspect is drawn from the distribution of that sentiment over aspects. Each
word of the sentence is then generated based on the selected aspect and sentiment.

Another work around the BOW assumption of LDA is Phrase-LDA (PLDA).
PLDA assumes a bag-of-phrase model of reviews (Fig. 5). An opinion phrase is a
pair of $<h_n, m_n>$ which leads to two observed variables head term h_n and modifier
term m_n. In [61] PLDA is applied to extract topics from reviews with the further
assumption that each sentence of the review is related to only one topic. Compared
to PLDA, separate PLDA (SPLDA) model introduces a separate rating variable which
is conditionally independent from the aspect. In this model a review is generated by
first drawing a distribution over aspect/rating pairs θ, and then repeatedly sampling N
aspects and ratings as well as opinion phrases $<h_n, m_n>$ conditioned on the chosen
value of θ.

In [42] a set of guidelines for designing LDA-based models are presented by
comparing a series of increasingly sophisticated probabilistic graphical models based
on LDA. In order to get familiar with some aspects of LDA-based methods and
compare the impact of various design decisions such as adding latent and observed
variables and dependencies we refer the readers to [42].

Latent Dirichlet Allocation is a member of a larger family of methods: latent variables models. Although there are so many works on aspect-level sentiment analysis
based on LDA, there are other latent variable model approaches based on conditional
random fields [10, 36] and Hidden Markov Models [30, 60]. Interested readers should
look into them.

4   Conclusions

Computational intelligence methods play a central role in sentiment analysis, and
have proven to be powerful tools in helping to understand customers' perceptions
related to products and services. While there have been many advances in the short
history of this field, there is still much work to be done. Most of the work thus far

has been focused on deciphering the semantics of written text, and this research is impacted by various linguistic challenges. Nevertheless, we see that past research has been able to propose methods that uncover sentiments, opinions, and aspects and that correlate very well with customer satisfaction scores. What remains less clear is the generalizability of the methods across contexts or domains in order to really test to what extent the probabilistic computational intelligence methods are generalizable. Thus, the opportunities for continued research is large, and this research area will in turn bring about a sea-change in how organizations understand their customers, and possibly in how customers understand and evaluate products and services.

5   Questions for Further Research

While research will continue to improve on methodologies for sentiment discovery and aspect identification, there is also a need for bringing in other types of context specific information into the analysis. For example, one could imagine including more information about the reviewer, as such information often is readily available (many reviewing systems contain potentially useful reviewer information). Also, one could also incorporate location specific information (geotagging) to improve on understanding sentiments and opinion, and possibly quantitative or qualitative information related to non-textual media (such as pictures or videos) for further enhance the analytical capabilities of sentiment analysis. Further, we believe that the dynamic aspect of sentiment analysis will need to be better addressed. That is, sentiments, opinions, and aspects are dynamic in nature, and change with changing competition, technology, use, and so on. We stress that the computation efforts associated with sentiments analysis will have to be efficient enough to handle large data in a dynamic fashion. Finally, the we believe that sentiment analysis as discussed herein will be strongly dependent on better understanding information quality. That is, garbage-in may result in garbage-out for sentiment analysis. As such, Information quality as a research field, also holds many opportunities for future research, and those may be strongly coupled to sentiment analysis.

References

1. Andreevskaia, A. Bergler, S.: Mining wordnet for a fuzzy sentiment: Senti-ment tag extraction from wordnet glosses. In: EACL, vol. 6, pp. 209–216, (2006)
2. Bates, M.J.: Subject access in online catalogs: a design model. J. Am. Soci. Inf. Sci. **37**(6), 357–376 (1986)
3. Bezdek, J.C. Sankar, K.: Fuzzy models for pattern recognition. Technical report, USDOE Pittsburgh Energy Technology Center, PA (United States); Oregon State University, Corvallis, OR (United States). Department of Computer Science; Naval Research Lab., Washington, DC (United States); Electric Power Research Institute, Palo Alto, CA (United States); Bureau of Mines, Washington, DC (United States) (1994)

4. Blair-Goldensohn, S., Hannan, K., McDonald, R., Neylon, T., Reis, G.A., Reynar, J.: Building a sentiment summarizer for local service reviews. In: WWW Workshop on NLP in the Information Explosion Era (2008)
5. Blei, D.M.: Probabilistic topic models. Commun. ACM **55**(4), 77–84 (2012)
6. Blei, D.M., Ng, A.Y., Jordan, M.I.: Latent dirichlet allocation. J. Mach. Learn. Res. **3**, 993–1022 (2003)
7. Brody, S., Elhadad, N.: An unsupervised aspect-sentiment model for online reviews. In: Human Language Technologies: The 2010 Annual Conference of the North American Chapter of the Association for Computational Linguistics, Association for Computational Linguistics, pp. 804–812 (2010)
8. Cambria, E., Havasi, C., Hussain, A.: Senticnet 2: a semantic and affective resource for opinion mining and sentiment analysis. In: FLAIRS Conference, pp. 202–207 (2012)
9. Cambria, E., Schuller, B., Xia, Y., Havasi, C.: New avenues in opinion mining and sentiment analysis. Intell. Syst. IEEE **28**(2) (2013)
10. Choi, Y., Cardie, C.: Hierarchical sequential learning for extracting opinionsand their attributes. In: Proceedings of the ACL 2010 conference: short papers, pp. 269–274. Association for Computational Linguistics (2010)
11. Deerwester, S.C., Dumais, S.T., Landauer, T.K., Furnas, G.W., Harshman, R.A.: Indexing by latent semantic analysis. JASIS **41**(6), 391–407 (1990)
12. Duchi, J., Hazan, E., Singer, Y.: Adaptive subgradient methods for online learning and stochastic optimization. J. Mach. Learn. Res. **12**, 2121–2159 (2011)
13. Dumais, S.T.: Latent semantic analysis. Annu. Rev. Inf. Sci. Technol. **38**(1), 188–230 (2004)
14. Esuli, A., Sebastiani, F.: Sentiwordnet: a publicly available lexical resource for opinion mining. In: Proceedings of LREC, vol. 6, pp. 417–422. Citeseer (2006)
15. Farhadloo, M., Rolland, E.: Multi-class sentiment analysis with clustering and score representation. In: 2013 IEEE 13th International Conference on Data Mining Workshops (ICDMW), pp. 904–912. IEEE (2013)
16. Feldman, R.: Techniques and applications for sentiment analysis. Commun. ACM **56**(4), 82–89 (2013)
17. Fellbaum, C.: WordNet. Wiley Online Library, (1998)
18. Filatova, E.: Irony and sarcasm: Corpus generation and analysis using crowdsourcing. In: LREC, pp. 392–398 (2012)
19. Finn, A., Kushmerick, N., Smyth, B.: Genre classification and domain transfer for information filtering. In: Advances in Information Retrieval, pp. 353–362. Springer (2002)
20. Furnas, G.W., Landauer, T.K., Gomez, L.M., Dumais, S.T.: The vocabulary problem in human-system communication. Commun. ACM **30**(11), 964–971 (1987)
21. Gamon, M., Aue, A., Corston-Oliver, S., Ringger, E.: Pulse: mining customer opinions from free text. Advances in Intelligent Data Analysis VI, pp. 121–132. Springer, Berlin (2005)
22. Gibbs, R.W.: On the psycholinguistics of sarcasm. J. Exp. Psychol. Gen. **115**(1), 3 (1986)
23. Gibbs, R.W., Colston, H.L.: Irony in language and thought: a cognitive science reader. Psychology Press (2007)
24. González-Ibáñez, R., Muresan, S., Wacholder, N.: Identifying sarcasm in twitter: a closer look. In: Proceedings of the 49th Annual Meeting of the Association for Computational Linguistics: Human Language Technologies: short papers, vol. 2, pp. 581–586. Association for Computational Linguistics (2011)
25. Griffiths, T.L., Steyvers, M., Tenenbaum, J.B.: Topics in semantic representation. Psychol. Rev. **114**(2), 211 (2007)
26. Hatzivassiloglou, V. McKeown, K. R.: Predicting the semantic orientation of adjectives. In: Proceedings of the 35th annual meeting of the association for computational linguistics and eighth conference of the european chapter of the association for computational linguistics, pp. 174–181. Association for Computational Linguistics (1997)
27. Hatzivassiloglou, V., Wiebe, J.M.: Effects of adjective orientation and gradability on sentence subjectivity. In: Proceedings of the 18th conference on Computational linguistics, vol. 1, pp. 299–305. Association for Computational Linguistics (2000)

28. Hofmann, T.: Probabilistic latent semantic indexing. In: Proceedings of the 22nd annual international ACM SIGIR conference on Research and development in information retrieval, pp. 50–57. ACM (1999)
29. Hu, M., Liu, B.: Mining and summarizing customer reviews. In: Proceedings of the tenth ACM SIGKDD international conference on Knowledge discovery and data mining, KDD'04, pp. 168–177, ACM, New York, USA (2004). ISBN:1-58113-888-1
30. Jin, W., Ho, H.H., Srihari, R.K.: Opinionminer: a novel machine learning system for web opinion mining and extraction. In: Proceedings of the 15th ACM SIGKDD international conference on Knowledge discovery and data mining, pp. 1195–1204. ACM (2009)
31. Jo, Y., Oh, A.H.: Aspect and sentiment unification model for online review analysis. In: Proceedings of the fourth ACM international conference on Web search and data mining, pp. 815–824. ACM (2011)
32. King, R.: Sentiment analysis gives companies insight into consumer opinion. Business Week: technology (2011)
33. Lakkaraju, H., Bhattacharyya, C., Bhattacharya, I., Merugu, S.: Exploiting coherence for the simultaneous discovery of latent facets and associated sentiments. In: SDM, pp. 498–509. SIAM (2011)
34. Lakkaraju, H., Socher, R., Manning, C.: Aspect specific sentiment analysis using hierarchical deep learning. In: Deep Learning and Representation Learning Workshop, NIPS (2014)
35. Lancaster, F.W.: Vocabulary control for information retrieval (1972)
36. Li, F., Han, C., Huang, M., Zhu, X., Xia, Y.-J., Zhang, S., Yu, H.: Structure-aware review mining and summarization. In: Proceedings of the 23rd international conference on computational linguistics, pp. 653–661. Association for Computational Linguistics (2010)
37. Liu, B.: Sentiment analysis and subjectivity. In: Handbook of natural language processing vol. 2, pp. 627–666 (2010)
38. Liu, B.: Sentiment analysis and opinion mining. Synth. Lect. Hum. Lang. Technol. 5(1), 1–167 (2012)
39. Lovins, J.B.: Development of a stemming algorithm. MIT Information Processing Group, Electronic Systems Laboratory (1968)
40. Lu, Y., Zhai, C., Sundaresan, N.: Rated aspect summarization of short comments. In: Proceedings of the 18th international conference on World wide web, pp. 131–140. ACM (2009)
41. Maynard, D., Greenwood, M.A.: Who cares about sarcastic tweets? investi-gating the impact of sarcasm on sentiment analysis. In: Proceedings of LREC (2014)
42. Moghaddam, S., Ester, M.: On the design of lda models for aspect-based opinion mining. In: CIKM'12: Proceedings of the 21st ACM international conference on Information and knowledge management, pp. 803–812, ACM, New York, USA (2012). ISBN:978-1-4503-1156-4, http://doi.acm.org/10.1145/2396761.2396863
43. Pang, B., Lee, L.: Opinion mining and sentiment analysis. Found. Trends Inf. Retr. 2(1–2), 1–135 (2008)
44. Pang, B., Lee, L., Vaithyanathan, S.: Thumbs up?: sentiment classification using machine learning techniques. In: Proceedings of the ACL-02 Conference on Empirical Methods in Natural Language Processing, vol. 10, pp. 79–86 (2002)
45. Poria, S., Gelbukh, A., Hussain, A., Das, D., Bandyopadhyay, S.: Enhanced senticnet with affective labels for concept-based opinion mining. Intell. Syst. IEEE, 28(2) (2013)
46. Porter, M.F.: An algorithm for suffix stripping. Program 14(3), 130–137 (1980)
47. Resnik, P., Hardisty, E.: Gibbs sampling for the uninitiated. Technical report, Maryland University College Park Institute For Advanced Computer Studies, DTIC Document (2010)
48. Steyvers, M., Griffiths, T.: Probabilistic topic models. Handb. Latent Sem. Anal. 427(7), 424–440 (2007)
49. Subasic, P., Huettner, A.: Affect analysis of text using fuzzy semantic typing. IEEE Trans. Fuzzy Syst. 9(4), 483–496 (2001)
50. Svenonius, E.: Unanswered questions in the design of controlled vocabularies. J. Am. Soc. Inf. Sci. 37(5), 331–340 (1986)

51. Tatemura, J.: Virtual reviewers for collaborative exploration of movie reviews. In: Proceedings of the 5th international conference on Intelligent user interfaces, pp. 272–275. ACM (2000)
52. Turney, P.D.: Thumbs up or thumbs down?: semantic orientation applied to unsupervised classification of reviews. In: Proceedings of the 40th annual meeting on association for computational linguistics, pp. 417–424. Association for Computational Linguistics (2002)
53. Utsumi, A.: Verbal irony as implicit display of ironic environment: distinguishing ironic utterances from nonirony. J. Pragmat. 32(12), 1777–1806 (2000)
54. Wallach, H.M.: Topic modeling: beyond bag-of-words. In: Proceedings of the 23rd international conference on Machine learning, pp. 977–984. ACM (2006)
55. Wang, H., Lu, Y., Zhai, C.: Latent aspect rating analysis on review text data: a rating regression approach. In: Proceedings of the 16th ACM SIGKDD international conference on Knowledge discovery and data mining, pp. 783–792. ACM (2010)
56. Wang, X., McCallum, A., Wei, X.: Topical n-grams: phrase and topic discovery, with an application to information retrieval. In: Seventh IEEE International Conference on Data Mining, 2007. ICDM 2007, pp. 697–702. IEEE (2007)
57. Ward, K.F., Rolland, E., Patterson, R.A.: Improving outpatient health care quality: understanding the quality dimensions. Heal. Care Manag. Rev. 30(4), 361–371 (2005)
58. Wiebe, J.: Learning subjective adjectives from corpora. In: AAAI/IAAI, pp. 735–740 (2000)
59. Willett, P.: The porter stemming algorithm: then and now. Program 40(3), 219–223 (2006)
60. Wong, T.-L., Bing, L., Lam, W.: Normalizing web product attributes and discovering domain ontology with minimal effort. In: Proceedings of the fourth ACM international conference on Web search and data mining, pp. 805–814. ACM (2011)
61. Zhan, T.-J., Li, C.-H.: Semantic dependent word pairs generative model for fine-grained product feature mining. In: Advances in Knowledge Discovery and Data Mining, pp. 460–475
62. Zhu, X., Ghahramani, Z.: Learning from labeled and unlabeled data with label propagation. Technical report, Technical Report CMU-CALD-02-107, Carnegie Mellon University (2002)

Fundamentals of Sentiment Analysis: Concepts and Methodology

A.B. Pawar, M.A. Jawale and D.N. Kyatanavar

Abstract Internet has opened the new doors for information exchange and the growth of social media has created unprecedented opportunities for citizens to publicly raise their opinions, but it has serious bottlenecks when it comes to do analysis of these opinions. Even urgency to gain a real time understanding of citizens concerns has grown very rapidly. Since, the viral nature of social media which is fast and distributed one, some issues get rapidly distributed and unpredictably become important through this word of mouth opinions expressed online which in turn has known as sentiments of the users. The decision makers and people do not yet realized to make sense of this mass communication and interact sensibly with thousands of others with the help of sentiment analysis. To understand thoroughly use of sentiment analysis in today's business world, this chapter covers the brief about sentiment analysis including introduction of sentiment analysis, early history of sentiment analysis, problems of sentiment analysis, basic concepts of sentiment analysis with mathematical treatment, sentiment and subjectivity classification comprises of opinion mining and summarization, past scenarios of opinion or sentiment collection and their analysis. Methodologies like Sentiment Analysis as Text Classification Problem, Sentiment analysis as Feature Classification with mathematical treatment are explored. Also, Economic consequences of sentiment analysis on individual, society and organization with the help of social media sentiment analysis are provided as supporting component.

Keywords Feature extraction · Sentiment analysis · Opinion mining

A.B. Pawar (✉)
Department of Computer Engineering, S.R.E.S', College of Engineering,
Kopargaon 423 603, India
e-mail: anil.pawar1983@gmail.com

M.A. Jawale
Department of Information Technology, S.R.E.S', College of Engineering,
Kopargaon 423 603, India
e-mail: jawale.madhu@gmail.com

D.N. Kyatanavar
Savitribai Phule Pune University, Pune, Maharashtra, India
e-mail: kyatanavar@gmail.com

© Springer International Publishing Switzerland 2016
W. Pedrycz and S.-M. Chen (eds.), *Sentiment Analysis and Ontology Engineering*,
Studies in Computational Intelligence 639, DOI 10.1007/978-3-319-30319-2_2

25

1 Introduction

Opinion mining also termed as sentiment analysis is the mining of opinions of individuals, their appraisals, and feelings in the direction of certain objects, facts and their attributes. As stated by Liu [10, 11] and Jawale et al. [6], in the latest years, opinion mining has attracted great deal of concentration from both the academicians and industry persons because of various challenging research issues and support of sentiment analysis for a broad set of applications. Opinions play a very important role in making a proper decision. As it is wise to get or listen to the opinions from other people while we make a choice. This scenario is not only true in the case of individual choice but today it is useful and right for organizations also. Very little computational study was carried out on opinions prior to the introduction of World Wide Web (WWW) due to limited availability of opinionated text for such analysis. In earlier days, when the person used to go for taking a decision, she usually used to ask for opinion from different sources either friends or relatives.

As stated by Liu [10] and Jawale et al. [7, 8] when an organization is in need of opinions about their product or service from customer or general public, they generally conduct surveys or opinion polls from a group. Because of increase in the contents on the WWW through social media, the world has changed and became wealthy in data through advancement of Web. Currently people can put their reviews about products on respective business organizational sites and can express their opinions on nearly everything in various blogs and discussion forums. If anyone wants to purchase a product, there is no need to ask about the product to someone from friends and family. As many user reviews are easily available on the Web. So for the industries perhaps it is not required to conduct surveys to get customer opinions about their product or service or about their competitors because ample of information about the same is publicly available.

As stated by [8], due to growth of internet and data contents on Web, searching the sites where opinions are available and then monitoring such sites on the internet is quite an intensive job because the opinion contents are available on different sites, and in turn every site may also have large amount of opinions. It is possible that, in most of the cases, opinions or judgments about particular thing are not directly expressed. So it becomes difficult for a user to identify such opinion related sites, read and extract opinion related contents, analyze such contents, get useful summary out of it, and use this summary to form an opinion. Thus, to do all these tasks there is need of automated opinion discovery and summarization system in the field of business decision making process to enhance the business strategies and business profit in the competitive world.

It is observed in Liu [11] that, the opinion contents which are available online on internet as well as off line are containing mostly textual information used by the customer to provide relevant product feedback. The information available in textual format can be generally classified as either facts or opinions. An objective expression about entities, objects or events and their attributes is known as facts. On the other hand, opinion is a subjective expression that describes sentiments of an individual, assessment of performance or emotions about entities, things or events and their attributes.

Opinion is a broad concept. Majority of the current research study from Liu [9, 11, 14] based on opinion mining concentrated on textual information processing. In addition, it pays attention on opinion mining and retrieval of factual information provided and expressed in opinionated text. It also supports information retrieval, text classification, text mining and natural language processing. Processing of opinions was very little focused and studied concept to get exact opinions of the customer than factual information in recent research of sentiment analysis.

Following section introduces the problem of sentiment analysis and its challenges in bringing automation in opinion mining system.

Problems with Sentiment Analysis

Research in the field of sentiment analysis began with the study of the problem of subjectivity classification and sentiment classification, which mainly considered the sentiment analysis problem as a text classification problem. The subjectivity classification is the field which identifies whether a given text document contains factual information or opinionated information. Then the sentiment classification is responsible for categorizing an opinion into either positive opinion or negative opinion from the set of opinionated documents. In reality, it is required to have in-depth analysis of these opinionated documents because the user is interested to know what opinion have been expressed on certain product, service or on individual. From the survey of a product, one wants to know what features of the product have been praised or criticized by user of the product.

Let us consider an example of opinion expressed by a user on fruit Guava:

"(1) I purchased 2 kg of Guava day before yesterday. (2) They appeared very fresh when I bought them home. (3) The shape of each Guava was very appropriate. (4) The color of some of the pieces was bright. (5) I prepared a tasty salad of it. (6) According to my mother I bought them for higher rate."

In this context, main issue is exactly what we want to extract from this review? We may notice that there are several opinions in this provided input review. In this one, sentences (2), (3) and (4) express three positive opinions on fruit-Guava, while sentences (5) is expressing a fact. Sentence (6) expresses negative opinion about the same. It is also observed that, all the opinions presented by user here have some targets on which the opinions are expressed or given. In this sense, the opinion in sentence (2) is on the appearance of the fruit while the opinions expressed in sentences (3) and (4) are on the features such as 'shape' and 'color' respectively. The opinion given in sentence (6) is expressed on the cost of the fruit, but the opinion expressed in sentence (5) is on consumer herself i.e. on 'me', not on the fruit. Finally, it is also noticed that the source or holder of opinions is also part of this opinionated text. The source or holder of the opinions in sentences (2), (3) (4) and (5) is the author of the expressed review, where as in sentence (6) it is 'my mother'. With the help of this example, one can realize that challenges in sentiment analysis or opinion mining are part of intensive computing and require huge data analysis to get accurately analyzed business opinion.

Before proceeding further, this chapter highlights various basic concepts which
are related with opinion mining field, like:

What is an opinion or sentiment?

What is an emotion?

What is an object?

What is a feature?

What are the types of opinions?

What is an opinion orientation?

Without the knowledge of these factors, it would be nearly impossible for any
business organization decision maker to understand the trend of customer choice,
individual expectations and decisions for improvement measures to be carried out in
organization product development procedure. This chapter also highlights the growth
of the product contents on Web and their diversified formats of representations,
the language contents used by the consumers while expressing the reviews on the
product and challenges in sentiment analysis. The need of automated sentiment or
opinion mining system has been illustrated. The consideration of sentiment analysis
as opinion classification problem rather than text classification is also brought out.
This chapter also gives the details about motivation, issues in sentiment analysis
as well as scope and objective of the proposed research work, contributions of the
proposed research work and detailed outline of the report of research work.

2 Sentiment Analysis: Basic Concepts

In sentiment analysis, the opinionated text is important for decision making based on
its analysis. So while collecting the opinionated text and treating it as an input for the
opinion mining systems, one has to understand the basic terminologies associated
with the opinion mining. The following section highlights the basic concepts related
to this opinionated text and its further processing for decision making process.

As stated by Liu [10], information available in textual format can be classified
into two main things: Facts and Opinions.

An objective expression made by user regarding certain objects, entities or events
and their attributes is known as facts. In the similar way, a subjective expression which
describes emotions of a person, her sentiments and performance assessment about
objects, entities and events and their characteristics is known as opinion. Generally,
opinions can be expressed on anything in this world including a product, person,
business industries, or a topic. It indicates that mostly opinions are expressed on
target entities which are having their own components and attributes. So in opinion
mining, an object can be divided into the hierarchical levels based on the part-of
relation. So, exactly object can be defined as:

Object

An object is mainly any entity which can be anything in real world i.e. person,
organization, event, product, topic etc. Liu [11]. Let, *Obj* represents object here. So,

it can be represented by a pair, *Obj(L, A)*, where L represents hierarchy of levels of components and different sub levels of components. A indicates attribute set of *Obj*. Similarly, every component in the set can have the sub-component set and their respective attribute set. Following example explained below illustrates the definition of object*Obj*, levels of components *L*, and attributes *A*.

Consider phone as general class. So a particular brand of phone can be considered as object. Then this phone has a set of components like battery, speaker and screen, and also its attributes will be quality of voice, slimness of the phone i.e. size and heaviness of the phone. Then component named battery also has separate set of attributes, e.g., battery time, and battery dimension.

So with the help of this definition, an object forms a tree or hierarchy. So phone object can be represented as tree or hierarchy shown in Fig. 1.

Here, the root of the tree represents the object itself (In Fig. 1. it is 'Phone'). After this, every node which is not a root represents the component (for this example, the components of Phone object are Battery, Speaker, Screen) or sub-component of the object (in this case sub-components are represented in {...} brackets i.e. for Battery component, the sub-components are {Battery Life}, {Battery Size}, for Speaker component, the sub-components are {Type}, {Volume Level}, and for Screen component, the sub-components are {Screen Size}, {Screen Color}) and every connection i.e. link represents the available part-of relation. Additionally, a set of attributes defined for that node are in connection with the node. It is important because an opinion is given for any node and on attribute of any node which is associated with the object.

While expressing the opinion, one can comment on object i.e. the phone, which is the root in Fig. 1. These opinions may be like "I don't like this phone", or one can express opinion on one of its attributes as given in Fig. 1 like "The quality of voice of this phone is poor". Similarly, opinions can be expressed on any component of the Phone or any property of that component as shown in Fig. 1.

In theory context, it is sufficient to define an opinionated text object as described earlier, but in practice to understand this opinion concept clearly, we can simplify this definition. The question can be raised here about this simplification in object definition. Basically, there are two main reasons behind it which are illustrated as follows:

Reason 1: Opinions are generally expressed by user in the natural language and this natural language processing is really difficult task as it requires extensive

Fig. 1 Tree structure of an object in opinion mining

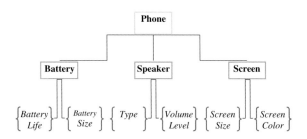

computational text processing. So, to do this text processing, defining an object is extremely challenging.

Reason 2: As defined earlier, object in opinionated text is represented by using hierarchical representation of an object. For normal users; it may become complicated to utilize this form of representation of an object while expressing the opinion about an object. So, instead of this tree kind of hierarchical representation of the object, we will use the concept feature to represent the components and attributes of opinionated text object.

Simplification of Object into Feature Form

Object itself can acts as a feature and becomes a special feature, which is represented as the root in the tree representation.

So, the opinion expressed on this object is then known as general opinion on the object. For this example as shown in Fig. 1 the general opinion statement can be "I like Nokia Phone". When the opinion is expressed on specific feature of the object, then it is known as specific opinion which is given on some feature of that object. For Phone object, specific opinion can be expressed as like [11] "Nokia Phone is having a very good touch screen", here Nokia Phone has a feature 'touch screen'. In real world, people are often habitual with product feature term than using feature term for an object. Even, at the point the objects are considered as topics or events, the feature term may not sound common and simpler one. Additionally, it is observed that researchers also used the terms either topic or aspect to get meaning as feature.

In this research work, we have chosen the term feature along with the term object. Since, both of these terms are required because in majority of the applications, the essential concern of the client or user is a set having interesting objects i.e. a set of products which are related to each other. After this, user is required to know about each feature which are expressing about an opinion document belonging to which object. This feature term is somewhat confusing if one considers the field of machine learning, where a feature indicates data characteristic or attribute. So to evade this conflict, the term which will be used further in this thesis is object feature to indicate feature of an object. To elaborate it in clearer manner, the following mathematical notations will help in this context.

Let, an object O has feature f in an opinionated text. Let d be a document, containing reviews about the products. Now, this opinionated document is collection of expressed opinion by user in statements. So, document d can be described with the help of series of sentences.

$$d = <s_1, s_2, \ldots, s_m>$$

Opinion Passage

An opinion passage is generally expressed on an object O about its feature f in document d which is a collection of sentences. These sentences can be positive opinions or negative opinions on feature f.

From this, it is understood that opinion can be expressed as collection of statements in document d on features f of an object O. Even possibility is that a single sentence may consist of opinions provided on multiple features of an object. The same is given in following sentence.

"The color of this guava fruit is nice but it does not taste that good".

This statement expresses opinion on features i.e. 'color' and 'taste' of an object 'fruit'. In recent studies, it is observed that current research focuses on sentences in opinionated text.

Explicit and Implicit Feature

When user expresses the opinion on any object by considering the features of an object, then one has to carefully categorize these opinions based on whether these are expressed explicitly or implicitly by using features of the object. So based on this expression, such features are called as either explicit or implicit features. The definition of these features is given below:

Explicit feature in sentence s of any document d is defined as follows: The feature f is called explicit feature if the feature or its alternatives is present in a sentence s. For example, in the next statement 'color' is an explicit feature expressed in opinionated text.

"The color of this Guava fruit is pleasant".

Similarly, in the following statement, the opinion is expressed on the 'size' feature of the fruit, but it is not appeared in the sentence directly but it is implied, so it is an implicit feature. "This fruit is big". In this sentence, big is not synonym of feature 'size', but it indicates opinion towards this feature, so such indications are known as a feature indicator. Mainly adjectives and adverbs are used as feature indicators. Some of the adverbs and adjectives are very commonly used like nice, good, bad or worst. Apart from these, other adjectives adverbs may be used for indicating specific feature, like beautiful can be used to indicate appearance, etc. These kinds of feature indicators can be mapped directly with their respective features.

Opinion Holder and or Opinion Sources

Mainly opinions on certain objects are expressed by users. Users may be individual person, group, and organization. It means that these users are authors of the opinions. In the field of sentiment analysis, such users are known as holder of an opinion. These holders of opinion are also known as opinion sources. Whenever feedback or review about particular thing is published or a blog is written, the opinion is given by the author of that review or blog. Additionally opinion holders are considered vital in the news articles as they provide explicit view about a situation, place, organization or person which is published in news articles. To understand it, consider the sentence, "Sarita expressed her disagreement on the purchased car." The opinion holder in this sentence is 'Sarita' since she is the opinion source in this sentence as 'Sarita' is mentioned explicitly in this sentence.

Opinion and Opinion Orientation

With the help of all above concepts related to opinion in sentiment analysis, opinion can be defined as, a positive or negative view, appraisal, emotion or feeling about object O with the help of feature f given by an opinion holder. These views on opinion lead us towards the next concept i.e. opinion orientation. Opinion orientation decides orientation of the expressed opinion on an object O for a feature f into positive, negative or neutral form. The term opinion orientation is alternatively used in many research papers as 'sentiment orientation', 'polarity of opinion', or even 'semantic orientation'.

Based on these concepts, the opinion mining system can be expressed in theory with the help of Object Model for opinion mining. It is described in the next section in depth.

Object Model

The object model can be derived as follows. Let, object O is a set of finite features as given below,

$$F = \{f_1, f_2, \ldots, f_n\}$$

This feature set F may include the object as a special feature. As described earlier, these features are the attributes of object. So, each feature $f_i \in F$ can be represented as a finite set of words or phrases where,

$$W_i = \{W_{i,1}, W_{i,2}, \ldots, W_{im}\}$$

Which are the feature synonyms, and finite set of feature indicators is expressed as of the feature.

Opinionated Document Model

A general document d containing opinion on set of objects as $\{Obj_1, Obj_2, \ldots, Obj_n\}$ from set of opinion holders represented as $\{OH_1, OH_2, \ldots, OH_n\}$. Now, these opinions expressed on each object Obj_1 are a subset F_i of features.

There are mainly two types of opinions, called as direct opinion and comparative opinion which are defined as below.

Direct Opinion

A direct opinion is defined in the form of quintuple $(Obj_i, f_{ik}, Obj_{ijk1}, h_i, t_l)$, where an object is represented as Obj_i, the feature is denoted as f_{ik} of an object obj_i, and the orientation or polarity of the opinion on feature f_{ik} of object obj_i is denoted as obj_{ijkl}, opinion holder is represented as h_i and t_l denotes the time when an opinion is commented by opinion holder h_i. The orientation of an opinion obj_{ijkl} can be identified as positive, negative or neutral. For feature, f_{ik}, the opinion holder h_i selects words or sentences from the corresponding alternative W_{ik}, or a sentence from the feature indicator set as I_{ik} to explain the feature, and then it is articulated as a positive, negative or neutral opinion on the feature in the opinionated text.

More specifically, the direct opinions are divided into two sub-types based on direct opinion expression on an object and on its effect on some other object. The first case considers opinions which are directly expressed on an object or its features; the next example clarifies this concept. "The color of Guava fruit is really fresh." Here, the object is Guava fruit and its directly stated feature in this sentence is 'color'. In the latter case, opinions expressed for an object are commented based on the effect of an object on some other objects. It is very easily adopted and frequently used in the medical field where patients expresses their opinions on drugs or describe drug side effects very frequently. If patient expresses opinion like "After taking this medicine, my leg pain reduced." Then it describes an expected effect of the specified drug on the leg, and so it implies a positive opinion about that medicine. For simplicity, we are considering these both sub types under the direct opinion category in the further discussion.

Comparative Opinion

A comparative opinion as the name itself explores that it is generally expressed on two or more objects based on their similarities or differences, and the object preferences are given by the opinion holder based on some of the shared features or attributes of these objects. Comparative opinion is described with the help of comparative or superlative form of an adverb or adjective, but it is not true for all comparative opinions.

Strength of Opinion

Strength of an opinion is mainly defined by opinion orientation obj_{ijkl} as described in quintuple of opinion. Each opinion comes in different strengths where few are very strong and few are weak. The strong opinion may be like opinion expressed in the following statement "This phone is rubbish" and weak opinion may be expressed like, "As per my thinking this phone is nice". So, it is possible to determine power of opinions. A feeling of contentedness, happiness, joyfulness, or delightedness is expressed with the help of positive opinion. In practice, as per the application, one can select scale to measure the strength of opinion. The more depth of this discussion indirectly leads towards the emotions.

Emotions

Emotions are human beings subjective feelings and thoughts expressed by them. The significant study has been done on the emotions and studied in many fields like biology, social science, psychology, philosophy, etc. Still it is found that researchers are not agreed on basic set of emotions. Mainly, there are six types of emotions, namely joy, love, surprise, anger, fear and sadness. These emotions further can be classified into their own subtypes.

Though emotions are influencing the strength of opinion and are exhibiting close relationship with opinion, there is no equivalence between emotion and opinion. So when the impact of emotions or opinions is discussed, it is important to differentiate between these two things i.e. feelings and language expressions used to describe feelings. As stated earlier, there are only six types of primary emotions; but on other side there are many expressions of the language that can be used to convey emotions. In the same sense, there are also many opinion expressions used to explain positive sentiments or negative sentiments by users. Therefore, mainly sentiment analysis attempts to conclude individual sentiments based on expressions provided with the help of any language to portray their emotions.

Extensive Purpose of Sentiment Analysis

The main purpose of sentiment analysis, aims to conclude positive sentiments or negative sentiments from given opinionated text. But this objective can be extended to discover the additional information which is significant for use of opinions in the practical decision making process. So the objectives of direct opinion mining can be illustrated as: In given opinionated document d, firstly, identify the quintuple of opinion i.e. $(obj_i, f_{ik}, obj_{ijkl}, h_i, t_l)$ and secondly, find out all the feature indicators I_{ik} and synonyms (W_{ik}) of each feature f_{ik} in d.

Issues present in the Feature based Opinion Mining

Here we will discuss the different issues related to feature based opinion mining [2].

Issue 1

In the feature based opinion mining, each time it is expected that the quintuple containing five pieces of information about an opinion should be related to each other. It indicates that, the opinion have to be expressed on feature of object at specified time by opinion holder. It can be concluded from this condition that, sentiment analysis is really challenging problem since in opinionated text; even discovering each piece of information itself is very complicated. Even it becomes worse when in opinionated text, sentence does not explicitly mention some pieces of information required in quintuple, and may be stated with the help of pronouns, language expressions, and the context of opinionated text.

The following blog example highlights this issue with the help of further illustration. Each statement is identified with the number for simplicity and understanding in blog opinionated text.

"(1) Last Monday, I purchased a 'Samsung' phone and my granny got her gift as an 'Asus' phone. (2) She called me to inform me the same. (3) I could not listen to her as voice of her phone was not clearly audible. (4) The picture taken from camera was fine. (5) I am really happy and comfortable with my phone. (6) I got the phone as per my expectations with good voice. (7) So the phone was really displeasure for us. (8) My granny was not at all satisfied with her gift and sent it back to the retailer day before yesterday."

The identified objects in this given blog contain opinionated text describing 'Samsung phone' and 'Asus phone', which is really difficult to identify in practical way. To get an idea about what does 'my phone' indicate and what does 'her phone' shows in sentence (3) and sentence (5) is actually difficult. Sentence (4) is really conflicting one as it is not having mention of either 'Samsung phone' or 'Asus phone' and it also does not contain a pronoun. Then the question raised here is that 'the camera' belongs to which phone. Sentence (6) directly gives a positive opinion about a phone and quality of its voice. In sentences (7) and (8), it is very difficult to get information about which 'phone' it is and what 'the gift' is. Here, the opinion holder of all the opinions is the author himself of the opinionated blog text except in sentence (8) the opinion holder is 'my granny' instead of author of the blog.

Issue 2

It is not needed to identify the quintuple with all five pieces of information every time for each application in practice. Since some time some of them may be unknown and unwanted. In case of reviews about a product, the reviewed object i.e. product is assessed in each review. When the review is written, the author of the review always knows that a review site typically records and displays review information. Though this information is not required every time, one needs to take out such review information from the Web pages of review sites because of structured data extraction issue.

The example stated earlier in issue1 additionally revealed another issue, known as, subjectivity. It states that, in any document which is having collection of opinions, some sentences always give opinions where as some does not. As seen in earlier example, sentences (1), (2) and (8) do not comment on any opinions. It gives the clue for sentence subjectivity concept which is defined and described in the next section.

Sentence Subjectivity

Generally, factual information about the world is always represented with the help of objective sentences, while subjective sentences are used to express personal emotions or beliefs. As example given in issue1, sentences (1), (2) and (8) are categorized as objective sentences, whereas all remaining sentences are called as subjective sentences. Many forms of subjective expressions are like desires, opinions, speculations, beliefs, suspicions, and allegations. Therefore, a subjective sentence may or may not contain an opinion every time. For example, the following sentence indicates the sentence subjectivity but it does not state either positive opinion or negative opinion on any precise smart phone object. "I expected a smart phone with good reception signal power." In the same way, not every objective sentence contains any opinion.

Explicit Opinion and Implicit Opinion

As discussed earlier a feature which is specified explicitly and a feature which is implied in the sentence, an explicit opinion is defined as an opinion which is expressed explicitly in a sentence on feature f in a subjective sentence of the opinionated text [2]. Similarly, an implicit opinion is defined as an opinion on feature f which is implied in an objective sentence of the opinionated text. The sentence as presented below is an example of an explicit positive opinion in subjective sentence. "Clarity of this TV is excellent." and in the similar way the next sentence expresses an implicit negative opinion in an objective sentence. "The earphone failed working in two days." It is noted that, this sentence presents a fact with the help of objective sentence; but still it is a negative opinion on an 'earphone' which is indicated implicitly.

Opinionated Statement

An opinionated statement is defined as a statement that explicitly expresses positive or negative opinion or implicitly implies positive or negative opinions. This opinionated sentence may be either of these two categories viz., subjective sentence or objective sentence. One needs to understand that subjective sentences and opinionated sentences are two different concepts, though these sentences which are opinionated are most of the time the subset of subjective sentences. The ways with which opinionated and subjective sentences are determined are also similar. In this research work to avoid the reader confusion we will use these terms alternative to each other. The function that is used for distinguishing subjective or objective sentences is known as subjectivity classification. The next section describes in depth about this concept.

Sentiment and Subjectivity Classification

The term sentiment classification is used to classify an opinionated document based on whether it expresses the positive opinion or negative opinion. This is also called as the sentiment classification done at document level because it takes the complete document as the basic information unit used as an input for performing opinion mining [10]. The more details about sentiment classification at document level are illustrated in the following section. The most of the existing researcher done the study based on the document is known to be opinionated as the basic information unit. However, this sentiment analysis can be applied to the individual sentences of the document too. But one has to note that, we cannot assume each sentence to be opinionated as part of the opinionated document. The function that classifies a given sentence into either opinionated sentence or non-opinionated sentence is known as subjectivity classification. Even the resultant sentences which are opinionated are also devised into positive sentence or negative sentence, which is further known as sentence-level sentiment classification [1, 13].

Sentiment Classification at Document Level

Document level sentiment classification is used to determine whether the document d is expressing a positive or negative opinion when given a set of opinionated documents D and each document $d \in D$.

More specifically, sentiment classification done at the document level can be illustrated better in the following way. Consider, a document d containing opinionated sentences which provides an opinion on object O, then identify the orientation or polarity of the opinion expressed on an object O, i.e. determine what is the orientation of that opinion on feature f. It can be represented in the quintuple (o, f, s, h, t), where $f = O$ and h, t, O are assumed to be known or not relevant one.

During this study it is observed that the traditional research work done on classification of sentiments is based on the theory that the document d which is opinionated contains usually the product reviews on an object O and these opinions are usually expressed by only one opinion holder h. Such kind of assumption can be useful for user reviews on products and services. But it cannot be applicable for the opinionated text generated by blog posts and a forum and since in these posts, the opinion holder may provide her opinions on multiple targets and may compare those opinions using superlative or comparative sentences.

Additionally, sentiment classification done at document-level can be carried out using supervised learning and unsupervised learning methods. These methods are described below.

Sentiment Classification at Document Level Based on Supervised Learning

Sentiment classification can be done as a supervised learning problem with two class labels which are positive and negative. In existing research studies, the training and testing data used are mostly reviews about a product of certain organization.

Sentiment classification is similar with classic topic-based text classification. But distinction can be made between two as sentiment classification classifies the documents into either positive or negative whereas text classification using classic topic based methods classifies documents into already defined topic classes like movies, sports, food, education, etc. Additionally, in classification using topic-based methods, the main emphasis is given on topic related words. On the other hand, in sentiment classification importance is not given to topic-related words during the classification. Instead, it puts emphasis on identifying expressed sentiment, feeling, emotion or opinion words which expresses either positive opinions or negative opinions like good, bad, outstanding, nice, etc.

It is found that present supervised learning classification techniques can be directly used to perform sentiment classification; such techniques are Naïve Bayesian Classification Algorithms, Nearest Neighbor Algorithm, Decision Tree Classifier, Artificial Neural Networks and Support Vector Machines, etc. It is proved earlier that features in classification are considered using unigrams (also called as a bag of individual words) have shown the expected results with supervised learning techniques either Naïve Bayesian Algorithm or Support Vector Machine. The main thing to be clarified here is about use of features. Here, the features are considered as data attributes as described in machine learning, instead of object features which are mentioned in earlier section. Following are the examples of such features used in existing research work [1]:

Term (T) and Term Frequency (TF)

The feature considered as individual word or word n-grams is called as term and its occurrence count in the document is known as term frequency. During the classification word positions may be another issue. So from the field of information retrieval, the feature Term Frequency—Inverse Document Frequency weighting scheme can be applied. These features are common part of traditional topic-based text classification, but it is found that these are quite effective and useful one in sentiment classification also.

Part of Speech Tags (POS)

It is the method used to assign a Part-of-Speech to every word present in the sentence. Every word in the sentence is assigned a tag like, verb, noun, prepositions and adjective etc. In English language, mainly adjectives are used to identify subjectivity and opinions. So, the earlier researchers used these adjectives as significant indicators of either subjectivities or opinions and counted these adjectives as the special features in the field of opinion mining.

Opinion Words and Opinion Phrases

In the opinionated text positive or negative sentiments which are commonly used to express emotions of the opinion holder are called as opinion words. We can consider the beautiful, good, amazing, etc. as positive opinion words and negative opinion words like, bad, weak, poor, etc. Additionally, as stated earlier, adjectives, adverbs, nouns like rubbish, junk, and crap and even verbs like hate, like, etc. are used as opinion words.

Moreover, instead of such individual opinion words, there are also idioms and phrases which can be used to indicate opinions. Consider an example, "These opera tickets cost us an arm and a leg". Here, 'cost someone an arm and a leg' is a phrase which means having a negative impact of something. Therefore, opinion phrases and opinion words play a vital role in performing sentiment analysis.

Syntactic Dependency

It is defined as words dependency which is based on features generated from dependency trees or parsing. Several researchers can use it in their research in the field of sentiment analysis.

Negation

This term is vital in sentiment analysis since its appearance normally change the opinion orientation results. In the next example, the sentence "I am not happy with this camera" is negative. Handling of negation words requires careful analysis because not all occurrences of such words are used to indicate negation like in the sentence 'not' in 'not only ... but also' does not change the orientation status.

Sentiment Classification at Document Level based on Unsupervised Learning.

In sentiment classification opinion phrases and opinion words are the most important indicators. Hence use of unsupervised learning methods for sentiment analysis which is based on such opinion phrases and opinion words can be obvious in the research of sentiment analysis. Mainly, these methods perform classification by using fixed syntactic phrases that are mostly used by the user to state the opinions. Generally, these algorithms consist of the following steps.

Step 1

Initially, from the given document, it does the extraction of phrases containing adverbs or adjectives. Because as subjectivity and opinions can be indicated from adverbs and adjectives as described earlier. However, it is observed that subjectivity is indicated with the help of adjective, but there may be insufficient relative information to state or decide its opinion orientation results based on such isolated adjectives. So, these algorithms extract two successive words, out of these two one member represents either an adjective or adverb and the other gives context dependent word. If the Part-of-Speech tag of these two successive words conform to any of the patterns as shown in Table 1 then only these words will be extracted. As shown in line 2 of Table 1 two successive words are extracted if there is match for the pattern as like first word is an adverb and the second word is an adjective, but the third word cannot be a noun.

Table 1 POS patterns for two words extraction

First word		Second word	Third word (Not Extracted)
1.	JJ	NN or NN	Anything
2.	RB, RBR, or RBS	JJ	not NN nor NNS
3.	JJ	JJ	not NN nor NNS
4.	NN or NNS	JJ	not NN nor NNS
5.	RB, RBR, or RBS	VB, VBD, VBN, or VBG	Anything

3 Sentiment and Subjectivity Classification

The term sentiment classification is used to classify an opinionated document based on whether it expresses the positive opinion or negative opinion [10]. This is also called as the sentiment classification done at document level because it takes the complete document as the basic information unit used as an input for performing opinion mining. The more details about sentiment classification at document level are illustrated in the following section. The most of the existing researcher done the study based on the document is known to be opinionated as the basic information unit. However, this sentiment analysis can be applied to the individual sentences of the document too. But one has to note that, we cannot assume each sentence to be opinionated as part of the opinionated document. The function that classifies a given sentence into either opinionated sentence or non- opinionated sentence is known as subjectivity classification. Even the resultant sentences which are opinionated are also devised into positive sentence or negative sentence, which is further known as sentence-level sentiment classification [10, 13].

Sentiment Classification at Document Level

Document level sentiment classification is used to determine whether the document d is expressing a positive or negative opinion when given a set of opinionated documents D and each document $d \in D$.

More specifically, sentiment classification done at the document level can be illustrated better in the following way. Consider, a document d containing opinionated sentences which provides an opinion on object O, then identify the orientation or polarity of the opinion expressed on an object O, i.e. determine what is the orientation of that opinion on feature f. It can be represented in the quintuple (o, f, S, h, t), where $f = O$ and h, t, O are assumed to be known or not relevant one.

During this study it is observed that the traditional research work done on classification of sentiments is based on the theory that the document d which is opinionated contains usually the product reviews on an object O and these opinions are usually expressed by only one opinion holder h. Such kind of assumption can be useful for user reviews on products and services. But it cannot be applicable for the opinionated text generated by blog posts and a forum and since in these posts, the opinion holder may provide her opinions on multiple targets and may compare those opinions using superlative or comparative sentences.

Additionally, sentiment classification done at document-level can be carried out using supervised learning and unsupervised learning methods. These methods are described below.

Sentiment Classification at Document Level Based on Supervised Learning

Sentiment classification can be done as a supervised learning problem with two class labels which are positive and negative. In existing research studies, the training and testing data used are mostly reviews about a product of certain organization.

Sentiment classification is similar with classic topic-based text classification. But distinction can be made between two as sentiment classification classifies the documents into either positive or negative whereas text classification using classic topic based methods classifies documents into already defined topic classes like movies, sports, food, education, etc. Additionally, in classification using topic-based methods, the main emphasis is given on topic related words. On the other hand, in sentiment classification importance is not given to topic-related words during the classification. Instead, it puts emphasis on identifying expressed sentiment, feeling, emotion or opinion words which expresses either positive opinions or negative opinions like good, bad, outstanding, nice, etc.

It is found that present supervised learning classification techniques can be directly used to perform sentiment classification; such techniques are Naïve Bayesian Classification Algorithms, Nearest Neighbor Algorithm, Decision Tree Classifier, Artificial Neural Networks and Support Vector Machines, etc. It is proved earlier that features in classification are considered using unigrams (also called as a bag of individual words) have shown the expected results with supervised learning techniques either Naïve Bayesian Algorithm or Support Vector Machine. The main thing to be clarified here is about use of features. Here, the features are considered as data attributes as described in machine learning, instead of object features which are mentioned in earlier section. Following are the examples of such features used in existing research work:

Term (T) and Term Frequency (TF)

The feature considered as individual word or word n-grams is called as term and its occurrence count in the document is known as term frequency. During the classification word positions may be another issue. So from the field of information retrieval, the feature Term Frequency- Inverse Document Frequency weighting scheme can be applied. These features are common part of traditional topic-based text classification, but it is found that these are quite effective and useful one in sentiment classification also.

Part of Speech Tags (POS)

It is the method used to assign a Part-of-Speech to every word present in the sentence. Every word in the sentence is assigned a tag like, verb, noun, prepositions and adjective etc. In English language, mainly adjectives are used to identify subjectivity and opinions. So, the earlier researchers used these adjectives as significant indicators of either subjectivities or opinions and counted these adjectives as the special features in the field of opinion mining.

Opinion Words and Opinion Phrases

In the opinionated text positive or negative sentiments which are commonly used to express emotions of the opinion holder are called as opinion words. We can consider the beautiful, good, amazing, etc. as positive opinion words and negative opinion words like, bad, weak, poor, etc. Additionally, as stated earlier, adjectives, adverbs, nouns like rubbish, junk, and crap and even verbs like hate, like, etc. are used as opinion words.

Moreover, instead of such individual opinion words, there are also idioms and phrases which can be used to indicate opinions. Consider an example, "These opera tickets cost us an arm and a leg". Here, 'cost someone an arm and a leg' is a phrase which means having a negative impact of something. Therefore, opinion phrases and opinion words play a vital role in performing sentiment analysis.

Syntactic Dependency

It is defined as words dependency which is based on features generated from dependency trees or parsing. Several researchers can use it in their research in the field of sentiment analysis.

Negation

This term is vital in sentiment analysis since its appearance normally change the opinion orientation results. In the next example, the sentence "I am not happy with this camera" is negative. Handling of negation words requires careful analysis because not all occurrences of such words are used to indicate negation like in the sentence 'not' in 'not only ... but also' does not change the orientation status.

Sentiment Classification at Document Level based on Unsupervised Learning

In sentiment classification opinion phrases and opinion words are the most important indicators. Hence use of unsupervised learning methods for sentiment analysis which is based on such opinion phrases and opinion words can be obvious in the research of sentiment analysis. Mainly, these methods perform classification by using fixed syntactic phrases that are mostly used by the user to state the opinions. Generally, these algorithms consist of the following steps.

Step 1

Initially, from the given document, it does the extraction of phrases containing adverbs or adjectives. Because as subjectivity and opinions can be indicated from adverbs and adjectives as described earlier. However, it is observed that subjectivity is indicated with the help of adjective, but there may be insufficient relative information to state or decide its opinion orientation results based on such isolated adjectives. So, these algorithms extract two successive words, out of these two one member represents either an adjective or adverb and the other gives context dependent word. If the Part-of-Speech tag of these two successive words conform to any of the patterns as shown in Table 1 then only these words will be extracted. As shown in line 2 of

Table 1 two successive words are extracted if there is match for the pattern as like first word is an adverb and the second word is an adjective, but the third word cannot be a noun.

Step 2

The orientation estimation of the phrases which are extracted can be calculated with the measure of Point-wise Mutual Information (PMI) as given in equation:

$$PMI(t1, t2) = log_2 \left(\frac{Pr(t1 \wedge t2)}{Pr(t1)Pr(t2)} \right)$$

In this equation, the term $Pr(t1 \wedge t2)$ is the co-occurrence probability of t1 and t2, and another term $Pr(t1)Pr(t2)$ represents the probability of the occurrences of these two terms if and only if they are statistically not dependent on each other. Therefore, this degree is a measure of the level of statistical dependence between them and log of this proportion gives the measure of information that we are getting about the occurrence of one of the words when we see its connection with other word.

Now, based on this we can calculate the individual phrase opinion orientation based on its connection with the words like outstanding which is a positive reference and based on its connection with the words like unfortunate which is a negative reference. So the individual phrase opinion orientation can be calculated using the following equation.

$$OO(phrase) = PMI(phrase, "excellent") - PMI(phrase, "poor")$$

The probabilities can be calculated by targeting the queries to any information retrieval tool and based on its response in terms of hits i.e. count. For every targeted request, information retrieval tool provides the response in terms of documents which are relevant to the request made by user, which represents the total count. So, it is easy to calculate the probability of the two terms together as well as separately as per the PMI calculation equation.

Step 3

Based on given collection of reviews, the algorithm can compute the average object orientation value of all the provided phrases and categorize the provided reviews as important and useful if average of this object orientation is positive; otherwise it is not recommended as important and useful.

Sentiment Classification at Sentence Level

Now for the similar task we can compute the sentence-level classification. Suppose the task is given as below. For a sentence s, perform the two important sub-tasks which are given below [1].

Perform Subjectivity Classification

In this one can find out sentence s is either an objective sentence or a subjective sentence as per the opinion expressed.

Perform Sentiment Classification at Sentence Level

In this, if sentence s is subjective sentence, then one can find out sentence s is either a positive opinion or negative opinion.

In this case, it is observed that the quintuple (o, f, OO, h, t) is not utilized for describing the task since the classification at sentence level is a middle step. Then again, the two sub-assignments of the classification done at sentence-level are still very significant because (1) it sort out those sentences which are not having any opinion, and (2) after the identification which are the objects and which are the features of that objects are expressed in these sentences, it provides help to decide that the provided opinions on the underlying objects and also on the features of that objects are either positive or negative. It is found that significant study has been done on these problems, but in separate directions as classification problem, so, we can easily apply the supervised learning methods for their solution. The work reported by earlier researchers was focused on subjectivity classification. Classifier used for doing subjectivity classification is the Naïve Bayesian Classifier.

The main drawback of such supervised learning is, it requires extensive manual efforts to do annotation of a big number of training examples. Another approach to reduce the manual classification attempt is a bootstrapping method which was proposed for automatically labeling the training data. In this case, the algorithm considers the uses of two classifiers with high precision namely, HP-Subject and HP-Object which can categorize automatically subjective sentences and objective sentences. These classifiers list out lexical tokens such as words and n-grams words which give clues for doing better subjectivity classification. The HP-Subject determines a subjective sentence if it indicates more than two well-built subjective clues and HP-Object identifies objective sentence if it does not show strongly subjective clues. The merit of these classifiers is that precision is very high but demerit is that its recall is low. These extracted sentences are then added to the training data in order to learn about the patterns. These patterns are then utilized to perform identification of other sentences which are either subjective sentences or objective sentences. These classified sentences are further added in the training data set, and algorithm continues with its next continues loop. For pattern learning, we can provide syntactic template set to limit the types of patterns to be learned by this classifier. Few examples of syntactic templates and their respective patterns as example are given in Table 2.

Assumptions made for Sentiment Classification at Sentence Level

Table 2 Syntactic template pattern with examples

Syntactic template pattern	Example
<subject> passive-verb	<subject> was satisfied
<subject> active-verb	<subject> complained
Active-verb <dobject>	endorsed <dobject>
noun aux <dobject>	fact is <dobject>
passive-verb prep <np>	was worried about <np>

It is most of the time assumed that the sentence is provided by a single opinion holder to express a single opinion. This hypothesis is suitable only when a simple sentence expresses a single opinion. Consider the example, "Clarity of this TV is excellent." But in fact, compound sentences can provide one or more opinions. Consider the same example with opinion expressed as, "Quality and Clarity of this TV is excellent but its SMPS is poor." provides both the types of positive as well as negative opinions in mixed form. This sentence gives positive opinion for 'quality' and 'clarity', but expresses negative opinion for 'SMPS', even this whole compound sentence is positive for the camera. At the last, it should be kept in mind that subjective sentences are subset of opinionated sentences, and additionally, many objective sentences are also used to express the opinions. So, we need to identify opinions from given textual information with the consideration of subjective as well as objective sentences.

4 Opinion Mining and Summarization

The Internet has changed every individual's life to a great extent which was very sophisticated and simpler one previously. Generally, people are using the Web technology for their functionalities like information retrieval, to perform communication, to enjoy their relaxation and to have an entertainment, to done online shopping, etc. Even, Internet forums are also used as medium of information and online resource exchange. The commercial review [14] Web sites are used to express opinions on the products, individuals, and on organizations. These commercial review websites allow users to express their opinions in their own ways, so the collection of such reviews for specific product is available enormously. Hence it becomes challenge for the customers to observe all these reviews to make a decision on certain product choice and its selection.

So, with the help of opinion mining, an extraction technique can be developed to score the reviews and summarize the opinions to end user. From this, user can decide whether to recommend the product or not based on these opinions mined.

Past Scenario of Collecting Opinions and Their Analysis

During the decision making process we always take in account that "What is the opinion or thinking of other persons?" This type of view always had been a vital part of information for most of us during the decision making to get the idea or views of people in the decision. When we decide to make a decision we want to hear other persons for their opinions. This is applicable not only for individual's performance assessment but also for organizations. In recent days, when WWW has grown largely and easily available, many of us used to ask our friends to recommend a certain product or vote poll for local elections, reference for individual, etc. In past when an organization requires finding general public opinions about its products and related services, they mainly go for surveys and their center of attention is groups. As there is a tremendous growth of the data contents on the Web due to social media in the last few years, the world has been changed completely. Internet and the Web made it very simpler to discover the opinions of users and share their knowledge and their feedback in the massive group of unknown people that are never been our individual good wishers or professional criticizer. Generally, these are people we have never heard of. Today, it is observed that many users are expressing their opinions on Internet without any constraints.

Due to this explosive growth of Web, finding such sites which have opinions and monitoring such sites on the Web became really challenging job. Since a large numbers of such opinion sites are available and each one structured differently, and each site may contain large amount of text which is opinionated. Most of the times, opinions are indirectly expressed by the user in their posts and blogs. It is not possible for the human expert to locate opinionated Web sites, then take out opinionated text, understand it, summarize it, and organize it into usable forms for the decision making. So, it demands the need of automated opinion discovery and summarization system which is the main theme of this research work.

Sentiment Analysis as Text Classification Problem

In sentiment analysis field research is done mostly by academicians. In this study, it is found that, sentiment analysis is treated as a problem of text classification. Additionally, two associate research areas have been comprehensively studied in this study which are: (1) classification of document which is opinionated based on either a document expresses a positive opinion or negative opinion, and (2) classification of a sentence either subjective sentence or objective sentence, and for the sentences which are subjective, again classify it on the basis of either it expresses a positive opinion, negative opinion or neutral opinion. First research area is universally known as sentiment classification at document level, which focuses on finding the broad feeling of the opinion holder in the given opinionated text. With the help of given product reviews; it helps to recognize and understand whether the opinion holder is positive or negative on the subject of that product. The second area concentrates on the sentences individually to identify whether this sentence is responsible for contributing any opinion or not which is known as subjectivity classification, and also it is known as sentence-level sentiment classification.

Sentiment Analysis as Feature Classification

During the study, it is observed that in most of the cases it is important to done the classification of the opinionated texts is either at the sentence level or at the document level. However, such classification is not giving the necessary details required for other different types of applications. In general, a positive opinion on a certain feature of an object in the opinionated document does not indicate that the person has positive opinions about all the features of that object. Similarly, a negative opinion on a certain feature of an object in the opinionated document does not indicate that the person has negative opinions about all the features of that object. Writer may put in writing both aspects of the object in an opinionated text as positive and negative. These types of details are not given by the classification at document-level and classification at sentence-level. So, for such information, there is a need to identify features of an object that means there is a requirement of extraction of all five pieces of quintuple i.e. full model of an object. As declared previously, at the feature level, the mining task deals with discovery of every quintuple $(obj_i, f_{ik}, obj_{ijkl}, h_i, t_l)$ and determine all the synonyms W_{ik} and feature indicator is I_{ik}.

Here, we basically perform two important mining tasks as given below:

(1) Determine all the features of an object that have been expressed. Consider the example sentence, "The taste of tea is very good." and here, the feature of an object is 'taste'.
(2) Identify the expressed opinions on the features of an object is a positive remark, negative remark or neutral one.

Here in the above example sentence, the opinion expressed on the feature 'taste' of a tea object is positive.

This model identifies the opinion sentences and opinion objects on which opinions are given in those sentences, and subsequently classifies the opinions as positive, negative or neutral kind of. Generally, the targeted objects are real world entity and their features are identified. Here, in opinion mining, an object can be manufactured goods, any types of service, human being, any institute, any firm, occasion, theme, etc. on which opinions are expressed. Consider the case of product, in review sentences of product, it finds features of an object which is product that has been expressed by the user and identifies the comments expressed are either positive type or negative type. In the sentence, for example, "The battery of this Laptop is giving less backup," here, the opinion is expressed on 'battery' of the considered object 'Laptop' and the expresses negative kind of opinion. This level of thorough investigation is the need of real existence applications during the decision making process. Since for the improvement of product one must be familiar with which are the components or which are the features of the product are positively recommended or not recommended by the users. Generally, this type of detailed investigation of information is not acknowledged by sentiment classification and subjectivity classification.

5   Economic Consequences of Sentiment Analysis—Impact on Individual, Society and Organization

Many business analysts of online reviews believe that these reviews considerably control their marketing strategies as well as user purchasing decisions. Today, users are always searching for and dependent upon online advice and recommendations, suggestions rather than asking particular individual about his or her choice. Due to the growth and easy availability of Internet, it is easily possible to discover about the opinions expressed and shared feedbacks of people from all the corners of the world regardless of their professions and expertise in the particular domain. And conversely, many users are putting their opinionated text available to strangers via the Internet and social media sites. The key findings about impact on individual, society and organizations of opinion mining are stated below as per the two surveys carried out in America on 2000 American people in year 2008; details are reported by Horrigan [3].

Online investigation about a product minimum once is performing by 81% of Internet users before purchasing it; 20% Internet users perform such investigation on a usual day the majority of the time; During selecting the various daily services, most of the readers of online reviews of hotels, restaurants, and a variety of services like hospitals, among 73% and likely 87% users reported that reviews had a major influence while doing selection of such services and products; Consumers report says that users are easily ready to pay starting from 20 to 99% additional for a 5-star-rated products than that of a 4-star-rated products based on reviews given by various users.Out of all users, 32% users have given and commented an evaluation about a person, a product, or a service via an online evaluation system, and 30% users counting 18% of online senior general public have posted and expressed an online observation or comment about the service or product during their online purchase.

It is noticed that utilization of services and goods is not the merely inspiration and foundation behind citizens expressing opinions online. Even, there are so many factors involved in this posting of reviews online. Mostly, for political information, it is used extensively.

Following example illustrate this fact more clearly. Survey report of more than 2500 Americans reported by Lee and John [12] pointed that the 31% of Americans were 2006 election campaign Internet users; 28% users believed that a most important issue for these online actions was to reach and to know judgments and perspectives from within their society, and 34% users said that a most important cause was to get judgments and perspectives from outside their society and to see its impact during election; 27% users had stared online for the approval of outside organizations posted online; 28% users declare that the majority of the sites they utilize to share their point of observation on certain public issues, but 29% users said that most of the sites they use challenge their point of view against others views; 8% users posted their own political expressed comments, opinions online without considering any public issues.

The mentioned above two report findings are just representative example that shows user expectations and dependency based on online recommendations and advice is major cause at the back the surge of attention in innovative systems that concerned in a straight way with opinions as a primary rank object and gives them opinion analysis as early as possible. This suggests an unambiguous necessity to assist customers by constructing improved information analysis systems as compared to those that are presently in way of life.

6 Conclusion and Future Scope

In this chapter, detailed and clear suggestions about findings and conclusions of opinion into the present and current research study are studied for analysis of an opinion. Current research studies major contribution is to distinguish either opinion is positive, negative or neutral on the basis of explicit opinions and its explicit feature recognitions only. Not provided a great deal of attentions on implicit opinion and its implicit feature extraction and identifications. So the planned research studies center of attention will be on recognizing and identifying explicit as well as implicit opinions with the help of their explicit and implicit features. This sensible and large requirement in major applications and the technological challenges will stay the field lively and active.

References

1. Dalal, M.K., Zave, M.A.: Automatic text classification: a technical review. Int. J. Comput. Appl. (0975–8887) **28**(2), 37–40 (2011)
2. Eirinaki, M., Pisal, S., Singh, J.: Feature-based opinion mining and ranking. J. Comput. Syst. Sci. 1175–1184 (2012)
3. Horrigan, J.A.: Online Shopping (2008) http://www.pewinternet.org/2008/02/13/online-shopping/
4. http://nishithsblog.wordpress.com/2013/09/27/social-media-facts-and-statistics-for-2013/
5. http://www.academia.edu/6723240/Mining_Opinion_Features_in_Customer_Reviews
6. Jawale, M.A., Dr., Kyatanavar, D.N., Pawar, A.B.: Design of automated sentiment or opinion discovery system to enhance its performance. In: Proceedings of International Conference on Advances in Information Technology and Mobile Communication 2013 and In: Journal of ACEEE 2013, pp. 48–53 (2013) http://searchdl.org/public/book_series/LSCS/2/79.pdf
7. Jawale, M.A., Dr., Kyatanavar, D.N., Pawar, A.B.: Development of automated sentiment or opinion discovery system: review. In: Proceedings of ICRTET 2013 (2013)
8. Jawale, M.A., Dr., Kyatanavar, D.N., Pawar, A.B.: Implementation of automated sentiment discovery system. In: Proceedings of IEEE International Conference on Recent Advances and Innovations in Engineering (ICRAIE- 2014), pp. 1–6 (2014). ISBN: 978-1-4799-4041-7
9. Leong, C.K., Lee, Y.H., Mak, W.K.: Mining sentiments in SMS texts for teaching evaluation. In: Expert Systems with Applications, pp. 2584–2589 (2012)
10. Liu, B.: Sentiment analysis and subjectivity. In: Handbook of Natural Language Processing, 2nd edn. pp. 1–38 (2012)
11. Liu, B.: Sentiment analysis: a multi-faceted problem. In: IEEE Intelligent Systems, pp. 1–5 (2010)
12. Rainie, L., Horrigan, J.: Election 2006 online, Pew Internet & American Life Project Report, Jan 2007
13. Tang, H., Tan, S., Cheng, X.: A survey on sentiment detection of reviews. In: Science Direct, Expert Systems with Applications, pp. 10760–10773 (2009)
14. Yin, C., Peng, Q.: Sentiment analysis for product features in Chinese reviews based on semantic association. In: International Conference on Artificial Intelligence and Computational Intelligence, pp. 82–85 (2009)

The Comprehension of Figurative Language: What Is the Influence of Irony and Sarcasm on NLP Techniques?

Leila Weitzel, Ronaldo Cristiano Prati and Raul Freire Aguiar

Abstract Due to the growing volume of available textual information, there is a great demand for Natural Language Processing (NLP) techniques that can automatically process and manage texts, supporting the information retrieval and communication in core areas of society (e.g. healthcare, business, and science). NLP techniques have to tackle the often ambiguous and linguistic structures that people use in everyday speech. As such, there are many issues that have to be considered, for instance slang, grammatical errors, regional dialects, figurative language, etc. Figurative Language (FL), such as irony, sarcasm, simile, and metaphor, poses a serious challenge to NLP systems. FL is a frequent phenomenon within human communication, occurring both in spoken and written discourse including books, websites, fora, chats, social network posts, news articles and product reviews. Indeed, knowing what people think can help companies, political parties, and other public entities in strategizing and decision-making polices. When people are engaged in an informal conversation, they almost inevitably use irony (or sarcasm) to express something else or different than stated by the literal sentence meaning. Sentiment analysis methods can be easily misled by the presence of words that have a strong polarity but are used sarcastically, which means that the opposite polarity was intended. Several efforts have been recently devoted to detect and tackle FL phenomena in social media. Many of applications rely on task-specific lexicons (e.g. dictionaries, word classifications) or Machine Learning algorithms. Increasingly, numerous companies have begun to leverage automated methods for inferring consumer sentiment from online reviews and other sources. A system capable of interpreting FL would be extremely beneficial to a wide range of practical NLP applications. In this sense, this chapter aims at evaluating how two

L. Weitzel (✉)
Departamento de Computação Rio das Ostras Rj,
Univerisidade Federal Fluminense (UFF), Rio de Janeiro, Brazil
e-mail: leila_weitzel@id.uff.br

R.C. Prati · R.F. Aguiar
Centro de Matemática, Computação e Cognição,
Universidade Federal do ABC (UFABC), Santo André, SP, Brazil
e-mail: ronaldo.prati@ufabc.edu.br

R.F. Aguiar
e-mail: f.raul@ufabc.edu.br

© Springer International Publishing Switzerland 2016
W. Pedrycz and S.-M. Chen (eds.), *Sentiment Analysis and Ontology Engineering*,
Studies in Computational Intelligence 639, DOI 10.1007/978-3-319-30319-2_3

specific domains of FL, sarcasm and irony, affect Sentiment Analysis (SA) tools. The study's ultimate goal is to find out if FL hinders the performance (polarity detection) of SA systems due to the presence of ironic context. Our results indicate that computational intelligence approaches are more suitable in presence of irony and sarcasm in Twitter classification.

Keywords Sentiment analysis · Irony · Sarcasm · Figurative language · Machine learning · Natural language processing

1   Introduction

The Internet consolidated itself as a very powerful platform that has changed forever the way we do business, and the way we communicate. The Internet, as no other communication medium, has given an International or, a "Globalized" dimension to the world. As such, it has become the Universal source of information for millions of people, on the go, at home, at school, and at work. From 2000 to 2015, the overall global population of people using Internet grew more than 753 % to more than 3 billion people. Furthermore, online search technology is barely 20 years old, yet has profoundly changed how we behave and get things done at work or at home [1].

Undoubtedly, Internet has become an essential component of everyday social and business people lives. Two things have marked Internet evolution recently: the social web and mobile technology [9].

On the one hand, with a very low investment, almost anybody have access and a presence in the World Wide Web. In this sense, "blogging" has consolidated the social media: people everywhere are expressing and publishing their ideas, and opinions like never before. Consumers now search for opinions online before, during, and after a purchase [37].

On the other hand, the enterprises globally have stored exabytes of new data. This trend has the potential of not only drive competitiveness in the private sector but also fundamentally transform government operations, health care and education. The large datasets generated from every customer interaction, every wired object, and every social network represents one example of stored data [9]. Thus, due to the growing volume of available textual information, there is a great demand for Natural Language Processing (NLP) techniques that can automatically process and manage large collections of texts. The term NLP encompasses a broad set of techniques for automated (or semi-automated) generation, manipulation and analysis of natural (both spoken and written) human languages.

NLP has to tackle the ambiguous and linguistic structures people often use in everyday speech. As such, there are many issues that have to be considered, for instance slangs, grammatical errors, regional dialects, figurative languages, etc. Figurative Language (FL) is a frequent phenomenon within human communication, occurring both in spoken and written discourse, including books, websites, fora, chats, Twitter messages, Facebook posts, news articles and product reviews. Examples of FL devices are: simile, metaphor, personification, and irony or sarcasm [70].

Irony and sarcasm poses several issues for NLP techniques. The Oxford Dictionary defines irony as, "the expression of one's meaning by using language that normally signifies the opposite, typically for humorous or emphatic effect". For instance, "It has been a warm summer" is a sarcastic sentence when it has been cold and raining almost every day in the summer season. Sarcasm is praise, which is really an insult; sarcasm generally involves malice, the desire to put someone down, e.g., "This is my brilliant son, who failed out of college". When an individual is sarcastic or ironic, he or she is expressing a sentiment (often superficially positive) but intending the opposite.

When people are engaged in an informal conversation, they almost inevitably use irony (or sarcasm) to express something else or different than stated by the literal sentence meaning. Generally, verbal irony is often stated in the form of a metaphor or simile. This fact not only has an influence on the human understanding; it goes beyond the Text Classification [70].

Opinion-oriented information extraction is categorized as a Text Classification task. Text Classification (TC) is the process of categorizing documents into different classes using predefined category labels. One of the strands of this area is Sentiment Analysis (SA), also called Opinion Mining. The aim of the Opinion Mining is to recognize the subjectivity and objectivity of a text and further classify the opinion orientation of subjective text. The essential issue in SA is to identify how sentiments are expressed in texts and whether the expressions indicate positive (favorable) or negative (unfavorable) opinions toward the subject. It is difficult task because of the complexity and ambiguity of natural language, where a word may have different meanings or multiple phrases can be used to express the same idea [48].

Language that cannot be taken literally (or should not be taken literally only) becomes an enormous challenge for NLP, especially to SA tools. In the presence of words that have a strong polarity, but are used sarcastically, which mean that the opposite polarity was intended, thus FL can easily mislead SA tools.

In this sense, this chapter aims at evaluating how two specific domains of FL, sarcasm and irony, affect Sentiment Analysis (SA) tools. The study's ultimate goal is to find out if FL hinders the performance (polarity detection) of SA systems due to the presence of ironic context. Our results indicate that computational intelligence approaches are more suitable in presence of irony and sarcasm in Twitter classification.

The rest of the chapter proceeds as follows: In Sect. 2, we highlight the basic concepts related to NLP research. Section 3 briefly introduces Online Social Media (OSM), followed by the challenges in processing user generated content in OSM Sect. 4. Section 5 provides a brief description of the related work. Our approach and the experimental results are presented in Sects. 6 and 7. Section 8 highlighted the results obtained. We conclude in Sect. 9.

2   Natural Language Processing

Natural Language Processing (NLP) is a large and multidisciplinary field, where most techniques inherit largely from Linguistics and Computational Intelligence. Relatively newer areas such as Machine Learning, Computational Statistics and Cognitive Sciences also have influenced some of these techniques. They are aimed at developing computer algorithms capable of human-like activities related to understanding or producing texts or speech in a natural language [35].

There are many NLP applications developed over the years. They can be mainly divided into two parts as follows [35]:

Text-based applications: This part involves applications such as searching for a certain topic or a keyword in a database; extracting information from a large document; translating from one language to another or summarizing a text for different purposes; among other tasks.

Dialogue based applications: Some of the typical examples in this part includes question-answering systems; services that can be provided over a telephone without an operator; teaching-tutoring systems; voice controlled machines (that take instructions by speech); among other tasks.

A incomplete, but useful list of NLP systems includes: spelling and grammar checking; optical character recognition (OCR); screen readers for blind or partially sighted users; automatic report generation (possibly multilingual); machine translation; plagiarism detection tools; email understanding and dialogue systems [5].

NLP has to tackle the often-ambiguous linguistic structures. In fact, in natural languages (as opposed to programming languages), ambiguity is ubiquitous. Thus, exactly the same string might have different meanings, depending upon the context they are user. As an example of ambiguity, consider the sentence "At last, a computer that understands you like your mother". One can consider the intended meanings as, for instance "the computer understands you as well as your mother do" or "the computers understands (that) you cherish your mother".

When using Literal Language (LL), a writer is expressing in words exactly what she/he means. This means that, when writers are being literal, they are not exaggerating, using sarcasm (irony), metaphor or implying anything else is written. A word (isolated or within a context) conveys one single meaning [57].

Although the field of NLP has made considerable strides in the automated processing of standard language, figurative (i.e., non-literal) language still poses great difficulty [56]. Indeed, FL is one of the most arduous topics facing NLP research [61].

Indeed, the writer may take advantage of numerous linguistic figures such as metaphors, analogy, ambiguity, irony, and so on, in order to project more complex meanings to a sentence. This usually represents a real challenge, not only for computers, but sometimes for humans as well. Each figure can exploit different linguistic strategies to be able to produce the desirable effect (e.g., ambiguity and alliteration regarding humor; similes regarding irony) [58]. Sometimes the strategies are similar (e.g., the use of satirical or sarcastic utterances to express a negative attitude). These figures suppose cognitive capabilities to abstract and meta-represent meanings out of the "physical" words [57].

According to [24, 41] the ability to identify sarcasm and irony has gotten a lot of attention recently. With the proliferation of Web based applications such as micro blogging, forums and social networks, Internet users generate numerous reviews, comments, recommendations, ratings and feedbacks daily. One of the major issues within the task of irony identification is the absence of an agreement among researchers (linguists, psychologists, computer scientists) on how one can formally define irony or sarcasm and their structure. On the contrary, many theories that try to explain the phenomenon of irony and sarcasm agree that it is impossible to come up with a formal definition of these phenomena.

Moreover, there exists a belief that these linguistic figures are not static but undergo changes and that sarcastic sentences even has regional variations. Thus, it is not possible to create a definition of irony or sarcasm for training annotators to identify ironic utterances following a set of formal criteria [7].

3   Online Social Media

The Web 2.0 is the ultimate manifestation of User-Generated Content (UGC) systems. The UGC can be virtually about anything including politics, products, people, events, etc. Popular UGC systems domains include blogs and web forums, social bookmarking sites, photo and video sharing communities, as well as Online Social Media (OSM) such as Twitter, Facebook and MySpace. The most basic feature is the ability to create and share a personal profile. This profile page typically includes a photo, some basic personal information (name, age, sex, and localization) and extra space for listing your preferences, like favorite music bands, books, TV shows, movies, hobbies and Web sites. Social media sites also have a timeline feed, where people update their statuses.

Twitter constitutes a very open social network space, whose lack of barriers to access. Even non-registered users are able to use Twitter to track public feeds by, e.g., following breaking news on their chosen topics, from "World Economic Crisis" to "European Football Championship", for instance. Twitter users communicate with each other by posting tweets on his timeline, that his followers (in a private account) or other users (in a public account) can access and comment, allowing for a public interactive dialogue. On Twitter, the kind of messages users often post or update are short messages referred to as tweets, describing one's current status within a limit of 140 characters [74].

Microblogging sites such as Twitter and Sina Weibo (a Chinese microblogging website similar to Twitter) has rapidly established as an emerging global communication service due to its timeliness, convenience as a lightweight and easy way of communication and information sharing. Beyond merely displaying news and reports, the Twitter itself is also a large platform where different opinions are presented and exchanged [75].

Sentiment Analysis over Twitter offers organizations and individuals a fast and effective way to monitor the publics' feelings towards them and their competitors. In the current global market, designers, developers, vendors and sales representatives of new products need to carefully study whether and how their products offer competitive advantages. Twitter, with over 500 million registered users and over 400 million messages per day (as from 2014), has become a gold mine for organizations to monitor their reputation and brands by extracting and analyzing the sentiment of the tweets posted by the public about them, their markets, and competitors [34].

In just a few years, the use of social technologies has become a sweeping cultural, social, and economic phenomenon. OSM has literally changed how millions of people live. Businesses are changing their behaviors as well. In these few short years, OSM has evolved from simply another "new media" platform to an increasingly important business tool, with wide-ranging capabilities. Thousands of companies have found that OSM can generate rich new forms of consumer insights, at lower cost and faster than conventional methods. Moreover, companies are monitoring what consumers do and say to one another on OSM, which provides unfiltered feedback and behavioral data (e.g. do people who like this wine also like that brand of vodka?) [27].

A while ago, consumers had few options for expressing their dissatisfaction with a product or service. They may have told their friends and family about their experience. Maybe they filed in complaints with governmental regulatory agencies or other agencies. Today, social network users are using channels such as Twitter and Facebook for discussing shopping decisions and experiences with their peers. Indeed, consumers spread negative or positive word-of-mouth globally through online forums, online reviews, video sharing sites, and personal blogs. Furthermore, many Internet sites encourage consumers to share their product and service experiences with consumers worldwide. Given that, individuals are able to freely express their opinions [26].

Thus, it is almost certain that customers provide a genuine opinion about products and services. As consequence of that, it might be crucial to monitor such opinionated documents. As stated earlier, uncovering how customers feel about specific products or brands and detecting purchase habits and preferences has traditionally been a costly and highly time-consuming task, which involved the use of methods such as focus groups and surveys. OSM sites are commonly known for information dissemination, personal activities posting, product reviews, online pictures sharing, professional profiling, advertisements and opinion/sentiment expression. OSM has also given users the privilege to give opinions with very little or no restriction. Social media analytics is the practice of gathering data from OSM and analyzing that data in order to transform social media data into knowledge for decision makers and e-marketers. The most common use of social media analytics is to mine customer sentiment with aim of increasing competitive advantage and effectively assesses the competitive environment of businesses. Companies need to monitor and analyze not only the customer-generated content on their own social media sites, but also the textual information on their competitors' social media sites. Enormous sums are being spent on customer satisfaction surveys and their analysis. Yet, the effectiveness of such surveys is usually very limited in spite of the amount of money and effort spent on them, both because of the sample size limitations and the difficulties of making effective questionnaires [21]. In order to avoid these issues, numerous companies have begun to leverage automated methods for inferring consumer sentiment from online reviews.

4 Challenges Sentiment Analysis for Online Social Media

Sentiment analysis and opinion mining are very growing topics of interest over the last few years due to the large number of texts produced through Web 2.0. A common task in opinion mining is to classify an opinionated document as having a positive or a negative opinion. Subjectivity in natural language refers to aspects of language used to express opinions and evaluations. Opinion is an expression of sentiment by an author about something or an aspect of something. The purpose of sentiment analysis is to determine the attitude or inclination of a communicator through the contextual polarity of their speaking or writing. Their attitude may be reflected in their own judgment, emotional state of the subject, or the state of any emotional communication they are using to affect a reader or listener. In simple words, it is used to track the mood of the public. A comprehensive review of both sentiment analysis and opinion mining as a research field for Natural Language Processing (NLP) is presented in [49].

There are different types of polarities in texts: positive evaluation (e.g., "it is a fascinating tale"), negative evaluation (e.g., "the software is horrible") and objective (or neutral) evaluation (e.g., "Bell Industries Inc. increased its quarterly to 10 cents from 7 cents a share"). Sentiment classification also looks, for instance, at emotional states such as anger, sadness, and happiness [56, 70].

Sentiment classification mainly falls into three levels: document-level, sentence-level and aspect-level. Document-level analysis is based on the sentiments represented on the overall sentiments expressed by authors in the documents. Documents are classified according to the sentiments instead of topics. Sentence-level sentiment classification models are used for the extraction of the sentences contained in the opinionated terms, opinion holder and opinionated object. Aspect level assumes that a document contains opinion on several entities and their aspects. Aspect-level analysis requires the discovery of these entities, as well as aspects, and sentiments associated to each of them. In this approach, positive or negative opinion is identified from the already extracted entities. It is very useful in summarizing parts of the whole document as having positive or negative polarity about any object (e.g., a camera, a fridge, a mobile phone, a car, a movie, or a politician) [49].

The demand for applications and tools to accomplish sentiment classification tasks has attracted the researcher's attention in this area. Hence, sentiment analysis applications have spread to many domains: from consumer products, healthcare and financial services to political elections and social events, to name a few.

Opinions are easy to understand by human beings, but it is not that easy for a computer to achieving the same level of understanding. We must always keep in mind that SA is a NLP problem and, consequently, many of the issues in NLP are also problems that must be addressed when dealing with SA problems. According to [40], some of the sub-problems that still are the object of further research attention by the NLP community are:

Coreference resolution: Given a sentence or larger chunk of text, determine which words ("mentions") refer to the same objects ("entities"). Anaphora resolution is a specific example of this task, and is specifically concerned with matching up pronouns with the nouns or names that they refer to. For example, in a sentence such as "He entered John's house through the front door", "the front door" is a referring expression and the bridging relationship to be identified is the fact that the door being referred to is the front door of John's house (rather than of some other structure that it might also be referred to).

Negation handling: This is typically done by inverting the polarity of negated words. For instance, the negative sentiment associated with the word "bad" would typically be inverted into a positive sentiment for the phrase "not bad". Negation identification is not a simple task and its complexity increases, since negation words such as not, nor etc., (syntactic negation) are not the only criterion for negation calculation. The linguistic patterns—prefixes (e.g., un-, dis-, etc.) or suffixes (e.g., -less) also introduce the context of negation in textual data.

Word sense disambiguation: This is the task of determining the meaning of an ambiguous word in its context. For example, the word "cold" has several senses and may refer to a disease, a temperature sensation, or an environmental condition. The specific sense intended is determined by the textual context in which an instance of the ambiguous word appears.

Meaning extraction: Rhetorical devices/modes such as sarcasm, irony, implication, etc.

Optimized parsing Understanding natural language is much more than parsing sentences into their individual parts of speech and looking those words up in a dictionary. NLP involves a series of steps that make text understandable (or computable).

There are challenges that need to be faced to implement sentiment analysis tools. For example, opinion words that are considered to be positive in one situation may be considered negative in another situation; people do not always express opinions in the same way; people can be contradictory in their statements. Orientation of opinion of words could be different according to different situation. For example, "The camera size of the mobile phone is small". Here, the adjective small was used in a positive sense but if another customer had replied that "the battery time is also small", the word "small" represent a negative orientation to the battery of the phone. To identify the polarity of the same adjective words in different situation is also a challenging task.

Another aspect in respect to the SA challenges is the importance of highlighting Figurative Language. As described early, FL is a term used to describe author's use of language to extend meaning. This is achieved by deviating from the literal meaning of words or by deviating from the usual arrangement of words. According to [7, 19, 41, 56, 68, 70] FL is one of the most arduous issues facing SA. Metaphor, analogy, ambiguity, irony are common examples of FL. Irony is, basically, a communicative act that expresses the opposite of what was literally said. Unlike a simple negation, an ironic message typically conveys a negative opinion using only positive words or even intensified positive words [7, 56], e.g.:

> "Yes. That's JUST what I needed: a fever and a sore throat".
>
> "What lovely weather we're having, I just loved being harassed on the street".
>
> "It is such a compliment. Nothing like as a sore throat to start the day. Yay".
>
> "So grateful for being ill again, can't get enough of it".

As depicted in the above examples, some lexical phenomena can cause the valence of a lexical item to shift from one pole to the other or, less forcefully, to modify the valence towards a more positive or negative position. Irony behaves as polarity reversers, which "flip" or "reverse" the polarity of an expression. Therefore, a major obstacle for automatically determining the polarity of a (short) text is constructions in which the literal meaning of the text is not the intended meaning of the writer [38].

5   Related Work

This section reviews some recent studies on Sentiment Analysis applied to Twitter texts, focusing on handling irony and sarcasm. We begin by analyzing the SemEval2015 Task 11 (Sect. 5.1) followed by some recent related studies (Sect. 5.2).

5.1   A Review of the SemEval2015 Task 11: Sentiment Analysis of Figurative Language in Twitter

The studies outlined next show an analysis of the Task 11 at the SemEval2015 (The Semantic Evaluation). SemEval2015 is the Ninth International Workshop on Semantic Evaluation, held in Denver, Colorado, collocated with the 2015 Conference of the North American Chapter of the Association for Computational Linguistics – Human Language Technologies (NAACL HLT 2015). In the workshop, the Task 11 concerns with the classification of tweets containing irony and metaphors, aiming to evaluate the degree to which conventional sentiment analysis can handle FL, and to determine whether systems that explicitly model these phenomena demonstrate a remarked increase in performance [10, 21].

The main goal of the task 11 is:

> "…given a set of tweets that are rich in metaphor, sarcasm and irony, the goal is to determine whether a user has expressed a positive, negative or neutral sentiment in each, and the degree to which this sentiment has been communicated".

The task concerns itself with the classification of tweets containing irony and metaphors. The trial, training and test data contain a concentrated amount of these phenomena to evaluate to which degree conventional sentiment analysis can handle creative language, and to determine whether systems that explicitly model these phenomena demonstrate a marked increase in competence. The participants can use a fine-grained sentiment scale to capture the effect of irony and figurative language on the perceived sentiment of a tweet [10, 21].

Three broad classes of figurative language are present in the task: irony, sarcasm and metaphor. The participants must use an eleven-point interval scale, ranging from −5 (very negative, for tweets with highly critical meanings) to +5 (very positive, for tweets with flattering or very upbeat meanings). The point zero on this scale is used for neutral tweets. Queries for hashtags such as #sarcasm, #sarcastic and #irony, and for words such as "figuratively" were used to generate the dataset. Only URLs have been removed from the tweets, all other content has been left in place. Annotators were asked to use ± 5, ± 3 and ± 1 as scores for tweets presenting strong, moderate or weak sentiment, and to use ± 4 and ±2 for tweets with nuanced sentiments that fall between these gross scores. An overall sentiment score for each tweet was calculated as a weighted average of all seven annotators. Annotators were explicitly asked to consider all of a tweet's content when assigning a score, including any hashtags (such as # sarcasm, #irony, etc.). The tweets are hand annotated. In order to avoid disagreements, the SemEval organizers adopted the following methodology. If 60 % or more of an annotator's judgments are judged to be outliers, thus the annotator is considered a scammer and dismissed from the task.

Seven human annotators (internal team) marked a trial dataset, consisting of 746 sarcasm, 81 irony and 198 Metaphor tweets (1025 in total). The training and test datasets contains about 8000 and 4000 tweets, respectively. The training and test set was annotated on the crowd-sourcing platform CrowdFlower.com, following the same annotation scheme as for the trial dataset. A total of 15 teams participated in Task 11: CLaC (Concordia University); UPF (Universitat Pompeu Fabra); LLT_PolyU (Hong Kong Polytechnic University); LT3 (Ghent University); elirf (Universitat Politècnica de València); ValenTo (Universitat Politècnica de València); HLT (FBK-Irst, University of Trento); CPH (Københavns Universitet); PRHLT (PRHLT Research Center); DsUniPi (University of Piraeus); PKU (Peking University); KELabTeam (Yeungnam University); RGU (Robert Gordon University); SHELLFBK (Fondazione Bruno Kessler); BUAP (Benemèrita Universidad Autónoma de Puebla) [10, 21].

Most teams performed well on the detection of sarcasm and irony tweets, but the Metaphor and other categories prove to be more challenging. Most approaches were based on computational intelligence methods, using supervised learning methods (SVMs and regression models).

The best-reported system was by CLaCSentiPipe (CLaC team) [46], which achieves the score of 0.758 using the cosine similarity measure, and a score of 2.117 using the mean squared error (MSE) measure. The CLaC team used different sentiment lexica. After a set of steps (tokenization, POS tagging, named entity recognition, etc.), the resulting feature space is grouped into subsets of features in order to create feature combinations and processed with decision tree regressor to predict continuous values for Task 11.

LLT_PolyU and ELiRF [22, 77] achieve the best performance on the classification of sarcasm and irony tweets, and ranked 3rd and 4th in the combined task, respectively. The score of each system are f 0.059 and 0.758 using cosine similarity and 11.274 and 2.117 using MSE.

The ELiRF [22] team used lexicons (Afinn, Pattern, SentiWordNet, Jeffrey and NRC) as features for training their systems. For example, using the bag-of-words approach, they considered the number of capitalized words and the number of words with elongated characters, they removed the stop words, and others terms. The team classified tweets using the SVM algorithm (using scikit-learn toolkit—a machine learning library for the Python programming language) with a linear kernel for regression. They used a cross-validation technique (10-fold cross validation) on training set. The authors developed a framework to define functional classification models (a set of steps comprising: preprocess, mining, feature vectorization, and classification). This framework receives one to n models. Hence, a tweet was classified using the most voted category or using the mean of predictions in the case of using regression. The best performance was achieved by the model using: 3-grams to 9-grams of characters; a set of Lexicons: Afinn, Pattern and NRC; and none of figurative hashtag has been removed.

The LLT-PolyU [77] team used some heuristic rules to pre-process the texts. For example: removing repeated vowels, e.g. "loooove" to "love"; standardizing capital letters, e.g., "LOVE" to "love"; substitution of the combination of exclamation and question marks (e.g. "?!?!!") with the form "?!". An emoticon dictionary was used on training data and Internet resources. Maximal matching segmentation was used to segment all the out of vocabulary tokens through a maximal matching algorithm according to an English dictionary (e.g. the token "yeahright" would be segmented as "yeah right"). In order to get the POS tags and dependency structures of the normalized tweets, they used the Stanford parser. The polarity was tested using four Opinion Lexicon, Afinn, MPQA and SentiWordnet. Two different models are used: Decision Tree Regression model (RepTree) implemented in Weka and Support Vector Regression model (SVR) implemented in LibSVM. The best performance was obtained with the value of r^2 between 0.03 and 0.04 with the RepTree model.

5.2 Recent Studies on Handling Sarcasm in Twitter

In [53], experiments were conducted using document-level detection on Czech and English Twitter datasets. It was the first attempt at sarcasm detection in the Czech language. Their evaluation was performed using the Maximum Entropy (MaxEnt) and Support Vector Machine (SVM) classifiers. They gathered 140,000 Czech and 780,000 English tweets (130,000 sarcastic and 650,000 non-sarcastic), respectively. The #sarcasm hashtag was used as an indicator of sarcastic tweets.

Seven thousand of tweets in Czech were arbitrarily selected for annotation purposes. Only two annotators undertake this task, given the information that "A tweet is considered sarcastic when its content is intended ironically/sarcastically without anticipating further information. Offensive utterances, jokes and ironic situations are not considered ironic/sarcastic". In order to resolve some classification conflicts, they used a third annotator. Ptáček et al [53] created two scenarios based on the work of [57] using three different preprocessing pipelines, including a basic pipeline (tokenization and POS-tagging only). For each preprocessing pipeline they gathered various sets of features and used two classifiers. All experiments were conducted in the 5-fold cross validation. In most classification cases for English tweets, the MaxEnt classifier significantly outperforms the SVM classifier. Macro F-measure (Fm) with 95 % confidence interval (CI) was about 94 % for balanced distribution and 92 % for imbalanced distribution.

Reyes et al. [57] collect a training corpus of irony based on tweets that consist of the hashtag #irony in order to train classifiers on different types of features (signatures, unexpectedness, style and emotional scenarios) and try to distinguish #irony-tweets from tweets containing the hashtags #education, #humour, or #politics. They achieved F1-scores of around 70 %. In [57], the word focused on developing classifiers to detect verbal irony based on ambiguity, polarity unexpectedness and emotional cues derived from text. An evaluation corpus of 50,000 tweet texts automatically retrieved from Twitter was used to evaluate the patterns. Two goals were considered in the evaluation: representatively and relevance. Some of the results, apart from being satisfactory in terms of classification accuracy, precision, recall, and F-measure, confirmed their initial assumptions about the usefulness of this kind of information to characterize these FL.

In [31], the authors investigate how emotional signals (emoticons) can be used to improve unsupervised sentiment analysis. Authors investigate the use of indicated signals, as well as the correlation among the signals, into an unsupervised learning framework, comparing their results with lexical based approaches. In [64], authors developed an approach for learning sentiment-specific word embedding for twitter sentiment classification. Word embedding is a continuous vector-based representation of words. The approach developed by the authors focus on words expressing sentiments. This approach was very competitive with results from SemEval.

In [59], the authors explore the contrast the duality of some sarcastic texts, where in the same sentence both positive and negative sentiments occur together. They use NLP techniques for POS tagging, and aspect-level sentiment identification. They use machine learning techniques to learn patterns of positive/negative contrast that commonly occurs in tweets. Results show an improvement in terms of recall for sarcasm recognition. Frequent patterns in sarcastic tweets was also explored in [13], associated with a semi-supervised approach, in order to classify sarcastic sentences in Twitter texts and Amazon Review data set.

The network structure of the social network and endogenous information was explored in [4, 32, 54, 63] aiming to improve the detection of sarcasm. In [32], authors explore a sociological theory to develop an algorithm approach to explore the social relations among users. The authors of [54] also use an behavioral theory to analyze past users tweets to improve sarcasm detection. The motivation is that sarcasm is a user personality characteristic, and users that used sarcasm in the past are more likely to user sarcasm in the future. In [63], the authors use both the network structure associated with the information of the users' mentions for sentiment analysis. This information was incorporated with text content using a semi-supervised transductive learning approach. Results show in improvement over using Support Vector Machines that access only the textual data. In [4], an approach that incorporates information of the profile of the user, as well as the intended audience, can improve sarcasm detection.

6   Sentiment Analysis Pipeline

The SA analysis process involves two stages: pre-processing and sentiment classification. As Twitter allows only 140 characters per each post, the messages generally heavily use abbreviations, irregular expressions, slang, and emoticons. This phenomenon increases the level of data sparsity, affecting the performance of Twitter sentiment classification.

6.1   Pre-processing Phase

The pre-processing stage generally involves the phases [3, 72]:

Extraction (or tokenization): this method is used to tokenize the file content into individual words or tokens. A text document is split into a stream of words, which may be accompanied by the removal of all punctuation marks and by replacing tabs and other non-text characters by single white spaces. The set of different tokens produced by merging all text documents of a collection is called the dictionary of a document collection [45].

Filtering methods: these methods remove words from the dictionary and thus from the documents. A standard filtering method is stop word removal. The idea of stop word filtering is to remove words that bear little or no content information, like articles, conjunctions, prepositions, etc. This method is based on the idea that discarding non-discriminating words reduces the feature space of the classifiers and helps them to produce more accurate results [47, 62].

Normalization: this process involves transforming text to ensure consistency. Some examples of this process include converting upper case letters to lower case, Unicode conversion, and removing diacritics from letters, punctuation, or numbers.

Stemming: this process is used to find out the root/stem of a word. For example, the words user, users, used, using all can be stemmed to the word "use". The purpose of this method is to remove various suffixes, to reduce number of words, to have exactly matching stems, to save memory space and time.

6.2   Sentiment Classification Approaches

Sentiment classification is commonly categorized in two basic approaches [70]:

- Machine Learning:
  - Supervised Learning (e.g., [36, 69]);
  - Unsupervised Learning (e.g., [76]);
- Lexicon-based:
  - Dictionary-based approach (e.g., [29]);
  - Corpus-based Approach: Statistical and Semantic, (e.g., [78]).

Machine learning [20] is a sub-field of artificial intelligence that explores the development and study of algorithms that can build models (learn) and make predictions from data. Supervised learning generates a function, which maps inputs to desired outputs, also called as labels. In general, for the text classification problem, any supervised learning method can be applied, e.g., Naïve Bayes classification, and Support Vector Machines. Unsupervised learning algorithms do not have access to the labels. The main goal is to find out regularities in the data, e.g., by creating clusters.

In computational linguistics, the lexicon provides paradigmatic information about words, including part of speech tags, irregular plurals, and information about subcategorization of verbs. Traditionally, lexicons were quite small and were constructed largely by hand. A sentiment (opinion) lexicon is defined as a list of positive and negative opinion words or sentiment words. According to [18], the sentiment lexicon "is the most important resource for most crucial analysis algorithms". Weichselbraun et al. [71] highlight the importance of context when producing sentiment lexicons

For mining large document collections, it is necessary to pre-process the text documents and store the information in a suitable data structure, which is more appropriate for further processing than a plain text file. However, even though several methods that try to explore both the syntactic structure and semantics of text exist, most text mining approaches are based on the idea that a text document can be represented by a set of words, i.e. a text document is described based on the set of words contained in it (bag-of-words representation).

Moreover, in order to be able to define at least the importance of a word within a given document, usually a vector representation is used, where for each word a numerical "importance" value is stored. The currently predominant approaches based on this idea are the vector space model, the probabilistic model and the logical model [67, 72]. Recent work [39, 44] uses neural networks for learning distributed representations based on word co-occurrence. The Vector Space Model is described in Sect. 6.3, while the distributed vector representation is described in Sect. 6.4.

6.3   Vector Space Model

A text (document) is a sequence of words. For most SA techniques, a text document is often represented by a vector. The vectors have as many elements as words in the dictionary of the examined document collection. The vector space model represents documents as vectors in m-dimensional space, *i.e.*, each document d is described by a numerical feature vector $w(d) = (x(d, t_1), \ldots, x(d, t_m))$. Each element of the vector usually represents a word (or a group of words) of the document collection. Each word for the document can be encoded in several ways: word frequency that uses simple frequency of word incidence; binary frequency, which is a binary indicator of incidence; log frequency, which uses a logarithm function that provides a "damping" of word incidences by taking its logarithm. Term Frequency–Inverse Document Frequency (TF-IDF), which is a numerical statistic that accounts to what extent a word is important to a document in a collection. TF–IDF is the product of two statistics that are named frequency and inverse document frequency. Term Frequency (TF) is defined as the number of times a term occurs in a document. Inverse Document Frequency is a statistical weight used for measuring the importance of a term in a text document collection. The TF-IDF is often used as a weighting factor in information retrieval and text mining. The intuition was that a query term that occurs in many documents is not a good discriminator, and should be given less weight than one that occurs in few documents. The value of TF-IDF increases proportionally to the number of times a word appears in the document [3, 25].

6.4   Distributed Vector Representation

Although the Vector Space Model is quite common in text mining applications, it is known to produce sparse vectors. The dimension of the vectors is the size of the dictionary, but for each document, only words that actually appear in it have values different from zero. As, in general, the dictionary size is much larger than the length of each document. Therefore, the vectors tend to be sparse.

To avoid this sparsity, in [39, 44] a recent approach was proposed that uses unsupervised neural networks to learn a continuous vector representation of words, based on word co-occurrence. This model, called Word2Vec, has been attracting considerable attention in recent years [6, 43, 51, 52, 60]. In this model, each word has a vector representation in the space. The dimension of the vector is a parameter of the model. These vectors are trained by a neural network by two main approaches: CBOW and Skip-Gram. In CBOW (Continuous Bag-Of-Words), the inputs to the model are word windows preceding and following the current word we want to predict. For instance, if the windows size is five, the inputs are the words w_{i-2}, w_{i-1}, w_{i+1} and w_{i+2} and the output the word w_i , i.e., the two previous and the two following words of the current word. The main idea is "predicting the word given its context". The Skip-gram, on the way around, the input to the model is the word wi, and the output are the words

w_{i-2}, w_{i-1}, w_{i+1} and w_{i+2}, so the task is predicting the context given a word, hence the name "skip"-gram.

An interesting property about these word representations is that the learned vectors implicitly encode numerous linguistic regularity and patterns, and these patterns can be processed by means of linear vector translations in the vector space. For instance, the result of the vector calculation vector ("king") – vector ("man") + vector("woman") is very close to the vector ("queen"). Another example is the vector ("Madrid") – vector ("Spain") + vector ("France") is closest to the vector ("Paris"). Given sufficient data, several other regularities like these have been emerged and observed, like male-female, company-CEO, city-ZIP code, comparative-superlative, among others.

In [39], this representation was extended to larger chunks of texts, like sentences or paragraphs. This method is called Doc2Vec. A paragraph vector is a continuous vector representation for pieces of texts. These paragraph vectors share the word vector representation. The main idea is to try to infer some text structure by trying to predict the surrounding words in a paragraph. Thus, a paragraph vector captures some of the language structure by capturing patterns in words' ordering in the paragraph. A drawback of this technique is that it requires a large collection of documents to converge. This structure is lost in the vector space model based in the bag-of-words. Paragraph vectors can be used in substitution of bag-of-word vectors in the Vector Space Model, or concatenated to them.

7 Experimental Setup

In this section, we describe the experimental process we have used to evaluate SA approaches in the context of sarcastic tweets. We concentrate on approaches that use only the text. We analyze different approaches for Sentiment Analysis in the context of sarcastic and ironic sentences. The main idea is to analyze to what extent two different approaches correctly classify sarcastic tweets, and how these methods differ from each other.

7.1 Data Collection

To carry out the experiments we use the same data set used in [53]. It consists of a balanced sample containing 50,000 sarcastic tweet IDs and 50,000 non-sarcastic tweet IDs. As only the IDs were available, and some tweets were erased or the owners made their accounts private, we ware only able to download 40,262 sarcastic tweets and 41,636 non-sarcastic tweets. Thus, the final data set contains 81,898 tweets from the original 100,000 tweet ids. The tweets are written in English, and the presence of the hashtag #sarcasm was used the authors in [53] as an indicator of sarcastic tweets. Similar approaches were used in other studies on Twitter sarcasm detection [4, 54, 73]

7.1.1 Data Pre-processing

Tweets were tokenized using the tweetokenize package.[1] This package was developed to extract tokens from twitter messages, and can easily replace tweet features (such as usernames; URLs; phone numbers; date-times; etc.) with tokens in order to reduce feature set complexity and to improve performance of classifiers. It can also correctly separate most emoji, written consecutively, into individual tokens. Usernames mentions, URLs, and numbers were replaced by the general tags <USERNAME>, <URL> and <NUMBERS>, respectively. As hashtags may be a strong indicator of irony due to the way tweets were collected, they were removed from the tweets.

As it is common to use slangs in twitter messages, we also did an additional step of slang expansion. To this end, a slang dictionary was used. The process consists in consulting each token in the dictionary, and if the token matches a slang expression then it was replaced by the corresponding meaning in the dictionary. Frequent bigrams and trigrams were generated using the approach described in [42], using the Gensim software package [55].

We carried out experiments with both machine learning and lexicon based approaches, in order to evaluate their performances in the task of identifying sarcastic tweets. To conduct the experiments, we used the scikit-learn software package [50]. As described early, this package is a Python library containing implementations of several machine-learning algorithms. As input to these algorithms, we used the TF-IDF vector representation, and the TF-IDF representation combined with the vectors get from the distributed vector representation algorithm Word2Vec (W2V), available in the Genism Software package. Due to the highly sparsity and dimensionality of the TF-IDF representation, not all machine-learning algorithms can be applied to this task due to hardware constraints. Therefore, the two machine learning algorithms used in this study are:

Support Vector Machines (SVMs): SVMs [28] is supervised machine learning algorithm that aims to find the "maximal margin" hyperplane that separates the classes. This hyperplane is defined by the support vectors (examples near the class boundaries) so that the distance between these support vectors of different classes is maximized. We used the linear SVM algorithm implemented in the Scikit-package; build upon the Liblinear optimization library [17].

Logistic Regression (LogReg): Logistic Regression [30] is a regression model where the dependent variable is categorical. Logistic regression is also known in the literature as logit regression, maximum-entropy classification (MaxEnt) or the log-linear classifier. It constructs a probabilistic model, aiming to fit a logistic

---

[1] https://github.com/jaredks/tweetokenize.

function for the class distribution, according to the data. The scikit learn provides a regularized logistic regression using the Liblinear library [17].

All the experiments using SVM and LogReg were run using 5-fold stratified cross validation, using default parameters of the learning algorithms.

We also used a set of SA tools; they are described briefly as follows. These methods are incorporated in the iFeel toolkit [23].

SentiWordNet [16]: is an approach that is widely used in opinion mining, and is based on an English lexical dictionary WordNet. This lexical dictionary groups adjectives, nouns, verbs, and other grammatical classes into synonym sets called synsets. To assign polarity, the method considers the average scores of all associated synsets of a given text. The relative strength of positive and negative synsets affects the polarity of a given text.

Happiness Index:[15]: is a sentiment scale based on the Affective Norms for English Words (ANEW) [8], which is a collection of 1,034 commonly used words associated with their affective dimensions of valence, arousal, and dominance. Happiness Index is scaled between 1 and 9, depending on the amount of happiness inferred from text. To assign polarities, the method consider a range between 1 and 4 as negative and a range from 5 to 9 as positive.

SenticNet [12]: is based on artificial intelligence and semantic Web techniques. The approach uses Natural Language Processing (NLP) techniques to create a polarity for nearly 14,000 concepts and was evaluated in measuring the level of polarity in opinions of patients about the National Health Service in England. SenticNet uses the affective categorization model Hourglass of Emotions [11] that provides an approach that classify messages as positive and negative of a given text.

SentiStrength [66] is a mix of supervised and unsupervised classification methods, including simple logistic regression, SVM, J48 classification tree, JRip rule-based classifier, SVM regression, and Naïve Bayes. This method extends the existing LIWC dictionary [65], which is made for structured text, to include a wide range of Online Social Network contexts. LIWC is a text analysis tool that evaluates, among cognitive and structural components, the emotional (positive and negative affects) of a given text.

Pattern.en [14]: is a web mining module for the Python programming language. Its main functionalities are web mining, natural language processing, machine learning, network analysis and data visualization. The SA function returns a polarity for the given sentence, based on the adjectives it contains, where polarity is a value between -1 and 1.

Vader [33] is a method for sentiment analysis based on validated human judges. It is developed for twitter and social media contexts. VADER incorporates a "gold-standard" sentiment lexicon, which is especially attuned to microblogs-like contexts. The Vader sentiment lexicon is sensitive both to the polarity and to the intensity of sentiments expressed in social media contexts, and it is generally applicable to sentiment analysis in other domains besides microblogs.

8 Results and Discussion

We begin by analyzing the classification performance of the methods and tools described in Sect. 6.3. To this end, we compute four different classification performance measures:

Accuracy (ACC): this measure compute the percentage of instances correctly classified in the test set folds.

F1-measure (F1): Also known as, F-score, this measure is the harmonic mean between precision and recall. Precision is the number of correct positive results divided by the number of all positive results, and recall is the number of correct positive results divided by the number of positive results that should have been returned.

Mattews correlation coefficient (MCC): this measure is in essence a correlation coefficient between the observed and predicted binary classifications; it returns a value between -1 and +1. A coefficient of +1 represents a perfect prediction, 0 no better than random prediction and -1 indicates total disagreement between prediction and observation.

Area under the ROC curve (AUC): a ROC curve is a graphical plot about the relationship between true positive and false positive rates of a binary classifier system as its discrimination threshold is varied. The Area under the Curve is a summary statistic. It measures the probability that a positive example is given a high classification score than a negative example.

Table 1 shows these four performance measures for the classifiers induced by SVM and Logistic Regression. The performances are very similar, although the Logistic regression was a little better. Furthermore, incorporating the vectors from the Doc2Vec also improved a little the results for all four performance measures and for both classifiers.

Although the average performance measures are similar, the two methods may have differences in classifying sarcastic tweets that do not reflect in the average performance. To further characterize this, Fig. 1 plots the joint distribution between the test scores produced by the models produced by SVMs and Logistic Regression, factored by the class of the tweets (whether it is sarcastic or not). The "S" shape of the plot came from the sigmoid shape of the logistic function. As can be seen

Table 1 Classification performances using machine-learning methods

	SVM		LogReg	
	TF-IDF	TF-IDF + W2V	TF-IDF	TF-IDF + W2V
Accuracy	0.8116	0.8157	0.8155	0.8167
F1	0.8112	0.8153	0.8144	0.8160
MCC	0.6239	0.6320	0.6317	0.6342
AUC	0.8962	0.9002	0.8989	0.9002

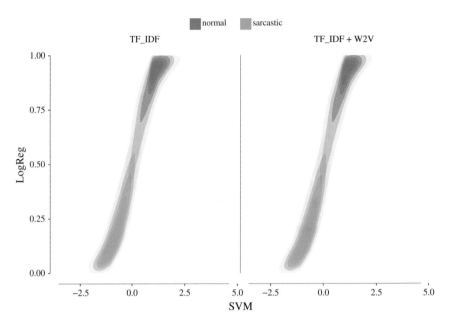

Fig. 1 Joint plot of SVM and logistic regression scores for sarcastic and non-sarcastic tweets

from the plots, both models strongly agree, although the overlap region indicates a difference in performance.

To gain some insight in this overlapping region, we applied the Apriori association rule algorithm [2]. This algorithm was developed for the basked analysis problem, with the objective of identifying correlations between items shop by the customers. In our case, we considered items as the words (tokens) in the tweets, and we want to identify which words prejudice the performance of the methods. Figure 2 gives a graphical summary of the rules. In this figure, the columns indicate the words and the rows the machine-learning configuration, which misclassify tweets with that word. The bullet sizes indicate frequency (support) and the gray scale the degree of correlation (lift).

Analyzing the plot, one can be seen that words that common words (which probably appears in both in sarcastic and non-sarcastic context) such as laughing, good and love prejudice all models. On the other hand, the vectors from Word2Vec helped in reducing errors for the words best and like, for the SVM classifier, and the word today for the logistic regression. A possible explanation to this is that the vectors from Word2Vec were able to identify different contexts in which these two words were used, and thus helped in avoiding mistakes for these frequent words. This may also give a hint on why the difference in performance between the models induced using only TF-IDF models and TF-IDF concatenated with Doc2Vec vectors is small. Doc2Vec is known to need a large collection to converge for infrequent words, which may not be the case from our tweet sample. A larger tweet sample may help in strengthening the influence of vector from Doc2Vec in the models.

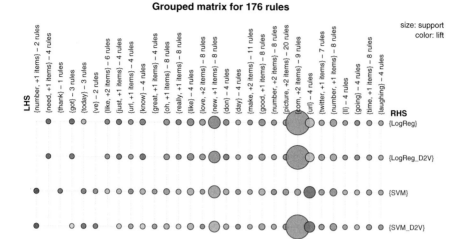

Fig. 2 Word associations with common errors by the models

Table 2 Classification performances using lexicon-hybrid methods

	Happiness	Senticnet	Vadder	Sentstrength	Sentiwordnet	Pattern.en
AUC	0.5290	0.5844	0.5263	0.5283	0.5513	0.5524
MCC	0.0484	0.0554	0.0734	0.0862	0.0920	0.1322
Accuracy	0.5241	0.5186	0.5331	0.5375	0.5442	0.5653
F1	0.5217	0.6193	0.5849	0.2936	0.5731	0.5770

Table 2 shows the results of lexicon based methods. Overall, results are quite poor, when compared to Machine Learning approaches. For most methods, the area under the curve is just about 0.5 (which indicates random performance). The Matthews Correlation Coefficients for all methods are also all close to 0 (which also indicates random performance) and accuracy is close to the majority error.

As the results for lexical based methods are quite poor, and motivated by the work reported in [59], we did a modification in these methods. The rationale behind this modification is that sarcastic texts presents a contrast between positive and negative, and calculating the polarity as the difference between these two polarities would indicate a polarity close to neutral. Therefore, we calculate the sum of absolute positive and negative valences, so that sarcastic tweets would have a larger sum. Results are presented in Table 3.

Comparing these results with the ones reported in Table 2, we can see an improvement in performance for all methods and metrics. This is an indicator that standard SA tools based on lexical approaches can be easily misled by sarcastic and ironic tweets. However, results are still quite poor when compared with machine learning techniques. This fact poses computational intelligence approaches as more indicate candidates, when considering applications that may contain sarcastic tweets.

Table 3 Classification performance using lexicon-hybrid methods, using the absolute sum of sentiment valences

	Happiness	Senticnet	Vadder	Sentstrength	Sentiwordnet	Pattern.en
AUC	0.5414	0.6046	0.5894	0.6057	0.5419	0.5749
MCC	0.0684	0.1599	0.5894	0.1938	0.0825	0.1700
Accuracy	0.5345	0.5801	0.5685	0.5839	0.5408	0.5723
F1	0.5658	0.6584	0.6496	0.6592	0.6543	0.6521

9   Concluding Remarks

Figurative Language is a frequent phenomenon within human communication, occurring both in spoken and written discourse including books, websites, fora, chats, Twitter messages, Facebook posts, news articles and product reviews. Knowing what people think can, in fact, help companies, political parties, and other public entities in strategizing and decision-making. With the increase in reviews, ratings, recommendations and other forms of online expression, online opinion has turned into a kind of virtual currency for businesses looking to sell their products, identify new opportunities and manage their reputations.

In this sense, when people are engaged in an informal conversation, they almost inevitably use irony (or sarcasm) to express something more or different than stated by the literal sentence meaning. Sentiment analysis methods can be easily misled by the presence of words that have a strong polarity but are used sarcastically, which means that the opposite polarity was intended. Several efforts have been recently devoted to detect and tackle FL phenomena in social media. Many of applications rely on task-specific lexicons (e.g. dictionaries, word classifications) or Machine Learning algorithms.

This chapter aims at showing how two specific domains of Figurative Language, sarcasm and irony, affect Sentiment Analysis tools. The study's ultimate goal is to find out if figurative language hinders the performance (polarity detection) of sentiment analysis systems due to the presence of ironic context.

The classifications were conducted with Support Vector Machine and lexicon based approaches. The results and conclusions of the experiments raise remarks and new questions. A first remark to be made is that all experiments are performed with English texts. Consequently, the result cannot directly be generalized to other languages.

Another interesting observation is whether very similar results are obtained when the experiments are carried out on data from the machine learning algorithms SVM and LogReg. Word embedding further improved these results. Although the improvements are small, our analysis indicates interesting fine granulated improvements in the sarcasm detection. Furthermore, results based on lexical based approaches are worse when compared to computational intelligence approaches. Our work indicates that the machine learning algorithms were reasonably able to detect irony in twitter messages.

In this work, we concentrate on off-the-shelve SA approaches. We did not use the network structure, as this network is difficult to process in, e.g., the classification of tweets in a stream real-time feed. It would be interesting to concentrate on specific methods developed for detecting iron, such as learning specific ironic patterns, as in [59]. Such approaches, however, tends to improve precision, but have lower recall. An interesting venue for future research is to learn sarcasm-specific vector embedding, in an approach similar to sentiment-specific vector developed in [64]

References

1. Internet World Stats. http://www.internetworldstats.com (2015). Accessed October 2015
2. Agrawal, R., Imieliński, T., Swami, A.: Mining association rules between sets of items in large databases. In: ACM SIGMOD International Conference on Management of Data, pp. 207–216. ACM (1993)
3. Baeza-Yates, R., Ribeiro-Neto, B., et al.: Modern information retrieval, vol. 463. ACM press, New York (1999)
4. Bamman, D., Smith, N.A.: Contextualized sarcasm detection on twitter. In: Ninth International AAAI Conference on Web and Social Media (2015)
5. Bikel, D., Zitouni, I.: Multilingual Natural Language Processing Applications: From Theory to Practice. IBM Press (2012)
6. Bogdanova, D., dos Santos, C., Barbosa, L., Zadrozny, B.: Detecting semantically equivalent questions in online user forums. CoNLL **2015**, 123 (2015)
7. Bowes, A., Katz, A.: When sarcasm stings. Discourse Process. **48**(4), 215–236 (2011)
8. Bradley, M.M., Lang, P.J.: Affective norms for english words (anew): instruction manual and affective ratings. Technical report, Technical Report C-1, The Center for Research in Psychophysiology, University of Florida (1999)
9. Bughin, J., Corb, L., Manyika, J., Nottebohm, O., Chui, M., de Muller Barbat, B., Said, R.: The impact of internet technologies: search. Technical report, McKinsey & Company, High Tech Practice (2011)
10. Calzolari, N., Choukri, K., Declerck, T., Loftsson, H., Maegaard, B., Mariani, J., Moreno, A., Odijk, J., Piperidis, S. (eds.): Proceedings of the Ninth International Conference on Language Resources and Evaluation (LREC-2014), European Language Resources Association (ELRA), Reykjavik, Iceland, 26–31 May 2014. http://www.lrec-conf.org/lrec2014
11. Cambria, E., Livingstone, A., Hussain, A.: The hourglass of emotions. In: Cognitive Behavioural Systems, pp. 144–157. Springer (2012)
12. Cambria, E., Speer, R., Havasi, C., Hussain, A.: Senticnet: a publicly available semantic resource for opinion mining. In: AAAI Fall Symposium: Commonsense Knowledge, vol. 10, p. 02 (2010)
13. Davidov, D., Tsur, O., Rappoport, A.: Semi-supervised recognition of sarcastic sentences in twitter and amazon. In: Proceedings of the Fourteenth Conference on Computational Natural Language Learning, pp. 107–116. Association for Computational Linguistics (2010)
14. De Smedt, T., Daelemans, W.: Pattern for python. J. Mach. Learn. Res. **13**(1), 2063–2067 (2012)
15. Dodds, P.S., Danforth, C.M.: Measuring the happiness of large-scale written expression: songs, blogs, and presidents. J. Happiness Stud. **11**(4), 441–456 (2010)
16. Esuli, A., Sebastiani, F.: Sentiwordnet: a publicly available lexical resource for opinion mining. In: Proceedings of LREC, vol. 6, pp. 417–422. Citeseer (2006)
17. Fan, R.E., Chang, K.W., Hsieh, C.J., Wang, X.R., Lin, C.J.: LIBLINEAR: a library for large linear classification. J. Mach. Learn. Res. **9**, 1871–1874 (2008)

18. Feldman, R.: Techniques and applications for sentiment analysis. Commun. ACM **56**(4), 82–89 (2013)
19. Fellbaum, C.: Wordnet. In: Theory and Applications of Ontology: Computer applications (2010)
20. Flach, P.: Machine Learning: The Art and Science of Algorithms that Make Sense of Data. Cambridge University Press (2012)
21. Ghosh, A., Li, G., Veale, T., Rosso, P., Shutova, E., Reyes, A., Barnden, J. (eds.): Semeval-2015 task 11: sentiment analysis of figurative language in twitter. In: International Workshop on Semantic Evaluation (SemEval-2015). Denver, Colorado (2015)
22. GIménez, M., Pla, F., Hurtado, L.: Elirf: a svm approach for sa tasks in twitter at semeval-2015. In: Proceedings of the 9th International Workshop on Semantic Evaluation (SemEval 2015), pp. 574–581. Association for Computational Linguistics, Denver, Colorado (2015)
23. Gonçalves, P., Araújo, M., Benevenuto, F., Cha, M.: Comparing and combining sentiment analysis methods. In: Proceedings of the First ACM Conference on Online Social Networks, pp. 27–38. ACM (2013)
24. González-Ibánez, R., Muresan, S., Wacholder, N.: Identifying sarcasm in twitter: a closer look. In: Proceedings of the 49th Annual Meeting of the Association for Computational Linguistics: Human Language Technologies: short papers, vol. 2, pp. 581–586. Association for Computational Linguistics (2011)
25. Grossman, D.A., Frieder, O.: Information Retrieval: Algorithms and Heuristics, vol. 15. Springer Science & Business Media (2012)
26. Gu, B., Ye, Q.: First step in social media: measuring the influence of online management responses on customer satisfaction. Prod. Oper. Manage. **23**(4), 570–582 (2014)
27. Hayta, A.B.: A study on the of effects of social media on young consumers' buying behaviors. Management **65**, 74 (2013)
28. Hearst, M.A.: Trends & controversies: support vector machines. IEEE Intell. Syst. **13**(4), 18–28 (1998). http://dx.doi.org/10.1109/5254.708428
29. Heerschop, B., Goossen, F., Hogenboom, A., Frasincar, F., Kaymak, U., de Jong, F.: Polarity analysis of texts using discourse structure. In: Proceedings of the 20th ACM International Conference on Information and Knowledge Management, pp. 1061–1070. ACM (2011)
30. Hosmer, D.W. Jr., Lemeshow, S.: Applied Logistic Regression. Wiley (2004)
31. Hu, X., Tang, J., Gao, H., Liu, H.: Unsupervised sentiment analysis with emotional signals. In: Proceedings of the 22nd International Conference on World Wide Web, pp. 607–618. International World Wide Web Conferences Steering Committee (2013)
32. Hu, X., Tang, L., Tang, J., Liu, H.: Exploiting social relations for sentiment analysis in microblogging. In: Proceedings of the Sixth ACM International Conference on Web search and Data Mining, pp. 537–546. ACM (2013)
33. Hutto, C., Gilbert, E.: Vader: a parsimonious rule-based model for sentiment analysis of social media text. In: Eighth International AAAI Conference on Weblogs and Social Media (2014)
34. Ingle, A., Maheshwari, N., Sutrave, N., Akumarthi, S., Bhitre, T.: Sentiment analysis: Sarcasm detection of tweets. B.Sc, Disssertation, May 2014
35. Jurafsky, D., Martin, J.H.: Speech and Language Processing, 2nd edn. Prentice-Hall Inc. (2008)
36. Kang, H., Yoo, S.J., Han, D.: Senti-lexicon and improved naïve bayes algorithms for sentiment analysis of restaurant reviews. Expert Syst. Appl. **39**(5), 6000–6010 (2012)
37. Kende, M.: How the internet continues to sustain growth and innovation. Technical report 20, Analysys Mason Limited and The Internet Society (ISOC) (2012)
38. Kunneman, F., Liebrecht, C., van Mulken, M., van den Bosch, A.: Signaling sarcasm: from hyperbole to hashtag. Inf. Process. Manage. (2014)
39. Le, Q.V., Mikolov, T.: Distributed representations of sentences and documents. In: Proceedings of the 31th International Conference on Machine Learning, ICML 2014, pp. 1188–1196, Beijing, China, 21–26 June 2014. http://jmlr.org/proceedings/papers/v32/le14.html
40. Liu, B.: Sentiment analysis and subjectivity. In: Handbook of Natural Language Processing, vol. 2, pp. 627–666 (2010)

41. Maynard, D., Greenwood, M.A.: Who cares about sarcastic tweets? investigating the impact of sarcasm on sentiment analysis. In: Proceedings of 9th Language Resources and Evaluation Conference (2014)
42. Mikolov, T., Chen, K., Corrado, G., Dean, J.: Efficient estimation of word representations in vector space (2013). arXiv preprint arXiv:1301.3781
43. Mikolov, T., Le, Q.V., Sutskever, I.: Exploiting similarities among languages for machine translation (2013). arXiv preprint arXiv:1309.4168
44. Mikolov, T., Yih, W., Zweig, G.: Linguistic regularities in continuous space word representations. In: Human Language Technologies: Conference of the North American Chapter of the Association of Computational Linguistics, Proceedings, pp. 746–751. Westin Peachtree Plaza Hotel, Atlanta, Georgia, USA, 9–14 June 2013. http://aclweb.org/anthology/N/N13/N13-1090. pdf
45. Munková, D., Munk, M., Vozár, M.: Data pre-processing evaluation for text mining: transaction/sequence model. Procedia Comput. Sci. **18**, 1198–1207 (2013)
46. Ozdemir, C., Bergler, S.: Clac-sentipipe: semeval 2015 subtasks 10 b, e, and task 11. In: Proceedings of the 9th International Workshop on Semantic Evaluation (SemEval 2015), pp. 479–485. Association for Computational Linguistics, Denver, Colorado (2015)
47. Pak, A., Paroubek, P.: Twitter as a corpus for sentiment analysis and opinion mining. In: International Conference on Language Resources and Evaluation, vol. 10, pp. 1320–1326 (2010)
48. Pang, B., Lee, L.: Opinion mining and sentiment analysis. Found. Trends Inf. Retrieval **2**(1–2), 1–135 (2008)
49. Pang, B., Lee, L., Vaithyanathan, S.: Thumbs up?: sentiment classification using machine learning techniques. In: Proceedings of the ACL-02 Conference on Empirical methods in Natural Language Processing, vol. 10, pp. 79–86. Association for Computational Linguistics (2002)
50. Pedregosa, F., Varoquaux, G., Gramfort, A., Michel, V., Thirion, B., Grisel, O., Blondel, M., Prettenhofer, P., Weiss, R., Dubourg, V., Vanderplas, J., Passos, A., Cournapeau, D., Brucher, M., Perrot, M., Duchesnay, E.: Scikit-learn: machine learning in Python. J. Mach. Learn. Res. **12**, 2825–2830 (2011)
51. Pennington, J., Socher, R., Manning, C.D.: Glove: Global vectors for word representation. In: Proceedings of the Empiricial Methods in Natural Language Processing (EMNLP 2014), vol. 12, pp. 1532–1543 (2014)
52. Perozzi, B., Al-Rfou, R., Skiena, S.: Deepwalk: online learning of social representations. In: Proceedings of the 20th ACM SIGKDD International Conference on Knowledge Discovery and Data Mining, pp. 701–710. ACM (2014)
53. Ptáček, T., Habernal, I., Hong, J.: Sarcasm detection on czech and english twitter. In: Proceedings of the 25th International Conference on Computational Linguistics: Technical Papers, COLING 2014, pp. 213–223. Dublin City University and Association for Computational Linguistics, Dublin, Ireland, August 2014. http://www.aclweb.org/anthology/C14-1022
54. Rajadesingan, A., Zafarani, R., Liu, H.: Sarcasm detection on twitter: a behavioral modeling approach. In: Proceedings of the Eighth ACM International Conference on Web Search and Data Mining, pp. 97–106. ACM (2015)
55. Řehůřek, R., Sojka, P.: Software framework for topic modelling with large corpora. In: Proceedings of the LREC 2010 Workshop on New Challenges for NLP Frameworks, pp. 45–50. ELRA, Valletta, Malta, May 2010. http://is.muni.cz/publication/884893/en
56. Reyes, A., Rosso, P.: On the difficulty of automatically detecting irony: beyond a simple case of negation. Knowl. Inf. Syst. **40**(3), 595–614 (2014)
57. Reyes, A., Rosso, P., Buscaldi, D.: From humor recognition to irony detection: the figurative language of social media. Data Knowl. Eng. **74**, 1–12 (2012)
58. Reyes, A., Rosso, P., Veale, T.: A multidimensional approach for detecting irony in twitter. Lang. Resour. Eval. **47**(1), 239–268 (2013)
59. Riloff, E., Qadir, A., Surve, P., Silva, L.D., Gilbert, N., Huang, R.: Sarcasm as contrast between a positive sentiment and negative situation. In: Proceedings of the 2013 Conference on Empirical

Methods in Natural Language Processing, EMNLP 2013, pp. 704–714. Grand Hyatt Seattle, Seattle, Washington, USA, A meeting of SIGDAT, a Special Interest Group of the ACL, 18–21 October 2013. http://aclweb.org/anthology/D/D13/D13-1066.pdf

60. Santos, C.D., Zadrozny, B.: Learning character-level representations for part-of-speech tagging. In: Proceedings of the 31st International Conference on Machine Learning (ICML-14), pp. 1818–1826 (2014)

61. Sikos, L., Brown, S.W., Kim, A.E., Michaelis, L.A., Palmer, M.: Figurative language: "meaning" is often more than just a sum of the parts. In: AAAI Fall Symposium: Biologically Inspired Cognitive Architectures, pp. 180–185 (2008)

62. Silva, C., Ribeiro, B.: The importance of stop word removal on recall values in text categorization. In: Proceedings of the International Joint Conference on Neural Networks, 2003, vol. 3, pp. 1661–1666. IEEE (2003)

63. Tan, C., Lee, L., Tang, J., Jiang, L., Zhou, M., Li, P.: User-level sentiment analysis incorporating social networks. In: Proceedings of the 17th ACM SIGKDD International Conference on Knowledge Discovery and Data Mining, pp. 1397–1405. ACM (2011)

64. Tang, D., Wei, F., Yang, N., Zhou, M., Liu, T., Qin, B.: Learning sentiment-specific word embedding for twitter sentiment classification. In: Proceedings of the 52nd Annual Meeting of the Association for Computational Linguistics, vol. 1: Long Papers, pp. 1555–1565. Association for Computational Linguistics, Baltimore, Maryland, June 2014. http://www.aclweb.org/anthology/P14-1146

65. Tausczik, Y.R., Pennebaker, J.W.: The psychological meaning of words: liwc and computerized text analysis methods. J. Lang. Soc. Psychol. **29**(1), 24–54 (2010)

66. Thelwall, M.: Heart and soul: sentiment strength detection in the social web with sentistrength. In: Proceedings of the CyberEmotions, pp. 1–14 (2013)

67. Van Rijsbergen, C.J.: A non-classical logic for information retrieval. Comput. J. **29**(6), 481–485 (1986)

68. Vanin, A.A., Freitas, L.A., Vieira, R., Bochernitsan, M.: Some clues on irony detection in tweets. In: Proceedings of the 22nd International Conference on World Wide Web companion, pp. 635–636. International World Wide Web Conferences Steering Committee (2013)

69. Walker, M.A., Anand, P., Abbott, R., Tree, J.E.F., Martell, C., King, J.: That is your evidence?: classifying stance in online political debate. Decis. Support Syst. **53**(4), 719–729 (2012)

70. Wallace, B.C., Do Kook Choe, L.K., Charniak, E.: Humans require context to infer ironic intent (so computers probably do, too). In: Proceedings of the Annual Meeting of the Association for Computational Linguistics (ACL), pp. 512–516 (2014)

71. Weichselbraun, A., Gindl, S., Scharl, A.: Extracting and grounding context-aware sentiment lexicons. IEEE Intell. Syst. **28**(2), 39–46 (2013)

72. Weiss, S.M., Indurkhya, N., Zhang, T., Damerau, F.: Text mining: predictive methods for analyzing unstructured information. Springer Science & Business Media (2010)

73. Weitzel, L., Freire, R.A., Quaresma, P., Gonçalves, T., Prati, R.C.: How does irony affect sentiment analysis tools? In: Progress in Artificial Intelligence—Proceedings of the 17th Portuguese Conference on Artificial Intelligence, EPIA 2015, pp. 803–808. Coimbra, Portugal, 8–11 Sept 2015

74. Weitzel, L., de Oliveira, J.P.M., Quaresma, P.: Measuring the reputation in user-generated-content systems based on health information. Procedia Comput. Sci. **29**, 364–378 (2014)

75. Weitzel, L., de Oliveira, J.P.M., Quaresma, P.: Exploring trust to rank reputation in microblogging. In: Database and Expert Systems Applications. Lecture Notes in Computer Science, vol. 8056, pp. 434–441. Springer (2013)

76. Xianghua, F., Guo, L., Yanyan, G., Zhiqiang, W.: Multi-aspect sentiment analysis for chinese online social reviews based on topic modeling and hownet lexicon. Knowl.-Based Syst. **37**, 186–195 (2013)

77. Xu, H., Santus, E., Laszlo, A., Huang, C.: Llt-polyu: identifying sentiment intensity in ironic tweets. In: Proceedings of the 9th International Workshop on Semantic Evaluation (SemEval 2015), pp. 673–678. Association for Computational Linguistics, Denver, Colorado (2015)

78. Zirn, C., Niepert, M., Stuckenschmidt, H., Strube, M.: Fine-grained sentiment analysis with structural features. In: IJCNLP, pp. 336–344 (2011)

Probabilistic Approaches for Sentiment Analysis: Latent Dirichlet Allocation for Ontology Building and Sentiment Extraction

Francesco Colace, Massimo De Santo, Luca Greco, Vincenzo Moscato and Antonio Picariello

Abstract People's opinion has always driven human choices and behaviors, even before the diffusion of Information and Communication Technologies. Thanks to the World Wide Web and the widespread of On-Line collaborative tools such as blogs, focus groups, review web sitesorums, social networks, millions of messages appear on the web, which is becoming a rich source of opinioned data. *Sentiment analysis* refers to the use of natural language processing, text analysis and computational linguistics to identify and extract subjective information in documents, comments and posts. The aim of this work is to show how the adoption of a probabilistic approach based on the Latent Dirichlet Allocation (LDA) as Sentiment Grabber can be an effective Sentiment Analyzer. Through this approach, for a set of documents belonging to a same knowledge domain, a graph, the Mixed Graph of Terms, can be automatically extracted. This graph, which contains a set of Mixed Graph of Terms, can be transformed in a Sentiment Oriented Terminological Ontology thanks to a methodology that involves the introduction of annotated lexicon as Wordnet. The chapter shows how the obtained ontology can be discriminative for sentiment classification. The proposed method has been tested in different contexts: standard datasets and comments extracted from social networks. The experimental evaluation shows how the proposed approach is effective and the results are quite satisfactory.

F. Colace (✉) · M. De Santo · L. Greco
Dipartimento di Ingegneria dell'Informazione, Ingegneria Elettrica
e Matematica Applicata, Università degli Studi di Salerno,
Via Ponte Don Melillo 1, 84084 Fisciano, Salerno, Italy
e-mail: fcolace@unisa.it

M. De Santo
e-mail: desanto@unisa.it

L. Greco
e-mail: lgreco@unisa.it

V. Moscato · A. Picariello
Dipartimento di Ingegneria elettrica e delle Tecnologie dell'Informazione,
Università degli Studi di Napoli Federico II, via Claudio 21, 80125 Napoli, Italy
e-mail: vmoscato@unina.it

A. Picariello
e-mail: picus@unina.it

© Springer International Publishing Switzerland 2016
W. Pedrycz and S.-M. Chen (eds.), *Sentiment Analysis and Ontology Engineering*,
Studies in Computational Intelligence 639, DOI 10.1007/978-3-319-30319-2_4

Keywords Sentiment analysis · Ontologies · Latent Dirichlet Allocation · Natural language processing

1 Introduction

Millions of messages are shared daily on the internet thanks to blogs, microblogs, social networks or reviews collector sites. These textual contexts can be grouped in two main categories: facts and opinions. Facts are objective statements while opinions reflect people's sentiments about products, other person and events and are extremely important when someone needs to evaluate the feelings of other people before taking a decision [38].

Before the wide diffusion of the Internet and Web 2.0, people used to share opinions and recommendations with traditional approaches: asking friends, talking to experts and reading documents. The Internet and web made possible to find out opinions and experiences from people being neither our personal acquaintances nor well known professional critics. The interest, that potential customers show in online opinions and reviews about products, is something that vendors are gradually paying more and more attention to. Companies are interested in what customers say about their products as politicians are interested in how different news media are portraying them. Therefore there is a lot of information on the web that have to be properly managed in order to provide vendors with highly valuable network intelligence and social intelligence to facilitate the improvement of their business.

In this scenario, a promising approach is the *sentiment analysis*: the computational study of opinions, sentiments and emotions expressed in a text [27, 45]. Its main aim is the identification of the agreement or disagreement statements that deal with positive or negative feelings in comments or reviews.

In this chapter we propose an alternative method to sentiment extraction relying on *Sentiment Oriented Terminological Ontologies* (SOTO). Such ontologies can be built automatically starting from Mixed Graph of Terms, that have proven to represent the most discriminative sentiment oriented words in a collection of user comments [12]. The ontology formalism not results reulsts but is important for our aims because it allows to easily share the obtained knowledge about positive and negative terms for other kinds of systems and applications.

The rest of the chapter is organized as follows: Sect. 2 shows related work in the field of sentiment extraction and terminological ontology building; Sect. 3 contains an overview of the proposed architecture; in Sect. 4 some details on the main involved methodologies are provided, with particular emphasis on Mixed Graph of Terms extraction and Terminological Ontology Building; in Sect. 5 an algorithm for Sentiment extraction with SOTO is presented; in Sect. 6 some experiments are reported and then some conclusions and future works about the proposed work are finally discussed.

2   Related Work

A very broad overview of the existing work on sentiment analysis was presented in [33]. The authors describe in detail the main techniques and approaches for an opinion oriented information retrieval. Early work in this area was focused on determining the semantic orientation of documents.

In particular some approaches attempt to learn a positive-negative classifier at a document level. Turney [43] introduces the results of review classification by considering the algebraic sum of the orientation of terms as respective of the orientation of the documents. Starting from this approach other techniques have been developed by focusing on some specific tasks as finding the sentiment of words [46]. Baroni [3] proposed to rank a large list of adjectives according to a subjectivity score by employing a small set of manually selected subjective adjectives and computing the mutual information of pairs of adjectives using frequency and co-occurrence frequency counts on the web. The work of Turney [41] proposes an approach to measure the semantic orientation of a given word based on the strength of its association with a set of context insensitive positive words minus the strength of its association with a set of negative words. By this approach sentiment lexicons can be built and a sentiment polarity score can be assigned to each word [22, 31]. Sentiment polarity score means the strength or degree of sentiment in a defined sentence pattern.

Artificial Intelligence and probabilistic approaches have also been adopted for sentiment mining. In [34] three machine learning approaches (Naïve Bayes, Maximum Entropy and Support Vector Machines) has been adopted to label the polarity of a movie reviews datasets. A promising approach is presented in [36] where a novel methodology has been obtained by the combination of rule based classification, supervised learning and machine learning. In [39] a SVM based technique has been introduced for classifying the sentiment in a collection of documents. In [32], instead, a Naïve Bayes classifier is used for the sentiment classification of tweets' corpora. Other approaches are inferring the sentiment orientation of social media content and estimate sentiment orientations of a collection of documents as a text classification problem [18].

More in general, sentiment related information can be encoded within the actual words of the sentence through changes in attitudinal shades of word meaning using suffixes as discussed in [19]. This has been investigated in [31] where a lexicon for sentiment analysis has been obtained.

In [50] a probabilistic approach to sentiment mining is adopted. In particular this work uses a probabilistic model called Sentiment Probabilistic Latent Semantic Analysis (S-PLSA) in which a review, and more in general a document, can be considered as generated under the influence of a number of hidden sentiment factors [24]. The S-PLSA is an extension of the PLSA where it is assumed that there are a set of hidden semantic factors or aspects in the documents related to documents and words under a probabilistic framework.

In [5] an approach combining the ontological formalism and a machine learning technique has been introduced. In particular the proposed system uses domain ontology to extract the related concepts and attributes starting from a sentence and then labels the sentence itself as positive, negative or neutral by means of the Support Vector Machine (SVM) classifier.

In this chapter, we investigate the adoption of a similar approach based on the Latent Dirichlet Allocation (LDA). In LDA, each document may be viewed as composed by a mixture of various topics. This is similar to probabilistic latent semantic analysis, except that in LDA the topic distribution is assumed to have a Dirichlet prior. By the use of the LDA approach on a set of documents belonging to a same knowledge domain, a Mixed Graph of Terms can be automatically extracted [17, 30], which we demonstrated to be discriminative also for sentiment classification [12].

The main reason of such discriminative power is that LDA-based topic modeling is essentially an effective conceptual clustering process and it helps discover semantically rich concepts describing the respective "sentimental" relationships. By means of applying these semantically rich concepts, that contain more useful relationship indicators to identify the sentiment from messages and by using a Terminological Ontology Builder which allows to identify the kind of semantic relationship between word pairs in mGT, the proposed system can accurately discover more latent relationships and make less errors in its predictions.

In the last decade many researchers have been involved in the development of methodologies for ontology definition, building, learning and population, since ontologies are considered as an effective answer to the need of semantic interoperability among modern information systems: it is well known, in fact, that ontologies are considered as the backbone of the Semantic Web, being fundamental components for sharing, reusing as well as reasoning over knowledge domains. Several theories have been developed, in different application domains and especially in the semantic web framework. However the process of learning and populating ontologies is generally not trivial and very time consuming: it still remains an open research challenge.

The term "ontology learning" has been introduced in [28] to describe the acquisition of a domain model from data. This process is historically connected to the introduction of the Semantic Web and calls for new capabilities to learn from input data relevant concepts for a given domain, their definitions as well as the relations holding between them. In particular, ontologies can be learnt from different sources, such as databases, structured and unstructured documents or even existing linguistic resources like dictionaries and taxonomies. Then, with the explosion of available textual data due to the Read/Write Web, ontology learning from text is becoming one of the most investigated issue in the literature: it is the process of identifying terms, concepts, relations, and axioms from textual information and using them in order to construct and maintain an ontology [37, 48].

Such a process is generally composed by five phases that aim at returning five main outputs: *terms*, *concepts*, *taxonomic relations*, *non-taxonomic relations* and *axioms* [7].

To obtain each output, some tasks have to be accomplished and the techniques employed for each task may change among different systems. In this sense the ontology learning process is really modular: in [48] the corresponding tasks and the plethora of employed techniques for each output are described.

The extraction of terms from text usually needs a *preprocessing phase* to convert a textual stream into a target format to feed an ontology learning system: this generally involves text analysis methods [2]. In particular, the terms extraction begins with a *tokenization* or *part of speech tagging* to break texts into smaller components. In this phase, statistical or probabilistic measures are adopted for determining the "unithood"—i.e. the strength or stability of syntagmatic combinations or collocation of terms - and the "termhood"—i.e. the relevance or the specificity of a term with respect to the domain.

The relations model the kind of interactions among the concepts in ontology; in general, two types of relations are exploited: taxonomic and non-taxonomic relations. Taxonomic relations—hypernym—are useful to build concepts' hierarchies and can be labeled as "is-a" relations [10]. This kind of relations can be recognized in several ways such as using predefined relations from existing background knowledge, using statistical subsumption models, relying on semantic similarity between concepts and utilizing linguistic and logical rules or patterns. In tur1n, the non-taxonomic relations represent the remaining kind of interactions and their extraction is a very challenging task: verbs play a significant role together with the support of domain experts.

Axioms are propositions or sentences that always hold a true value and are the starting point for deducing other truth, verifying the correctness of the ontological elements and defining constraints. The process of learning axioms is still complex and there are few examples in the literature.

Ontology learning problem can be faced using techniques coming from different fields such as information retrieval, machine learning, and natural language processing [11]. These techniques can be generally classified into *statistics-based, linguistic-based, logic-based* or *hybrid* [48].

Statistics-based techniques work at a syntactical level and are exploited in the early stages of ontology learning, such as term extraction and hierarchy construction [6]. The most common techniques include clustering [47], Latent Semantic Analysis [42], term subsumption [20] and contrastive analysis [44]. The main idea behind these techniques is that the co-occurrence of words provides a reliable estimate about their semantic identity. In this way the ontology concepts can be inferred.

The linguistics-based techniques can support all tasks in ontology learning and are based on natural language processing tools. In general, they include Part of Speech (POS) tagging, such as [4], syntactic structure analysis [23] and dependency analysis [9]. Other adopted techniques are related to semantic lexicon [35], lexico-syntactic patterns [7, 40], subcategorization frames [21] and seed words [49].

The logic-based techniques are the least common in ontology learning and are mainly adopted for more complex tasks involving relations and axioms. The two main employed techniques are based on inductive logic programming [26] and logical inference [38]. In the inductive logic programming, rules are derived from existing collection of concepts and relations which are divided into positive and negative samples. In logical inference, implicit relations are derived from existing ones using particular rules (transitivity and inheritance).

As previously said, each phase of the ontology learning process can exploit one of these approaches in order to maximize the related effectiveness. In particular the terms and concepts extraction can be performed by the use of the statistics-based techniques, while the inference of relations can be obtained by the use of linguistic and logic based techniques. Indeed, the hybrid approach is mainly used in the existing studies and achieves the best results [11, 48].

Differently from several approaches in ontology learning and population, where concept hierarchies are usually obtained by means of statistical and/or proba-bilistic methods, we enrich our terminological ontologies with the semantic fea-tures contained in general purpose lexical ontologies, such as WordNet. The use of both statistical and semantic techniques allows to have suitable and effective domain ontologies for a number of applications such as topic detection and track-ing, opinion and sentiment analysis, text mining/classification and recommendation [1, 13, 15, 16].

3   System Architecture

Figure 1 describes at a glance the proposed system architecture. The system ana-lyzes a set of documents—positive/negative comments coming from different web sources or from specific collections—to produce a Sentiment Oriented Terminolog-ical Ontology (SOTO). The system is formed by two main components:

- *Mixed Graph of Terms* building component, that uses *Latent Dirichlet Analysis* (LDA) on the input documents and produces a *Mixed Graph of Terms* (mGT) representation containing the most relevant domain terms and their co-occurrence values (relations) in the analyzed set.
- *Sentiment Oriented Ontology* building component, that using general purpose or domain-specific lexical databases, refines the previous discovered terms, exploit-ing their lexical relationships, adding hidden concepts, and producing the final ontology schema and population.

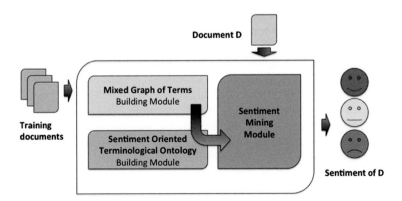

Fig. 1 System architecture

Once the positive/negative SOTOs have been built, they can be used as a sentiment filter by applying an particular algorithm as described further.

In the following we will discuss into details the basic components of the proposed architecture, first focusing on the terminological ontology building phase.

4 Terminological Ontology Building

4.1 Ontology Learning: Concepts and Relation Extraction

In this section we explain how a mGT structure (Mixed Graph of Terms) can be extracted from a corpus of documents.

The Feature Extraction module (FE) is represented in Fig. 2. The input of the system is the set of documents:

$$\Omega_r = (\mathbf{d}_1, \cdots, \mathbf{d}_M)$$

The pre-processing phase provides tokenization, stopwords filtering and stemming; then a Term-Document Matrix is built to feed the Latent Dirichlet Allocation (LDA) [5] module. The LDA algorithm, assuming that each document is a mixture of a small number of latent topics and each word's creation is attributable to one of the document's topics, provides as output two matrices—Θ and Φ—which contain probabilistic relations between topic-document and word-topic respectively.

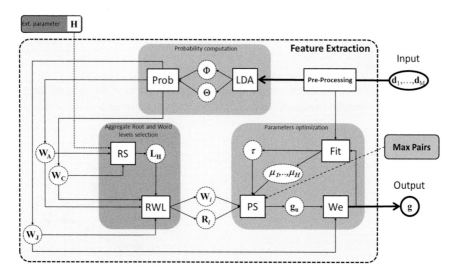

Fig. 2 Proposed feature extraction method. A *mGT* **g** structure is extracted from a corpus of training documents

Under a given set of assumptions [14], LDA module's results can be used to determine: the probability for each word v_i to occur in the corpus $W_A = \{P(v_i)\}$; the conditional probability between word pairs $W_C = \{P(v_i|v_s)\}$; the joint probability between word pairs $W_J = \{P(v_i, v_j)\}$. Details on LDA and probability computation are discussed in [5, 14, 17].

Defining *Aggregate roots* (AR) as the words whose occurrence is most implied by the occurrence of other words of the corpus, a set of H aggregate roots $\mathbf{r} = (r_1, \ldots, r_H)$ can be determined from W_C:

$$r_i = \operatorname{argmax}_{v_i} \prod_{j \neq i} P(v_i|v_j) \tag{1}$$

This phase is referred as Root Selection (RS) in Fig. 2. A weight ψ_{ij} can be defined as a degree of probabilistic correlation between AR pairs: $\psi_{ij} = P(r_i, r_j)$. We define an *aggregate* as a word v_s having a high probabilistic dependency with an aggregate root r_i. Such a dependency can be expressed through the probabilistic weight $\rho_{is} = P(r_i|v_s)$. Therefore, for each aggregate root, a set of aggregates can be selected according to higher ρ_{is} values. As a result of the Root-Word level selection (RWL), an initial mGT structure, composed by H aggregate roots (R_l) linked to all possible aggregates (W_l), is obtained. An optimization phase allows to neglect weakly related pairs according to a fitness function discussed in [14]. Our algorithm, given the number of aggregate roots H and the desired max number of pairs as constraints, chooses the best parameter settings τ and $\mu = (\mu_1, \ldots, \mu_H)$ defined as follows:

1. τ: the threshold that establishes the number of *aggregate root/aggregate root* pairs. A relationship between the aggregate root v_i and aggregate root r_j is relevant if $\psi_{ij} \geq \tau$.
2. μ_i: the threshold that establishes, for each aggregate root i, the number of *aggregate root/word* pairs. A relationship between the word v_s and the aggregate root r_i is relevant if $\rho_{is} \geq \mu_i$.

Note that a mGT structure can be suitably represented as a *graph* $\mathbf{g}$ of terms (Fig. 3). Such a graph is made of several clusters, each containing a set of words v_s

Fig. 3 Graphical representation of a *mGT* structure

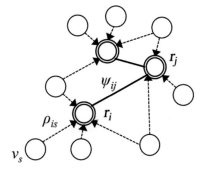

(*aggregates*) related to an *aggregate root* (r_i), the centroid of the cluster. *Aggregate roots* can be also linked together building a centroids subgraph.

For sentiment analysis goals, we consider mGTs having as aggregate roots positive or negative adjectives and as aggregates adjectives or adverbs.

4.2 Ontology Refinement

The main goal of such a component is to transform, for each topic, the obtained mGT graphs into a particular terminological ontology—by using information related to *adjectives* and *adverbs* coming from comments and coded in the shape of mGT structures—in order to represent and manage the knowledge coming from the document corpus in a more effective way. In the following, we will introduce some preliminary definitions and describe an algorithm for automatically building the described ontologies.

A *Sentiment Oriented Terminological Ontology* is a labeled graph $(\mathscr{V}, \mathscr{E}, \rho)$, where: (i) $\mathscr{V} = \{v_1, \ldots, v_n\}$ is a finite set of *nodes*, formed by a set of *aggregates* and the related *aggregate roots* together with a common *root*; (ii) $\mathscr{E}$ is a subset of $(\mathscr{V} \times \mathscr{V})$; (iii) ρ is a function that associates to each couple of nodes a label λ indicating the kind of *relationship* between them. In addition, we assume that the code of a given vocabulary synset s is associated to each ontology node.

We implemented an algoirthm to build the local terminological ontology by means of: the set of (positive/negative) adjectives with the related adverbs for a single topic; a set of relationships among ontology nodes, using a general lexical vocabulary.

The input of the algorithm is a mGT graph: in particular, a set of aggregate root nodes (adjectives) and a set of aggregated words (other adjectives or adverbs) for a given topic/domain are considered.

As an example, we extracted from the following user positive comments (about the "Million Dollar Baby" movie):

1. "fantastic...really good movie..."
2. "very good movie, fantastic interpretation of hilary swanks...she played suggestively.."
3. "really fantastic, wonderful, extraordinary movie..."

{good, fantastic, wonderful, suggestive, extraordinary} and {really, very, suggestively} as lists of adjectives and adverbs, which we then combined in the mGT graph in Fig. 4 using the approach previously described.

In a first phase, all the adjective nodes related to some aggregate roots are opportunely linked to the ontology common root. Then, the aggregate root nodes with the same synset are linked (each one to other ones) by a *synonym edge*, while nodes having a *similar meaning* are linked creating a *similarity edge*.

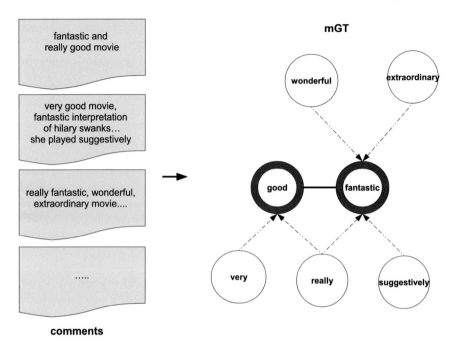

Fig. 4 mGT example

In a second step, the adjective nodes (for aggregates) are linked to the related aggregate roots by a similarity edge, if they express a similar meaning; otherwise, they are removed from the ontology.

In the final step, all the adverb nodes are linked to the related aggregate root nodes with a *correlation edge*. If an adverb derives from an adjective (not corresponding to any node present in the ontology), a new node for the adjective is created and a *pertinence edge* is generated between the new term and the aggregate root.

The algorithm can be used to create both the ontologies for negative and positive comments. Figure 5 shows an example of SOTO building process during the described steps.

In our implementation, we exploit:

- *WordNet* as lexical database;
- *Neo4j* graph database to store and query the different ontologies;
- a combination of *see_also* and *similar_to* properties of WordNet database to verify if two adjectives are similar;
- *pertainym* property of WordNet database to verify if an adverb derives from a given adjective.

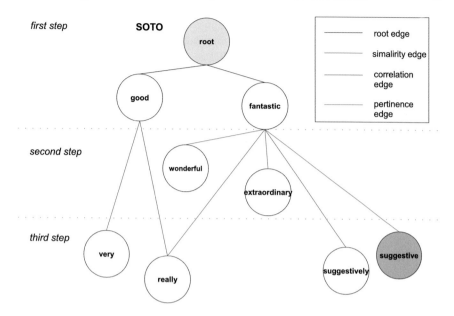

Fig. 5 SOTO building process

5 Sentiment Extraction

In [12], the LDA approach and the mGT formalism have been used for the detection of sentiment in tweets. The approach aims at using the mGT, obtained by LDA based analysis of tweets, as a filter for the classification of the sentiment in a tweet.

Here, we want to improve that methodology by applying the sentiment mining algorithm to terminological ontologies built from a set of positive and negative comments.

- *Input of the algorithm*: (i) A subset $\hat{F}^k$ of the feature space F^k corresponding to the set of adverbs, adjectives or $\langle adverb, adjective \rangle$ pairs coming from comments in a specific knowledge domain. The Sentiment Oriented Terminological Ontologies SOTO$^+$ and SOTO$^-$ built from (positively and negatively) opinionated comments related to a knowledge domain.
- *Output of the algorithm*: (i) The sentiment extracted from the comment.
- *Description of the main steps*:

  1. For each comment the scores $S_{c_i}^+$ and $S_{c_i}^-$ are determined as:

  $$S_{c_i}^{\pm} = \alpha * N + \beta * M$$

  with N being the number of matched adjective or adverb nodes in the ontology and M the number of matched pairs of nodes (in our implementation a pair $\langle adverb, adjective \rangle$ from a comment is matched by a pair of nodes of the ontology, if there exists a ontology graph path with length $k > 1$ that contains the input pair and not containing the root node); $\alpha, \beta \in [0, 1]$ are proper normalizing factors. The values of $\alpha = 0.3$ and $\beta = 0.7$ have been chosen since they have proven to achieve the best results.[1]

2. For each comment $\frac{S_{c_i}^+}{S_{c_i}} > \delta$, the comment is labelled as positive; if $\frac{S_{c_i}^-}{S_{c_i}^+} > \delta$ the comment is labelled as negative; otherwise it is neutral. The value δ has been chosen in order to maximize accuracy.

The use of Neo4j graph database technology permits to deal with very large ontologies (that are eventually distributed on different computing nodes) and to easily and efficiently implement the required search of paths to determine the number of matched pairs of adjective/adverb. Thus, our approach can be practically applied in a variety of domain and contexts related to sentiment and opinion analysis topics.

6 Experimental Stage

In a previous work we have shown the effectiveness of a terminological ontology building method with standard measures [16]. Here we are interested in evaluating the performances of a SOTO based method for sentiment classification. A standard dataset has been used for the experiments, the *Movie Reviews Dataset* provided by Pang et al. in [34]. Such a dataset consist of 1000 positive and 1000 negative reviews from the Internet Movie Database. Positive labels were assigned to reviews that had a rating above 3.5 stars and negative labels were assigned to the rest. Positive and negative documents were random selected and divided into three equal-sized folds in order to maintain balanced class distributions in each fold.

We compare the proposed method with the state of art approaches for sentiment classification through three-fold cross-validation accuracies; we also show the improvements of the SOTO based method with respect to the standard mGT based approach proposed in [12]. Table 1 shows the results of the comparison. The approach has proven to obtain effective results also in real time scenarios.

An additional experimental campaign on posts coming from social networks has been conducted. A custom dataset built for the work in [12] was used to this aim; it is composed by tweets collected from twitter along a period of about two weeks and contains the following topics:

- Foreign News
- Italian Politics
- Show

---

[1]The chosen value of β permites to weight in a more significant manner the presence of pairs adjectives/adverbs in a comment rather than the simple occurrence of the positive/negative adjectives or adverbs.

Table 1 The average three-fold cross-validation accuracy obtained by the various methods on the standard dataset

Reference paper	Methodology	Accuracy (%)
Pang et al. [34]	Support Vector Machines	82.90
	Naive Bayes	81.50
	Maximum Entropy	81.00
Kennedy and Inkpen [25]	Support Vector Machines	86.20
Chaovalit and Zhou [8]	Ontology supported polarity mining	72.20
Melville et al. [29]	Bayesian classification with the support of lexicons	81.42
Chaovalit and Zhou [39]	Formal concept analysis	77.75
mGT Approach	LDA	88.50
SOTO	LDA with support of lexicons	93.45

Table 2 Dataset structure

Topic	Language	Tweets
Topic	Language	# Tweets
Foreign News	English	500
Italian Politics	English	500
Show	English	450
Sport	English	400
Technology	English	500
Foreign News	Italian	450
Italian Politics	Italian	450
Show	Italian	500
Sport	Italian	500
Technology	Italian	450

- Sport
- Technology

The dataset was divided in two different subsets according to their original language (Italian and English). The Dataset structure is described in Table 2. As ground truth, we used the same judgements from two groups of experts in Computational Linguistics that were considered in [12]. A first group of 15 experts classified 100 different tweets and at the end each tweet was labeled according to a majority vote rule. After this phase the same unlabelled tweets were submitted to a second group of 15 experts to be judged according to a majority vote rule. If both classifications were consistent, tweets could be introduced into the dataset. Otherwise they would be discarded. The Naïve Bayes approach, the SVM approach, our standard mGT approach and the

Table 3 Obtained results (accuracy percentage) on the custom dataset (English language)

	Foreign News	Italian Politics	Show	Sport	Technology
Naïve Bayes	88.67	93.98	95.43	93.23	83.35
SVM	90.32	89.53	94.75	91.76	95.43
mGT	97.45	95.08	97.63	95.43	94.83
SOTO	98.01	95.93	98.22	96.71	95.96

Table 4 Obtained results (accuracy percentage) on the custom dataset (Italian language)

	Foreign News	Italian Politics	Show	Sport	Technology
Naïve Bayes	89.47	92.67	91.56	89.32	89.64
SVM	89.43	91.37	92.56	97.43	92.32
mGT	96.92	95.56	95.98	96.34	93.45
SOTO	97.74	96.87	96.48	97.84	94.85

SOTO based method have been tested on the dataset obtaining the results, in terms of accuracy, depicted in Tables 3 and 4. Also in this case, we observe that the SOTO approach shows the best results.

7   Conclusion

In this chapter we propose the use of Sentiment Oriented Terminological Ontologies, obtained by Mixed Graph of Terms (LDA based structure), as tool for the sentiment classification of documents. The method relies on extracting complex structures called mGTs from documents labeled according their sentiment and then build Sentiment Oriented Terminological Ontologies according to a novel algorithm. The sentiment classification of a new document can then be performed from SOTO through a sentiment mining module.

Our method was compared to the state of art methods in literature and our standard mGT approach by the use of standard and real datasets. In both cases obtained results reveal a better accuracy.

Future work will be devoted to enlarge our experimentation to more large datasets, also considering data form different kinds of social networks (OSN and multimedia sharing systems). From the other hand, we are planning to compare our approach not only with the classical approaches but only with the most recent ones in the literature.

Once obtained a number of significant ontologies for several languages, topics and application domains, a further goal of our project will be to share such knowledge in a standard format (e.g. RDF) in order to allow other systems to use SOTO structures for the related aims.

References

1. Albanese, M., d'Acierno, A., Moscato, V., Persia, F., Picariello, A.: A multimedia semantic recommender system for cultural heritage applications. In: 2011 Fifth IEEE International Conference on Semantic Computing (ICSC), pp. 403–410. IEEE (2011)
2. Amato, F., Mazzeo, A., Moscato, V., Picariello, A.: Semantic management of multimedia documents for e-government activity. In: International Conference on Complex, Intelligent and Software Intensive Systems, 2009. CISIS'09, pp. 1193–1198. IEEE (2009)
3. Baroni, M., Vegnaduzzo, S.: Identifying subjective adjectives through web-based mutual information. In: Proceedings of the 7th Konferenz zur Verarbeitung Natrlicher Sprache (German Conference on Natural Language Processing KONVENS04, pp. 613–619 (2004)
4. Bird, S., Klein, E., Loper, E., Baldridge, J.: Multidisciplinary instruction with the natural language toolkit. In: Proceedings of the Third Workshop on Issues in Teaching Computational Linguistics, TeachCL'08, pp. 62–70 (2008)
5. Blei, D.M., Ng, A.Y., Jordan, M.I.: Latent dirichlet allocation. J. Mach. Learn. Res. **3**, 993–1022 (2003)
6. Brewster, C., Jupp, S., Luciano, J., Shotton, D., Stevens, R., Zhang, Z.: Issues in learning an ontology from text. BMC Bioinformatics **10**(Suppl 5), S1 (2009)
7. Buitelaar, P., Magnini, B.: Ontology learning from text: an overview. In: Paul Buitelaar, P., Cimiano, P., Magnini, B. (eds.) Ontology Learning from Text: Methods, Applications and Evaluation, pp. 3–12. IOS Press (2005)
8. Chaovalit, P., Zhou, L.: Movie review mining: a comparison between supervised and unsupervised classification approaches. In: Proceedings of the Proceedings of the 38th Annual Hawaii International Conference on System Sciences (HICSS'05)—Track 4—HICSS'05, vol. 04, p. 112.3. IEEE Computer Society, Washington, DC, USA (2005)
9. Ciaramita, M., Gangemi, A., Ratsch, E., Šaric, J., Rojas, I.: Unsupervised learning of semantic relations between concepts of a molecular biology ontology. In: Proceedings of the 19th international joint conference on Artificial intelligence, IJCAI'05, pp. 659–664 (2005)
10. Cimiano, P., Pivk, A., Schmidt-Thieme, L., Staab, S.: Learning taxonomic relations from heterogeneous evidence
11. Cimiano, P., Völker, J., Studer, R.: Ontologies on demand a description of the state-of-the-art, applications, challenges and trends for ontology learning from text (2006)
12. Colace, F., De Santo, M., Greco, L.: Sentiment mining through mixed graph of terms. In: 2014 17th International Conference on Network-Based Information Systems (NBiS), pp. 324–330, Sept 2014
13. Colace, F., De Santo, M., Greco, L.: A probabilistic approach to tweets' sentiment classification. In: 2013 Humaine Association Conference on Affective Computing and Intelligent Interaction (ACII), pp. 37–42 (2013)
14. Colace, F., De Santo, M., Greco, L., Napoletano, P.: Weighted word pairs for query expansion. Inf. Process. Manage. (2014)
15. Colace, F., De Santo, M., Greco, L.: An adaptive product configurator based on slow intelligence approach. Int. J. Metadata Semant. Ontol. **9**(2), 128–137, 01 (2014)
16. Colace, F., De Santo, M., Greco, L., Amato, F., Moscato, V., Picariello, A.: Terminological ontology learning and population using latent dirichlet allocation. J. Vis. Lang. Comput. **25**(6), 818–826 (2014)
17. Colace, F., De Santo, M., Greco, L., Napoletano, P.: Text classification using a few labeled examples. Comput. Human Behav. **30**, 689–697 (2014)
18. Colbaugh, R., Glass, K.: Estimating sentiment orientation in social media for intelligence monitoring and analysis. In: 2010 IEEE International Conference on Intelligence and Security Informatics (ISI), pp. 135–137, May 2010
19. Esuli, A., Sebastiani, F.: Sentiwordnet: a publicly available lexical resource for opinion mining. In: Proceedings of the 5th Conference on Language Resources and Evaluation (LREC06), pp. 417–422 (2006)

20. Fotzo, H.N., Gallinari, P.: Learning generalization/specialization relations between concepts—application for automatically building thematic document hierarchies
21. Gamallo, P., Agustini, A., Lopes, G.P.: Learning subcategorisation information to model a grammar with "co-restrictions" (2003)
22. Gamon, M., Aue, A.: Automatic identification of sentiment vocabulary: exploiting low association with known sentiment terms. In: Proceedings of the ACL Workshop on Feature Engineering for Machine Learning in Natural Language Processing, pp. 57–64. Association for Computational Linguistics, Ann Arbor, Michigan, June 2005
23. Hippisley, A., Cheng, D., Ahmad, K.: The head-modifier principle and multilingual term extraction. Nat. Lang. Eng. **11**(2), 129–157 (2005)
24. Hofmann, T.: Probabilistic latent semantic analysis. In: Proceedings of Uncertainty in Artificial Intelligence, UAI99, pp. 289–296 (1999)
25. Kennedy, A., Inkpen, D.: Sentiment classification of movie reviews using contextual valence shifters. Comput. Intell. **22**, 2006 (2006)
26. Wrobel, S.: Inductive logic programming for knowledge discovery in databases. In: Dzeroski, S., Lavrač, N. (eds.) Relational Data Mining. Springer, pp. 74–101 (2001)
27. Liu, B.: Sentiment Analysis and Subjectivity. In: Handbook of Natural Language Processing, 2nd edn. Taylor and Francis Group, Boca (2010)
28. Maedche, A., Staab, S.: Ontology learning for the semantic web. IEEE Intell. Syst. **16**(2), 72–79 (2001)
29. Melville, P., Gryc, W., Lawrence, R.D.: Sentiment analysis of blogs by combining lexical knowledge with text classification. In: Proceedings of the 15th ACM SIGKDD International Conference on Knowledge Discovery and Data Mining, KDD'09, pp. 1275–1284. ACM, New York, NY, USA (2009)
30. Napoletano, P., Colace, F., De Santo, M., Greco, L.: Text classification using a graph of terms. In: 2012 Sixth International Conference on Complex, Intelligent and Software Intensive Systems (CISIS), pp. 1030–1035, July (2012)
31. Neviarouskaya, A., Prendinger, H., Ishizuka, M.: Sentiful: a lexicon for sentiment analysis. IEEE Trans. Affect. Comput. **2**(1), 22–36, jan–june 2011
32. Pak, A., Paroubek, P.: Twitter as a corpus for sentiment analysis and opinion mining. In: Proceedings of the Seventh conference on International Language Resources and Evaluation (LREC'10). European Language Resources Association (ELRA), Valletta, Malta, May 2010
33. Pang, B., Lee, L.: Opinion mining and sentiment analysis. Found. Trends Inf. Retr. **2**(1–2), 1–135 (2008)
34. Pang, B., Lee, L., Vaithyanathan, S.: Thumbs up? sentiment classification using machine learning techniques. In: Proceedings of EMNLP, pp. 79–86 (2002)
35. Pedersen, T., Patwardhan, S., Michelizzi, J.: Wordnet::similarity: measuring the relatedness of concepts. In: Demonstration Papers at HLT-NAACL 2004, HLT-NAACL-Demonstrations'04, pp. 38–41 (2004)
36. Prabowo, R., Thelwall, M.: Sentiment analysis: a combined approach. J. Informetrics **3**, 143–157 (2009)
37. Sebastiani, F.: Machine learning in text categorization. ACM Comput. Surv. **34**, 1–47 (2002)
38. Shamsfard, M., Barforoush, A.A.: The state of the art in ontology learning: a framework for comparison. Knowl. Eng. Rev. **18**(4), 293–316 (2003)
39. Shein, K.P.P.: Ontology based combined approach for sentiment classification. In: Proceedings of the 3rd International Conference on Communications and Information Technology, CIT'09, pp. 112–115. World Scientific and Engineering Academy and Society (WSEAS), Stevens Point, Wisconsin, USA (2009)
40. Snow, R.: Semantic taxonomy induction from heterogenous evidence. In: Proceedings of COLING/ACL 2006, 801–808 (2006)
41. Turney, P., Littman, M.: Unsupervised learning of semantic orientation from a hundred-billion-word corpus. Technical report NRC technical report ERB-1094, Institute for Information Technology, National Research Council Canada (2002)

42. Turney, P.D.: Mining the web for synonyms: PMI-IR versus LSA on TOEFL. In: Proceedings of the 12th European Conference on Machine Learning, EMCL'01, pp. 491–502 (2001)
43. Turney, P.D.: Thumbs up or thumbs down?: semantic orientation applied to unsupervised classification of reviews. In: Proceedings of the 40th Annual Meeting on Association for Computational Linguistics, ACL'02, pp. 417–424. Association for Computational Linguistics, Stroudsburg, PA, USA (2002)
44. Velardi, P., Navigli, R., Cucchiarelli, A., Neri, F.: Evaluation of ontolearn, a methodology for automatic learning of ontologies. In: Buitelaar, P., Cimiano, P., Magnini, B. (eds.) Ontology Learning from Text: Methods, Evaluation and Applications. IOS Press, pp. 92–105
45. Wang, C., Xiao, Z., Liu, Y., Yanru, X., Zhou, A., Zhang, K.: Sentiview: sentiment analysis and visualization for internet popular topics. IEEE Trans. Hum. Mach. Syst. **43**(6), 620–630 (2013)
46. Wilson, T., Wiebe, J., Hwa, R.: Just how mad are you? finding strong and weak opinion clauses. In: Proceedings of the 19th national conference on Artifical intelligence, AAAI'04, pp. 761–767. AAAI Press (2004)
47. Wong, W., Liu, W., Bennamoun, M.: Tree-traversing ant algorithm for term clustering based on featureless similarities. Data Min. Knowl. Discov. **15**(3), 349–381 (2007)
48. Wong, W., Liu, W., Bennamoun, M.: Ontology learning from text: a look back and into the future. ACM Comput. Surv. **44**(4), 20:1–20:36, Sept 2012
49. Yangarber, R., Grishman, R., Tapanainen, P.: Automatic acquisition of domain knowledge for information extraction. In: Proceedings of the 18th International Conference on Computational Linguistics, pp. 940–946 (2000)
50. Yu, X., Liu, Y., Huang, X., An, A.: Mining online reviews for predicting sales performance: a case study in the movie domain. IEEE Trans. Knowl. Data Eng. **24**(4), 720–734 (2012)

Description Logic Class Expression Learning Applied to Sentiment Analysis

Alberto Salguero and Macarena Espinilla

Abstract Description Logic (DL) Class Expression Learning (CEL) is a recent research topic of interest in the field of machine learning. Given a set of positive and negative examples of individuals in an ontology, the learning problem consists of finding a new class expression or concept such that most of the positive examples are instances of that concept, whereas the negatives examples are not. Therefore, the class expression learning can be seen as a search process in the space of concepts. In this chapter, the use of CEL algorithms is proposed as a tool to find the class expression that describes as much of the instances of positive documents as possible, being the main novelty of the proposal that the ontology is focused on inferring knowledge at syntactic level to determine the orientation of opinion. Furthermore, the use of CEL algorithms can be an alternative to complement other types of classifiers for sentiment analysis, incorporating such description classes as relevant new features into the knowledge base. To do so, an ontology-based text model for the representation of text documents is presented. The process for the ontology population and the use of the class expression learning of sentiment concepts are also described. To show the usefulness and effectiveness of our proposal, we use a set of documents about positive feedback focused on films to learn the positive sentiment concept and to classify the documents, comparing the results obtained against the result obtained by a C4.5 decision tree classifier, using the standard bag of words structure. Finally, we describe the problems that have arisen and solutions that have been adopted in our proposal.

Keywords Description logic class expression learning · Ontology-based text model · Text mining · Sentiment analysis

A. Salguero (✉)
Computer Sciences Department, University of Cádiz, Cádiz, Spain
e-mail: alberto.salguero@uca.es

M. Espinilla
Computer Sciences Department, University of Jaén, Jaén, Spain
e-mail: mestevez@ujaen.es

© Springer International Publishing Switzerland 2016
W. Pedrycz and S.-M. Chen (eds.), *Sentiment Analysis and Ontology Engineering*,
Studies in Computational Intelligence 639, DOI 10.1007/978-3-319-30319-2_5

1 Introduction

Sentiment analysis, also called opinion mining, is a research area that analyzes people's opinions or sentiments such as products, services, organizations, individuals, issues, events, topics, and their attributes Refs. [8, 18]. In recent years there is a huge amount of research related to the extraction and analysis of the opinions like subjectivity detection, opinion extraction, irony detection, etc. However, among these areas, the process to determine opinions, which express or imply positive or negative sentiments, is becoming an area increasingly important Refs. [9, 22].

Several proposals have been presented in the literature to determine positive or negative opinions, which can be classified into two different groups Ref. [18]. On the one hand, supervised methodologies that use machine learning algorithms, specific to text mining, in which training data exist. This kind of methodology requires large datasets for training and learning that, due to the novelty of the research area, it is sometimes difficult to obtain. On the other hand, unsupervised methodologies that use resources such as dictionaries or lexical ontologies. Therefore, it is necessary linguistic resources, which generally depend on the language, that are used to determine the orientation of opinion.

Furthermore, there are hybrid methodologies that combine both groups. The main advantage of the hybrid methodology is that takes advantage of the strengths of both methodologies without allowing them to get in each others way Ref. [20]. Usually, the hybrid methodologies are focused on to apply learning algorithms (supervised methodology) with a set of attributes that are obtained by means of linguistic resources (unsupervised methodology) in order to improve the results.

The standard Bag Of Words (BOW) is typically used to classify textual data when using supervised learning techniques. With the standard BOW approach, each document is transformed into a vector containing a set of features, which usually represent the frequency of each term. There exist many algorithms that try to improve the quality of the classifiers based on this technique by learning more features Ref. [26]. However, those new features are obtained using an *ad hoc* application that usually works for a very limited domain.

The algorithms for Class Expression Learning (CEL) are mainly used in the field of ontology engineering. They can be used to suggest new class descriptions that are relevant for the problem while the ontologies are being developed. So, CEL is a recent research topic of interest in the field of machine learning that can be used as an learning tool in order to improve the obtained results. In this chapter, we propose the use of CEL algorithms as a useful tool to find the class expression that has as much of the instances of positive documents or sentences as possible. So, given a set of positive and negative examples of individuals in an ontology, the learning problem consists on finding a new class expression or concept such that most of the positive examples are instances of that concept, whereas the negatives examples are not.

Therefore, we propose the use of the Description Logic (DL) Class Expression Learning technique for sentiment classification with the main advantage that offers a very flexible approach that can be used to classify textual data in many domains without the need of developing *ad hoc* applications. To do so, first, we present an ontology-based text model that contains the concepts and properties needed in order to represent the text documents, presenting also a procedure to generate the ontology population. Finally, the CEL technique is used to find the class expression that represents as many of the positive documents as possible.

The main novelty of the proposal, in contrast to most of the ontology-based text analyzers, is that our proposal is focused on inferring knowledge at syntactic level to learn an equivalent description for the positive or negative opinions. So, the reasoning capabilities of ontologies are used in this contribution to infer hidden features and relationships among terms with the advantage of that not all the relationships among terms need to be explicitly recorded due to the fact that they can be inferred while traversing the space of concepts.

To illustrate our proposal, we apply the CEL technique to a set of documents containing opinions about films, so we get as results DL class descriptions that describe documents expressing positive opinions about films, being used to classify the documents. Furthermore, we compare the results obtained by the CEL-based classifier against the result obtained by a C4.5 decision tree classifier that uses the usual BOW structure.

The chapter is structured as: Sect. 2 provides a brief introduction to ontologies as well as some related concepts and tools that will be used throughout the content of this chapter. Section 3 introduces the CEL problem. Section 4 proposes a procedure for applying CEL techniques in the field of sentiment analysis. Section 5 presents the results obtained by the CEL-based classifier applied to sentiment analysis and compares the results obtained by the CEL-based classifier with respect to a C4.5 decision tree classifier that uses the standard BOW approach. Section 6 the encountered problems and the solutions that have been adopted are discussed in this section. Finally, in Sect. 7, conclusions and future works are pointed out.

2 Foundations of Ontologies

Ontologies are used to provide structured vocabularies that explain the relationship among terms, allowing an unambiguous interpretation of their meaning. So, ontologies are formed by concepts (or classes) which are usually organized into a hierarchy of concepts Refs. [1, 27], being the ontologies more complex than taxonomies because they not only consider type-of relations, but they also consider other relations, including part-of or domain-specific relations Ref. [10].

The main advantage of the ontologies is that they codify knowledge and make it reusable by people, databases, and applications that need to share information Refs. [10, 28]. Due to this fact, the construction, the integration and the evolution of ontologies have been critical for the so-called Semantic Web Refs. [4, 6, 11, 19]. However, obtaining a high quality ontology largely depends on the availability of well-defined semantics and powerful reasoning tools.

The ontologies have been applied to sentiment analysis, obtaining successful results in several applications. So, in Ref. [13], the aspect-oriented sentiment analysis is studied, learning fuzzy product ontologies. In Ref. [17] was proposed an ontology-based sentiment analysis of network public opinions by applying semantic web technology. An ontology-based linguistic model was presented in Ref. [29] to identify the basic appraisal expression in Chinese product by mapping product features and opinions to the conceptual space of the domain ontology. In Ref. [12] was proposed an ontology-based sentiment analysis of twitter posts, computing a sentiment grade in the post. Furthemore, in Ref. [3] was presented an approach that shows how to automatically mine positive and negative sentiments with an ontological filtering. So, we can see that the proposals presented in the literature are focused on the representation and the analysis only at semantic level instead of our proposal that will be focused on syntactic level to learn class description for positive or negative opinions.

Regarding semantic web, a formal language is OWL Refs. [7, 24] that was developed by the World Wide Web Consortium (W3C) in the Web Ontology Working Group. Originally, OWL was designed to represent information about categories of objects and how objects are related. OWL inherits characteristics from several representation languages families, including the Description Logics and Frames basically. Furthermore, OWL shares many characteristics with Resource Description Framework (RDF), the W3C base for the Semantic Web. The major extension over RDF Schema (RDFS) is that OWL has the ability to impose restrictions on properties for certain classes.

The design of OWL is greatly influenced by Description Logics (DL), particularly in the formalism of semantics, the choice of language constructs and the integration of data types and data values. In fact, OWL DL and OWL Lite (subsets of OWL) is seen as expressive DL, offering a DL knowledge base equivalent ontology. They are in fact extensions of the DL "Attributive Concept Language with Complements" ($\mathcal{ALC}$). More formally, let N_C, N_R and N_O be (respectively) sets of "concept names", "role names" (also known as "properties") and "individual names". The semantics of DL are defined by interpreting concepts as sets of individuals and roles as sets of ordered pairs of individuals.

The Manchester OWL Syntax is derived from the OWL Abstract Syntax, but is less verbose and minimises the use of brackets. This means that it is quicker and easier to read and write by humans than DL formal syntax Ref. [5]. We introduce the Manchester OWL Syntax in Definition 2.1 because the solutions found by the tool we will use to solve the CEL problem are expressed in such language.

Definition 2.1 (*Terminological interpretation*) A "terminological interpretation" $\mathcal{I} = (\Delta^{\mathcal{I}}, \cdot^{\mathcal{I}})$ over a "signature" (N_C, N_R, N_O) for $(\mathcal{ALC})$ consists of the following concepts:

- A non-empty set $\Delta^{\mathcal{I}}$ called the "domain".
- A "interpretation function" $\cdot^{\mathcal{I}}$ that maps:

Table 1 Semantics of DL constructions

DL Syntax	Manchester syntax	Semantics
$\top^{\mathcal{I}}$	*Thing*	$\Delta^{\mathcal{I}}$
$\perp^{\mathcal{I}}$	*Nothing*	$\emptyset$
$(C \sqcup D)^{\mathcal{I}}$	C or D	$C^{\mathcal{I}} \cup D^{\mathcal{I}}$
$(C \sqcap D)^{\mathcal{I}}$	C and D	$C^{\mathcal{I}} \cap D^{\mathcal{I}}$
$(\neg C)^{\mathcal{I}}$	not C	$\Delta^{\mathcal{I}} \setminus C^{\mathcal{I}}$
$(\forall R.C)^{\mathcal{I}}$	R only C	$\{x \in \Delta^{\mathcal{I}} \vert \text{for every } y, (x, y) \in R^{\mathcal{I}} \text{ implies } y \in C^{\mathcal{I}}\}$
$(\exists R.C)^{\mathcal{I}}$	R some C	$\{x \in \Delta^{\mathcal{I}} \vert \text{there exists } y, (x, y) \in R^{\mathcal{I}} \text{ and } y \in C^{\mathcal{I}}\}$

- every "individual" a to an element $a^{\mathcal{I}} \in \Delta^{\mathcal{I}}$
- every "concept" to a subset of $\Delta^{\mathcal{I}}$
- every "role name" to a subset of $\Delta^{\mathcal{I}} \times \Delta^{\mathcal{I}}$

such that the semantics in Table 1 holds.

OWL extends $\mathcal{ALC}$ with role hierarchies, value restrictions, inverse properties, cardinality restrictions and transitive roles. One of the main advantages of high formalization of the OWL language is the possibility of using automated reasoning techniques. In 2009, the W3C proposed the OWL 2 recommendation in order to solve some usability problems detected in the previous version, keeping the base of OWL. So, OWL 2 adds several new features to OWL, some of the new features are syntactic sugar (e.g., disjoint union of classes) while others offer new expressivity, including: increased expressive power for properties, simple metamodeling capabilities, extended support for datatypes, extended annotation capabilities, and other innovations and minor features Ref. [30].

Sometimes, it is difficult to express certain kind of knowledge in OWL. In such cases, it is possible to use OWL extensions such as the Semantic Web Rule Language (SWRL) in order to include a high-level abstract syntax for Horn-like rules in OWL.[1] The SWRL is used to provide more powerful deductive reasoning capabilities than OWL alone Ref. [2]. It is important to note that several OWL reasoners, which are well known open source, support the SWRL language as Hermit[2] and Fact++.[3]

Another key tool that can be used in conjunction with OWL is SPARQL,[4] which is a query language able to retrieve and manipulate data stored in RDF triples. The main difference with SWRL is that the SPARQL language has been designed to work with RDF triples, at a lower abstraction level. While, the SWRL has been built on top of OWL, extending the set of its axioms, SPARQL is designed to work with individuals. Therefore, it is mainly used to retrieve individuals meeting certain conditions, moreover, it can also be used to add new knowledge to the ontology

---

[1] http://www.w3.org/Submission/SWRL.

[2] http://hermit-reasoner.com.

[3] http://owl.man.ac.uk/factplusplus.

[4] http://www.w3.org/TR/sparql11-query.

through the CONSTRUCT clause. By using this clause, it is possible to establish new relations among individuals or among individuals and classes.

Due to the fact that OWL is heavily based on formal semantics, there are some situations in which SPARQL is highly useful because it overcomes some of the limitations of OWL when, for example, open-world assumption issues arise. Finally, it also supports aggregations, which are very useful on the extraction of information from ontologies.

3   Class Expression Learning in Ontologies

In the field of ontologies, we can find two different types of statements. On the one hand, we have the set of statements that define the classes and properties in the scheme of the knowledge base. Those statements are typically used by reasoners in order to obtain new knowledge about the classes and properties already defined in the ontology. Normally, the task of automatic reasoners consists on determining the subsumption relationship between classes in the knowledge base, extending the asserted hierarchy of concepts. On the other hand, it is possible to define instances of the classes defined in the schema. In this case, the task of reasoners consists on classifying individuals as instances of the classes defined in the scheme of the ontology. Following, we provide a typical example in order to illustrate the reasoning.

Example The *GrandParent* class is a subclass of *Parent* class, assuming the following statements are defined in the ontology.

$$Grandparent \equiv \exists\, hasChild.(\exists\, hasChild.\top)$$
$$Domain(hasChild) : Father$$

The reasoners would classify all the individuals doubly related to other individuals through the *hasChild* property as instances of the *GrandParent* class.

In both cases, the reasoners cannot modify the description of a class or suggest the existence of new classes. For this reason, the usefulness of ontologies is usually limited to verify the consistency of the knowledge base or to extend the hierarchy of concepts.

On the contrary, the objective of CEL algorithms is to determine new class descriptions for concepts that may be used to classify individuals in an ontology according to some criterion. More formally, given a class C, the goal of CEL algorithms is to determine a class description A such that $A \equiv C$.

Let suppose an ontology O that has enough individuals defined in it. The set of individuals in $\Delta^{\mathcal{I}}$ is the search space S. CEL algorithms search in S, trying to find a description for class A such that $A^{\mathcal{I}}$ contains the same individuals in $C^{\mathcal{I}}$.

Definition 3.1 (*Pos(C)*) $Pos(C) \subseteq \Delta^{\mathcal{I}}$ is the set of individuals in O such that $x \in Pos(C) \implies x \in C^{\mathcal{I}}$.

Definition 3.2 (*Neg(C)*) $Neg(C) \subseteq \Delta^{\mathcal{I}}$ is the set of individuals in O such that $x \in Neg(C) \implies x \notin C^{\mathcal{I}}$.

As it can be seen, the CEL problem may be defined as a supervised learning problem but unlike the usual supervised learning problems the number of features for each instance is not fixed. They are dynamically generated as the CEL algorithm moves along the search space S. To navigate through the space S the CEL algorithms usually apply a refinement operator to existing classes in the knowledge base Ref. [15].

Definition 3.3 (*Refinement operator*) A refinement operator ρ is a mapping from S to 2^S such that for any $C \in S$ we have that $C' \in \rho(C)$ implies C' is a generalisation or a specialisation of C.

In addition to this operator, it is also necessary to establish a search strategy in S that maximizes the searched area and avoid the analysis of already visited areas. In literature, we can find several proposed search strategies Refs. [15, 25]. Most of them are based on graph exploration algorithms and some of them are based on computational intelligence Ref. [23] like a genetic algorithms Ref. [16] where the refinement operator consists on the combination of existing classes in the knowledge base.

The Algorithm 1 represents a very basic implementation of a CEL algorithm. First, the algorithm gets the current class description that best fit the POS(C) and NEG(C) sets. This class description is combined, using a selected refinement operator, with all of the other class descriptions that are present in the ontology and only the valid descriptions are added to the ontology. The process is restarted until the stopping condition is met. In this case the algorithm stops when an number of class description are evaluated.

Example Let suppose for example the existence of a family ontology O, having a sufficient number of individuals, where the concepts *Male*, *Female*, *Parent* and *Child* and the property *hasChild* are defined conveniently. Let suppose we want to automatically find a description for a new class *Father*.

1. First, the individuals in the *Pos(Father)* and *Neg(Father)* sets have to be identified. The *Pos(Father)* set contains individuals in O that should be classified as *Father*. The *Neg(Father)* set contains individuals that should be classified as $\neg Father$.
2. Using a refinement operator ρ the search space S is travelled. During this travel a set of class descriptions $D \subseteq S$ is generated, where the classes and properties in O are combined using the DL operators. Following the example of the family ontology, the following class descriptions may be eventually generated: $\neg Male$, $Male \sqcap Female$, $\exists hasChild.\top$, $\forall hasChild.(\neg(Parent \sqcup Child))$.
3. For each class description $d_i \in D$, the sets $d_i^{\mathcal{I}}$ and $(\neg d_i)^{\mathcal{I}}$ are calculated. The process stops when $d_i^{\mathcal{I}} = Pos(Father)$ and $(\neg d_i)^{\mathcal{I}} = neg(Father)$. Depending on the complexity of the concept that we want to learn and the number of individuals in O, it may be complex to travel the entire search space S and find

---

Algorithm 1 CEL algorithm

Require: C is the set of class descriptions in the ontology. C_{pos} and C_{neg} are the $Pos(C)$ and $Neg(C)$ sets, respectively. P is the set of refinement operators that is used to generate new class descriptions. n is the maximum number of class description the algorithm generates in the search process. α is a constant float value that indicates the importance of negative samples classification accuracy.

```
 1: function CEL(C, C_pos, C_neg, P, n)
 2:     while |C| < n do
 3:         best ← BEST-DESCRIPTION(C, C_pos, C_neg)
 4:         C' ← ∅
 5:         for all ρ_i ∈ P do
 6:             for all c_i ∈ C do
 7:                 d ← ρ_i(best, c_i)
 8:                 if valid(d) then
 9:                     C' ← C' ∪ d
10:             C ← C ∪ C'
11:     return best
12:
13: function BEST- DESCRIPTION(C, C_pos, C_neg)
14:     v ← −∞
15:     c ← c_0
16:     for all c_i ∈ C do
17:         v_pos ← |c_i^I ∩ C_pos^I|/|C_pos^I|
18:         v_neg ← |(¬c_i)^I ∩ C_neg^I|/|C_neg^I|
19:         if v_pos − α · v_neg > v then
20:             v ← v_pos − α · v_neg
21:             c ← c_i
22:     return c
```

---

a solution within a reasonable time. So, CEL algorithms usually give the class descriptions that best approximate the *Father* concept as result.

If the process runs for enough time, the CEL algorithm will eventually found a description $d_i = Male \sqcap \exists hasChild.\top$ in the second stage. Assuming the individuals in O are correctly annotated, $d_i^{I} = Pos(Father)$ and $(\neg d_i)^{I} = Neg(Father)$, so the process will stop and d_i will be proposed as a solution. In some cases, the CEL algorithm may continue searching for other alternative solutions.

Therefore, one of the main advantages of the use of ontologies in the field of machine learning is that the information is perfectly structured, so it is possible to define refinement operators and search strategies in S regardless of the scope of the problem. Therefore, our proposal overcome the problem related to create *ad hoc* applications for this task every time. Furthermore, another advantage of using ontologies is that invalid solutions can be discarded quickly, without the need of evaluating them. Not all descriptions in D are valid. Some of them can be discarded because they produce contradictions in the knowledge base, reducing strongly the search space.

4   Class Expression Learning Applied to Sentiment Analysis

In this section we describe the procedure to find a class description that represents documents that express a positive or negative opinion based on the use of CEL algorithms in the field of sentiment analysis.

The procedure for obtaining a solution requires the steps that are illustrated in Fig. 1. First, all text documents need to be transformed in form of ontologies with the help of a POS tagger. The stop words may be taken into account by the POS tagger but for efficiency they are not included in the resulting ontologies. The text documents that are used to train the classifier are merged in a global ontology. This global ontology is used as the input of the CEL algorithm in order to find a class description that best fits *POS(Positive)*. The resulting expression is used by the reasoner to classify any other text document, which needs to be also expressed in form of ontology.

So, CEL algorithms require the knowledge base to be expressed in the form of ontology. Therefore, first, we present an ontology-based text model to transform the information contained in text documents to an ontology. Then, we present a procedure to built an ontology population based on a set of documents. Finally, we propose the use of a CEL algorithm to find a class description that describes the positive concepts in the set of documents.

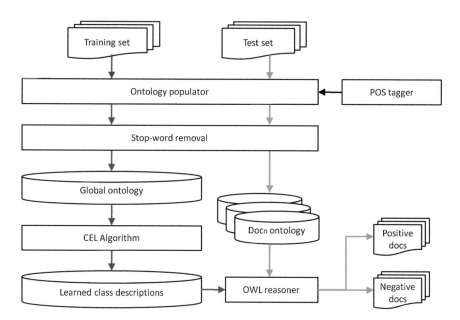

Fig. 1 Functional architecture

4.1 Ontology-Based Text Model

In this section, the ontology-based text model for ontology representation of text documents is presented. The kernel of the proposed ontology-based text model offers a reusable basis to the analysis of textual data. Here, the relevant DL axioms of the kernel are presented that include list patterns as well as the most important basic concepts.

4.1.1 List Pattern

In order to identify the most popular entities in a text and the relations among them, our proposal is based on a list structure. The basic concepts $List \sqsubseteq \top$ and $Item \sqsubseteq \top$ with the following relations among them are defined in the proposed model:

$$hasNext \sqsubseteq isFollowedBy \sqsubseteq itemProperty$$

hasNext is defined as a functional, asymmetric and irreflexive property. Because it has been defined as a functional property just one item can follow to an item. The inverse property is also defined as functional, forcing an item to be directly preceded by an unique item. The transitive property *isFollowedBy* is defined as a superproperty of *hasNext*. The property *hasNext* is referred to the item that is just after another item, while the property *isFollowedBy* is refereed to the set of items following an item in the list. Furthermore, the property *hasItem* establishes the membership of an item in its list.

$$hasItem \sqsubseteq listProperty$$
$$hasNext \sqsubseteq inTheContextOf$$
$$hasNext^- \sqsubseteq inTheContextOf$$

The symmetric property *inTheContextOf* is defined as a superproperty of both the *hasNext* property and *hasNext⁻*, which is the inverse of property *hasNext*. This property relates an item in the list with any of the elements that are immediately before or after of this item.

There is a set of concepts and relations that do not need to be explicitly defined in the model because they can be expressed using DL operators. However, in order to simplify the design of new concepts and relations, these are explicitly defined in the model. So, *hasPrevious*, *isPrecededBy* and *isPartOf* are defined as inverse properties of *hasNext*, *isFollowedBy* and *hasItem*, respectively.

The concepts *First* and *Last* identifies the starting and ending items of the list. Due to open-world assumption in OWL, reasoners cannot automatically infer the individuals that belong to these concepts. Therefore, it is necessary to annotate these individuals when the text is processed.

$$\top \sqsubseteq \forall\, has I D.Datatype\#long$$

Finally, for practical reasons, a functional property *hasID* is used to identify all of the individuals in the model with an unique code. In this way, it is easier the addition of new items to the ontology without the need of asserting that all of them are different from the existing individuals.

4.1.2 Basic Concepts

Once the list structure has been defined, it is possible to describe the most popular entities that can be found in a text such as sentences, documents and terms. Following, we present the relevant axioms of these entities in the ontology-based text model.

In some applications, it is also desirable to take *Punctuation marks* into account. For this reason, the concept *Token* is defined that encompasses both the *Term* and the *Punctuation marks* concepts. A *Sentence* is defined as a list containing tokens and a *Document* is defined in turn as a list containing sentences. The concepts *Document*, *Sentence*, *Term* and *PunctuationMark* are established as disjoint concepts.

$$Token \equiv PunctuationMark \sqcup Term$$
$$\top \sqsubseteq \forall\, has Lexeme.Datatype\#string$$
$$Term \sqsubseteq\ \leq 1\, has Lexeme$$
$$Sentence \equiv \exists\, has Element.Token$$
$$Document \equiv \exists\, has Element.Sentence$$

The terms in a sentence are classified according to their Parts of Speech (POS). Each of them must be represented as an unique individual in the ontology regardless of their lexeme. Therefore, the property *hasLexeme* is used to link a term with its lexeme.

Depending on the objectives of the text analyzer, the sentences may be excessively large information units Ref. [28]. For this reason, sentences are sometimes split into shorter segments that are called *Contexts*. *Contexts* could be a powerful tool for text analyzers to identify patterns in the text. To do so, arbitrarily long sequences of tokens connected by the property *hasNext* should be established. To find a particular pattern in the text, it is only necessary to apply some restrictions to the items in the context. For example, to find contexts of four elements length, being the initial item a punctuation mark and the ending term a noun, the following class description may be defined:

$$PunctuationMark \sqcap \exists\, has Next.(\exists\, has Next.(\exists\, has Next.Noun))$$

By default the relation *inTheContextOf* assumes contexts of three tokens length. If longer contexts are needed, it is necessary to increase the length of the contexts by means of rules such as the one shown below (in SWRL), for example, which sets contexts of five tokens length.

$$hasNext(token_i, token_j) \wedge hasNext(token_j, token_k)$$
$$\rightarrow inTheContextOf(token_i, token_k)$$

Although it might seem useful to define concepts such as *Positive*, *Neutral* or *Negative* to identify the different types of documents, it is not the aim. Due to the fact that the proposal is based on learning equivalent description for the concepts *Positive*, *Neutral* or *Negative*. Furthermore, if the concept *Positive* were included in the scheme of the ontology and the corresponding individuals annotated, the CEL problem would be trivially solved by giving the class expression *Positive* as result. Due to the fact that additions of these concepts to the core of the model for describing text documents would make it not very reusable. So, the model could be used in the field of sentiment analysis and, moreover, other purposes. Therefore, one of the advantages of using ontologies is kept due to the fact that they can be easily combined to form a global ontology.

4.2 Ontology Population

In this section, the construction of the ontology population is described for a set of text documents.

First, each sentence in a document is divided in terms. For each term in the sentence, a new individual is created in the ontology. The lexemes of the terms are related to the term individual by mean of the data type property *hasLexeme*. It is not necessary to assert that all of the terms are individuals of the concept *Term*. Any reasoner will identify these individuals as terms because the concept *Term* has been defined to be the domain of the property *hasLexeme*.

As was indicated in Sect. 4.1.2, each token (*Term* or *PunctuationMark*) has to be associated to an unique identifier through the property *hasID*, which has been defined as a functional relation. For this purpose, a sequence generator function has to be created with the aim that reasoners identify all tokens in the documents as different individuals.

The order among elements in a sentence is established by relating two consecutive tokens by mean of the property *hasPrevious*. The reasoners can always get the next token using the property *hasNext*, which has been defined as the inverse property of *hasPrevious*.

In order to relate all tokens to their sentences, the property *isPartOf* is used. In our proposal, it is not necessary to relate all tokens to the sentence they belong to because the tokens of a sentence are part of the same sentence the previous tokens are part of. Therefore, just the first token of the sentence needs to be associated to the sentence of which it is a member. It is important to note that in our proposal this kind of reasoning is based on the SWRL language. By relating the tokens to their sentences, it is very easy to analyze the sentences according to their elements with the property *hasItem* that was defined as the inverse of the property *isPartOf*. So, for example, an interrogative sentence may be defined as a sentence having a question mark.

$$QuestionMark \sqsubseteq PunctuationMark$$
$$InterrogativeSentence \equiv \exists\, hasItem.QuestionMark$$

4.3 Class Expression Learning of Sentiment Concepts

Once the documents are expressed in the ontology-based model proposed in Sect. 4.1 by building the ontology population presented in Sect. 4.2, the CEL algorithm is proposed to find a class description that describes the positive documents or negative documents. To do so, in this chapter, we propose the use of the DL-Learner tool that was presented in Ref. [14].

The process begins with the identification of *Pos(Positive)* and *Neg(Positive)* individuals, where *Positive* is a new empty concept representing the positive documents. For the creation of these groups of individuals, we have developed a tool, which is also responsible for transforming text documents in form of ontologies. In addition, this tool is able to generate a configuration file for the DL-Learner tool where individuals belonging to the *Pos(Positive)* and *Neg(Positive)* sets are pointed out. The tool assigns individuals to either *Pos(Positive)* or *Neg(Positive)* sets depending on the location of the text file in the directory structure, so it can be used in other text classification problems without having to be modified.

In this case, we are trying to find the sentiment polarity of the whole document but in some cases a finer detail is needed and the polarity of each paragraph of sentence may be calculated. The CEL algorithm can also be applied to those cases. The *Pos(Positive)* and *Neg(Positive)* sets just need to be populated accordingly. In the later case, for instance, the application that processes the text documents just need to populates both sets with individuals of type *Sentence*. In this case, the class description obtained as result describes the sentences expressing a positive opinion.

5 Results

In order to show the usefulness and effectiveness of our proposal, results obtained by the CEL-based classifier applied to sentiment analysis are shown. Furthermore, the results obtained by the CEL-based classifier with respect to a C4.5 decision tree classifier that uses the standard BOW approach are compared. To do so, we have made use of the freely available documents in Ref. [21] that are a list of two thousand annotated documents containing opinions about films.

5.1   Class Expression Learning of Positive Documents

We have used some of the documents in Ref. [21] in order to learn the concept
of document giving positive feedback about a film. The first twenty five positive
documents have been selected as *Pos(Positive)* whereas the first twenty five negative
documents have been selected as *Neg(Positive)*. As was shown in Fig. 1, all documents
have been processed and a single global ontology has been generated as result,
following the model presented in Sect. 4.1. This ontology, along with the list of
individuals in *Pos(Positive)* and *Neg(Positive)* are used as the input for the DL-Learner
tool, obtaining the following two class descriptions as result (in Manchester syntax),
among others.

$$hasItem\ some\ (hasItem\ min\ 2\ (PastTense\ and\ (isPrecededBy\ some\ Thing)))$$
$$(1)$$

$$hasItem\ some\ (hasItem\ some(PossessiveEnding\ and\ (inTheContextOf\ some$$
$$(Noun\ and\ (isFollowedBy\ some\ Thing)))\ and\ (isPrecededBy\ some\ (Adverb\ and$$
$$(hasPrevious\ some\ ((Noun\ or\ (hasPrevious\ some\ Verb))$$
$$and\ (isFollowedBy\ some\ (Adjective\ and$$
$$(isFollowedBy\ some\ Singular)))))))))$$
$$(2)$$

The class description (1), with 60.42 % of accuracy when classifying positive
documents, represents those documents that contain a sentence with at least two
verbs in the past tense, not being the first terms of the sentence. Furthermore, the
class description (2), illustrated in Fig. 2, with 81.25 % of efficacy, is the best solution
found by the CEL algorithm. It represents documents that contain any sentence having
a possessive ending that is in the context of a noun that is not the final term of that
sentence. In addition, the possessive ending must be preceded in the sentence by some
adverb, which must be immediately preceded in turn by a noun or some other term
immediately preceded by a verb. That name or term preceded by a verb should be

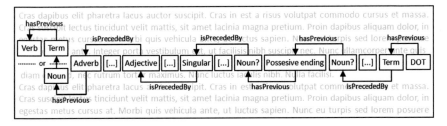

Fig. 2   Best class description found by CEL algorithm

followed by an adjective or a term followed, in turn, by a singular term. The following sentence is an example of a sentence represented by the this class description.

> "Even so, the strengths of election rely upon its fantastic performances from Broderick, Witherspoon, and newcomer Jessica Campbell, as Paul's anti-social sister, Tammy".

Let C be the class description that best describes text documents with a positive opinion. It is possible to use the class C to classify a single document d without using the DL-Learner tool. For this, the text document d has to be transformed in the form of an ontology, following the model proposed in Sect. 4.1, and check if the document complies with the class description C, that is, if $d \in C^{\mathcal{I}}$. To do this, we just need to check if $C^{\mathcal{I}} = \varnothing$, that is, if there is some individual in d that fits the class description C.

5.2 Comparative Analysis

A comparative analysis of the results obtained by the CEL-based classifier and a C4.5 tree-based classifier that use the usual BOW representation of the text documents are presented here. The DL-Learner tool is used as the CEL-based classifier. Default options for Weka's implementation of a C4.5 tree-based classifier (J48) have been used in all tests, which have been made in one of the nodes of the computing cluster at the University of Cádiz (2xIntel Xeon E5 2670, 2.6 GHz, 256 GB RAM). The memory of the Java Virtual Machine is limited to a maximum of 16 GB per process.

The efficacy percentages given in Table 2 are those indicated by the DL-Learner tool and they refer to the efficacy of the classifier for the training set, that is, for the case of the first fifty documents in Ref. [21], which were used to train both classifiers. DL-Learner makes use of an own approximate incomplete reasoning procedure for Fast Instance Checks (FIC) which partially follows a closed world assumption. When using a reasoner that complies with the OWL standard, such as Hermit, the accuracy rate downs to 78 %. A classifier based on a C4.5 decision tree has a 98 % accuracy for the same data set.

The next hundred documents in Ref. [21] are used as the test set. Fifty of them are documents with a positive opinion, while the remaining fifty documents have a negative opinion. Using the test data set the classifier based on a C4.5 decision tree offers an accuracy of 51 %, whereas the version of the classifier using the class description (2), in Sect. 4.3, is able to correctly classify 63 % of the documents. In this case, we have no information about the efficacy on the test data set for the classifier

Table 2 Accuracy of classifiers

	FIC (%)	Hermit (%)	C4.5 (%)
Training set	81.25	78.00	98.00
Test set	–	63.00	51.00

that uses the FIC of the DL-Learner tool because the reasoner runs out of memory. The learned class description has been tested on each document individually, using the Hermit reasoner, as described at the end of the previous section.

6 Discussion

In this section we describe the problems that we have found while building the former and the solutions we have adopted are discussed.

The DL-learner tool implements several CEL algorithms Ref. [16]. Some of them, like the Class Expression Learning for Ontology Engineering (CELOE), are focused on the process of building ontologies and are biased towards simple class descriptions. This type of algorithms sacrifices accuracy for the sake of simplicity of class descriptions so that the developer can easily understand them and incorporate to the ontology scheme, if appropriate. This is not the type of algorithms that should be used in our proposal, unless the learned class description is intended to be incorporated to the ontology. In this paper we have chosen to use the Ontology Class Expression Learning (OCEL) algorithm. Although it generates more complex class descriptions they offer the key advantage that are more effective for classifying text documents.

The complexity of the class description obtained by the algorithm is not only a problem from the point of view of the ontology developer. As the number of concepts and DL operators increase, the time required for the reasoners to classify the instances of ontology increases. In fact, since this task must be repeated for each new class description proposed in the learning process, the high processing time required for this task may be considered the main drawback. For example, the class description (2) shown in Sect. 4.3, was found after 155 minutes of executions of the DL-Learner tool, of which 145 were employed in the classification of instances of the ontology.

Despite the large amount of time required for the DL-Learner tool, this could have been much higher if the FIC were not used. The FIC implementation provides better reasoning performance when classifying individuals but uses an approximate reasoning. This means that the FIC classifies instances into classes that are excluded by standard OWL reasoners. Therefore, some of the class descriptions given as result of the CEL algorithm do not offer the same results when evaluated in OWL standard reasoners. For example, the class description shown in description (3) always represents an empty set of individuals when open world assumptions and the methodology proposed in Sect. 4.3 are followed, for example. For this reason we have decided to disable the inclusion of the complement operator in the class descriptions generated by the DL-Learner tool.

$$not \ (has Item \ some \ PastTense) \tag{3}$$

A similar problem occurs with the semantics of the *allValuesFrom* constraint for a property r. The standard option is to also include all those individuals that have no value for property r at all, whereas the FIC in DL-Learner includes only explicitly related individuals through the property r. To mitigate this problem an option has been activated in the DL-Learner tool that only includes the *allValuesFrom* constraints in the class descriptions for a property r when there is already a cardinality and/or existential restriction on r.

As shown above, the developers of the DL-Learner tool, aware of the problem of the high processing time required by the reasoners, have opted for the development of an approximate, incomplete reasoning procedure for fast instance classification. However, there are other problems arising from the use of reasoners for instance classification. The property *isFollowedBy* is a composite property because it is defined to be transitive.[5] To maintain the decidability of OWL, it is not allowed to be used in combination with *ObjectMinCardinality* constraint. Therefore, some of the class descriptions proposed by the CEL algorithm cannot be tested. Although it is possible to disable the use of cardinality restrictions in DL-Learner, it has not been deactivated because not all the properties in the model have been defined as transitive roles. The class description (2), which was illustrated in Sect. 4.3, is an example of a valid class description that includes cardinality restrictions.

To improve reasoning performance, we also decided to ignore all those terms providing little information about the content of the text. We have set a list of stop words that are ignored by the tool that transform text documents in ontologies, following the scheme of the model proposed in Sect. 4.1. Actually, as shown in Fig. 1, the stop words may be taken into account by the POS tagger but they are not included in the resulting ontologies. The list structure of sentences is maintained through the *hasNext* property, but only among non-stop words.

In view of the results obtained, it is shown how the accuracy of the CEL-based classifier for the sentiment classification is somewhat higher than the classifier based on the usual BOW structure for the test set. However, we have to keep in mind that the C4.5 decision tree classifier has not been optimized and its accuracy could probably be increased by tuning its parameters. So, its accuracy may also be improved by increasing the number of individuals in the training set. However, this would cause the reasoner of DL-Learner runs out of memory, so we could not compare the efficacy of both types of classifiers for the same training set data.

7   Conclusions and Future Works

In this chapter, we have presented the development of a text document classifier based on class descriptions in DL. The objective of the classifier is to identify text documents expressing a positive and negative opinions. To do so, an ontology-based text documents model has been proposed to represent text documents as ontologies, containing all the necessary concepts and properties for describing text documents.

---

[5]http://www.w3.org/TR/2009/REC-owl2-syntax-20091027/#Property_Hierarchy_and_Simple_Object_Property_Expressions.

A procedure to transform the text documents to this ontology-based text model has also been proposed. Following this procedure, some of the documents have been expressed in form of ontology and the CEL technique has been used in order to find the class expressions that represent different types of documents.

The main proposal of this chapter have been to show that CEL-based classifiers are an excellent option for sentiment analysis and, moreover, they can complement other types of classifiers. Class descriptions obtained by the CEL algorithm can be used by the developers to incorporate relevant new features into the knowledge base. Those features add valuable information for the classifiers that is usually difficult to be identified by people.

The DL-Learner tool, which implements several CEL algorithms, has been used in order to find a class description that represents documents expressing positive opinions. Some of the annotated documents have been converted to the ontology-based text model and used as the training set of the CEL-based classifier. The class description given as result has a significant lower classification accuracy for the training set than the accuracy achieved by the classifier based on a C4.5 decision tree, which uses the usual BOW approach. However, the class description based classifier obtains a slightly higher accuracy when the test set is used.

Furthermore, we have pointed out the main advantages and disadvantages of using CEL-based classifiers for text classification. Those disadvantages are related with the large number of resources required by reasoners, mainly. Some of the solutions we have adopted to mitigate this problem have been proposed, but it remains an open research area to obtain improved results.

References

1. Chandrasekaran, B., Josephson, J.R., Benjamins, V.R.: What are ontologies, and why do we need them? IEEE Intell. Syst. Appl. **14**(1), 20–26 (1999)
2. Chen, R.C., Huang, Y.H., Bau, C.T., Chen, S.M.: A recommendation system based on domain ontology and swrl for anti-diabetic drugs selection. Expert Syst. Appl. **39**(4), 3995–4006 (2012)
3. Colace, F., De Santo, M., Napoletano, P., Becchi, C., Chang, S.K.: Ontological filtering for sentiment analysis. In: DMS'2012, pp. 60–66 (2012)
4. Gomez-Perez, A., Corcho, O.: Ontology languages for the semantic web. IEEE Intell. Syst. Appl. **17**(1), 54–60 (2002)
5. Horridge, M., Drummond, N., Goodwin, J., Rector, A., Wang, H.H.: The manchester owl syntax. In: Proceedings of the 2006 OWL Experiences and Directions Workshop (OWL-ED2006) (2006)
6. Horrocks, I.: Ontologies and the semantic web. Commun. ACM **51**(12), 58–67 (2008)
7. Horrocks, I., Patel-Schneider, P.F., Van Harmelen, F.: From SHIQ and RDF to OWL: the making of a web ontology language. Web Semant. **1**(1), 7–26 (2003)
8. Kaur, A., Gupta, V.: A survey on sentiment analysis and opinion mining techniques. J. Emerg. Technol. Web Intell. **5**(4), 367–371 (2013)
9. Kearney, C., Liu, S.: Textual sentiment in finance: a survey of methods and models. Int. Rev. Fin. Anal. **33**, 171–185 (2014)
10. Knijff, J., Frasincar, F., Hogenboom, F.: Domain taxonomy learning from text: the subsumption method versus hierarchical clustering. Data Knowl. Eng. **83**, 54–69 (2013)

11. Kohler, J., Philippi, S., Specht, M., Ruegg, A.: Ontology based text indexing and querying for the semantic web. Knowl.-Based Syst. **19**(8), 744–754 (2006)
12. Kontopoulos, E., Berberidis, C., Dergiades, T., Bassiliades, N.: Ontology-based sentiment analysis of twitter posts. Expert Syst. Appl. **40**(10), 4065–4074 (2013)
13. Lau, R.Y.K., Li, C., Liao, S.S.Y.: Social analytics: learning fuzzy product ontologies for aspect-oriented sentiment analysis. Decis. Support Syst. **65**(C), 80–94 (2015)
14. Lehmann, J.: DL-learner: learning concepts in description logics. J. Mach. Learn. Res. **10**, 2639–2642 (2009)
15. Lehmann, J., Auer, S., Bhmann, L., Tramp, S.: Class expression learning for ontology engineering. J. Web Seman. **9**(1), 71–81 (2011)
16. Lehmann, J., Hitzler, P.: Concept learning in description logics using refinement operators. Mach. Learn. **78**, 203–250 (2010)
17. Li, S., Liu, L., Xiong, Z.: Ontology-based sentiment analysis of network public opinions. Int. J. Digit. Content Technol. Appl. **6**(23), 371–380 (2012)
18. Liu, B.: Sentiment Analysis and Opinion Mining. In: Synthesis Lectures on Human Language Technologies. Morgan & Claypool Publishers (2012)
19. Maedche, A., Staab, S.: Ontology learning for the semantic web. IEEE Intell. Syst. Appl. **16**(2), 72–79 (2001)
20. Martin-Valdivia, M.T., Martinez-Camara, E., Perea-Ortega, J.M., Urena-Lopez, L.A.: Sentiment polarity detection in spanish reviews combining supervised and unsupervised approaches. Expert Syst. Appl. **40**(10), 3934–3942 (2013)
21. Pang, B., Lee, L.: A sentimental education: sentiment analysis using subjectivity summarization based on minimum cuts. In: Proceedings of the ACL (2004)
22. Pang, B., Lee, L.: Opinion mining and sentiment analysis. Found. Trends Inf. Retr. **2**(1–2), 1–135 (2008)
23. Pedrycz, W.: Computational Intelligence: An Introduction. CRC Press Inc, Boca Raton (1997)
24. Sirin, E., Parsia, B., Grau, B.C., Kalyanpur, A., Katz, Y.: Pellet: a practical owl-dl reasoner. Web Seman. **5**(2), 51–53 (2007)
25. Tran, A., Dietrich, J., Guesgen, H., Marsland, S.: An approach to parallel class expression learning. In: Bikakis, A., Giurca, A. (eds.) Rules on the Web: Research and Applications. Lecture Notes in Computer Science, vol. 7438, pp. 302–316. Springer, Berlin (2012)
26. Uguz, H.: A two-stage feature selection method for text categorization by using information gain, principal component analysis and genetic algorithm. Knowl.-Based Syst. **24**(7), 1024–1032 (2011)
27. Uschold, M., Gruninger, M.: Ontologies: principles, methods and applications. Knowl. Eng. Rev. **11**(2), 93–136 (1996)
28. Wei, T., Lu, Y., Chang, H., Zhou, Q., Bao, X.: A semantic approach for text clustering using wordnet and lexical chains. Expert Syst. Appl. **42**(4), 2264–2275 (2015)
29. Yin, P., Wang, H., Guo, K.: Feature-opinion pair identification of product reviews in chinese: a domain ontology modeling method. New Rev. Hypermedia Multimedia **19**(1), 3–24 (2013)
30. Zhang, F., Ma, Z.M., Li, W.: Storing owl ontologies in object-oriented databases. Knowl.-Based Syst. **76**, 240–255 (2015)

Capturing Digest Emotions by Means of Fuzzy Linguistic Aggregation

C. Brenga, A. Celotto, V. Loia and S. Senatore

Abstract Distilling sentiments and moods hidden in the written (natural) language is a challenging issue which attracts research and commercial interests, aimed at studying the users behavior on the Web and evaluating the public attitudes towards brands, social events, political actions. The understanding of the written language is a very complicated task: sentiments and opinions are concealed in the sentences, typically associated to adjectives and verbs; then the intrinsic meaning of some textual expressions is not amenable to rigid linguistic patterns. This work presents a framework for detecting sentiment and emotion from text. It exploits an affective model known as Hourglass of Emotions, a variant of Plutchik's wheel of emotions. The model defines four affective dimensions, each one with some activation levels, called 'sentic levels' that represent an emotional state of mind and can be more or less intense, depending on where they are placed in the corresponding dimension. Our approach draws from the Computational Intelligence area to provide a conceptual setting to sentiment and emotion detection and processing. The novelty is the fuzzy linguistic modeling of the Hourglass of Emotions: dimensions are modeled as fuzzy linguistic variables, whose linguistic terms are the sentic levels (emotions). This linguistic modeling naturally enables the use of fuzzy linguistic aggregation operators (from Computing with Words paradigm), such as LOWA (Linguistic Ordered Weighted Averaging) that inherently accomplishes an aggregation of the emotions

C. Brenga · A. Celotto
CORISA (Consorzio di Ricerca Sistemi ad Agenti), Universitá degli Studi di Salerno,
84084 Fisciano, SA, Italy
e-mail: cbrenga@gmail.com

A. Celotto
e-mail: acelotto@unisa.it

V. Loia
Dipartimento di Scienze Aziendali - Management & Innovation Systems,
Universitá degli Studi di Salerno, 84084 Fisciano, SA, Italy
e-mail: loia@unisa.it

S. Senatore (✉)
Dipartimento di Ingegneria Dell'Informazione ed Elettrica e Matematica Applicata,
Universitá degli Studi di Salerno, 84084 Fisciano, SA, Italy
e-mail: ssenatore@unisa.it

© Springer International Publishing Switzerland 2016
W. Pedrycz and S.-M. Chen (eds.), *Sentiment Analysis and Ontology Engineering*,
Studies in Computational Intelligence 639, DOI 10.1007/978-3-319-30319-2_6

113

in order to get an emotional expression that synthesizes a set of emotions associated with different sentic levels and activation intensities. The whole process for the emotion detection and synthesis is described through its main tasks, from the text parsing up to emotions extraction, returning a predominant emotion, associated with each dimension of the Hourglass of Emotions. An ad-hoc ontology has been designed to integrate lexical information and relations, along with the Hourglass model.

Keywords Fuzzy linguistic modeling · Ontology · Sentiment analysis · Sentic computing · Linguistic Ordered Weighted Averaging (LOWA) · Computing with words · Fuzzy logic

1 Introduction

During recent years, the study of the sentiments, opinions and emotions has attracted many research fields, such as psychology cognitive science, social sciences, and decision-making. Particularly, the automatic detection of emotions in the written text has drawn the attention of researchers in area of computer science and it has been becoming increasingly important at applicative level.

At the same time, commercial and marketing companies analyze customer feedback for their products and services: blogs and review sites are used to collect the customer opinion in order to evaluate the actual utility value of the products [15] and estimate the reputation of the company in the market, product positioning, and financial market prediction.

In an era where people express their opinions about everything and everywhere on the Web, the understanding of the written language targeted at interpreting human moods is a very complicated task. Sentiments and opinions indeed are concealed in the sentences, typically associated to adjectives and verbs; then the intrinsic meaning of some textual expressions is not amenable to rigid linguistic patterns.

Our approach exploits an affective-based model known as Hourglass of Emotions [7], that is a variant of Plutchik's wheel of emotions. This model is based on four affective dimensions: *Pleasantness, Attention, Sensitivity and Aptitude* and each dimension has some levels of activation, called 'sentic levels' which represent an emotional state of mind and can be more or less intense, depending on the position on the corresponding dimension. Since users' opinions are naturally expressed in the speaking language rather than numeric scales, they fit well to be modeled by techniques from the Fuzzy Logic. From this perspective, Fuzzy Logic assumes a central role to deliver a conceptual and methodological setting to deal inherently with the imprecision, the subjectivity and the vagueness, hidden in concepts, sentiments, emotions of the natural language. Particularly, emotions are strictly related to the human feeling, they are often tricky to grab in the written language. Our approach exploits indeed, the fuzzy linguistic modeling to represent the emotions that are for their own nature, imprecise and vague. The novelty is the modeling of the emotional dimensions of the Hourglass model as fuzzy linguistic variables, whose fuzzy linguistic terms are the sentic levels (emotions). In order to individuate the main emotions, and in some way, to provide the digest emotions expressed in an analyzed text, linguistic terms are collapsed together by a linguistic aggregation operator, LOWA (Linguistic Ordered Weighted Averaging) that inherently accomplishes an aggregation of emotions in order to get a emotional expression that synthesizes a set of emotions associated with different sentic levels and activation intensities.

In nutshell, this work describes an approach to detect relevant emotions from textual data. Precisely, our contribution is twofold.

- A modeling of the emotional dimensions of the Hourglass of Emotions through fuzzy linguistic variables and the use of a fuzzy aggregator to transform several sentic levels in a few "digest" emotions.
- The design of a lexical ontology to support the text analysis and parsing. The ontology, called SentiWordSKOS, reproduces the well-known lexicon, WordNet [20], but extends it, adding to each (sense of a) word, the polarity and the emotion. Moreover, SentiWordNet encloses the ontological modeling of the Hourglass of Emotions, by linking words and sentic levels.

The work shows the whole process from the text parsing up to the emotions extraction, and the returning of the digest ones. Its structure is as follows.

Section 2 introduces a brief overview of the main related literature; then, in order to have a clearer idea of the approach, Sect. 3 provides a high-level step-by-step outlook on the whole process of emotion detection and extraction emotions; then, before describing the process, the ontology SentiWordSKOS and related external resources are presented in Sect. 4. Section 5 introduces the theoretical fundamentals on which the process is built. Then, in Sect. 6 through a simple example, the whole process is instantiated, across all its steps. Finally, experimental evaluations and conclusion close the chapter.

2   Related Work

The Web is becoming a big data repository with a a huge volume of opinionated data to analyze. Capturing emotions in texts, such as reviews, forum discussions, blogs, tweets, etc. is assuming a considerable importance with the growth of the social Web. Opinion mining is a wide area of interest that gathers all the methodologies and techniques aimed at extracting opinions, sentiments, attitudes from the written language.

Sentiment analysis and opinion mining often refer interchangeably to the same area of study. More specifically, the sentiment analysis relates to techniques to automatically classify sentiments (in terms of positive or negative polarity) for entire documents. The opinion mining considers approaches for capturing opinions, moods and human feelings for the webposts, reviews or documents in general. Sentiment Analysis (SA) has been extensively studied in recent years. Existing work mainly concentrates on the use of three types of features: lexicon features, POS features, and microblogging features for sentiment analysis and opinion mining [26]. Some approaches combine these three types of features, giving major emphasis to POS tags with or without word prior polarity involved [1], others explore the use of microblogging features [4, 16].

Opinion mining techniques have been devised and evaluated on many domains [9, 30], focusing on the categorization of overall sentiment [21], or recognizing the sentiments and/or the opinions at the feature level [2, 9, 17].

In the recent years, research trends are studying the affective information embedded in the natural language. Emotions are immediately perceived from facial expressions or voice inflection, but their extraction from the text needs major attention. In [27], the automatic analysis of emotions in text has been built on a large data set annotated with six basic emotions: anger, disgust, fear, joy, sadness and surprise. The study described in [10] detects emotions and their intensity from both speech and textual data: a set of emotional keywords and (like in our approach) emotion modification words has been taken into account even though they are manually built. For instance, words such as *very* and *not* are examples of modifiers of the following adjectives, affecting the emotional intensity.

SenticNet [8] is a publicly available semantic resource for opinion mining. It is built using common sense reasoning techniques together with an emotion categorization model. It generates a collection of common sense concepts with a positive or negative polarity. SenticNet exploits AffectiveSpace [11], a n-dimensional vector space that, through the Singular Vector Decomposition, achieves the dimensionality reduction of the emotional space in order to perform emotive reasoning on it.

In general, the existing resources detect sentiments and emotions from text, exploiting the lexicons, also defining ontologies to model and classify emotions in categories; yet, it is missing an integrated view that translates the terms and glosses from (emotion-based) lexicons into ontological concepts. A contribution that partially addresses this issue is YAGO [29]: it is a light-weight and extensible ontology that merges the information from WordNet and Wikipedia, although it is not targeted to describe emotions. Our ontology SentiWordSKOS, herein described, is defined for a similar goal: providing a shared representation of the knowledge by integrating more information sources. As far as we know, there are no other ontological resources which expand the meaning of a word with the sentiment and the emotion associated with its own sense.

Emotions play a crucial role in interpreting the opinions: between "like" and "dislike", there are a number of alternatives that describe the state of mind. Fuzzy techniques are often used in literature to model imprecision: the nature of emotions introduces some imprecision, in addition to the natural language ambiguity and complexity. In [12], the dynamics of 14 emotions has been achieved through fuzzy logic-based rules; the final emotion is represented as the average of all emotional intensities. A synergic approach based on the natural-language processing and the fuzzy logic techniques is presented in [28], where emotions are extracted from free text by defining a degree of relatedness with the some affective categories and an intensity value which represents the strength of the affect level described by that word.

3 Process Overview

Our proposal has been inspired by the work described in [18]. Figure 1 describes the pipeline of whole process of text analysis, across a fine grained (word-based) to a coarse grained (whole text-based) analysis. For each step of the pipeline, any involved resource is listed above, whereas the tasks accomplished in that step are listed under it.

A tweet, a document, or a paragraph extracted from any textual resource represent an acceptable input. The predominant emotions, selected among the four dimensions of the Hourglass of Emotion are returned as an output.

Precisely, the main steps are sketched as follows.

- *Syntactic Analysis*: the most traditional parsing activities are applied: stop-word removal and part of speech (POS) tagging are accomplished by using the Stanford CoreNLP. All words, considered irrelevant for the text analysis are furthermore eliminated. The remaining words will be processed in the next steps.
- *Word Analysis*: this step is in charge of getting the sentiments (polarity) and emotions associated with each selected word. This information is coded in SentiWord-SKOS, which associates with each word, a polarity and a set of sentic levels, according to the fuzzy linguistic modeling of the Hourglass of Emotion. To synthesize the emotions associated with a word, a further task of the step achieves the fuzzy aggregation to raise more relevant emotions.
- *Pattern Analysis*: once discovered emotions for each word, the analysis is extended to the linguistic patterns. Since the patterns are collections of two or three words, probably, they can better represent the emotions. They are often composed of sequences of adverb and adjective that can act increasing or reducing the intensities of the emotions expressed by those words.
- *Sentence Analysis*: emotions calculated for the words and patterns are further aggregated in order to get a comprehensive representation of the emotions expressed in the sentence.
- *Text Analysis*: in this step, the aggregation is computed at level of the whole given text.

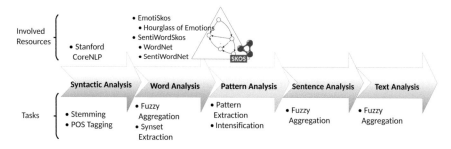

Fig. 1 Text processing pipeline

The process (mainly, the involved step is the *Word Analysis*) exploits SentiWord-SKOS that encloses the ontology EmotiSKOS, modeling the Hourglass of Emotion. Next section provides additional details on these ontological resources.

4 SentiWordSKOS and the Related Lexical Resources

This section introduces our ontological resources SentiWordSKOS and EmotiSKOS and provides details about how they are structured, and/or integrated to each others. As stated, SentiWordSKOS is based on the lexicon WordNet, but extends it, by associating each word with an emotion. Furthermore, it uses EmotiSKOS, ad-hoc built for coding the Hourglass. For the sake of the completeness, a brief description of the technological background will be given: the external resources involved in the process will be presented, before introducing our ontologies. Moreover, since both SentiWordSKOS and EmotiSKOS are coded by the SKOS language, a short introduction on semantic web technologies and SKOS will be given as well.

4.1 Technological Background

4.1.1 WordNet

WordNet [20] is a lexical database for the English language (even though there is a multilingual version). It associates each word with one or more synsets, i.e., sets of synonyms, each expressing a distinct concept. WordNet can individuate nouns, verbs, adjectives and adverbs, according to the defined grammatical rules. Each synset has a unique index and provides also the sense and a gloss (or dictionary) definition. For instance the word "cat" belonging to the syntactical category of noun, has offset 02124272 and sense "domestic cats".

WordNet includes lexical and semantic relations. Lexical relations are, however, relations between meanings. Examples of lexical relations are synonymy and antonymy. WordNet provides a semantic network that links the senses of words to each other, maintaining the semantic relations among these synsets. The main types of relations among synsets are the super-subordinate relations, that are hyperonymy, hyponymy or ISA relations. Other relations are the meronymy and the holonymy. Since a semantic relation is a connection between meanings, and since meanings can be represented by synsets, it is natural to think of semantic relations as pointers between synsets. Examples of semantic relations are hypernymy, hyponymy.

4.1.2 SentiWordNet

SentiWordNet[1] is strictly related to WordNet. It is a lexical resource designed for supporting sentiment classification and opinion mining applications. It is the result of automatic annotation of all the synsets of WordNet, by means of three numerical scores, $Pos(s)$, $Neg(s)$, and $Obj(s)$, that are associated with each synset. These scores represent how "objective", "positive", and "negative" are the terms associated to the synset. These values are in the range [0, 1], and their sum is 1.0 for each synset. Thus, a synset may have non-zero score for all the three classes. In other words, each term of the synset has the same opinion-related property with a certain degree. Different senses of a term may have different opinion-related properties [3]. SentiWordNet has been developed by exploiting the quantitative analysis of the glosses associated with synsets and the use of the resulting vectorial term representations for semi-supervised synset classification. The triplet of scores is computed by combining the results produced by a committee of eight ternary classifiers [3], each of which has demonstrated similar accuracy although presented different characteristics in terms of classification behaviour.

4.1.3 Sentic Levels of the Hourglass of Emotions

Classifying the emotional states of human being is a study that has been fascinating many researchers, first among all, Plutchik [24] that identified eight primary emotions—anger, fear, sadness, disgust, surprise, anticipation, trust, and joy. Our approach relies on an affective-based model presented in [7], that is a variant of Plutchik's wheel of emotions, whose target is to develop emotion-sensitive systems, able to answer to questions regarding the satisfaction of a user with a services provided, the information supplied, the interface used, etc. This model emulates Marvin Minsky's conception of emotions, that considers each emotional state as a result of turning some resources (which compound our mind) on and turning off some others. For example, the state of anger turns on a set of resources that help us react with more speed and strength while suppressing some other resources that usually make us act prudently. The model considers four affective dimensions: *Pleasantness, Attention, Sensitivity and Aptitude*. Table 1 describes these dimensions and their different levels of activation, the *sentic levels*, which measure the strength (in the range [−3, 3]) of the perceived emotion. Each level represents an emotional state of mind and can be more or less intense, depending on the position on the corresponding dimension. For instance, for the Pleasantness dimension, ecstasy and grief are two opposite emotions at the sentic levels 3 and −3 respectively.

---

[1] http://sentiwordnet.isti.cnr.it/.

Table 1 Sentic levels for the four dimensions of Hourglass of Emotions [7]

	Pleasantness	Attention	Sensitivity	Aptitude
+3	Ecstasy	Vigilance	Rage	Admiration
+2	Joy	Anticipation	Anger	Trust
+1	Serenity	Interest	Annoyance	Acceptance
0	Limbo	Limbo	Limbo	Limbo
−1	Pensiveness	Distraction	Apprehension	Boredom
−2	Sadness	Surprise	Fear	Disgust
−3	Grief	Amazement	Terror	Loathing

4.1.4 SKOS

Simple Knowledge Organisation System (SKOS) is a language for supporting the use of knowledge organization system. It is part of the new emerging web technologies aimed at web-based information that would be understandable and reusable by both humans and machines. Particularly, ontologies represent a way to conceptualize information by a well-defined, machine interpretable mark-ups. Ontologies provide many benefits in representing and processing data, by sharing of common knowledge, and reusing it for different applications. Nowadays, the Semantic Web encourages the annotation of documents or general web resource content with metadata descriptions, using semantic information from domain ontologies. Semantic Web annotations go beyond familiar textual annotations, which are basically aimed at inserting keywords for use by the document creator. Instead, semantic annotation aims at individuating concepts and relations between concepts in web resources, intended for use by humans but especially by machines, in a way that is cost-effective and consistent with adopted schemas and ontologies.

SKOS is a data model for sharing and linking knowledge organization systems (such as thesauri, classification schemes and taxonomies) via the Semantic Web. SKOS Core is an RDF vocabulary for expressing the basic structure and content of concept schemes. A concept scheme is "a set of concepts, optionally including semantic relationships between those concepts" [19]. Thesauri, classification schemes, taxonomies, terminologies, glossaries and other types of controlled vocabulary are all examples of concept schemes. SKOS Vocabulary considers the "Concept" as the fundamental unit of any concept scheme.

The SKOS Core Vocabulary consists of a set of RDF-based properties and classes that can be used to express the content and structure of a concept scheme as an RDF graph. It may be used on its own, or in combination with formal knowledge representation languages such as the Web Ontology language (OWL). Each SKOS concept has a single *skos:prefLabel*; while it may have unlimited alternative descriptions (*skos:altLabel*); then groups of concepts are organized into concept schemes (*skos:ConceptScheme*). The membership to a concept scheme is defined by the property *skos:inScheme*. One of the key features of SKOS is the natural extensibility and interoperability with other vocabularies.

4.2 EmotiSKOS

EmotiSKOS is one of the ontology modeled in our approach and represents a SKOS-compliant translation of the Hourglass of Emotions. Each dimension of the Hourglass of Emotions is an instance of *skos:ConceptScheme* and includes a set of *skos:Concept* (representing the emotions associated with a dimension) and relations between these concepts. The possible relations among the emotions of a dimension are listed as follows:

- **hasHigherIntensity** is the relation that connects two adjacent emotions of a dimension; the domain of the relation is the emotion that has a sentic level higher than the other emotion that plays the role of the range. For instance, in the Pleasantness dimension, *ecstasy hasHigherIntensity* than *joy*.
- **hasLowerIntensity** is the inverse relation of *hasHigherIntensity*.
- **oppositeEmotion** is the relation that connects two opposing emotions within a same dimension. The opposite emotions are at the same sentic level, but they have a different sign, viz, one is positive while the other negative. For instance, in the Pleasantness dimension, *joy* and *sadness* are *oppositeEmotions*.

All the relations are coded as sub-properties of the native property *skos:related*.

Figure 2 shows a portion of the ontology EmotiSKOS, which describes the *Pleasantness* dimension through its own emotions. Let us notice that *ecstasy, joy*, and *serenity* are positive emotions, whereas *grief, sadness* and *pensiveness* are the opposite negative emotions, respectively.

Then, *serenity hasLowerIntensity* than *joy*, since, as shown in Table 1, *serenity* has a sentic level $+1$, that is lower than the level of *joy* that has $+2$; finally, *joy* is an *oppositeEmotion* of *sadness* (with value $+2$ and -2, respectively).

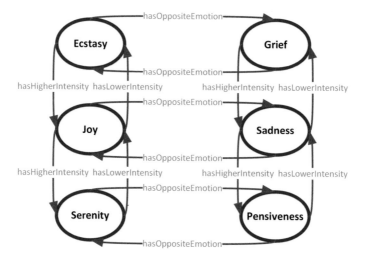

Fig. 2 EmotiSKOS ontology—the schema of the Pleasantness dimension

An example of the machine-readable representation in RDF/XML of the *joy* emotion of the *Pleasantness* dimension is shown in Listing 1.

Listing 1 The SKOS-coded description of the emotion *joy*

```
<Emotion rdf:about="#joy">
        <hasOppositeEmotion rdf:resource="#sadness"/>
        <skos:inScheme rdf:resource="#Pleasantness"/>
        <hasLowerIntensity rdf:resource="#ecstasy"/>
        <hasHigherIntensity rdf:resource="#serenity"/>
</Emotion>
```

4.3 SentiWordSKOS

SentiWordSKOS is our light-weight ontology, developed by integrating WordNet and SentiWordNet. Specifically, each WordNet synset has been modeled as an *owl:Class* called *Synset*, which is subclass of *skos:Concept*. Each WordNet syntactic category (noun, adjective, adverb and verb) is a subclass of the *Synset* class.

Figure 3 shows the SentiWordSKOS schema. For each WordNet synset, there are the following *owl:DatatypeProperty* properties, associated with the *Synset* class:

- **offset** is a unique identifier composed of a character indicating the syntactic category and the WordNet offset.
- **prefLabel** is one of the words belonging to the synset and is modeled by using skos:prefLabel.
- **altLabel**, is a label associated with the synset label; it has been modeled by the native skos:altLabel.
- **polarity** is a numeric value representing the sentiment of the synset. It is derived from the numerical scores *Pos(s)* and *Neg(s)* associated with a synset *s*, extracted from SentiWordNet.

Then, each synset has an *owl:ObjectProperty* **hasOpposite** that joins a synset with its antonym, as defined in WordNet antinomy relation.

Another *owl:ObjectProperty* is **derivedFrom** which reflects the WordNet's lexical relations: starting from a syntactic category of a word, it returns the nominal form (i.e., noun) of the synset associated with the word. The SentiWordSKOS scheme is joined to the EmotiSKOS ontology by a relation defined between the syntactic category *Noun* and the class *Emotion* (in Fig. 3, it appears with prefix *emoti* to outline that it is defined in EmotiSKOS). The *owl:ObjectProperty* **hasEmotionX** connecting them, states that the noun expresses a certain emotion with intensity value X. This value is calculated by exploiting the WordNet similarity measure [23], and it is in the range [0, 1]. This measure (often called also relatedness) says how much two nouns are related to each other [18] . The *hasEmotionX* relation, once instantiated, holds the similarity value that exists between the noun and the emotion. We add all the *hasEmotionX* relations whose value X is greater than 0.6 (lower values are considered not meaningful at emotional level). For the sake of a simple representation, this value has been coded in the corresponding range [60,100], with step 10. Let us

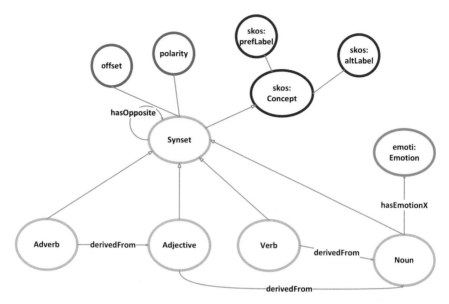

Fig. 3 SentiWordSKOS schema

remark that the *hasEmotionX* is just calculated between noun categories; this is in accord with the WordNet Similarity measure that applies only among nouns.

Listing 2 An example of synset

```
<Adjective rdf:about=''#a1148283''>
    <skos:prefLabel>happy_a_1</skos:prefLabel>
    <offset>a1148283</offset>
    <polarity>0.875</polarity>
    <hasOpposite rdf:resource=''#a1149494''/>

    ...
    <derivedFrom rdf:resource=''#n13987423''/>
</Adjective>
<Noun rdf:about=''#n13987423''>
    <skos:prefLabel>happiness_n_1</skos:prefLabel>
    <skos:altLabel>felicity_n_2</skos:altLabel>
    <offset>n13987423</offset>
    <hasEmotion80 rdf:resource=''&emoti;joy''/>
...
</Noun>
<Adjective rdf:about=''#a1149494''>
    <skos:prefLabel>unhappy_a_1</skos:prefLabel>
...
</Adjective>
```

Listing 2 shows an example of the synset *happy*. It is an adjective, whose off-set is *a*1148283 and the polarity value is 0.875 (positive); it has as an antonym *unhappy* (whose offset is *a*1149494) and is *derivedFrom* the noun *happiness* (with offset *n*13987423). In turn, the noun *happiness* is related to the emotion *joy*, with an intensity value equals to 0.80.

Table 2 SentiWordSKOS—classes and relations instances

Classes			
Synset	117659	Noun	82,115
		Verb	13,767
		Adjective	18,156
		Adverb	3621
Relations			
hasEmotionX	29,075	hasEmotion60	19,967
		hasEmotion70	7011
		hasEmotion80	1768
		hasEmotion90	298
		hasEmotion100	31
derivedFrom		34,776	
hasOpposite		7604	

4.3.1 SentiWordSKOS'size

Populating SentiWordSKOS was built automatically by means of SPARQL queries using Apache Jena library[2] by iterating on each WordNet synset through Java Word-Net Library[3] (JWNL). For each WordNet synset, a SentiWordSKOS individual has been created by adding the polarity extracted from SentiWordNet and the emotions from EmotiSKOS. Table 2 shows a synthetic view of the whole ontology population. The size of SentiWordSKOS ontology is about 65 Mb. It is composed of 117,659 *Synset* instances, whereas the relation instances (viz., the number of relations which have been instantiated) are in total 71,452.

Table 3 shows the numbers of existing *hasEmotionX* relations between a synset and each one of emotions, associated with the four dimensions. Precisely, the numbers of relations *hasEmotionX* generated in SentiWordSKOS, with different intensity values (viz., $X = 60$, $X = 70$, etc.), are listed. Last column gives all the instantiated relations, i.e., the number of nouns that are connected to each emotion, by *hasEmotionX* relation.

4.3.2 Ontology Querying and Exploration

SentiWordSKOS is very easy to integrate into any web application that needs to extract sentiments and emotions from a text in natural language, thanks to the ontology-based representation. The translation of WordNet into an ontology allows getting the meaning and/or the sense of a word along with the information on the

---

[2]https://jena.apache.org/.

[3]http://sourceforge.net/projects/jwordnet/.

Table 3 SentiWordSKOS—number of *hasEmotionX* relation instances associated with each sentic level (emotion), by varying the intensity value X of the relation

Emotion	Intensity					
	60	70	80	90	100	Total
Ecstasy	283	112	38	5	1	439
Joy	1187	407	106	11	1	1712
Serenity	469	34	46	2	1	552
Pensiveness	1092	147	20	1	1	1261
Sadness	560	156	8	3	3	730
Grief	451	157	52	5	1	666
Vigilance	2064	695	25	3	1	2788
Anticipation	376	159	54	2	1	592
Interest	268	77	17	10	3	375
Distraction	527	147	24	10	1	709
Surprise	539	212	99	2	1	853
Amazement	2218	903	265	36	1	3423
Rage	0	0	8	10	1	19
Anger	0	0	0	18	1	19
Annoyance	1748	839	438	28	1	3054
Apprehension	36	284	106	19	1	496
Fear	0	0	4	20	1	25
Terror	0	0	12	12	1	25
Admiration	1189	419	70	12	1	1691
Trust	1720	215	38	37	1	2011
Acceptance	2783	354	117	17	1	3272
Boredom	584	1009	22	5	4	1624
Disgust	1449	524	110	16	1	2100
Loathing	424	161	89	14	1	689

polarity and the emotions. Once the application has identified the synset associated with the word extracted from the text indeed, it is easy to recover the polarity and emotions associated with it, simply running a few SPARQL queries.

To explore the population of SentiWordSKOS graphically, a graphical tool, called LODLive [6] has been used. LodLive is a browser of RDF resources that draws an ontology-based graph where nodes and edges are respectively concepts and relations extracted by querying an associated SPARQL endpoint; it combines the capabilities of a RDF browser with the effectiveness of the graph representation.

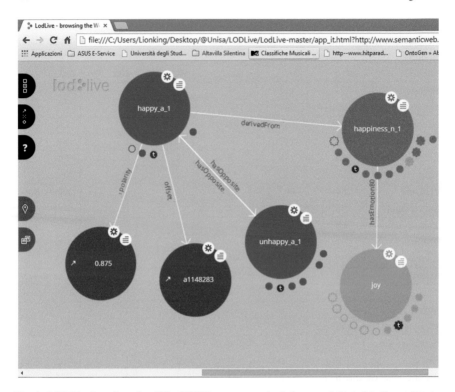

Fig. 4 LODLive Interface: SentiWordSKOS concepts and relations can be graphically explored

To easily interface LODLive[4] with our resources, we exploited the Fuseki triple-store[5] by loading our ontology and its population; we configured the endpoint of Fuseki so that LODLive can query, load and display the SentiWordSKOS triples.

Figure 4 shows a graphical representation of a portion of SentiWordSKOS obtained by LODLive. In particular, the information describing the synset containing the first sense of *happy* is shown: its polarity, its opposite, the noun from which it derives and one of the most relevant emotions. LODLive represents the various types of entities with different colours: we have in dark green the nouns, in violet the adjectives, in purple literal values and in light green, the emotions.

5   Theoretical Foundation

The natural language deals inherently with imprecision, subjectivity and vagueness in defining concepts, sentiments, perception. Particularly, emotions are strictly related

---

[4]http://lodlive.it/.

[5]http://jena.apache.org/documentation/serving_data/index.html.

to the human feeling, they are often hidden in the written language and tricky to grab. All these reasons motivate the choice to modeled emotions by techniques from the Fuzzy Theory, introduced by Zadeh as a means to model the uncertainty of natural language. The fuzzy linguistic modeling especially, fits to model fuzzy and qualitative aspects of an application domain by using users' opinions, judgments, etc. Our idea aims at using these fuzzy techniques to model the emotions.

Fuzzy linguistic modeling allows describing linguistic information by means of linguistic terms supported by linguistic variables [31]. The linguistic variables are defined by means of a syntactic rule and a semantic rule [5] and their values are not numbers but words or sentences in a natural or artificial language. In [13], the ordinal fuzzy linguistic approach is preferred, motivated by the fact that it is very useful for modeling the linguistic aspects in problems and then simplifies the definition of the semantic and syntactic rules of the fuzzy linguistic modeling.

In the light of these observations, the ordinal fuzzy linguistic approach seems an adequate formal model to map our emotional model. Figure 5 shows the four linguistic variables (associated with each dimension) which are designed to describe the 24 emotions (six for each dimensions). Each label represents an emotion described by the triangular membership function. For instance, the linguistic variable *Pleasantness* is described by the set of labels-emotion *Grief, Sadness, Pensiveness, Limbo, Serenity, Joy* and *Ecstasy*. This mapping meets the definition of the finite and ordered label set: $S = \{s_i\} i \in \{0, \ldots, \tau\}$ with $s_i \geq s_j$ if $i \geq j$ and odd cardinality for S (in our case, 7 labels). The mid term represents an assessment of "approximately 0.5", and the rest of the terms being placed symmetrically around it. *Limbo* is indeed mapped as the medium value, and it is neutral in our model.

According to [13], the Hourglass model holds the constraints that the semantics of the label set is established from the ordered structure of the label set by considering that each label for the pair $(s_i, s_{\tau - i})$ is equally informative.

This modeling fits to use tools of Computing With Words (CWW), such as the linguistic aggregation operators. An important aggregation operator of ordinal linguistic values based on symbolic computation is the Linguistic Ordered Weighted Averaging (LOWA) operator [14]. LOWA is an operator used to aggregate non-weighted

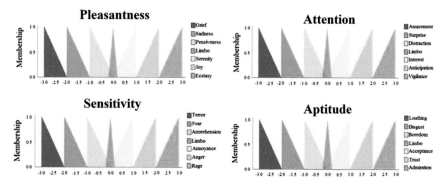

Fig. 5 Linguistic variables associated with the four dimensions of the Hourglass of Emotions

ordinal linguistic information, i.e., linguistic information values with equal importance [14]. LOWA aggregator allows us to get a consensus (synthesis) when more
than one sentic levels are activated for each dimension.

Usually, the fuzzy linguistic ordinal model for CWW is defined by setting the
following operators.

- A negation operator:

$$Neg(s_i) = s_j \mid j = \tau - i \qquad (1)$$

- Comparison operators:

  – Maximization operator:

$$MAX(s_i, s_j) = s_i \ if \ s_i \geq s_j \qquad (2)$$

  – Minimization operator:

$$MIN(s_i, s_j) = s_i \ if \ s_i \leq s_j \qquad (3)$$

- Aggregate operators: LOWA operator is idefined as follows.

The LOWA is an operator used to aggregate non-weighted ordinal linguistic information. Recalling the definition of LOWA [13]:

LOWA Definition: Let $A = \{a_1, \ldots, a_m\}$ be a set of labels to aggregate, then the
LOWA operator is defined as:

$$
\begin{aligned}
\phi(a_1, .., a_m) = W \cdot B^T &= C^m \{w_k, b_k, k = 1, \ldots, m\} = \\
&w_1 \odot b_1 \oplus (1 - w_1) \odot C^{m-1} \{\beta_h, b_h, h = 2, \ldots, m\}
\end{aligned}
\qquad (4)
$$

where: $W = [w_1, \ldots, w_m]$ is the weighting vector where $w_i \in [0, 1]$ and $\sum_i w_i =
1 \ \beta_h = \frac{w_h}{\sum_2^m w_k}$. $B = \{b_1, \ldots, b_m\}$ is the permutation of A defined as $B = \sigma(A) =
\{a_{\sigma(1)}, \ldots, a_{\sigma(m)}\}$ where $a_{\sigma(j)} \leq a_{\sigma(i)} \forall i \leq j$. C^m is a convex combination of m
labels and if m=2:

$$C^2 = \{w_i, b_i, i = 1, 2\} = w_1 \odot s_j \oplus (1 - w_1) \odot s_i = s_k \qquad (5)$$

such that $k = \min \{\tau, i + round(w_1 \cdot (j - i))\}$ and $s_i, s_j \in S, (j \geq i)$, round function is rounding operator with $b_1 = s_j, b_2 = s_i$. If $w_j = 1$ and $w_i = 0$ with $i \neq j \ \forall i$
then the convex combination and defined as: $C^m \{w_i, b_i, = 1, \ldots, m\} = b_j$.

The weighting vector W is important in LOWA calculation. According to [13], a
possible solution to define W by means of a fuzzy linguistic non-decreasing quantifier
Q to represent the concept of fuzzy majority:

$$w_i = Q\left(\frac{i}{m}\right) - Q\left(\frac{i-1}{m}\right) \forall i = 1, \ldots, m; \qquad (6)$$

The memberships function of Q is defined as:

$$Q(r) = \begin{cases} 0 & \text{if } r < a \\ \frac{r-a}{b-a} & \text{if } a \leq r \leq b \\ 1 & \text{if } r > b \end{cases} \tag{7}$$

where $a, b, r \in [0, 1]$, some examples of non-decreasing quantifier Q: "most" (0.3, 0.8); "at least half" $(0, 0.5)$ and "as many as possible" $(0.5, 1)$. When a fuzzy linguistic quantifier Q is used to compute the weights of LOWA operator ϕ, it is symbolized by ϕ_Q.

6 A Closer Look to the Step-by-Step Process

This section details the main steps of the text processing pipeline (Fig. 1). An illustrative example has been developed through all the steps of the process.

6.1 Syntactic Analysis

The input text is analyzed and refined through the typical activities of text parsing: POS tagging, lemmatization and stemming are carried out by Stanford POS tagger. After the stop word removal, the sentences are reduced to just words marked as noun (subject and object), verb, adverb and adjective. These four syntactic categories will be taken into account in our analysis. Let us consider the sentence portion:

"[...] and my guests were very happy for the very appealing restaurant and the excellent food."

After the POS tagging has been carried out, each word is grammatically tagged as shown:

"and/CC my/PRP guests/NNS were/VBD very/RB happy/JJ for/IN the/DT very/RB appealing/JJ restaurant/NN location/NN and/CC the/DT excellent/JJ food/NN./."

Once applied the stop word removal and the stemming, the sentence is reduced to "guest be very happy very appealing restaurant excellent food".

6.2 Word Analysis

The first task of this step is to associate a WordNet synset with each word remaining from the previous step. The selection of the appropriate synset is quite tricky, because it is related to the right sense of words that is often related to the linguistic context in which the word appears. Herein, for simplicity, a first sense is always selected. Each word is compared with the emotions in the Hourglass model in order to discover a semantic relation. To this purpose, Java WordNet Similarity Library

(JWSL)[6] has been taken into account. JSWL implements the most commons similarity and relatedness measures between words. The JSWL similarity selected for this approach is Wu and Palmer [22]. Since the JSWL similarity can be computed only between nouns, for the syntactical categories such as verbs, adverbs, adjective is necessary to "substantivize", i.e., finding the nominal form. As stated, SentiWord-SKOS holds the nominal form associated with non-noun word (see Sect. 4.3). For instance, In SentiWordSKOS, it is coded that the adjective *happy* is *derivedFrom* the noun *happiness*. Only words, whose similarity calculated with the emotion labels of the Hourglass model is non-zero, have been considered. Moreover, as shown in the SentiWordSKOS schema, only similarity values greater than 0.6 are stored. Values below this threshold are considered irrelevant and set to zero.

In our example, the words to process are *very happy, very appealing* and *excellent*. The remaining words are discarded. The candidate words have some (at most, six) values (computed by the similarity, in $[0, \ldots, 1]$) for each one of the four dimensions in the Hourglass of Emotions. The reading of this data could be tricky: each word can have more than one sentic level activated for each dimension with a difficult identification of the predominant emotions. For this reason, getting a compact view, with a main emotion for each dimension is desirable. Mapping the dimensions as linguistic variables and the emotions as fuzzy terms (Figure 5) allows us to apply an effective fuzzy aggregation operator such as LOWA.

According to (4), the calculation of LOWA needs a weighing vector whose elements sum is equal to 1. This vector is computed by scaling the similarity values in that interval. Just to give an example, the word *happy* in the dimension *Pleasantness* has associated with the vector $(0.9, 0.8, 0.6, 0, 0, 0)$ for the emotions labeled *Ecstasy, Joy, Serenity, Pensiveness, Sadness, Grief*, by computing by the similarity between *happiness* and those labels. That means, the word *happy* actives the sentic levels *Ecstasy, Joy, Serenity* in the dimension *Pleasantness*. After the scaling the vector becomes $(0.39, 0.35, 0.26, 0, 0, 0)$. It will be exploited in (4) to aggregate emotions according to LOWA:

$$\phi(Ecstasy, Joy, Serenity, Pensiveness,$$
$$Sadness, Grief) = W \cdot B^T = [0.39, 0.35, 0.26, 0.0, 0.0, 0.0]\cdot$$
$$(Ecstasy, Joy, Serenity, Pensiveness,$$
$$Sadness, Grief) = Joy$$

The label *Joy* synthesizes the emotional state that better describes the word *happy* in the dimension *Pleasantness*. Similar calculations are applied to get the main emotions in each other dimension.

At this point, in order to iterate the method at level of sentence, and then, at level of the whole text, a weight must be also assigned at the other granularity levels, to apply the calculation. A simple way is to calculate a such weight as the average on

---

[6]Java WordNet Similarity Library: http://grid.deis.unical.it/similarity/.

Table 4 Emotional description of the word *happy*

Dimension	Emotion	Weight
Pleasantness	Joy	0.77
Attention	Anticipation	0.7
Sensitivity	Apprehension	0.8
Aptitude	Trust	0.66

the non-zero[7] elements of the vector associated with a dimension. In our example, as shown the word *happy* on the dimension *Pleasantness* has as predominant emotion *joy*, whose weight can be calculated as: $(0.9 + 0.8 + 0.6)/3 = 0.77$, obtained by summing elements of its own original weighing vector.

In nutshell, the word *happy* is described by the digest emotions of the Hourglass shown in Table 4, with the relative weights, calculated as just described.

6.3 Pattern Analysis

This step concerns with the identification of possible patterns in the text analyzed. Patterns are composite words, or phrases, often called opinionated words, because they have composed by adjectives/adverbs, verbs used to qualify nouns. Particularly, by analyzing the returned POS tagged text, the selected patterns are described in Table 5. They involves nouns (NN,NNS, ect.), adverbs (RB,RBR, RBS, etc.), adjectives (JJ,JJR,JJS) and verbs (VB,VBD,VBN, ect.). In our sentence for example, the *Pattern 2* can be identified in *very/RB happy/JJ* and *very/RB appealing/JJ*.

Only phrases that match the patterns are considered for the further processing.

The semantics behind a phrase pattern affects the semantics (and the weight) of the individual words that form it, by intensify/reducing or negating the meaning (and consequently, the corresponding emotion). Adverbs such as *very, extremely* can be used to intensify phrases as *very good, extremely bad* in positive or negative way.

Table 5 Extracted phrases patterns

Pattern	First word	Second word	Third word
Pattern 1	JJ	NN	NN
Pattern 2	RB	JJ/RB	NN
Pattern 3	RB	VB	–
Pattern 4	VB	NN	–
Pattern 5	RB	RB	JJ/VB

---

[7] Generalizing, the average calculation can be applied on all element whose value is greater than a fixed threshold.

Table 6 Adverbs list

Intensifying	Negation
very; extremely; really; always; absolutely;	not; no; never;
bitterly; too; truly; deeply; quite; highly;	without; rarely
only; much; strongly; overall	

Table 7 Emotional description of the pattern *very happy*

Dimension	Emotion
Pleasantness	Ecstasy
Attention	Vigilance
Sensitivity	Fear
Aptitude	Admiration

Table 6 shows the list of adverbs, taken into account for intensifying or negating phrases patterns.

The adverbs act on the emotional dimension, modifying the sentic level activated for a word. In order to intensify the emotion of the individual words to the phrases patterns, the operator intensifier ι applied on a certain sentic level s_i is defined as follows.

$$\iota(s_i) = s_j \mid j = \begin{cases} \tau & \text{if } i = \tau \\ i+1 & \text{if } \frac{\tau}{2} < i \leq \tau \\ \frac{\tau}{2} & \text{if } i = \frac{\tau}{2} \\ i-1 & \text{if } 0 < i < \frac{\tau}{2} \\ 0 & \text{otherwise} \end{cases} \tag{8}$$

When the adverb involved in the pattern negates the phrase pattern, the negation operator is applied as defined in (1) .

By considering our pattern *very/RB happy/JJ*, the intensifier operator given in (8) is applied, activating the higher sentic level, *Ecstasy*. Table 7 shows the emotional intensification obtained for the pattern *very/RB happy/JJ* (cfr. Table 4). The weights associated with the pattern are the same of for *happy*, for each dimension. The same approach is applied for the pattern *very/RB appealing/JJ*, whose emotions (before and after the intensification) and weight are shown in Table 8.

6.4 Sentence Analysis and Text Analysis

As stated, the aggregation operation is iteratively applied on the words/patterns that compound a sentence and then on the sentences compounding a paragraph or text.

Table 8 Emotional description of the word *appealing* and the pattern *very appealing*

Before intensification			After intensification
Dimension	Emotion	Weight	Emotion
Pleasantness	Serenity	0.6	Joy
Attention	Vigilance	0.7	Vigilance
Sensitivity	No Emotion	0.0	No Emotion
Aptitude	No Emotion	0.0	No Emotion

Table 9 Emotional description of the word *excellent*

Dimension	Emotion	Weight
Pleasantness	Joy	0.6
Attention	Vigilance	0.6
Sensitivity	Annoyance	0.6
Aptitude	Trust	0.6

Table 10 Emotional description of the sentence "*I was very happy, the restaurant was very appealing and the food was excellent.*"

Dimension	Emotion
Pleasantness	Joy
Attention	Vigilance
Sensitivity	Fear
Aptitude	Trust

The only slight difference is in the calculation of the weights. Fixed a dimension, the weight associated with a sentence is computed as an average on the (non-zero or greater than a given threshold) weights of words/patterns for that dimension.

Recall that in the sentence "*[…] and my guests were very happy for the very appealing restaurant and the excellent food.*" the only words expressing emotions are *very happy, very appealing* and *excellent*. Resulting emotions for the word *excellent* (in the step *Word Analysis*) are summarized in Table 9.

Finally applying the LOWA operator at sentence level, the emotional description of that sentence is synthesized in Table 10.

7   Experimental Results

Our experimentation aims at evaluating the efficiency performance of our approach in the emotional analysis. To this purpose the International Survey of Emotion Antecedents and Reactions (ISEAR[8]) dataset has been used. ISEAR dataset was

---

[8]http://www.affective-sciences.org/researchmaterial.

build in 1990s across 37 countries by a group of psychologists that collected data from three thousand respondents. The respondents were asked to describe situations (through a few sentences) in which they had experienced some emotion. The emotions considered were joy, fear, anger, sadness, disgust, shame, and guilt. The dataset contains 7,666 such statements (about one thousand associated with each emotion), which include 18,146 sentences, 449,060 running words. As stated in [25], the choice of ISEAR as the source of corpus-based information is mainly motivated by the fact that sentences the corpus are rich in emotion-related words as compared with more standard corpora used in natural language processing.

Our evaluation consists in processing the ISEAR sentences in order to assess possible matches between the emotions that our system assigned to the sentences and those given by the ISEAR respondents. It is evident that the Hourglass of Emotions includes 24 emotions on 4 dimensions, so an ad-hoc mapping of emotions labels is necessary. Precisely, the performance of our approach has been assessed only on the matching emotion labels: joy, sadness, fear, anger and disgust. No further correspondence with guilt and shame seems to rise in the remaining emotions. Tables 11, 12 and 13 show the values of recall and precision computed for each dimension. According to the Hourglass model, the emotions *joy* and *sadness* are opposite in the *Pleasantness* dimension. Similarly, *fear* is opposite to *ager*, with respect to the *Sensitivity* dimension.

The calculation of the recall (R) and the precision (P) reflects the nature of the model that defines opposite emotions for each dimension. Thus, for example, in the calculation of R/P for the emotion *joy* (Table 11), $text_{ret}$ is the number of sentences that have been retrieved considering the positive emotions of Pleasantness, compared with $text_{rel}$, the sentences that are relevant, i.e., classified *joy* in ISEAR. The sentences that our system does not associate with any emotion are discarded. Similar considerations hold for the opposite emotion *sadness*. Table 12 shows the R/P for the emotions labeled *fear* and *anger* in the dimension *Sensitivity*. Let us stress that the ISEAR dataset provides emotions perceived by humans, with an additional emotional load, often encapsulated in the intended meaning of the sentences, rather than in the meaning of individual words/patterns that are in the sentences. For instance, euphemisms in sentences might hide emotions that unlikely are detected by an automatic process. For these reasons the results can be considered satisfactory in terms of R/P for the dimensions *Pleasantness* and *Sensitivity*; the only exception is the emotion *disgust* which does not have an opposite among the ISEAR emotions. In order to get a plausible evaluation on this emotion, the calculation of R/P has been done on all the sentences associated to the remaining emotions (*disgust, guilty* and *shame*). Table 13 shows the results for *disgust*: as expected, the missing opposite emotions has required a different setting of the R/P calculation; consequently the value of the recall is worsened, although the precision preserved a fair value.

Our experimentation was mainly based on the evaluation of the performance of our framework on documents from well-described dataset. The results are a strong

Table 11 Results for the *Pleasantness* dimension

Emotion	$text_{ret}$	$text_{rel}$	$text_{ret} \cap text_{rel}$	Precision	Recall
Joy	798	754	481	0.603	0.638
Sadness	616	839	413	0.67	0.492

Table 12 Results for the *Sensitivity* dimension

Emotion	$text_{ret}$	$text_{rel}$	$text_{ret} \cap text_{rel}$	Precision	Recall
Fear	301	686	235	0.781	0.342
Anger	918	715	565	0.615	0.79

Table 13 Results for the *Aptitude* dimension

Emotion	$text_{ret}$	$text_{rel}$	$text_{ret} \cap text_{rel}$	Precision	Recall
Disgust	361	973	158	0.43	0.162

point of our framework, since they come from a comparison with the human-driven evaluations.

Comparisons with existing tools have been quite tricky: to the best of our knowledge indeed, there are no other tools that provide a synthesis of emotions expressed in a text. However, some experiments have been conducted, by comparing the performance of our framework with SenticNet. The choice is motivated by the use of the same affective model, the Hourglass of Emotions, even though the provided APIs allow evaluating the polarity and emotion at level of a single word or composite linguistic expressions, rather than whole sentences. In order to set a common configuration, the comparison has been accomplished at level of single word. SenticNet provides a set of 15,000 concepts, where only 4,414 are composed of single words. The SenticNet APIs provide for each word, a value in the range $[-1, 1]$ for each dimension. Our approach has been set to work on the set of SenticNet's single words. Table 14 shows the recall and precision evaluated for each dimension, grouped by the emotions with positive sentic levels and negative ones (see Sect. 4.1.3). Table 15 shows the recall and precision evaluated with respect to the polarity.

Although our framework has been designed to provide a synthesis of emotions captured from a text, the result obtained on the single words are fairly good, considering also that the automatic word sense extraction implemented in our approach does not take into account the context where the word appears. This can affect on the selection of the right sense of the word, and then influences the emotion elicitation.

Table 14 Recall and Precision calculated on the Hourglass dimensions, with respect to SenticNet's single words

	Emotion	$text_{ret}$	$text_{rel}$	$text_{ret} \cap text_{rel}$	Precision	Recall
Pleasentness	Ecstacy, Joy, Serenity	368	299	155	0.42	0.52
	Limbo	577	709	398	0.69	0.56
	Pensiveness, Sadness, Grief	413	356	180	0.44	0.51
Attention	Vigilance, Anticipation, Interest	398	759	265	0.66	0.35
	Limbo	331	559	170	0.51	0.3
	Distraction, Surprise, Amazement	635	46	35	0.06	0.76
Sensitivity	Rage, Anger, Annoyance	364	178	123	0.34	0.69
	Limbo	630	650	396	0.62	0.61
	Apprehension, Fear, Terror	370	536	277	0.75	0.52
Aptitude	Admiration, Trust, Acceptance	675	625	417	0.62	0.67
	Limbo	140	340	90	0.64	0.26
	Boredom, Disgust, Loathing	549	399	261	0.48	0.65

Table 15 Recall and Precision calculated on the polarity with respect to SentiNet's single words

Polarity	$text_{ret}$	$text_{rel}$	$text_{ret} \cap text_{rel}$	Precision	Recall
Positive	554	1116	402	0.73	0.36
Neutral	3058	2532	2075	0.68	0.82
Negative	802	766	399	0.5	0.52

8 Conclusion

The chapter presents a process for emotion detections. The use of the fuzzy linguistic modeling allows mapping the emotions and dimensions as linguistic information reflecting the inherent imprecision of natural language: the emotions are mapped as linguistic terms and through the fuzzy aggregation, a unique comprehensive emotion summarizes the emotional content expressed by different activated sentic levels. The novelty of this work is in the use of the ordinal fuzzy linguistic modeling to map emotions; there is a subtle difference from the standard approach to map the linguistic variables: usually this model is used for the quality evaluation, where linguistic terms are user's labels (i.e., linguistic values such as *very high, high, low, very low*, etc.) associated with a linguistic variable.

Herein the role of the linguistic variables is played by the dimension and the terms are the emotions (not the value assumed by the dimension) activated in that dimension.

Then, a digest way to interpret more emotions associated to a text is the fuzzy aggregation operator applicable on these labels.

Although this model presents interesting performance in terms of detected emotions, there are some open questions in the natural language processing for the sentiment and emotion mining. Natural language is very difficult to analyze; emotions are often not directly associated with words, since the intrinsic meaning is not coded in the words but in the figurative sense of the phrase. Automatic text processing will be not able to recognize emotions in a review sentence like: "The hotel [...] should at least be able to provide their guests with comfortable accommodations no matter what the outside conditions provide". The reviewer expresses anger and loathing in this sentence, but these emotions are not easy to capture: no word in the sentence has a meaning that is explicitly related to an emotion. Scientific research community work to this direction, trying to improve the automatic extraction of feelings and emotions, but the result is still far to be effective, due to several expressive nuances that characterize the natural language.

A future development is to improve the word disambiguation, by analyzing the context (i.e., surrounding text and sentences) where a sentence appears, in order to identify the correct meaning (or sense). Then, phrases or sentences will be processed with reference to the identified context, in order to discern the intended meaning: for instance, "comfortable accommodation" or "clean towel" are crucial key-phrases to understand the emotion in the 'hotel' context.

References

1. Agarwal, A., Xie, B., Vovsha, I., Rambow, O., Passonneau, R.: Sentiment analysis of twitter data. In: Proceedings of the Workshop on Languages in Social Media. LSM'11, pp. 30–38. Association for Computational Linguistics, Stroudsburg, PA, USA (2011)
2. Liu, B., Hu, M., Cheng, J.: Opinion observer: analyzing and comparing opinions on the web. In: Proceedings of International World Wide Web Conference (WWW-05), pp. 342–351. ACM, New York, NY, USA (2005). doi:10.1145/1060745.1060797
3. Baccianella, A.E.S., Sebastiani, F.: Sentiwordnet 3.0: an enhanced lexical resource for sentiment analysis and opinion mining. In: Proceedings of the Seventh conference on International Language Resources and Evaluation (LREC'10). European Language Resources Association (ELRA), Valletta, Malta (2010)
4. Barbosa, L., Feng, J.: Robust sentiment detection on twitter from biased and noisy data. Proceedings of the 23rd International Conference on Computational Linguistics: Posters. COLING'10, pp. 36–44. Association for Computational Linguistics, Stroudsburg, PA, USA (2010)
5. Cabrerizo, F.J., Martinez, M.A., Gijon, J.L., Esteban, B., Herrera-Viedma, E.: A new quality evaluation model generating recommendations to improve the digital services provided by

the academic digital libraries. Word Congress of International Fuzzy System Association, Surabaya-Bali, Indonesia (2012)

6. Camarda, D.V., Mazzini, S., Antonuccio, A.: Lodlive, exploring the web of data. In: Proceedings of the 8th International Conference on Semantic Systems, I-SEMANTICS'12, pp. 197–200. ACM, New York, NY, USA (2012). doi:10.1145/2362499.2362532. http://doi.acm.org/10.1145/2362499.2362532

7. Cambria, E., Hussain, A., Havasi, C., Eckl, C.: Sentic computing: Exploitation of common sense for the development of emotion-sensitive systems. In: Development of Multimodal Interfaces: Active Listening and Synchrony, Lecture Notes in Computer Science, vol. 5967, pp. 148–156. Springer (2010)

8. Cambria, E., Speer, R., Havasi, C., Hussain, A.: Senticnet: a publicly available semantic resource for opinion mining (2010)

9. Cataldi, M., Ballatore, A., Tiddi, I., Aufaure, M.A.: Good location, terrible food: detecting feature sentiment in user-generated reviews. Soc. Netw. Anal. Mining, 1–15 (2013). doi:10.1007/s13278-013-0119-7

10. Chuang, Z., Wu, C.: Multi-modal emotion recognition from speech and text. Int. J. Comput. Linguis. Chin. Lang. Process. 9(2), 1–18 (2000)

11. Hussain, E.C.A., Havasi, C., Eckl, C.: Affectivespace: blending common sense and affective knowledge to perform emotive reasoning. In: Proceedings of the Workshop on Opinion Mining and Sentiment Analysis (2009)

12. El-Nasr, M.S., Yen, J., Ioerger, T.R.: Flamefuzzy logic adaptive model of emotions. Auton. Agents Multi-Agent Syst. 3(3), 219–257 (2000). doi:10.1023/A:1010030809960

13. Heradio, R., Cabrerizo, F.J., Fernandez-Amoros, D., Herrera, M., Herrera-Viedma, E.: A fuzzy linguistic model to evaluate the quality of library 2.0 functionalities. Int. J. Inf. Manage. 33(4), 642–654 (2013)

14. Herrera, F., Herrera-Viedma, E., Verdegay, J.: Direct approach processes in group decision making using linguistic owa operators. Fuzzy Sets Syst. 79(2), 175–190 (1996)

15. Kannan, K., Goyal, M., Jacob, G.: Modeling the impact of review dynamics on utility value of a product. Soc. Netw. Anal. Mining 1–18 (2012). doi:10.1007/s13278-012-0086-4

16. Kouloumpis, E., Wilson, T., Moore, J.: Twitter sentiment analysis: the good the bad and the OMG! In: ICWSM (2011)

17. Liu, B.: Sentiment analysis and opinion mining. Synth. Lect. Human Lang. Technol. 5(1), 1–167 (2012). doi:10.2200/S00416ED1V01Y201204HLT016

18. Loia, V., Senatore, S.: A fuzzy-oriented sentic analysis to capture the human emotion in web-based content. Know.-Based Syst. 58, 75–85 (2014). doi:10.1016/j.knosys.2013.09.024

19. Miles, A., Matthews, B., Wilson, M., Brickley, D.: Skos core: simple knowledge organisation for the web. In: Proceedings of the 2005 International Conference on Dublin Core and Metadata Applications: Vocabularies in Practice, DCMI'05, pp. 1:1–1:9. Dublin Core Metadata Initiative (2005)

20. Miller, G.A.: Wordnet: a lexical database for english. Commun. ACM 38(11), 39–41 (1995)

21. Pang, B., Lee, L.: A sentimental education: sentiment analysis using subjectivity summarization based on minimum cuts. In: Proceedings of the 42nd Annual Meeting on Association for Computational Linguistics, ACL'04. Association for Computational Linguistics (2004)

22. Pedersen, T., Patwardhan, S., Michelizzi, J.: Wordnet: similarity: measuring the relatedness of concepts. In: Demonstration Papers at HLT-NAACL 2004, pp. 38–41. Association for Computational Linguistics (2004)

23. Pedersen, T., Patwardhan, S., Michelizzi, J.: WordNet: Similarity: measuring the relatedness of concepts. In: Demonstration Papers at HLT-NAACL 2004. HLT-NAACL-Demonstrations'04, pp. 38–41. Association for Computational Linguistics, Stroudsburg, PA, USA (2004)

24. Plutchik, R.: The nature of emotions. Am. Sci. 89(4), 344–350 (2001)

25. Poria, S., Gelbukh, A., Hussain, A., Howard, N., Das, D., Bandyopadhyay, S.: Enhanced senticnet with affective labels for concept-based opinion mining. IEEE Intell. Syst. 28(2), 31–38 (2013). doi:http://doi.ieeecomputersociety.org/10.1109/MIS.2013.4

26. Saif, H., He, Y., Alani, H.: Semantic sentiment analysis of Twitter. In: Proceedings of the 11th International Conference on The Semantic Web—Part I. ISWC'12, pp. 508–524. Springer-Verlag, Berlin, Heidelberg (2012)
27. Strapparava, C., Mihalcea, R.: Learning to identify emotions in text. In: Proceedings of the 2008 ACM Symposium on Applied Computing, SAC'08, pp. 1556–1560. ACM, New York, NY, USA (2008). doi:10.1145/1363686.1364052
28. Subasic, P., Huettner, A.: Affect analysis of text using fuzzy semantic typing. Trans. Fuzzy Syst. **9**(4), 483–496 (2001). doi:10.1109/91.940962
29. Suchanek, F.M., Kasneci, G., Weikum, G.: Yago: A large ontology from wikipedia and wordnet. Web Semant. **6**(3), 203–217 (2008)
30. Ye, Q., Law, R., Gu, B.: The impact of online user reviews on hotel room sales. Int. J. Hosp. Manage. **28**(1), 180–182 (2009). doi:10.1016/j.ijhm.2008.06.011
31. Zadeh, L.A.: The concept of a linguistic variable and its application to approximate reasoning—I. Inf. Sci. **8**(3), 199–249 (1975)

Hyperelastic-Based Adaptive Dynamics Methodology in Knowledge Acquisition for Computational Intelligence on Ontology Engineering of Evolving Folksonomy Driven Environment

Massimiliano Dal Mas

Abstract Due to the rapid growth of structured/unstructured and user-generated data (e.g., social media sites) volume of data is becoming too big or it moves too fast or it exceeds current processing capacity and so traditional data processing applications are inadequate. Computational Intelligence with Concept-based approaches can detect sentiments analyzing the concept based on text expressions without analyzing the singlef words as in the purely syntactical techniques. On human-centric intelligent systems Semantic networks can simulate the human complex frames in a reasoning process providing efficient association and inference mechanisms, while ontology can be used to fill the gap between human and Computational Intelligence for a task domain. For an evolving environment it is necessary to understand what knowledge is required for a task domain with an adaptive ontology matching. To reflect the evolving knowledge this paper considers ontologies based on folksonomies according to a new concept structure called "Folksodriven" to represent folksonomies. To solve the problems inherent an uncontrolled vocabulary of the folksonomy it is presented a Folksodriven Structure Network (FSN): a folksonomy tags suggestions built from the relations among the Folksodriven tags (FD tags). It was observed that the properties of the FSN depend mainly on the nature, distribution, size and the quality of the reinforcing FD tags. So, the studies on the transformational regulation of the FD tags are regarded to be important for an adaptive folksonomies classifications in an evolving environment used by Intelligent Systems to represent the knowledge sharing. The chapter starts from the discussion on the deformation exhibiting linear behavior on FSN based on folksonomy tags chosen by different user on web site resources. Then it's formulated a constitutive law on FSN investigating towards a systematic mathematical analysis on stress analysis and equations of motion for an evolving ontology matching on an environment defined by the users' folksonomy choice. The adaptive ontology matching and the elastodynamics are merged to obtain what we can call the elasto-adaptive-dynamics methodology of the FSN. Further-

M. Dal Mas (✉)
Department of Research and Development - Artificial Intelligence Unit
technologos, via G. Mameli 4, 20129 Milano, Italy
e-mail: me@maxdalmas.com
URL: http://www.technologos.it

© Springer International Publishing Switzerland 2016   141
W. Pedrycz and S.-M. Chen (eds.), *Sentiment Analysis and Ontology Engineering*,
Studies in Computational Intelligence 639, DOI 10.1007/978-3-319-30319-2_7

more it is shown the last development defining a hyperelastic dynamic considering the internal folksonomy behavior of the stress and strain from original to deformed configuration.

Keywords Computational intelligence · Artificial intelligence · Big data · Sentiment analysis · Sentic computing · Semantic web · Folksonomy · Ontology · Network · Elasticity · Plasticity · Natural language processing · Quasicrystal

1 Introduction

Computational Intelligence (CI) is a methodology to deal/learn with new situations for which there are no effective computational algorithms [5, 11, 23].

A fundamental prerequisite for an evolving environment is to decide the "knowledge" for a task domain: what kinds of things consists of, and how they are related to each other. Semantic networks can be used to simulate the human-level intelligence providing efficient association and inference mechanisms to simulate the human complex frames in reasoning. Ontology can be used to fill the gap between human and CI for a task domain with an adaptive ontology matching.

The main purpose of this chapter is the development of a constitutive model emphasizing the use of ontology-driven processing to achieve a better understanding of the contextual role of concepts (Sentic computing). To reflect the evolving knowledge this chapter considers an ontology-driven process based on folksonomies according to a new concept structure called "Folksodriven" [17] to represent folksonomies—a set of terms that a group of users tagged content without a controlled vocabulary. To solve the problems inherent an uncontrolled vocabulary of the folksonomy a *Folksodriven Structure Network* (*FSN*), built from the relations among the *Folksodriven tags* (*FD tags*), is presented as a folksonomy tags suggestions for the user. It was observed that the properties of the *FSN* depend mainly on the nature, distribution, size and the quality of the reinforcing *FD tags*. So, the studies on the transformational regulation of the *FD tags* are regarded to be important for adaptive folksonomies classifications in an evolving environment used by Intelligent Systems.

This work proposes to use an adaptive ontology matching to understand what knowledge is required for a task domain in an evolving environment, using the elastodynamics—a mathematical study of a structure deformation that become internally stressed due to loading conditions—to obtain an *elasto–adaptative–dynamic methodology* of the *FSN*.

1.1 Motivation

Traditional data processing applications are becoming inadequate due to the rapid growth of structured/unstructured and user-generated data (e.g., social media sites)

[7]. Big data is a buzzword to describe that scenario where volume of data is too big or it moves too fast or it exceeds current processing capacity [2, 33, 43, 51] despite these problems, big data has the potential to help companies to improve operations and make faster, more intelligent decisions.

Personalized information systems aim at giving the individual user support in accessing, retrieving and storing information. Knowledge management and the associated tools aim to provide an environment where people may create, learn, share, use and reuse knowledge. However, instead of helping users, many systems are just increasing the information overload.

"Syntactical approaches" can result ineffective considering opinions and sentiments through latent predetermined semantics.

While in CI with "concept-level approaches" it is possible to grasp the conceptual and affective information associated with natural language opinions performing a semantic text analysis using Web ontologies or semantic networks.

Concept-based approaches can detect sentiments analyzing the concept based on text expressions without analyzing the single words as in the purely syntactical techniques. Such approaches rely on the implicit meaning associated with natural language concepts based on large semantic knowledge instead of using keywords and word co-occurrence counts as in purely syntactical techniques.

Sentic computing, as concept-based approaches, is a multi-disciplinary approach to sentiment analysis and common sense computing, working with computer and social sciences to better recognize, interpret, and process sentiments and opinion over the Web. Sentic computing (meant as the feeling of common sense) is based on the analysis of text based on ontologies and common sense reasoning tools of text.

The semantics of data and information structures can be used to aid the process of information use. That knowledge can be expressed in an ontology to represent the dynamic evolution of the knowledge [16].

1.2   The Approach and Related Works

In the field of social tagging exists several methods that study the association between social tags (as folksonomy tags) and semantics in order to make explicit the meaning of those tags as shown by García-Silva et al. [24], overall Damme [21] proposed the term "folksontology" as a branch of ontology deals the study for turning Folksonomies into Ontologies.

For the study of folksonomy, tagging and social tagging systems also Trant [48] provided a framework on different methods where three broad approaches were identified, focusing on the folksonomy itself (and the role of tags in indexing and retrieval), on tagging (and the behavior of users) and on the nature of social tagging systems (as socio-technical framework).

Grandi [25] exhaustively summarized on the actual researches on temporal and evolution aspects in the Semantic Web. He reported an annotated bibliography on several papers addressing the time-varying and evolutionary nature of the phenomena they model and the activities they support in the World Wide Web.

In this paper we propose a methodology framework for reusability and share ability of information based on an elastic adaptive ontology matching based on a network of folksonomies defined by the users as proposed by the *Folksodriven Structure Network (FSN)* [17].

The network structure of "Folksodriven tags"—*Folksodriven Structure Network (FSN)*—was thought as a "Folsksonomy tags suggestions" to represent the knowledge sharing. *FSN* is based on *Natural Language Processing (NLP)*, for the user on a dataset built on chosen websites. The single tag is called *Folksodriven tag (FD tag)*, while the network structure—called *Folksodriven Structure Network (FSN)*—can be considered as an adaptive way to solve the ontology matching problem between *Folksodriven tags (FD tag)*—that are hard to categorize. A *FSN* determines the flow of information in folksonomy tags network and determines its functional and computational properties. By observation it was found how serendipitous discovery of content among users can be facilitated by *FSN* [14].

A challenging topic today is to develop new theoretical methods and algorithms for the implementations of dynamically evolving systems with a higher level of flexibility and autonomy. An evolving and adaptive intelligent systems targets nonstationary processes by developing the vision of a Semantic Web environment in which ontology-based Web services [14, 20] with intelligent capabilities are able to self-configure themselves. This theory, unlike its name, does not limit the actual ontology matching to an evolving environment, but rather it requests the applications conformant to that to correctly display and interpret the data in a set of user interface [13, 15, 18, 19].

The scope is to provide an adaptive Web-based system in which members of a community are able to browse information tailored to their needs, store content of interest into their own information repository, and are able to augment it with metadata for reuse in a human-level CI.

It is possible to cluster *FD tags* components into larger *FSN structures* by understanding the inter-tags interactions of the *FD tags*. This paper provides a critical examination of the various inter-tags forces that can be used in *FD tags* self-assembly processes of *FSN*.

This chapter extends to Hyperelastic-based Adaptive Dynamics Methodology the work depicted in the paper "Elasticity on Ontology Matching of Folksodriven Structure Network" [16] and "Elastic Adaptive Dynamics Methodology on Ontology Matching on Evolving Folksonomy Driven Environment" [13] that describes the evolution of the *FSN* data model as well as their "elastic" properties.

1.3 Structure of the Chapter

The backgrounds of related technologies are briefly described in the next sections (Sects. 2 and 3). Then, a system overview is presented (Sect. 4–11), focusing on the definition of the "elasto-adaptative-dynamics"—for an adaptive folksonomies classifications in an evolving environment used by Intelligent Systems to represent the knowledge sharing. Furthermore Sect. 12 shows the last development defining a "hyperelastic dynamic" considering the internal folksonomy behavior of the stresses for a deformation process depending on the initial and final configurations.

Section 13 describes the resulting elasto-adaptative-dynamics equations of the proposed ontology-driven processing to achieve a better understanding of the contextual role of concepts (Sentic computing). While Sect. 14 summarizes the methodology described on the chapter.

Section 15 discusses the *computational complexity* of the proposed methodology.

The verification of the proposed method is described in Sect. 16 considering *experimental observation* with *capability results*, *performance results*, *data collection analysis* and *results verification*.

Section 17 describes the *experimental setup*: *cloud infrastructure* and *implementation*. Lastly on Sect. 18 some *conclusions and future works* are introduced.

2 Ontology for a Dynamic Evolving and Adaptive Intelligence Systems

Web originates data that need to be analyzed at various levels of details and for various purposes. As in the previous works [13–20] we define the process through which relevant data are analyzed in such a way that prevents the loss of important event affecting the data. This section presents our goal in support analytic bases for ontologies used by adaptive intelligence systems.

Evolving systems should be grown as they operate. They are inspired by the idea of modeling a system evolution in an environment that can dynamically change and evolve [37].

Such systems should be able to refine their model through their interaction with the environment and update their knowledge.

Real-world problems are complex and dynamic, they need sophisticated methods and tools for building adaptive Intelligent Systems (IS).

Computational Intelligence (CI) is a bottom-up approach, where the algorithm—inspired by biological processes—learns the "knowledge" to solve the problem. Different branches of CI evolve in many directions: a common definition of the CI field is therefore impossible, because different people include or exclude different methods under CI [5, 23].

A fundamental prerequisite to using it is to decide the "knowledge" for a task domain of the CI: what kinds of things consists of, and how they are related to each other.

An "ontology" is a representation to what exists in a particular task domain describing the world in term of "things" (or individuals) and their relationships among them. That is the same base assumption made in natural language and logic. A "thing" is defined by the observer and can be anything, whether concrete or abstract (e.g.: people, colors, emotions, numbers, and times). The world can be divided up in different ways according to the observer, or the task pursuit by the same observer. For each task, the observer needs to identify specific things and relations that can be used to express a "task domain" representing what is true on the world considered.

Those could be even worse when involving complex knowledge structures for reasoning. The human semantic is so sophisticated that no artificial system, even the most complex, can have the same "human" performance, moreover on qualitative way. The only way to target the human level of CI is that the artificial system itself has to decide how to parcel out the task domain representing the world with the relative relationships [5, 11]. So, to make the artificial system evolving by itself, it should be necessary in advance to grasp and to decide what kind of knowledge is required for. In this way, semantic networks can be used to simulate the human complex frames in reasoning process providing efficient association and inference mechanisms for Intelligent Systems.

3   Related Work

This section introduces the notation and the necessary background for this article about folksonomy tags and their correlation with the semantic web.

3.1   Folksonomy Tag

Folksonomies are concentrations of user-generated categorization principles. The term is a neologism consisting of a combination of the word *folk* and *taxonomy*[1] Folksonomies are thus created by the people for the people on the basis of the premise that the categorizing people can create a categorization that will better reflect the people's conceptual model, contextualization and actual use of the data [48]. The distinctive feature of folksonomies is that it is not a classification in a strict sense, but they consist of disconnected and loosely related keywords only connected by associative relations. Here, there is no hierarchy between superior and subordinated concepts [32].

3.2   Semantic Web and Folksonomy Tags

An ontology is a formal specification of a conceptualization of an abstract representation of the world or domain we want to model for a certain purpose.

---

[1]"folks" derives from the Old English "folc" (meaning 'people'), of Old High Germanic origin "folc", related to German "volk"; while "taxonomy" derives from the Greek words "taxis" (meaning 'order', 'arrangement') and "nomos" ('law' or 'science').

Ontologies can capture the semantics of a set of terms used by some communities: but meanings change over time, and are based on individual experiences, and logical axioms can only partially reflect them. Web developers encodes all the information in an ontology filled with rules that say, essentially, that "Robert" and "Bob" are the same. But humans are constantly revising and extending their vocabularies, for instance at one times a tool might know that "Bob" is a nickname for "Robert", but it might not know that some people named "Robert" use "Roby", unless it is told explicitly [1, 31].

There is a perception of ontologies as top-down authoritarian constructs, not always related to the variety resources in a domain context of the ontologies. So it is understandable that a bottom-up structure like folksonomy seams more attractive as technology for the web self-organization.

Folksonomies are trying to bypass the process of creating ontologies for semantic networks, supporting the development of evolutionary terms (and relations within a subject area). Folksonomies involve well-known problems and defects, such as the ambiguity of the terms and use of synonyms. Skeptics think that the folksonomy represents the first step toward anarchy of the Web but the Web is already an anarchist, which are both its greatest strength and its greatest weakness. As the Web grows, the best way to generate the metadata for the resources of the Web will always be a central issue.

Folksonomies represent exactly one of the attempts in place to monitor and categorize Web content to limit the costs to stand still "anarchy of the Internet".

4 Folksodriven Notation

In this section is introduced the formal notation used in the paper for the *Folksodriven tags* (*FD tags*) and the related *Time Exposition* (*E*).

In a model of space-time (Fig. 1), every point in space has four coordinates (*x, y, z, t*), three of which represent a point in space, and the fourth a precise moment in time. Intuitively, each point represents an event that happened at a particular place at a precise moment. The usage of the *four-vector* name assumes that its components refer to a "standard basis" on a Minkowski space [12]. Points in a Minkowski space are regarded as events in space-time. On a direction of time for the time vector we have:

- past directed time vector, whose first component is negative, to model the "history events" on the folksodriven notation
- future directed time vector, whose first component is positive, to model the "future events" on the folksodriven notation

$$FD := (C, E, R, X) \tag{1}$$

Fig. 1 In a model of
space-time, every point in
space has four coordinates
(C, E, R, X), representing a
point in space and a precise
moment in time E

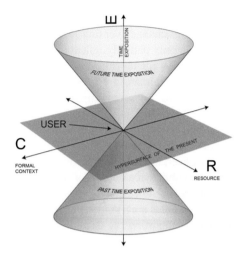

A Folksodriven will be considered as a tuple (1) defined by finite sets composed by
the *Formal Context (C)*, the *Time Exposition (E)*, the*Resource (R)* and the ternary
relation *X*—as in Fig. 1.

As stated in [16, 17] we consider a *Folksodriven tag (FD tag)* as a tuple (1) defined
by finite sets composed by:

- *Formal Context (C)* is a triple $C:=(T, D, I)$ where the *Topic of Interest T* and the
  Description D are sets of data and *I* is a relation between *T* and *D*;
- *Time Exposition (E)* is the clickthrough rate (CTR) as the number of clicks on a
  Resource (R) divided by the number of times that the *Resource (R)* is displayed
  (impressions);
- *Resource (R)* is represented by the URI of the webpage that the user wants to
  correlate to a chosen tag;
- *X* is defined by the relation $X = C \times E \times R$ in a Minkowski vector space [12]
  delimited by the vectors *C*, *E* and *R*.

5   Folksodriven as a Network

On scientific literature folksonomies are modelled in order to obtain a number of
model patterns and mathematical rules [4, 9]. The theoretical starting point considered is the emerging scientific network theory that aims to demonstrate mathe-
matically that all complex networks follow a number of general regularities. The
connection between the different vertices in the network forms a pattern. The pattern
of all complex systems is the power law that arises because the network expands
with the addition of new vertices which must be connected to other vertices in the
network. A network with a degree distribution that follows a power law is a "scale-
free network". An important characteristic of scale-free networks is the Clustering

Coefficient distribution, which decreases as the node degree increases following a power law [4].

We consider a Folksodriven network in which nodes are Folksodriven tags (*FD tags*) and links are semantic acquaintance relationships between them according to the SUMO (http://www.ontologyportal.org) formal ontology that has been mapped to the WordNet lexicon (http://wordnet.princeton.edu). It is easy to see that *FD tags* tend to form groups, i.e. small groups in which tags are close related to each one, so we can think of such groups as a complete graph. In addition, the *FD tags* of a group also have a few acquaintance relationships to *FD tags* outside that group. Some *FD tags*, however, are so related to other tags (e.g.: workers, engineers) that are connected to a large number of groups. Those *FD tags* may be considered the hubs responsible for making such network a scale-free network [4]. While the disambiguation on *FD tags* can be done using the relations with other *FD tags*. In a scale-free network most nodes of a graph are not neighbors of one another but can be reached from every other by a small number of hops or steps, considering the mutual acquaintance of *FD tags*.

The network structure of "Folksodriven tags" (*FD tags*)—*Folksodriven Structure Network (FSN)*—was thought as a "Folsksonomy tags suggestions" for the user on a dataset built on chosen websites.

Changes in folksonomy tags chosen cause structural changes on the *FSN* and its relative network connectivity (structural plasticity) of the sources considered (i.e. the website considered). As a consequence of structural changes, links between network tags may break and new links can be generated. Local structural changes at the single *FD tag* may entail alterations in global network topology of the *FSN*. Conversely, global topology can have impact on local chose on *FD tag* by the user on a new web source to correlate with a folksonomy tag.

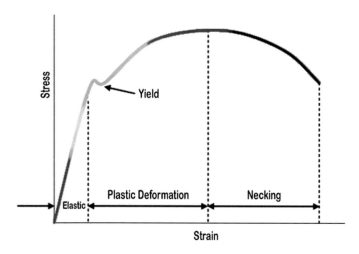

Fig. 2 Stress-strain curve

The *stress tensor* σe represents the set of forces that are imposed upon a system. Any system subjected to stress may deform, the deformation may change the *FSN* structure.

Strain εe is the change in shape of the *FSN* structure due to some inter-tags forces. The stress σe could be plotted against the strain εe to measure the ontological plasticity to obtain an engineering stress-strain curve such as that shown in Fig. 2. The curve is drawn with a rainbow grade colour from the red to indicate the elastic part and the green for the yield part, to the purple-black for the necking part [45].

6 Folksodriven Structure Network Plasticity

*FSN*s have regular elements, like normal networks. But these elements fit together in ways which never properly repeat themselves. The two-dimensional equivalent is known as Penrose tiling (Fig. 3), after Sir Roger Penrose, a British mathematician who put this form of geometry on a formal footing.

Penrose created "aperiodic" tiling patterns (Figs. 3 and 4) that never repeated themselves; work that he suspects was inspired by Kepler's drawings. Penrose tiling [27] has, however, been widely used in the past for decoration, as in the decoration tiles in the railway station of Toledo—Spain (Fig. 4) where we can see the correspondence with the relative Penrose tiles (Fig. 3)—see [36].

A three-dimensional Penrose tile equivalent is used in chemistry for the quasicrystals [6, 8, 39] having revolutionized materials science.

While *FSN* considers n-dimensions, virtually infinite dimensions, depending on the network connection. *FSN* deals mostly with how something evolves over time. Space or distance can take the place of time in many instances. *FSN* happens only

Fig. 3 Penrose "aperiodic" tiling patterns

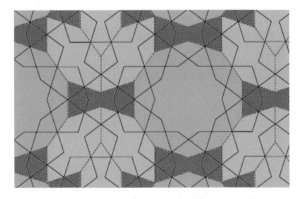

Fig. 4 Tails on the Toledo
(Spain) railway station

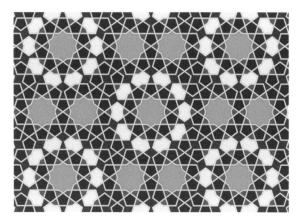

in deterministic, nonlinear,[2] dynamical systems.[3] For the conventional mathematical models the nonlinear behavior means mainly plasticity. In the study on the classical plasticity there are two different theories, one is the macroscopic plasticity theory and the other is based on the mechanism of motion of dislocation.

The so-called *FSN plasticity theory* here developed is based on classical theory [10] considering the mechanism of motion of dislocation, and in some extent can be seen as a "microscopic" theory. The difficulty for the Folksodriven plasticity lies in lack of both enough macro- and micro-data [28]. At present the macroscopic experiments have not, as yet, been properly undertaken.

Though there is some work on the mechanism in microscopy of the plasticity, the data are very limited. This leads to the constitutive law of *FSN* being essentially unknown. Due to that reason the systematic mathematical analysis on deformation and fracture for the ontology matching is not extensible available so far.

7 Macroscopic Behavior of Plastic FSN Lattice

The frame of the periodic arrangement of centers of *Time Exposition* (E) of *FD tags* is called a "lattice" [26]. Thus, the properties of corresponding points of different cells in a *FSN* are the same. The positions of these points can be defined in a Minkowsky vector space [12]. A *FSN* is an n-dimensional array of folksonomy forming a regular lattice.

---

[2]*Nonlinear* means that output isn't directly proportional to input, or that a change in one variable doesn't produce a proportional change or reaction in the related variable(s). In other words, a system's values at one time aren't proportional to the values at an earlier time.

[3]A *dynamical system* is anything that moves, changes, or evolves in time. Hence, chaos deals with what the experts like to refer to as dynamical-systems theory (the study of phenomena that vary with time) or nonlinear dynamics (the study of nonlinear movement or evolution).

The *FSN lattice* is designed considering its vertices and *FD tags* corresponding to the nodes of the mesh and its edges correspond to the ties between the nodes and so the *FD tags*. We observed the variations on the *FSN* using parameters defined on the *FD* structure (defined in Sect. 4 *Folksodriven Notation*). We consider the plasticity as the capacity of *FSN lattice* to vary in developmental pattern, by the interaction of the *FD tags* and the web source environment, or in user behavior according to varying user choose of the *FD tags*.

At medium and low time exposition (E), Folksodriven formal context (C) exhibit a dispersion with few points, but they present plasticity-ductility at high time exposition (E). Near the high stress concentration zone on Formal Context (C), e.g. around dislocation core or crack tip, plastic flow appears.

The important connection between plasticity and the structural defects in *FSN* was observed in experiments where plastic deformation of a Folksodriven Structure Network (*FSN*) was induced by motion of dislocations in the network. Salient structural features are presented in the *FSN*.

Some experimental data [13–15, 18–20] have been obtained observing tagging on news websites, whose posses a very high time exposition (E).

Figure 5 shows the stress-strain curves of the *FSN*: at strain rate of 10-5 time exposition (E) the ductile range sets on formal context (C) in at about 690 % corresponding to a homologous time exposition (E) of about 0.82. At lower strain rates of 10-6 E and below limited ductility can be observed down to formal context about 480 %. As another class of *FSN* the stress-strain curves [45] measured by experiments are similar to those given by Fig. 5. Combining relevant information, formula (1) can be used to well predict experimental curves, e.g. recorded by Fig. 5. The experimental data for stress-strain curves for *FSN* news websites are depicted in Fig. 5, it presents plastic properties. The reason for this is the distinguishing nature of *FSN lattice* between the two phases. Therefore, the influences of periodicity and quasiperiodicity (see Footnote 4) on the plastic deformation lead to an anisotropic behavior.

Fig. 5 Stress-strain curves of the *FSN* in the high formal context (C) range 32–145 at strain rate depicted as percentage respect the time exposition (E)

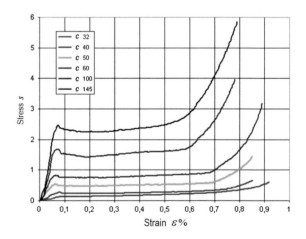

Fig. 6 Quasiperiodic
structure. Penrose tilings and
patterns

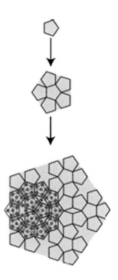

8 Quasiperiodic Order of FSN

FD tags are arranged in a way that we cannot define ordered, but it is not disorderly, we might say "almost" ordered or "nearly" messy. A quasiperiodic[4] structural shown in Fig. 6, or Penrose tile [3, 16, 34, 46], is a structure that is ordered but not periodic cause it lacks the invariance of its equations under any translation—translational symmetry. Quasiperiodic order seems to be much more complex respect periodic order. Hence we could think the formation and growth of quasiperiodic *FSN* to be a rather slow diffusion process. But the experimental observations indicate just the opposite. Quasiperiodic *FSN* can be easily obtained by rapid *FD tags* correlations [16] such as spinning consolidation[5] or collapse,[6] with *Time Exposition* (E) rates of $10^{-3}E$. At highest *Time Exposition* (E) rates periodic structures is formed, at intermediate *Time Exposition* (E) rates we observe quasiperiodic structure, while at lower *Time Exposition* (E) the process of formation of *FD tags* germs of periodic *FSN* takes longest. Some of the quasiperiodics *FSN* obtained by rapid link creations between *FD tags* are stable, however, most of them are metastable and transform into quasiperiodic phases during *Formal Context* (C) merging. The dimensionality seems to play a role in the *FSN* stability of no periodic structures. This view is also supported by the frequently observed toughen of quasiperiodic during the *FSN* construction. Sometimes the process of formation of *FD tags* germs is eased by epitaxial growth.

---

[4]*Quasiperiodicity* in the general definition also includes incommensurately modulated *FSN* as well as composite *FSN*. Here, we will not discuss these cases, which either can be seen as periodic modification of an underlying basic structure or as a kind of intergrowth of periodic structures.

[5]*Spinning consolidation*: the growing of *FD tags* connections around the original *FD tag*.

[6]*Collapse*: when links between *FD tags* shrink together abruptly and completely to a direct link with a main *FD tag*.

9 Dynamic Problems of FSN

In this paragraph we study the dynamic problems of the *Folksodriven Structure Network (FSN)*.

Elastodynamics of *FSN* could be a topic with different points of view due to the contradictions on the role of phason variables in the dynamic process. The phonon[7] field and the phason[8] field can play very different roles in the definition of the links towards *FD tags* of quasiperiodic structures [34]—see Sect. 7 *Macroscopic behavior of plastic FSN lattice.*

On this work we follow the Bak's argument [3] in dynamic study. According to Bak [3] the phason describes particular structure disorders or structure fluctuations and it can be formulated based on a six-dimensional space description. Since there are six continuous symmetries, there exist six vibration modes.

10 Elasticity

The theory of linear elasticity is a branch of continuum mechanics [3, 26, 46], it follows its basic assumptions:

1. *Continuity* A medium is continuous when it is filled the full space that it occupies. Because of this the field variables of the medium are continuous and differentiable functions of coordinates.
2. *Homogeneity* A medium is homogeneous when the physical constants describing the medium are independent of coordinates.
3. *Small deformation* Through small deformation, the Boundary Conditions are expressed respect to the boundaries before their deformation. This simplifies the solution making the problems to be linear.

10.1 Small Deformation Analysis on Ontology Matching

An elastic lattice *FSN* exhibits deformation is connected to the relative movement between points in it that is represented by the displacement field [44]. Considering an elastic lattice *FSN1*, refer to Fig. 7, a *FD tag* defined in a cell *i* turns into another cell *j* after deformation. The point *T* with radius vector r before deformation—respect

---

[7]A *phonon* is a quantum mechanical definition of the lattice vibration that uniformly oscillates at the same frequency. It is known as the "normal mode" in classical mechanics. According to it any arbitrary lattice vibration can be described as a superposition of the elementary vibrations described by the phonon (cfr. Fourier analysis—[12]).

[8]Similar to phonon, *phason* is associated with nodes of lattice motion, considered here as *FD tags*. However, whereas phonons are related to translation of *FD tags*, phasons are associated with *FD tags* rearrangements.

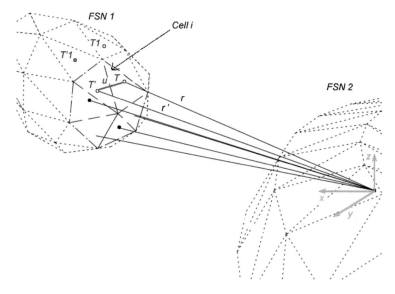

Fig. 7 Displacement of a *FD tag* from point *T* to point *T'* in an elastic lattice defined by the *FSN1* respect the *FSN2*

the ontology matching defined by another *FD tag* on *FSN2*—becomes point *T'* with radius vector *r'* after deformation, and *u* is the displacement vector of point *T* during the deformation process (see Fig. 7) as depict in (2).

$$\vec{r'} = \vec{r} + \vec{u} \tag{2}$$

In Fig. 7, x, y, z depicts the orthogonal coordinate system defined by rectilinear coordinate system from a point of reference in *FSN2*. The orthogonal coordinates x, y, z are dimensionally defined by the *FD tags* dimensions: *Formal Context (C)*, *Time Exposition (E)* and *Resource (R)*. Assume that $T1$ in *FSN1* is a point near the point *T*, the radius vector connecting them is $dr = dx_i$.

The point $T1$ becomes point $T'1$ in cell j' after deformation. The radius vector connecting points $T1$ and point $T'1$ is $dr = dx'i = dx_i + du_i$. The displacement of point $T1$ is u', thus (3).

$$\vec{u'} = \vec{u} + d\vec{u} \quad ; \quad du_i = u'_i - u_i \tag{3}$$

$$du_i = \frac{\partial u_i}{\partial x_i} dx_i; \tag{4}$$

We can express with (4) the Taylor expansion at point *T* taking the first order term only. Under the small deformation assumption, this reaches a very high accuracy.

$$\varepsilon_{ij} = \frac{1}{2}\left(\frac{\partial u_i}{\partial x_i} + \frac{\partial u_j}{\partial x_i}\right) \tag{5}$$

$$w_{ij} = \frac{1}{2}\left(\frac{\partial u_i}{\partial x_i} - \frac{\partial u_j}{\partial x_i}\right) \tag{6}$$

There you can define the partial derivatives as in (8) defined by the components ε_{ij} and ω_{ij}. The symmetric tensor ε_{ij}, where $\varepsilon_{ij} = \varepsilon_{ji}$, is defined by (5) and called the *"strain tensor"*; while (6) represents the asymmetric tensor ω_{ij}, which has only three independent components as depicted in (7).

The *"strain tensor"* ε_{ij} describes the volume and shape change of a cell; while ω_{ij} describes the rigid-body rotation of the cell that is independent respects the deformation.

$$\begin{cases} W_x = w_{yz} = \frac{1}{2}\left(\frac{\partial u_y}{\partial z} - \frac{\partial u_z}{\partial y}\right) \\ W_y = w_{zx} = \frac{1}{2}\left(\frac{\partial u_z}{\partial x} - \frac{\partial u_x}{\partial z}\right) \\ W_z = w_{xy} = \frac{1}{2}\left(\frac{\partial u_x}{\partial y} - \frac{\partial u_y}{\partial x}\right) \end{cases} \tag{7}$$

$$\frac{\partial u_i}{\partial x_i} = \varepsilon_{ij} + w_{ij} \tag{8}$$

$$\frac{\partial \sigma_{ij}}{\partial x_i} + f_i = 0 \tag{9}$$

The *"strain tensor"* ε_{ij} describes the volume and shape change of a cell; while ω_{ij} describes the rigid-body rotation of the cell that is independent respects the deformation. Subsequently, it is sufficient to consider only ε_{ij}. The volume change of a cell is described by the "normal strains", components ε_{xx}, ε_{yy} and ε_{zz}; while the shape change of a cell is described by the "shear strains", components $\varepsilon_{yz} = \varepsilon_{zy}$, $\varepsilon_{zx} = \varepsilon_{xz}$ and $\varepsilon_{xy} = \varepsilon_{yx}$.

10.2   Stress Analysis for Adaptive Matching

To analyze the "adaptive ontology matching" we consider the "stress" on *FSN* caused by internal forces per unit area as a result of deformation defined by σ_{ij}, and consequently it is zero if there is no deformation. It will be adequate for our purposes to consider a "unit cell" volume (Fig. 8) element aligned with the cartesian axes. Component σ_{ij} is defined as the force per unit area acting in the *ei* direction upon the surface normal to *ej*.

Fig. 8 Network that correlates different *FD tags* realized by "unit cell" smallest unit about a subject (e.g.: news websites on football matches, etc.)

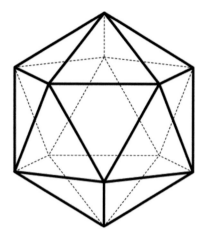

We consider a homogeneous strain in which the strain tensor εij is symmetric and it is the same at every point of the structure considered. When a *FSN* is in static equilibrium we can derive the equilibrium equations from the law of momentum conservation as follows in (9) that occurred for any infinitesimal volume element of the cells composing the *FSN* lattice. In (9) σij represents the components of the stress tensor, and suffix j indicates the component direction while i is the direction of external normal vector of the surface element where the stress component is applied; the density vector of the lattice force is designed by fi.

The "normal stresses" discloses the stress components of σij applied normally to the surfaces of the cells, depicted by σxx, σyy and σzz. While the "shear stresses" represents the components of σij that have the tangent directions of the surface elements, depicted by $\sigma yz, \sigma zy, \sigma zx, \sigma xz, \sigma xy$ and σyx. The *stress* tensor is a symmetric tensor due to the angular momentum conservation (10), which is defined as the "shear stress mutual equal law".

The internal stresses of a cell should be balanced with the strain—area force density. Those play a very important role for elasticity, defined by the stress Boundary Conditions with Eq. (11) where nj is the unit vector along the outside normal to the surface element of a cell.

$$\sigma_{ij} = \sigma_{ji};$$ (10)

$$\sigma_{ij} n_j = T_i$$ (11)

The *FSN* lattice can be drawn with a rainbow grade color from the red to indicate the elastic part and the green for the yield part, to the purple-black for the necking part (Fig. 9) as described in Sect. 5: *Folksodriven as a Network*.

Fig. 9 *Folksodriven*
Structure Network (FSN)
drawn with a *rainbow grade*
color from the *red* to indicate
the elastic part and the *green*
for the yield part, to the
purple-black for the necking
part

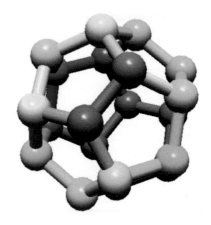

11   Generalized Hooke's Law

In physics *"Hooke's law"* of elasticity is an approximation that states that the extension of a spring is in direct proportion with the load applied to it. Materials for which
Hooke's law is a useful approximation are known as linear-elastic or "Hookean"
materials. Hooke's law[9] in simple terms says that strain εij is directly proportional
to stress σij.

$$\sigma_{ij} = \frac{\partial U}{\partial \varepsilon_{IJ}} = C_{ijkl}\varepsilon_{kl} \tag{12}$$

$$U = F = \frac{1}{2}C_{ijkl}\varepsilon_{ij}\varepsilon_{kl} \tag{13}$$

Between the stresses σij and the strains εij, there exists a certain relationship
depended upon the *"FSN link behavior"*, that for our study is composed by the
FD tags.

On this paper we consider only the linear elastic behavior of *FSN* without initial
stress. For our scope we can generalize the Hooke's law as in (12) where U denotes
the free energy, or the *"strain energy density"* (13). Due to the symmetry of stresses
σij and strains εij, each of them has 6 independent components only, such that there
are 36 independent components of the constant tensor $Cijkl$.

U is a quadratic function of εij, as in (13), considering the symmetry of εij,
we have $Cijkl = Cklij$. So the generalized Hooke's law (12) can describe anisotropic
elastic lattice with 21 independent elastic constants.

---

[9]According to classical physic the Hooke's law (law of elasticity), is depicted by $F = -kx$ Where
the movement of the end of the spring is expressed by **x** respect its equilibrium position. **F** depicts
the spring restoring force and **k** is the spring (or rate) constant.

12 Hyperelasticity Behavior of "malleable" and "frothy" FSN

Elastic deformation is described by an elastic strain tensor, based on the elastic deformation gradient tensor, and it is related to the stress tensor, by the hyperelastic constitutive equation, possessing the elastic strain energy function [29]. Then, the one-to-one correspondence between stress tensor and elastic strain tensor holds and the work done during a closed stress cycle is zero exactly when the plastic strain rate is not included. As cited by Hill [30], a natural generalization of Hook's law, Eq. (14) offers a framework for elastic constitutive laws. The concept of energy conjugated, first presented by Hill [30] states that a stress measure T—as external surface forces density—is said to be conjugate to a strain measure U—as free energy or the "strain" if (14) represents power or rate of change of internal energy per unit reference volume, $\dot{w}$

$$\dot{w} = T : \dot{U} \tag{14}$$

$$T^{(n)} = \partial w / \partial U^{(n)} \tag{15}$$

It was observed as in general, the response of a typical *Folksodriven Network* (*FSN*) is strongly dependent on *Time Exposition (E)* (see Sect. 4: *Folksodriven Notation)* from original to deformed configuration, strain history and loading rate (Fig. 10).

From observation we noted that Folksodriven Network has various kind of behavior that can be identified for a particular Folksodriven Network by applying a variation on stress to the *FSN* and measuring the resulting strain. The ratio of stress variation to strain variation can be determinate as correlated to *Time Exposition (E)*.

At a *Critical Time Exposition (E)* a Network undergoes a change in response with a different behavior.

Fig. 10 Deformation behavior of *Folksodriven Structure Network* (*FSN*)

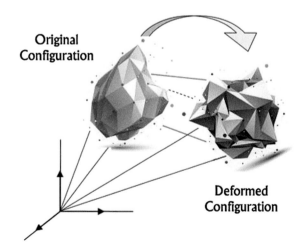

Original Configuration

Deformed Configuration

Above this *Critical Time Exposition (E)* the stress does not depend strongly on strain rate or strain history, and the modulus *increases* with *Time Exposition (E)*—the *Folksodriven Network* (*FSN*) shows "malleable" behavior for the elastic response.

While below this *Critical Time Exposition (E)* the stress depends strongly on strain rate or strain history, and the modulus *decreases* with *Time Exposition (E)*— the *Folksodriven Network* (*FSN*) shows "*frothy*" behavior for the elastic response.

All *FSNs* show this general trend, but the extent of each regime, and the detailed behavior within each regime, depend on the particular structure of the Network.

Hyperelastic constitutive law (15) is used to model *FSN* that respond elastically when subjected to very large strains to approximate this "*malleable*" behavior.

An ideal "malleable" behavior can be represented by heavily cross-linked *Folksodriven Networks* (*FSN*) been the most likely to show capability of being stretched or bent into different shapes.

Features of the behavior of a "*malleable*" *FSN* share some of these properties:

1. The *FSN* network is close to ideally elastic when deformed at constant.
2. *Time Exposition (E)* stress is a function only of current strain and independ- ent of history or rate of loading, or the behavior is reversible.
3. The *FSN* strongly resists volume changes—as increase of the network (*FSN*).
4. The *FSN* is isotropic; its stress-strain response is independent of network orien- tation.

Features of the behavior of a "*frothy*" *FSN* share some of these properties:

1. *FSN* are close to reversible, and show little dependence on strain history and loading rate.
2. Most "frothy" *FSN* are highly compressible, in contrast to "malleable".
3. Frothy have a may stress-true strain response—the strain response in compression deformation is quite different to that in tension.
4. The *FSN* can be anisotropic while with a random connection on a frothy *FSN* structure is isotropic.

Experiments described in the Sect. 16: *Experimental Observation* lead to define the general form of the strain energy density (13) and the power or rate of change of internal energy per unit reference volume (15)—those contains network properties to describe a particular FSN network.

13   Wave Motion of the Elastodynamics of FSN

The elastodynamic equations of the *FSN*, as extension to the classical elastodynamics formulation [3], can be deduced by considering the inertia effect (9) respect the time exposition (E) depicted as partial derivative as in (16). Where ρ is the mass density of the *FD tags* and e is the time exposition of the single *FD tag*. Coherently the adaptive ontology matching, formulated before, and the elastodynamics ontology matching (16) are merged to obtain what we can call the *elasto-adaptative-dynamics methodology* of the *FSN*.

$$\frac{\partial \sigma_{ij}}{\partial x_i} + f_i = \rho \frac{\partial^2 u_i}{\partial e^2} \tag{16}$$

Omitting the body forces inside the *FSN* lattice and considering the Eq. (16)— with the deformation history [13]—we can express the generalized Hooke's as in (17), where λ represents the modulated displacement.

$$\sigma_{ij} = 2\mu\varepsilon_{ij} + \lambda\varepsilon_{kk}\partial_{ij} \tag{17}$$

The strain εij and the stress σij are defined as in (18).

$$\begin{cases} \varepsilon_{kk} = \varepsilon_{11} + \varepsilon_{22} + \varepsilon_{33} = \varepsilon_{xx} + \varepsilon_{yy} + \varepsilon_{zz} \\ \delta_{ij} = \text{unit tensor} \end{cases} \tag{18}$$

The Eq. (19) is a simple form of a wave equation [22] concerning the time expo-sition e, the spatial variables xi and the scalar function u, whose values can model the height of a wave. The speeds of elastic longitudinal wave are represented by $c1$ while the transverse wave is represented by $c2$ as depicted in (20).

$$\left(c_1{}^2 - c_2{}^2\right)\frac{\partial^2 u_i}{\partial e^2} + c_2^2\frac{\partial^2 u_j}{\partial c_i^2} = \frac{\partial^2 u_j}{\partial e^2} \tag{19}$$

$$c_1 = \left(\frac{\lambda + 2\mu}{\rho}\right)^{1/2} ; c_2 = \left(\frac{\mu}{\rho}\right)^{1/2} \tag{20}$$

From (19) we can extrapolate the typical wave equations of mathematical physics (21) defining the scalar potential Φ, the vector potential Ψ and the relation between them—remembering the mathematical relation (22).

$$\nabla^2\Phi = \frac{1}{c_1^2}\frac{\partial^2\Phi}{\partial t^2} ; \nabla^2\Psi = \frac{1}{c_2^2}\frac{\partial^2\Psi}{\partial t^2} \tag{21}$$

The wave equations represent the relation between the adaptive ontology match-ing and the evolving folksonomy driven environment of the *FD tags* respect the *time exposition (E)* defined by the elastodynamics. We call that combination elasto-adaptative-dynamics of the *FSN*.

To define the problem the Boundary Conditions are defined by the strain εij and the stress σij, while the Initial Conditions are defined in (23).

$$\nabla^2 = \frac{\partial^2}{\partial x^2} + \frac{\partial^2}{\partial y^2} + \frac{\partial^2}{\partial z^2} \tag{22}$$

$$\begin{cases} u_i\,(x_i,0) = u_{i0}\,(x_i) \\ \dot{u}_i\,(x_i,0) = \dot{u}_{i0}\,(x_i) \end{cases} x_i \in \Omega \tag{23}$$

Ω denotes the region of the *FSN* lattice we studied.

14   Final Discussion

To sum up analyzing the elastic adaptive ontology matching on evolving folksonomy driven environment we can determine the initial Boundary Conditions represented by the strain εij and the stress σij on the *FD tags*. The advantages of that approach could be used on a new interesting method to be employed by a knowledge management system.

To develop the proposed methodology we used the classical theory of elasticity.

$$\varepsilon_{ij} = \tfrac{1}{2}\left(\frac{\partial u_i}{\partial x_i} + \frac{\partial u_j}{\partial x_i}\right) \quad ; \quad \frac{\partial \sigma_{ij}}{\partial x_i} = \rho\frac{\partial^2 u_j}{\partial e^2} - f_j$$
$$e > 0 \quad ; \quad x_i \in \Omega$$

where we set the *Initial Condition*:

$$u_i\left(x_j,0\right) = u_{io}\left(x_j\right) \qquad \sigma_{ij}n_j = T_i, t > 0, x_j \in \Gamma_t$$
$$\dot{u}_i\left(x_j,0\right) = \dot{u}_{io}\left(x_j\right) \qquad u_i = \overline{u}_i, t > 0, x_j \in \Gamma_u$$
$$x_j \in \Omega$$

While the *Boundaries Conditions* were depicted by the equilibrium equations of macro-plasticity of *FSN* [13]—see Sect. 7: *Macroscopic behavior of plastic FSN lattice*.

We consider $u_{i0}\left(x_j\right),\dot{u}_{i0}\left(x_j\right),T_i,\overline{u}_i$ as known functions.

Ω represents the region of the *FSN lattice* evaluated, Γt and Γu are parts of boundary on which the tractions and displacements are determined having: $\Gamma = \Gamma t + \Gamma u$.

We have a static problem as pure Boundary Condition problem if there are no initial conditions. $\partial^2 u_j/\partial t^2 = 0$.

The study of elasticity of *FSN* strengthens the link between computer science and different branches of physics (e.g.: elasticity, symmetry and symmetry-breaking, crystallography, hydrodynamics, etc.) and applied mathematics (e.g.: partial differential equations, complex variable function method, group theory, discrete geometry, numerical analysis, etc.) and in particular with some numerical methods in continuum mechanics. The study of the elasticity for the *FSN* also promotes the development of defect theories for the plasticity as developed in [16], *elasto-adaptive-dynamics* [13] and Hyperelastic-based Adaptive Dynamics as developed in this work. It is believed that extensive research on the new area will bring forth useful results of theoretical interest and practical importance.

We pointed out at the beginning of this chapter that in the study on the classical plasticity there are two different theories, one is the macroscopic plasticity theory, and the other is so-called crystal plasticity theory, in some extent the latter can be seen as a "microscopic" theory which is based on the mechanism of motion of dislocation.

15 Computational Complexity

We considered the complexity of the above procedure depending on the number of nodes in the kth "unit cells"—see Sect. 7: *Macroscopic behavior of plastic FSN lattice*—which can be also expressed by the number of iterations of the algorithm [13]. Even though it is possible to build examples where this number is large, we often observe in practice that the event where one variable leaves the active set is rare. The complexity also depends on the implementation [50]. The most widely technique used for variable selection is the Principal Component Analysis (PCA), which finds the corresponding projection by which the data show greater variability. For the square loss, the PCA remains the fastest algorithm for small and medium-scale problems on dimensionality reduction, since its complexity depends essentially on the size of the active sets. It computes the whole structure up to a certain scarcity level [49].

For mining large datasets the *FSN* could be used with the technique described by Reshef [41]—that is independent of any assumptions about the data.

16 Experimental Observation

For the purpose of verification and evaluation of the proposed method, we exploit the *Elastic Adaptive Ontology Matching* Algorithm

16.1 Experimental Setup

The experimental observation was done following a structure analysis that consists of three main parts: data collection, structure solution, and structure refinement. It includes a model of the underlying ideal structure and of the deviations from it, based on experimental observations. Such a model can serve as the basis for further modeling and for the derivation of *FSN* properties. Form the descriptions of the methods proposed researchers can carry out similar experiments to verify the results with experimental validation using publicly available test collections.

For the experimental observations it was considered the data collection acquired analyzing the hashtags of Twitter [42]. Hashtags are an example of folksonomies for social networks; they are used to identify groups and topics (the short messages called Tweet in Twitter). Hashtags are neither registered nor controlled by any one user or group of users; they are used to identify groups and topics to identify short messages on microblogging social networking services such as: Twitter, Google+, YouTube, Instagram, Pintrest… They can be used in theoretical perpetuity depending upon the longevity of the word or set of characters hashtags do not contain any set definitions, a single hashtag can be used for any number of purposes.

On December 2014 Twitter had surpassed 255 million monthly active users generating over 500 millions of tweets daily (source: Twitter).

For the Experimental Observation it was considered only 100 K tweets connected to "world news" topics.

For this paper the experimental evidence was carry out on the Twitter #hashtags collections chosen on five different topics thoughts as *Topics of interest* (T) on three different sets of #hashtags as attributes D:

- top fifty hashtags for the "world news" topics,
- group of wellknown ambiguous hashtags (#apple, #desert, #present, #refuse, #stream, #tube, …)
- subjective #hashtags chosen between the most popular ones (#jobs, #business, #britneyspears, #ladygaga, #obama, #oprah, #elearning, #humanrights, #poverty, #health, #green, …)

The Twitter #hashtags were chosen as attributes *Description D* on a *Formal Context C:=(T, D, I)* where the objects T is the *Topic of interest* for those #hashtags and I is a relation between T and D (see Sect. 4: *Folksodriven Notation*). For each #hashtag it was calculated the related *FD tag* and the connections with other *FD tags* to analyze the *Evolving Folksonomy Driven Environment* and the relative stress/strain values on the *FSN* to analyze the *Elastic Adaptive Ontology Matching*.

As described in Sect. 4 the *Folksodriven notation* was used for the content analysis on dynamic ontology matching for the Twitter #hashtags correlated to a topic T (for the evaluation on five different topics chosen).

The connection between the elastic adaptive ontology matching and the *FSN* was observed in experiments where plastic deformation of the network was induced by motion of dislocations in the network (Fig. 11).

In quasiperiodic structure we have to define the global and the local structure. The Local structure considers the *FD tags* arrangement inside the "unit cells" (i.e. the recurrent structural building unit tiles). While the Global structure considers the ordering of the "unit cells" (unit tiles) themselves on a higher hierarchy level. Another challenge for structure determination is disorder. In case of *FSN*, where at least two different "unit cells" make up the structure, disorder in the *FSN* can occur and is entropically favored.

Fig. 11 Here we see a
Penrose tiles structure
coloured with a stress-strain
scale of values during a
plastic deformation of the
FSN network induced by
motion of dislocations in the
network

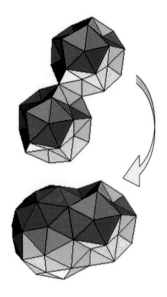

A satisfactory *FSN* is based on experimental data from complimentary methods
such as: degree distribution, clustering coefficient, path-based measures [42]. We
consider the complexity depending on the number of nodes in the kth "unit cells"—
which can be also expressed by the number of iterations of the algorithm, as proposed
by Merholz [38]. The Reshef [41] technique—independent of any assumptions about
the data—was used for mining large datasets of the *FSN*.

16.2 Capability Results

On Fig. 12 it is depicted the graphical representation of the stress-strain relation using
a chromatic scale for the five data-set chosen as *Topics of Interest T*—see Sect. 7
Macroscopic behavior of plastic FSN lattice. The experimental data were obtained
observing tagging for news topics as illustrated in Sect. 17: *Experimental Setup*. The
constitutive equation of dislocation density of formal context and dislocation time
exposition can be used for the structure refinement and so to predict experimental
curves and the *FD tags* that the users could choose.

The stress/strain values on the *FSN* of the *Elastic Adaptive Ontology Matching*
can be used as a folksonomy tags suggestions for the user to solve the problems
inherent in an uncontrolled vocabulary of the folksonomy.

On the experiment it was observed the growth of the quasiperiodic structure
defined by the *FD tags* during the evolving of the time exposition and it was observed
the merge between different kinds of *FSN* to create a new quasiperiodic structure.

Fig. 12 Stress-strain curves
of the *FSN* for 5 different
Topics of Interest T drawn
with a *rainbow grade* color
from the *red* to indicate the
elastic part and the *green* for
the yield part, to the
purple-black for the necking
part

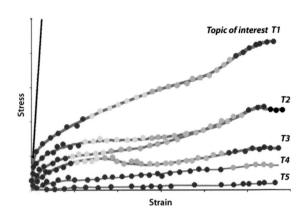

16.3 Performance Results

The system performance was measured in terms of efficiency of the analysis and
matching process.

To measure the efficiency of the *Elastic Adaptive Ontology Matching* a stress test
was done performing the two most expensive tasks occurring at run-time:

- the time needed by the algorithm to analyze a new set of 100K #hashtags from
  Twitter to be deployed and to generate the *Folksodriven tags* (*FD*),
- the time needed to automatically generate matching for the *Folksodriven Structure
  Network* (*FSN*)

The response time needed to perform the *Elastic Adaptive Ontology Matching*
depends mostly on the number of the Twitter #hashtags, chosen as attributes *D*,
respect a *Topic of interest* (*T*), with polynomial complexity as shown in Fig. 13
while, with fixed *T* and *D* (i.e. in a fixed schema), the number of *FD* creating
a *Folksodriven Structure Network* (*FSN*) influences the execution time linearly as
shown in Fig. 14.

Fig. 13 Polynomial
complexity on*Elastic
Adaptive Ontology Matching*
performance

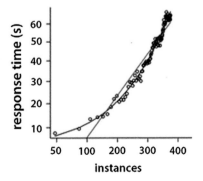

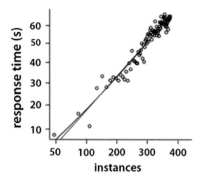

Fig. 14 Linear complexity on *Elastic Adaptive Ontology Matching* performance

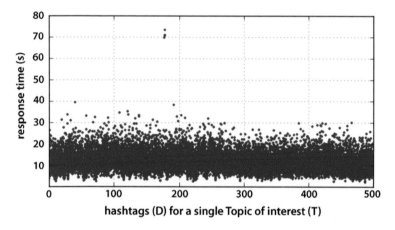

Fig. 15 Distribution of response time obtained by issuing 500 *#hashtags* (D) for a single *Topic of interest* (T)

Figure 15 shows the distribution of response time obtained by issuing 500 Twitter #hashtags (D) for a single *Topic of interest* (T).

16.4 Data Collection Analysis

We exemplify in this section the ability of the reference *FSN* implementation to adapt to the specific and dynamic user community. Using a *FSN* we were able to perform the largest Folksodriven data collection and analysis, to-date. We have gathered in a dataset of 2,543,546 users, based on a set of 3,987,595 users identifiers obtained manually through an accurate gathering process in 2012. We have repeated the process in 2015 using the cloud implementation described in the following Sect. 17 *Experimental Setup*. This time, a total of 4,324,874 unique users were gathered from the system and detailed data was successfully retrieved of 5,674,345 unique user were gathered from the system and detailed data was successfully retrieved for 4,335,756 of them.

16.5 Results Verification

In order to consider the effectiveness of the approach presented here, six recent references, as potential benchmarks, are now tabulated in Tables 1 and 2 to compare the results and consider the pros and cons of the proposed approach.

The potential benchmarks were chosen by the 2015 iteration of the SemEval-2015 Task 10 shared task on Sentiment Analysis in Twitter [42] that can be compared with the FSN algorithm proposed in this chapter. Three benchmarks were chosen by subtasks that ask to predict the sentiment towards a topic in a single tweet and the degree of prior polarity of a phrase (Subtask C of the SemEval-2015 Task 10). While other three benchmarks were chosen by the subtask on the degree of prior polarity of a phrase (Subtask E of the SemEval-2015 Task 10).

Now, by considering the whole of factors in the cases, in a careful manner, it is clear to note that the proposed approach FSN is well behaved on Topic-Level Polarity (Subtask C of the SemEval-2015 Task 10), see Table 1. This proved to be a hard subtask: given a message and a topic, decide whether the message expresses a positive, a negative, or a neutral sentiment towards the topic [35, 47, 52].

Table 1 The proposed approach results verification w.r.t. the potential benchmarks according to the Subtask C of the SemEval-2015: Topic-Level Polarity

	avgDiff	avgLevelDiff
KLUEless	0.202	0.810
FSN	**0.204**	**0.824**
Whu-Nlp	0.210	0.869
TwitterHawk	0.214	0.978

The systems are ordered by the official 2015 score

Table 2 The proposed approach results verification w.r.t. the potential benchmarks according to the Subtask E of the SemEval-2015: Degree of Prior Polarity

	Kendall's τ coefficient	Spearman's ρ coefficient
ONESC-ID	0.6251	0.8172
Ilsislif	0.6211	0.8202
FSN	**0.6109**	**0.8356**
ECNU	0.5907	0.7861

The systems are ordered by their Kendall's τ score, which was the official score

Moreover, on Table 2 is depicted the proposed approach FSN according the degree of prior polarity of a phrase (Subtask E of the SemEval-2015 Task 10)—determining Strength of Association of Twitter Terms with Positive Sentiment. Given a word/phrase, propose a score between 0 (lowest) and 1 (highest) that is indicative of the strength of association of that word/phrase with positive sentiment. If a word/phrase is more positive than another one, it should be assigned a relatively higher score.

17 Experimental Setup

To test a *FSN* in practice we follow the scenario introduced in Sect. 16 *Experimental Observation*. In this scenario, a *FSN* performs continuous analytics users resources leased from one of two clouds, the commercial Amazon Web Services and the Eucalyptus-based cloud installed locally.

17.1 Cloud Infrastructure

The *FSN* is built mainly as cloud architecture around the use of on-demand resource that can dynamically scale up or down depending on application workloads. In practice, we implemented an Infrastructure as a Service (IaaS) in a private cloud computing environment based on HP Helion Eucalyptus and on top of the Amazon Elastic Compute Cloud (EC2) and the Simple Storage Service (S3). A private cloud was set up on the open-source cloud middleware HP Helion Eucalyptus based on Ubuntu Enterprise Cloud. This cloud offers Amazon EC2 services-like allowing us to run the same experiments as on the Amazon cloud. While the Amazon EC2 provides scalable virtual private servers as computational resources to acquire and process Folksodriven data.

One of the main points on the implementation of the *FSN* is the ability to collect and process millions of small web pages. For this workload, the data collection time is dominated by Internet latencies. The latencies can be greatly reduced if the data server is located near the resource collecting data so through a location-aware resource selection. Amazon, like others cloud providers, have multiple site spread across the world that can perform that kind of data collection using on-demand resources near the resources of the web pages.

The Elastic IP Address (EIP) service provided by Amazon EC2 enables the *FSN* implementation to use cloud services that allows the IP address of a leased resource to change in time such as. We have already used several terabytes of data using Amazon S3, stored in labeled "buckets".

The data processing time can be reduced more by different kind of techniques as pipeling the use of data (i.e. through multi-threating), or by grouping data and storing it in larger chunks—as dome by Google' File System [40].

17.2   Implementation

The reference *FSN* implementation was written in Python, which makes it portable for a wide range of platforms but with a lower performance than can be archived in other programming languages, such as C. We have written specific web crawlers for the data collection process.

For enabling SP3 and EIPs support we have used Boto (http://code.google.com/p/boto), a Python interface to Amazon Web Services for the Elastic Computer Cloud (EC2) and Simle Storage Service (S3).

18   Conclusions and Future Development

The present research attempts to address an efficient approach in the area of Computational Intelligence for Sentiment Analysis and Ontology Engineering. The main purpose of this chapter is the development of a constitutive model emphasizing the use of ontology-driven processing to achieve a better understanding of the contextual role of concepts (Sentic computing). It is realized based on a hyperelastic-based algorithm to estimate elastic adaptive ontology matching on evolving folksonomy driven environment.

To date, ontology variance estimation has turned into an important aspect of Computational Intelligence process and continually is subject to many recent results. In the approach proposed here, firstly, a new concept structure called "Folksodriven" to represent folksonomies is accurately identified. To face the problem of an adaptive ontology matching, a *Folksodriven Structure Network* (*FSN*) is built to identify the relations among the *Folksodriven tags*.

The present algorithm finds the deformation on *FSN* based on folksonomy tags chosen by different user on web site resources; this is a topic which has not been well studied so far. For this, a constitutive law on *FSN* is investigated towards a systematic mathematical analysis on stress analysis and equations of motion for an evolving ontology matching on an environment defined by the users' folksonomy choice.

The outlines have utilized to present a low-complexity approach, as long as the investigated results guarantee that the proposed approach is able to estimate how the properties of the *FSN* depend mainly on the nature, distribution, size and the quality of the reinforcing Folksodriven tags (*FD tags*).

Through a series of experiments, the ability of the approach is shown to be able to estimate the relations among the *FD tags* for adaptive folksonomies classifications in an evolving environment used by Intelligent Systems.

Afterwards, the proposal is applied to use an adaptive ontology matching to understand what knowledge is required for a task domain for an evolving environment.

The discussion in the paper shows that the nonlinear elastic constitutive equation possesses some leaning for the investigation due to lack of plastic constitutive equation at present. An updated place for all sorts of information about the method presented is the Web site: http://www.folksodriven.com/hyperelastic-adaptive-dynamics-methodology where it's possible to see a video describing the features on the system proposed and the video recording of the presentation at the 2012 IEEE Conference on Evolving and Adaptive Intelligent System (EAIS 2012), Madrid, Spain

References

1. Antoniou, G., Groth, P., van Harmelen, F., Hoekstra, R.: A Semantic Web Primer. Cooperative Information Systems series (2012)
2. Asur, S., Huberman, B.: Predicting the future with social media. In: 2010 IEEE/WIC/ACM International Conference on Web Intelligence and Intelligent Agent Technology (WI-IAT), vol. 1. IEEE (2010)
3. Bak, P.: Symmetry, stability and elastic properties of icosahedral in commensurate crystals. Phys. Rev. B **32**(9), 5764–5772 (1985)
4. Barabási, A., Vazquez, A., Oliveira, J., Goh, K., Kondor, I., Dezso, Z.: Modeling bursts and heavy tails in human dynamics. Phys. Rev. E **73**(3) (2006)
5. Bezdek, J.C.: What is computational intelligence? Computational Intelligence Imitating Life, pp. 1–12. IEEE Press, New York (1994)
6. Born, M., Huang, K.: Dynamic Theory of Crystal Lattices. Oxford Clarendon Press (1954)
7. Burnap, P., Gibson, R., Sloan, L., Southern, R.: 140 Characters to Victory?: Using Twitter to predict the UK 2015 General Election (2015). arXiv preprint arXiv:1505.01511
8. Calliard, D.: Dislocation mechanism and plasticity of quasicrystals: TEM observations in icosahedral Al-Pd-Mn. Mater. Sci. Forum **509**(1), 50–56 (2006)
9. Cattuto, C., Schmitz, C., Baldassarri, A., Servedio, V.D.P., Loreto, V., Hotho, A., Grahl, M., Summe, G.: Network properties of folksonomies. In: Proceedings of the WWW2007 International World Wide Web Conference (2007)
10. Chakrabarty, J.: Theory of plasticity. Butterworth-Heinemann (2009)
11. Chen, Z.: Computational Intelligence for Decision Support. CRC Press (2000)
12. Courant, R., Hilbert, D.: Methods of Mathematical Physics, vol. 2. Interscience Wiley (2008)
13. Dal Mas, M.: Elastic adaptive dynamics methodology on ontology matching on evolving folksonomy driven environment. Evolving Syst. J. **5**(1), 33–48 (2014). (http://link.springer.com/article/10.1007%2Fs12530-013-9086-5)
14. Dal Mas, M.: Cluster analysis for a scale-free folksodriven structure network. Accepted for the International Conference on Social Computing and its Applications (SCA 2011) (2011). http://arxiv.org/abs/1112.4456
15. Dal Mas, M.: Elastic adaptive ontology matching on evolving folksonomy driven environment In: Proceedings of the IEEE Conference on Evolving and Adaptive Intelligent Systems EAIS 2012, pp. 35–40. Madrid, Spain (2012)
16. Dal Mas, M.: Elasticity on ontology matching of folksodriven structure network. Accepted for the CIIDS 2012—Kaohsiung Taiwan R.O.C. (2012). http://arxiv.org/abs/1201.3900
17. Dal Mas, M.: Folksodriven Structure Network. In: Ontology Matching Workshop (OM-2011) collocated with the 10th International Semantic Web Conference (ISWC-2011), CEUR WS, vol. 814 (2011). (http://arxiv.org/abs/1109.3138), (http://ceur-ws.org/Vol-814/)
18. Dal Mas, M.: Intelligent interface architectures for folksonomy driven structure net. In: Proceedings of the 5th International Workshop on Intelligent Interfaces for Human-Computer Interaction (IIHCI-2012), pp. 519–525. IEEE, Palermo, Italy (2012). doi:10.1109/CISIS.2012. 158, (http://ieeexplore.ieee.org/876xpl/articleDetails.jsp?arnumber=6245653)

19. Dal Mas, M.: Layered ontological image for intelligent interaction to extend user capabilities on multimedia systems in a folksonomy driven environment. In: Smart Innovation, Systems and Technologies, vol. 36, pp. 103–122. Springer (2015). doi:10.1007/978-3-319-17744-1_7, (http://link.springer.com/chapter/10.1007%2F978-3-319-17744-1_7)
20. Dal Mas, M.: Ontology Temporal Evolution for Multi-entity Bayesian Networks under Exogenous and Endogenous Semantic (2010). http://arxiv.org/abs/1009.2084
21. Damme, C.V., Hepp, M., Siorpaes, K.: FolksOntology. In: Proceedings of the ESWC Workshop 'Bridging the Gap between Semantic Web and Web 2.0, pp. 57–70 (2007)
22. Ding, D.H., Yang, W.G., Hu, C.Z.: Generalized elastic theory of quasicrystals. Phys. Rev. B 48(10), 7003–7010 (1993)
23. Duch, W.: Towards comprehensive foundations of computational intelligence. Challenges for Computational Intelligence. Springer, Berlin (2007)
24. García-Silva, A., Corcho, O., Alani, H., Gómez-Pérez, A.: Review of the state of the art: discovering and associating semantics to tags in folksonomies. J. Knowl. Eng. Rev. Arch. 27(1), 57–85 (2012)
25. Grandi, F.: Introducing an annotated bibliography on temporal and evolution aspects in the semantic web. ACM SIGMOD Rec. 41(4) (2012)
26. Grätzer, G.A.: General Lattice Theory. Birkhäuser, Basel (2003)
27. Grunbaum, B., Shephard, G.C.: Tilings and Patterns. Dover Publications (2011)
28. Han, W., Reddy, B.D.: Computational plasticity: the variational basis and numerical analysis. Comput. Mech. Adv. 2(4) (1995)
29. Hashiguchi, K.: Elastoplasticity Theory. Springer (2013)
30. Hill, R.: On constitutive inequalities for simple materials. Int. J. Mech. Phys. Solids 16(1968), 229–242 (1968)
31. Hitzler, P., Krötzsch, M., Rudolph, S.: Foundations of Semantic Web Technologies. Chapman & Hall/CRC Textbooks in Computing (2011)
32. Jacob, E.K.: Classification and categorization: a difference that makes a difference. Libr. Trends 52(3), 515–540 (2004)
33. Kitchin, R.: Big data, new epistemologies and paradigm shifts. Big Data & Society 1.1, 2053951714528481 (2014)
34. Levine, D., Steinhardt, P.J.: Quasicrystals: a new class of ordered structures. Phys. Rev. 53, 2477–2480 (1984)
35. Liu, B.: Sentiment Analysis: Mining Opinions, Sentiments, and Emotions. Cambridge University Press (2005)
36. Lu, P.J., Steinhardt, P.J.: Decagonal and quasi-crystalline tilings in medieval islamic architecture. Science 315(2007), 1106–1110 (2007)
37. McCarthy, J.: The future of AI-A manifesto. AI Magazine 26 (2005)
38. Merholz, P.: Metadata for the masses (2004). http://adaptivepath.com/ideas/e000361
39. Mussel, L.: Conformazione e struttura di bio-polimeri. Dip. di Scienze Chimiche—Università di Padova—Padova Digital University Archive (1964)
40. Ostermann, S., Iosup, A., Yigitbasi, M.N., Prodan, R., Fahringer, T., Epema, D.: An early performance analysis of cloud computing services for scientific computing. In: CloudComp, LNICST, vol. 34, pp. 1–10, (2009)
41. Reshef, D.N., Reshef, Y.A., Mitzenmacher, M., Sabeti, P.C.: Detecting novel associations in large data sets. Science 334(6062), 1518–1524 (2011). doi:10.1126/science.1205438
42. Rosenthal, S., Nakov, P., Kiritchenko, S., Mohammad, S.M., Ritter, A., Stoyanov, V.: SemEval-2015 task 10. In: Proceedings of the 9th International Workshop on Semantic Evaluation, SemEval '2015, Denver, Colorado, June 2015
43. Ruths, D., Pfeffer, J.: Social media for large studies of behavior. Science 346(6213), 1063–1064 (2014)
44. Sands, D.E.: Vectors and tensors in crystallography. Dover Publications (2002)
45. Simo, J.C., Ju, J.W.: Strain- and stress-based continuum damage models–II. Computational aspects Department of Mechanical Engineering, Stanford University Press (1986)
46. Steurer, W., Deloudi, S.: Crystallography of Quasicrystals. Springer (2009)

47. Thurner, S., Kyriakopoulos, F., Tsallis, C.: Unified model for network dynamics exhibiting nonextensive statistics. Phys. Rev. (2007)

48. Trant, J.: Studying social tagging and folksonomy: a review and framework. J. Digit. Inf. **10**(1) (2009)

49. van der Maaten, L.J.P., Postma, E.O., van den Herik, H.J.: Dimensionality reduction: a comparative review. Tilburg University Technical report, TiCC-TR 2009-005 (2009)

50. van Melkebeek, D.: Special Issue "Conference on Computational Complexity 2010". Computational Complexity vol. 20, no. 2, pp. 173–175. Springer, Basel (2010)

51. White, C.: Using big data for smarter decision making. BI research (2011)

52. Wilson, T., Wiebe, J., Hoffmann, P.: Recognizing contextual polarity in phrase-level sentiment Analysis. In: Proceedings of HLT-EMNLP-2005 (2005)

Sentiment-Oriented Information Retrieval: Affective Analysis of Documents Based on the SenticNet Framework

Federica Bisio, Claudia Meda, Paolo Gastaldo, Rodolfo Zunino
and Erik Cambria

Abstract Sentiment analysis research has acquired a growing importance due to its applications in several different fields. A large number of companies have included the analysis of opinions and sentiments of costumers as a part of their mission. Therefore, the analysis and automatic classification of large corpora of documents in natural language, based on the conveyed feelings and emotions, has become a crucial issue for text mining purposes. This chapter aims to relate the sentiment-based characterization inferred from books with the distribution of emotions within the same texts. The main result consists in a method to compare and classify texts based on the feelings expressed within the narrative trend.

Keywords Sentiment analysis · Text mining · Senticnet · SLAIR

1 Introduction

Sentiment analysis or opinion mining can be defined as a particular application of data mining, which aims to aggregate and extract emotions and feelings from different types of documents.

In recent years sentiment analysis has been applied especially to social networks, to analyze opinions on Twitter, Facebook or other digital communities in real time

F. Bisio (✉) · C. Meda · P. Gastaldo · R. Zunino
DITEN – University of Genoa, Via Opera Pia 11A, 16145 Genoa, Italy
e-mail: federica.bisio@edu.unige.it

C. Meda
e-mail: claudia.meda@edu.unige.it

P. Gastaldo
e-mail: paolo.gastaldo@unige.it

R. Zunino
e-mail: rodolfo.zunino@unige.it

E. Cambria
SCE – School of Computer Engineering, Nanyang Technological University, 50 Nanyang Ave, Singapore 639798, Singapore
e-mail: cambria@ntu.edu.sg

© Springer International Publishing Switzerland 2016   175
W. Pedrycz and S.-M. Chen (eds.), *Sentiment Analysis and Ontology Engineering*,
Studies in Computational Intelligence 639, DOI 10.1007/978-3-319-30319-2_8

[21, 39, 50]. The emerging field of big (social) data analysis deals with information retrieval and knowledge discovery from natural language and social networks. Graph mining and natural language processing (NLP) techniques contribute to distill knowledge and opinions from the huge amount of information on the World Wide Web.

Sentiment analysis can enhance the capabilities of customer relationship management and recommendation systems by allowing, for example, to find out which features customers are particularly interested in or to exclude from ads or recommendations items that have received unfavorable feedbacks. Likewise, it can be used in social communication to enhance anti-spam systems.

Business intelligence can also benefit from sentiment analysis. Since predicting the attitude of the public towards a brand or a product has become of crucial importance for companies, an increasing amount of money is invested in marketing strategies involving opinion and sentiment mining.

That scenario led to sentic computing [10], which tackles those crucial issues by exploiting affective common-sense reasoning, i.e., the intrinsically human capacity to interpret the cognitive and affective information associated with natural language. In particular, sentic computing leverages on a common-sense knowledge base built through crowdsourcing [16, 19]. Common-sense is useful in many different computer-science applications including data visualization [11], text recognition [64], and affective computing [9]. In this context, common-sense is used to bridge the semantic gap between word-level natural language data and the concept-level opinions conveyed by these [15].

To perform affective common-sense reasoning [4], a knowledge database is required for storing and extracting the affective and emotional information associated with word and multi-word expressions. Graph-mining [13] and dimensionality-reduction [7] techniques have been employed on a knowledge base obtained by blending ConceptNet [56], a directed graph representation of common-sense knowledge, with WordNet-Affect (WNA) [58], a linguistic resource for the lexical representation of feelings. Unlike WNA, SenticNet [14] exploits an ensemble of common and common-sense knowledge to go beyond word-level opinion mining and, hence, to associate semantics and sentics to a set of natural language concepts.

The task of extracting emotions from a large corpus of natural language texts can be viewed from a different point of view: instead of extracting feelings from data in order to assign different sentiment categories to the analyzed corpuses (e.g., tweets, posts, messages), one may try and identify the similarities between any text based on their sentiment distribution. This work presents both a *semantic descriptor* using the SenticNet emotional database [14] and a *sentiment distance metric* which can characterize the emotional states of different documents (i.e., books), and classify them based on their sentiment distributions.

The experiments were performed implementing these functionalities within a text-mining tool called SLAIR [51] in order to automate the extraction of emotional distributions from different categories of books and classify them based on their feelings trend.

To test the proposed approach on a dataset of documents dense of sentimental and affective contents, the experiments involved a dataset of books selected from different literary genres. This approach provided a group of texts able to convey clear feelings and emotions, while offering at the same time a novel field of analysis.

In order to clarify the scope, the following example sentences convey different emotions:

(A)—"Pain and suffering are always inevitable for a large intelligence and a deep heart. The really great men must, I think, have great sadness on earth."

(B)—"Seldom, very seldom, does complete truth belong to any human disclosure; seldom can it happen that something is not a little disguised or a little mistaken."

(C)—"Never forget what you are, for surely the world will not. Make it your strength. Then it can never be your weakness. Armour yourself in it, and it will never be used to hurt you."

Reading these sentences, it is possible to notice how excerpts from different books can convey similar emotions, even if they were written by different authors in different works. For example, one notes an emotional alignment between sentence (A) and (B), intuitively suggested by the presence of deep thoughts about humanity. Indeed, sentences (A) and (B) were drawn from the same book genre (XIX century novels); sentence (A) was extracted from "Crime and Punishment" by Dostoevskij, whereas sentence (B) was extracted from "Emma" by Jane Austen. On the contrary, sentence (C) was extracted from "Game of Thrones" by G.R.R. Martin and clearly conveys a heartening and encouraging tone, although yet involving again a reflection on humans. Details on the distance metric results related to these sentences will be given later in the experimental results section.

The rest of this chapter is organized as follows: Sect. 2 introduces related work in the field of sentiment analysis research; Sect. 3 describes the affective resource SenticNet; Sect. 4 illustrates the techniques employed for the analysis of the feelings distribution; Sect. 5 explains the experimental setup and the metrics used; Sect. 6 illustrates experimental results; finally, Sect. 7 offers some concluding remarks.

2 Sentiment Analysis and Opinion Mining

The Social Web has provided people with new content-sharing services that allow to create and share personal contents, ideas and opinions, in a time- and cost-efficient way, with virtually millions of other people connected to the World Wide Web. Since this amount of information is mainly unstructured, research has so far focused on online information retrieval, aggregation, and processing. Moreover, when it comes to interpreting sentences and extracting meaningful information, these tasks become very critical. NLP requires high-level symbolic capabilities [25], including:

- creation and propagation of dynamic bindings;
- manipulation of recursive, constituent structures;
- acquisition and access of lexical, semantic, and episodic memories;
- representation of abstract concepts.

All these capabilities are required to shift from mere NLP to what is usually called natural language understanding [1].

Therefore, opinion mining and sentiment analysis have recently emerged as a challenging and active field of research, due to many open problems and a wide variety of practical applications. The potential applications of concept-level sentiment analysis are indeed countless and span interdisciplinary areas, such as stock market prediction, political forecasting, social network analysis, social stream mining, and man-machine interactions.

Today, most of the existing approaches still rely on word co-occurrence frequencies, i.e., the syntactic representation of text. Therefore, computational models aim to bridge the cognitive gap by emulating the way the human brain processes natural language. For instance, by leveraging on semantic features that are not explicitly expressed in text, one may accomplish complex NLP tasks such as word-sense disambiguation, textual entailment, and semantic role labeling. Computational models are useful for both scientific purposes (such as exploring the nature of linguistic communication), and practical purposes (such as enabling effective human-machine communication).

Most existing approaches to sentiment analysis rely on the extraction of a vector representing the most salient and important text features, which is later used for classification purposes. Commonly used features include term frequency, presence and the position of a token in a text. An extension of this feature is the presence of n-grams, typically bi-grams and tri-grams. Other methods rely on the distance between terms or on the Part-of-Speech (POS) information: for example, certain adjectives may be good indicators of sentiment orientation. A drawback of these approaches is the strict dependency on the considered domain of application and the related topics.

Sentiment analysis systems aim to classify entire documents by associating contents with some overall positive or negative polarity [44] or rating scores (e.g., 1–5 stars) [42]. These approaches are typically supervised and rely on manually labelled samples. We can distinguish between knowledge-based systems [17], based on approaches like keyword spotting and lexical affinity, and statistics-based systems [18]. At first, the identification of emotions and polarity was performed mainly by means of knowledge-based methods; recently, sentiment analysis researchers have been increasingly using statistics-based approaches, with a special focus on supervised statistical methods.

Keyword spotting is the most straightforward, and possibly also the most popular, approach thanks to its accessibility and economy. Text is classified into affect categories based on the presence of fairly unambiguous "affect words" like 'happy', 'sad', 'afraid', and 'bored'. Elliott's Affective Reasoner [26], for example, watches for 198 affect keywords plus affect intensity modifiers and a handful of cue phrases. Other popular sources of affect words are Ortony's Affective Lexicon [40], which groups terms into affective categories, and Wiebe's linguistic annotation scheme [66]. The crucial issue of this approaches lies in the ineffectiveness at handling negations and in the structure based on the presence of obvious affect words.

Rather than simply detecting affect words, lexical affinity assigns each word a probabilistic 'affinity' for a particular emotion. These probabilities are usually learnt from linguistic corpora [48, 55, 57]. Even if this method often outperforms pure keyword spotting, it still works at word level and can be easily tricked by negations and different senses of the same word. Besides, lexical affinity probabilities are often biased by the linguistic corpora adopted, which makes it difficult to develop a reusable, domain-independent model.

Statistical based approaches, such as Bayesian inference and support vector machines (SVM), have been used on several projects [24, 27, 44, 65]. By feeding a machine learning algorithm [63] a large training corpus of affectively annotated texts, it is possible not only to learn the affective valence of affect keywords (as in the keyword spotting approach), but the one of other arbitrary keywords (like lexical affinity), punctuation, and word co-occurrence frequencies. Anyway, it is worth noticing that statistical classifiers work well only when a sufficiently large text is given as input. This is due to the fact that, with the exception of affect keywords, other lexical or co-occurrence elements possess a little predictive value individually.

For example, Pang et al. [44] used the movie review dataset to compare the performance of different machine learning algorithms: in particular, they obtained 82.90 % of accuracy employing a large number of textual features. Socher et al. [54] proposed a recursive neural tensor network (RNTN) and improved the accuracy (85 %). Yu and Hatzivassiloglou [68] identified polarity at sentence level using semantic orientation of words. Melville et al. [36] developed a framework exploiting word-class association information for domain-dependent sentiment analysis. Reference [3] tackled a particular aspect of the sentiment classification problem: the ability of the framework itself to operate effectively in heterogeneous commercial domains. The approach adopts a distance-based predictive model to combine computational efficiency and modularity.

Some approaches exploit the following fact: many short n-grams are usually neutral while longer phrases are well distributed among positive and negative subjective sentence classes. Therefore, matrix representations for long phrases and matrix multiplication to model composition can also be used to evaluate sentiment. In such models, sentence composition is modeled using deep neural networks such as recursive auto-associated memories. Recursive neural networks (RNN) predict the sentiment class at each node in the parse tree and try to capture the negation and its scope in the entire sentence.

Several unsupervised learning approaches have also been proposed and rely on the creation of the lexicon via the unsupervised labeling of words or phrases with their sentiment polarity or subjectivity [43]. To this aim, early works were mainly based on linguistic heuristics. For example, Hatzivassiloglou and McKeown [28] built a system based on opposition constraints to help labeling decisions, in the case of polarity classification.

Other works exploited the seed words, defined as terms for which the polarity is known, and propagated them to terms that co-occur with them in general text, or in specific WordNet-defined relations. Popescu and Etzioni [46] proposed an algorithm that, starting from a global word label computed over a large collection of generic topic text, gradually tried to re-define such label to a more and more specific corpus, until the one that is specific to the particular context in which the word occurs. Snyder and Barzilay [53] also exploited the idea of utilizing discourse information to aid the inference of relationships between product attributes.

Regression techniques are often employed for the prediction of the degree of positivity in opinionated documents since they allow for modeling classes that correspond to points on a scale, such as the number of stars given by a reviewer [43]. Works like [41] attempted to address the problem via incorporating location information in the feature set and underlined the importance of the last sentences of a review in the calculation of the overall sentiment of a document.

One problem is represented by the fact that contrary attitudes can be present in the same document: therefore, a fine-grained level analysis can be useful to distinguish sentimental from non-sentimental sections, e.g., by using graph-based techniques for segmenting sections of a document on the basis of their subjectivity [41], or by performing a classification based on some fixed syntactic phrases that are likely to be used to express opinions [62], or by bootstrapping using a small set of seed opinion words and a knowledge base such as WordNet [31] or, finally, using unsupervised methodolgies [29].

In recent works, text analysis granularity has been taken down to sentence level, e.g., by using presence of opinion lexical items to detect subjective sentences [32, 49], or by using semantic frames defined in FrameNet [2, 33]. Since an author usually does not switch too frequently between adjacent sentences, a certain continuity level is preserved and some works propose a classification of the document based on assigning preferences for pairs of nearby sentences [42, 69].

Concept-based approaches [6, 30, 47, 61] focus on a semantic analysis of text through the use of web ontologies or semantic networks, which allow the handling of conceptual and affective information associated with natural language opinions. By relying on large semantic knowledge bases, such approaches step away from blind use of keywords and word co-occurrence counts, but rather rely on the implicit meaning/features associated with natural language concepts. Unlike purely syntactical techniques, concept-based approaches are also able to detect sentiments that are expressed in a subtle manner, e.g., through the analysis of concepts that do not explicitly convey any emotion, but are implicitly linked to other concepts that do so.

More recent studies [20, 23, 34, 37] enhanced sentiment analysis of tweets by exploiting microblogging text or Twitter-specific features such as emoticons, hashtags, URLs, @symbols, capitalizations, and elongations. Tang et al. [59] developed a convolutional neural network based approach to obtain word embeddings for the words mostly used in tweets. These word vectors were then fed to a convolutional neural network for sentiment analysis. A deep convolutional neural network for sentiment detection in short text was also proposed by Santos et al. [52]. The approach based on Sentiment Specific Word Embeddings [60] considers word embeddings based on a sentiment corpora: this means including more affective clues than regular word vectors and producing a better result.

Finally recent researches faced the problem of identifying literary texts based on certain textual characteristics in common, for example [22], which limited the analysis to Dutch novels, or [38], that analyzed the narrative emotion related to the relationship of characters described into two different novels.

3   SenticNet

SenticNet[1] is a publicly available semantic and affective resource for concept-level sentiment analysis. The last release SenticNet 3 exploits 'energy flows' to connect different parts of both common and common-sense knowledge representations to one another, unlike standard graph-mining and dimensionality-reduction techniques. SenticNet 3 therefore models semantics and sentics (that is, the conceptual and affective information associated with multi-word natural language expressions).

To this aim, SenticNet 3 employs an energy-based knowledge representation [14] to provide the semantics and sentics associated with 30,000 concepts, thus enabling a fine-grained analysis of natural language opinions. SenticNet 3 contains both unambiguous adjectives as standalone entries (like 'good' and 'awful') and non-trivial multi-word expressions such as 'small room' and 'cold bed'. This is due to the fact that while unambiguous adjectives convey positive or negative polarities (whatever noun they are associated with), other adjectives are able to carry a specific polarity only when coupled with certain nouns.

SenticNet 3 focuses on the use of 'energy' or information flows to connect various parts of common and common-sense knowledge representations to one another. Each quantum of energy possesses a scalar magnitude, a valence (binary positive/negative), and an edge history, defined as a list of the edge labels that a particular quantum of energy has traversed in the past. Essentially, common and common-sense knowledge is broken down into 'atoms', thus allowing the fusing of data from different knowledge bases without requiring any ontology alignment.

3.1   Sources of Knowledge

SenticNet 3 embeds both common and common-sense knowledge, in order to boost sentiment analysis tasks such as feature spotting and polarity detection, respectively. In particular, it is generated by an ensemble of different methods. Regarding the common knowledge bases, the employed resources are either crafted by human experts or community efforts, such as DBPedia [5], a collection of 2.6 million entities extracted from Wikipedia, or automatically-built knowledge bases, such as Probase [67], Microsoft's probabilistic taxonomy counting about 12 million concepts learned iteratively from 1.68 billion web pages in Bing web repository.

Regarding common-sense knowledge, Open Mind Common Sense (OMCS) collects pieces of knowledge from volunteers on the Internet by enabling them to enter common-sense into the system with no special training or knowledge of computer science. OMCS has exploited these pieces of common-sense knowledge to build ConceptNet [56], a semantic network of 173,398 nodes.

---

[1]http://sentic.net/sentics.

3.2   Structure of SenticNet

The aggregation of common and common-sense knowledge bases is designed as a 2-stage process in which different pieces of knowledge are first translated into RDF triples and then inserted into a graph. Considering as an example 'Pablo Picasso is an artist', we obtain the RDF triple <Pablo Picasso-isA-artist> and, hence, the entry [artist—SUBSUME → Pablo Picasso]. In this way we obtain a shared representation for common and common-sense knowledge, thus performing a conceptual decomposition of relation types, i.e., the unfolding of relation types that are usually opaque in natural-language-based resources.

After low confidence score trimming and duplicates removal, the resulting semantic network (built out of about 25 million RDF statements) contains 2,693,200 nodes. Of these, 30,000 affect-driven concepts (that is, those concepts that are most highly linked to emotion nodes) have been selected for the construction of SenticNet 3.

SenticNet 3 conceptualizes the information as 'energy' and sets up pathways upon which this energy may flow between different semantic fragments. In this way, complex concepts can be built upon simpler pieces by connecting them together via energy flows. Once an element is reached by a certain quantum of energy flow, it is included in a wider concept representation, thus enabling simple elements to deeply affect larger conceptual connections. Such a representation is optimal for modeling domains characterized by nuanced, interconnected semantics and sentics (including most socially-oriented AI modeling domains).

Each quantum of energy possesses a scalar magnitude, a valence (binary positive/negative), and an edge history, defined as a list of the edge labels that a particular quantum of energy has traversed in the past. These three elements describe the semantics and sentics of the quantum of energy and they are extracted for each concept of the semantic network.

In particular, the extraction of semantics and sentics is achieved through multiple steps of spreading activation with respect to the nodes representing the activation levels of the Hourglass of Emotions [12], a brain-inspired model for the representation and the analysis of human emotions.

3.3   The Hourglass of Emotions

The Hourglass of Emotions is an affective categorization model developed starting from Plutchik's studies on human emotions [45]. The main advantage over other emotion categorization models is that it allows emotions to be deconstructed into independent but concomitant affective dimensions, whose different levels of activation make up the total emotional state of the mind. Such a modular approach to emotion categorization allows different factors (or energy flows) to be concomitantly taken into account for the generation of an affective state.

Table 1 The sentic levels of the Hourglass model

Interval	Pleasantness	Attention	Sensitivity	Aptitude
[G(1), G(2/3))	Ecstasy	Vigilance	Rage	Admiration
[G(2/3), G(1/3))	Joy	Anticipation	Anger	Trust
[G(1/3), G(0))	Serenity	Interest	Annoyance	Acceptance
(G(0), G(−1/3)]	Pensiveness	Distraction	Apprehension	Boredom
(G(−1/3), G(−2/3)]	Sadness	Surprise	Fear	Disgust
(G(−2/3), G(−1)]	Grief	Amazement	Terror	Loathing

The model can potentially synthesize the full range of emotional experiences in terms of four affective dimensions, Pleasantness, Attention, Sensitivity, and Aptitude, which determine the intensity of the expressed/perceived emotion as a $float \in [-1, +1]$. Each affective dimension is characterized by six levels of activation, termed 'sentic levels', which are also labeled as a set of 24 basic emotions (six for each affective dimension) (Table 1).

Previous works [8] already proved that a categorization model based on these four affective dimensions is effective in the design of an emotion categorization architecture.

The transition between different emotional states is modeled, within the same affective dimension, using the function $G(x) = -\frac{1}{\sigma\sqrt{2\pi}}e^{-x^2/2\sigma^2}$, for its symmetric inverted bell curve shape that quickly rises up towards the unit value. In particular, the function models how valence or intensity of an affective dimension varies according to different values of arousal or activation, spanning from null value (emotional void) to the unit value (heightened emotionality). Mapping this space of possible emotions leads to a hourglass shape (Fig. 1).

Complex emotions can be synthesized by using different sentic levels, as shown in Table 2.

3.4 Semantics and Sentics Representation

The RDF triples are encoded in a XML format, in order to represent SenticNet 3 in a machine-processable way. For each concept, semantics and sentics are provided.

Given the concept 'celebrate special occasion', for example, SenticNet 3 provides a set of semantically related concepts, e.g., 'celebrate holiday', 'celebrate occasion' or 'celebrate birthday'. The resource also provides a sentic vector specifying Pleasantness, Attention, Sensitivity, and Aptitude associated with the concept (for tasks such as emotion recognition), a polarity value (for tasks such as polarity detection).

Fig. 1 The 3D model of the Hourglass of emotions

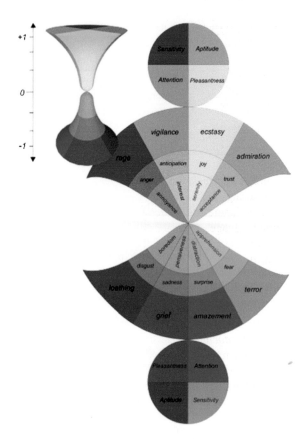

Table 2 The emotions generated by pairwise combination of the sentic levels of the Hourglass model

	Attention > 0	Attention < 0	Aptitude > 0	Aptitude < 0
Pleasantness > 0	Optimism	Frivolity	Love	Gloat
Pleasantness < 0	Frustration	Disapproval	Envy	Remorse
Sensitivity > 0	Aggressiveness	Rejection	Rivalry	Contempt
Sensitivity < 0	Anxiety	Awe	Submission	Coercion

4 Sentiment Distribution: A Different Point of View

From a sentiment analysis perspective, the present chapter aims to study and identify the best similarity metric able to describe the sentiment distribution of several types of books, establishing a different point of view on the interpretation of feeling extraction: the classification of documents based on an emotional distance (Fig. 2).

In order to implement this idea, an existing text miner application is employed: the SeaLab Advanced Information Retrieval—SLAIR—is a software, developed in

```
▼<rdf:RDF xmlns:rdf="http://www.w3.org/1999/02/22-rdf-syntax-ns#">
 ▼<rdf:Description rdf:about="http://sentic.net/api/en/concept/celebrate_special_occasion">
   <rdf:type rdf:resource="http://sentic.net/api/concept"/>
   <text xmlns="http://sentic.net/api">celebrate special occasion</text>
   <semantics xmlns="http://sentic.net/api" rdf:resource="http://sentic.net/api/en/concept/celebrate_holiday"/>
   <semantics xmlns="http://sentic.net/api" rdf:resource="http://sentic.net/api/en/concept/celebrate_occasion"/>
   <semantics xmlns="http://sentic.net/api" rdf:resource="http://sentic.net/api/en/concept/celebrate_birthday"/>
   <semantics xmlns="http://sentic.net/api" rdf:resource="http://sentic.net/api/en/concept/celebrate_wedding">
   <semantics xmlns="http://sentic.net/api" rdf:resource="http://sentic.net/api/en/concept/express_appreciation"/>
   <pleasantness xmlns="http://sentic.net/api" rdf:datatype="http://www.w3.org/2001/XMLSchema#float">0.93</pleasantness>
   <attention xmlns="http://sentic.net/api" rdf:datatype="http://www.w3.org/2001/XMLSchema#float">0.724</attention>
   <sensitivity xmlns="http://sentic.net/api" rdf:datatype="http://www.w3.org/2001/XMLSchema#float">0</sensitivity>
   <aptitude xmlns="http://sentic.net/api" rdf:datatype="http://www.w3.org/2001/XMLSchema#float">0</aptitude>
   <polarity xmlns="http://sentic.net/api" rdf:datatype="http://www.w3.org/2001/XMLSchema#float">0.551</polarity>
 </rdf:Description>
 </rdf:RDF>
```

Fig. 2 A sample of SenticNet

C++ programming language, dedicated to the analysis of large amount of documents for clustering and classification purposes [51]. In this context the word '*document*' is used to denote any source of data which carries information, e.g., text written in natural language, web pages, images [35]. SLAIR normalizes input documents into an internal representation and applies several metrics to compute distances between pair of documents; then, the document-distance takes into account a conventional content-based similarity metric, a stylistic similarity criterion and a semantic representation of the documents, applying machine learning algorithms for both cluster and classification purposes. After a pre-processing phase, in which language identification, stemming and stopword removal steps are carried out, a text document becomes a '*docum object*', deprived of useless information (e.g., articles, prepositions, punctuation, special characters). At this level, different types of semantic descriptors can be applied to the new object, e.g., the SenticNet semantic descriptor explained in the present work.

4.1 SenticNet Semantic Descriptor

A document can hold sentiment information and one may be interested in analyzing its sentiment distribution, in order to study and compare different documents from a sentiment point of view. In particular, this chapter adopts the SenticNet framework, which allows one to retrieve four different sentiment experiences associated with a specific word, in order to develop a sentiment semantic descriptor made up of a vector of four affective dimensions, i.e., Pleasantness, Attention, Sensitivity and Aptitude. Hence, each document is described by a sentiment vector having four values in the range $[-1, +1]$.

The SenticNet network has been integrated into a MySQL relational database management system, called MariaDB, and SLAIR interacts with it through MySQL C++ APIs. After SLAIR '*text processing phase*', the SenticNet Semantic Descriptor extracts the list of words that compose the document and submits each single word to MariaDB through a proper query. If the word is present in SenticNet, the database returns the corresponding four dimensions, which are later saved into a vector. If the word is not included in the database, the sentiment descriptor assigns a default vector, that will not be considered during the calculation of the sentiment distribution. The

entire SenticNet Semantic Descriptor Algorithm is described in Algorithm 1. After
the semantic descriptor step, a new process starts, during which the distance between
two document is calculated, using a specific metric.

Algorithm 1: SENTICNET SEMANTIC DESCRIPTOR

Input: A document $docum = \{(w)_i; i = 1, ..., N\}$,
where w is a *word* and N is the total number of words in *docum*.
1 **Initialization:** $sv[D]$, where sv is the sentiment vector that describes *docum* and $D = 4$ is
   the number of affective dimensions [pleasantness, attention, sensitivity, aptitude];
   $senticTemp[D]$ is a temporary vector used for calculation and wordCount = 0 is the number
   of valid SenticNet words.
2 **for(i=0; i<N; i++):** $SenticDB(w_i, senticTemp)$, where SenticDB is a function that
   makes a query to MariaDB and returns the filled senticTemp vector. If w_i is not present in
   MariaDB, the $senticTemp$ values are set to 2, which is a value out of the acceptable range
   [-1,+1], otherwise $wordCount + +$.
3   **for(j=0; j<D; j++)**
4   **if(senticTemp[j] != 2)**
5             $sv[j] = sv[j] + senticTemp[j]$
6 **for(i=0; i<wordCount; i++)**
7   **for(j=0; j<D; j++)**
8          $sv[j] = sv[j]/wordCount$
Output: *docum* is described by four normalized affective dimensions, *Pleasantness*,
     Attention, Sensitivity, Aptitude.

---

5 Experimental Setup

SLAIR framework is provided with a module able to calculate the distance between
two text documents, based on a specified metric. As shown in Algorithm 2, a *SenticNet Distance Metric Function* is implemented, and requires as arguments the two
documents to be compared and the metric. First a Semantic SenticNet Descriptor is
retrieved from each document. Once the two sentiment vectors *sv[D]* describing the
two documents are obtained, a specific metric is applied to *sv[D]*. It is possible to
choose one of the following different metrics:

- Manhattan norm $\|\mathbf{x}\|_1 :=$

$$\sum_{i=1}^{n} |x_i|$$

- Euclidean norm $\|\mathbf{x}\| :=$

$$\sqrt{x_1^2 + \cdots + x_n^2}$$

- Maximum norm $\|\mathbf{x}\|_\infty :=$

$$max(|x_1|, \ldots, |x_n|)$$

The *SenticNet Distance Metric Function* returns a distance which describes the closeness between the two documents; e.g., if the documents are very similar, the distance will be a positive number near zero, otherwise, if the documents are very different, the distance will be a number near one.

Algorithm 2: SENTICNET DISTANCE METRIC FUNCTION

Input: Two documents $docum A$, $docum B$, a $metric$ Φ and a distance Δ.
1 **Get sv[D]:** extract $sv[D]_A$ vector from $docum A$ and $sv[D]_B$ vector from $docum B$.
2 **Calculate SenticNet Distance**: Δ = SenticDistMetric($sv[D]_A$, $sv[D]_B$, Φ) where $SenticDistMetric$ is a function that calculates the distance between $sv[D]_A$ and $sv[D]_B$ based on Φ.
3 **Return Δ**

---

6   Experimental Results

This section reports on the experimental setup adopted to demonstrate the validity of the approach employing the *SenticNet Semantic Descriptor* and the *Sentiment Distance Metric*, implemented in SLAIR.

In order to take advantage of the use of the four affective dimensions, the choice of the documents was guided by the necessity of having texts long enough to convey clear feelings and emotions: therefore, the experiments of the present work have been performed on a dataset of books, written or translated in English language. In fact, on one hand, books are full of sentimental and affective contents and, on the other hand, the analysis of sentiment orientation applied to this type of dataset can inspire a new method of classification. For example, from a commercial perspective, the analysis of the sentiment distribution of a book may be useful within a search engine which suggests the book most emotionally similar to another one.

The aim of the present work is to classify different literary genres; to this purpose, we selected the following five distinct categories:

- XIX Century Novel
- Fantasy
- Horror
- XXI Century Novel
- Greek Tragedy

Table 3 Books dataset description

ID	Category	Author	Title
1	XIX century novel	F. Dostoevskij	Crime and Punishment
2	XIX century novel	J. Austen	Emma
3	XIX century novel	L.M. Alcot	Little Women
4	XIX century novel	L. Tolstoj	War and Peace
5	Fantasy	G.R.R. Martin	A Game of Thrones
6	Fantasy	M. Weis, T. Hickman	Dragonlance Tales
7	Fantasy	T. Brooks	The Sword of Shannara
8	Fantasy	David Gemmell	Winter Warriors
9	Horror fiction	H.P. Lovecraft	The Beast in the Cave
10	Horror fiction	D. Sharen	Blood Beast
11	Horror fiction	Robin Becker	Brains: a Zombie Memoir
12	Horror fiction	E. A. Poe	The Masque Of The Red Death
13	XXI century novel	J. Egan	A Visit from the Goon Squad
14	XXI century novel	E. Gilbert	Eat Pray Love
15	XXI century novel	J. Martel	Life of Pi
16	XXI century novel	D. Nicholls	One Day
17	Greek tragedy	Sophocles	Antogone
18	Greek tragedy	Euripides	Hecuba
19	Greek tragedy	Sophocles	Oedipus the King
20	Greek tragedy	Euripides	The Suppliants

The dataset used is shown in Table 3: it is made up of 20 extracts of books, 20.000 characters each, selected from a random line of each book. Therefore, each category is represented by four extracts of books.

Three different distance metrics are applied to the extracts, namely the Manhattan, the Euclidean and the Maximum norm. The experimental results underline some differences between the distances resulted from the use of the three norms: Tables 4 and 5 respectively show the most similar and distant categories based on the minimum and maximum distances of each book with respect to the others. For example, it is possible to notice a similarity between the XIX century novels and the XXI century novels according to all the norms, while both genres result distant from the fantasy novels; this evidence is completely coherent from an affective point of view, because even though XIX and XXI century novels may be set in different environments, mentality and social constraints, they can still convey similar types of feelings.

Table 4 Most similar categories with different norms

ID	Manhattan norm	Euclidean norm	Maximum norm
1	XXI century novel	XXI century novel	XXI century novel
2	XXI century novel	XXI century novel	XXI century novel
3	XXI century novel	XIX century novel	XXI century novel
4	XXI century novel	XXI century novel	XXI century novel
5	Fantasy	Fantasy	Fantasy
6	Horror	Horror	Horror
7	Horror	Horror	Horror
8	Fantasy	Fantasy	Fantasy
9	Horror	Horror	Horror
10	Horror	Fantasy	Horror
11	Fantasy	Fantasy	Horror
12	Horror	Horror	Horror
13	XIX century novel	XIX century novel	XIX century novel
14	XXI century novel	XXI century novel	XXI century novel
15	XIX century novel	XIX century novel	XIX century novel
16	XIX century novel	XIX century novel	XIX century novel
17	Tragedy	Tragedy	Tragedy
18	Horror	Horror	Horror
19	XXI century novel	XXI century novel	XXI century novel
20	Tragedy	Tragedy	Tragedy

We chose the Euclidean norm to perform a deeper analysis, since it better underlines the similarities and dissimilarities between different book categories distributions. Table 6 reports on the results of the distances between the 20 extracts, employing the *Euclidean norm*: in particular, the table represents a symmetric matrix, in which each row indicates an extract of a book compared with the other extracts. The zero values on the diagonal hence indicate the difference between an extract and itself. For each book in the rows, the corresponding black cell represents the minimum distance, while the grey one represents the maximum distance. It is therefore straightforward to use this information to extract not only the affective affinity or difference between books, but also the possible influence of a previous book or author in the following ones. For example, it can be noticed that the XIX century novel extracts are all emotionally close to the extracts of the XXI century and different from the Fantasy genre. Also, Fantasy and Horror extracts are both extremely distant from the XXI century novel.

In order to better clarify the book extracts sentiment distributions, the following pictures are shown: each figure reports on the emotional trend of each category, with four plots for each book of the category. The figures prove the effectiveness of

Table 5 Most distant categories with different norms

ID	Manhattan norm	Euclidean norm	Maximum norm
1	XIX century novel	XIX century novel	XIX century novel
2	Fantasy	Fantasy	Fantasy
3	Fantasy	Fantasy	Fantasy
4	Fantasy	Fantasy	Horror
5	Fantasy	Fantasy	XIX century novel
6	XXI century novel	XIX century novel	XXI century novel
7	XXI century novel	XIX century novel	XXI century novel
8	XIX century novel	Fantasy	XXI century novel
9	XXI century novel	XIX century novel	XXI century novel
10	XXI century novel	XIX century novel	XXI century novel
11	XXI century novel	XIX century novel	XXI century novel
12	XXI century novel	XIX century novel	XXI century novel
13	XXI century novel	XIX century novel	XXI century novel
14	Fantasy	Fantasy	Horror
15	Fantasy	Fantasy	Horror
16	Fantasy	Fantasy	Fantasy
17	XXI century novel	Fantasy	XXI century novel
18	XXI century novel	XIX century novel	XXI century novel
19	Fantasy	Fantasy	Horror
20	Fantasy	Fantasy	Horror

the proposed approach: each different category is in fact represented by a particular distribution, shared by the books of the same category; moreover, it is possible to use local minima in order to find similarity points between a certain category and specific books. Another important result is achieved when noticing that different books from different categories may show a similar distribution; for example, Crime and Punishment by Dostoevskij shows a similar emotional distribution compared to two fantasy novels, namely The Game of Thrones and Winter Warriors: this means that, even if the genre is different, these books arouse similar feelings.

Eventually, Table 7 with the sentiment distances of the example sentences named in the Introduction section is reported, in order to show the similarity between the Dostoevskij's book and the Austen's one, differently from Martin's fantasy novel.

Table 6 Books sentiment distribution

ID	1	2	3	4	5	6	7	8	9	10	11	12	13	14	15	16	17	18	19	20
1	0	3.373	0.865	1.255	0.26	0.839	2.698	0.323	1.195	2.022	1.422	1.504	0.19	2.823	1.669	0.644	1.198	2.229	1.303	1.259
2	3.373	0	0.878	0.601	3.073	3.691	7.954	2.947	4.654	5.742	5.667	7.35	2.359	0.561	0.546	1.198	2.405	5.172	1.279	1.928
3	0.865	0.878	0	0.156	0.732	1.155	4.163	0.684	1.825	2.678	2.354	3.302	0.381	1.081	0.406	0.092	0.808	2.547	0.503	0.621
4	1.255	0.601	0.156	0	1.227	1.924	5.482	1.091	2.697	3.921	3.397	4.505	0.835	0.458	0.09	0.136	1.108	3.301	0.252	0.786
5	0.26	3.073	0.732	1.227	0	0.59	3.261	0.126	1.378	2.121	1.233	1.721	0.308	2.83	1.631	0.589	1.237	2.389	1.206	1.137
6	0.839	3.691	1.155	1.924	0.59	0	1.598	0.354	0.295	0.75	0.222	0.835	0.554	4.228	2.742	1.122	0.535	0.724	1.78	0.703
7	2.698	7.954	4.163	5.482	3.261	1.598	0	2.943	0.617	0.331	0.976	0.51	2.396	8.958	6.819	4.273	2.477	1.107	5.659	3.354
8	0.323	2.947	0.684	1.091	0.126	0.354	2.943	0	1.008	1.925	0.941	1.596	0.329	2.746	1.595	0.46	0.665	1.631	0.849	0.591
9	1.195	4.654	1.825	2.697	1.378	0.295	0.617	1.008	0	0.281	0.255	0.499	0.869	5.339	3.709	1.824	0.68	0.313	2.655	1.128
10	2.022	5.742	2.678	3.921	2.121	0.75	0.331	1.925	0.281		0.499	0.536	1.48	7.004	5.076	2.933	1.621	0.845	4.227	2.241
11	1.422	5.667	2.354	3.397	1.233	0.222	0.976	0.941	0.255	0.499		0.34	1.245	6.314	4.462	2.275	1.14	0.623	3.108	1.509
12	1.504	7.35	3.302	4.505	1.721	0.835	0.51	1.596	0.499	0.536	0.34	0	1.588	7.714	5.627	3.189	2.097	1.238	4.375	2.69
13	0.19	2.359	0.381	0.835	0.308	0.554	2.396	0.329	0.869	1.48	1.245	1.588	0	2.396	1.294	0.383	0.757	1.781	1.11	0.83
14	2.823	0.561	1.081	0.458	2.83	4.228	8.958	2.746	5.339	7.004	6.314	7.714	2.396		0.176	1.021	2.804	6.02	0.815	2.198
15	1.669	0.546	0.406	0.09	1.631	2.742	6.819	1.595	3.709	5.076	4.462	5.627	1.294	0.176	0	0.373	1.813	4.466	0.418	1.364
16	0.644	1.198	0.092	0.136	0.589	1.122	4.273	0.46	1.824	2.933	2.275	3.189	0.383	1.021	0.373		0.7	2.448	0.227	0.479
17	1.198	2.405	0.808	1.108	1.237	0.535	2.477	0.665	0.68	1.621	1.14	2.097	0.757	2.804	1.813	0.7	0	0.621	0.882	0.079
18	2.229	5.172	2.547	3.301	2.389	0.724	1.107	1.631	0.313	0.845	0.623	1.238	1.781	6.02	4.466	2.448	0.621	0	2.908	1.072
19	1.303	1.279	0.503	0.252	1.206	1.78	5.659	0.849	2.655	4.227	3.108	4.375	1.11	0.815	0.418	0.227	0.882	2.908	0	0.502
20	1.259	1.928	0.621	0.786	1.137	0.703	3.354	0.591	1.128	2.241	1.509	2.69	0.83	2.198	1.364	0.479	0.079	1.072	0.502	0

Table 7 Sentiment distance metric example applied to sentences

Sentence ID	A	B	C
A	0	0.038827	0.11794
B	0.038827	0	0.035272
C	0.11794	0.035272	0

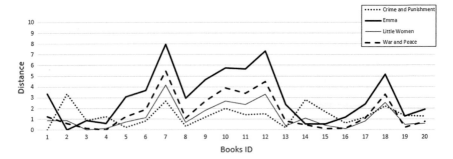

Fig. 3 XIX century novel books

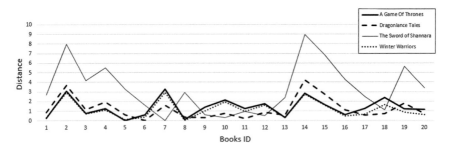

Fig. 4 Fantasy books

7 Conclusion

The present research has exploited a cognitive model for emotion recognition in natural language text. In particular, the aim of the proposed approach has involved the analysis of the sentimental and affective contents of some literary genres of books, in order to extract and study the distribution of each category and implement a method able to automatically detect similarities and dissimilarities between them.

In order to pursue this scope, SenticNet 3 has been used as a publicly available semantic and affective resource to obtain the values of Pleasantness, Attention, Sensitivity, and Aptitude of each book, which can potentially synthesize the full range of emotional experiences. The software application SLAIR has been employed to analyze the sentiment distribution of the books, in order to study and compare different documents from a sentiment point of view. In particular, a sentiment semantic

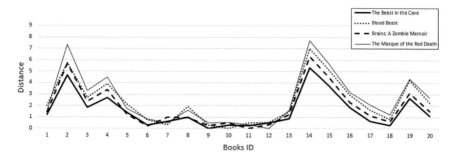

Fig. 5 Horror books

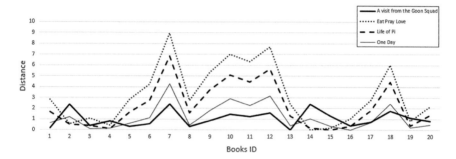

Fig. 6 XXI century novel books

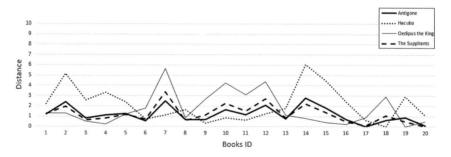

Fig. 7 Greek tragedy books

descriptor has been made up using a vector representing the four affective dimensions. Besides, different metrics have been proposed to calculate the distance between two documents (Figs. 3 and 4).

Experimental results have shown that the proposed approach is able to extract not only the affective affinity or difference between books, but also the possible influence of a previous book in the following ones. Moreover, we found out that each category is characterized by a certain distribution, and that local minima of this distribution can be exploited in order to find similarities with specific books (Figs. 5, 6 and 7).

References

1. Allen, J.: Natural Language Understanding. Benjamin/Cummings (1987)
2. Baker, C.F., Fillmore, C.J., Lowe, J.B.: The berkeley framenet project. In: Proceedings of the 17th international conference on Computational linguistics, vol, 1, pp. 86–90. Association for Computational Linguistics (1998)
3. Bisio, F., Gastaldo, P., Peretti, C., Zunino, R., Cambria, E.: Data intensive review mining for sentiment classification across heterogeneous domains. In: Advances in Social Networks Analysis and Mining (ASONAM), 2013 IEEE/ACM International Conference on. pp. 1061–1067. IEEE (2013)
4. Bisio, F., Gastaldo, P., Zunino, R., Cambria, E.: A learning scheme based on similarity functions for affective common-sense reasoning. In: IJCNN. pp. 2476–2481 (2015)
5. Bizer, C., Jens, L., Kobilarov, G., Auer, S., Becker, C., Cyganiak, R., Hellmann, S.: Dbpedia—a crystallization point for the web of data. Web Semant.: Sci. Serv. Agents World Wide Web **7**(3), 154–165 (2009)
6. Bosco, C., Patti, V., Bolioli, A.: Developing corpora for sentiment analysis and opinion mining: a survey and the Senti-TUT case study. IEEE Intell. Syst. **28**(2), 55–63 (2013)
7. Cambria, E., Fu, J., Bisio, F., Poria, S.: AffectiveSpace 2: enabling affective intuition for concept-level sentiment analysis. In: AAAI. pp. 508–514. Austin (2015)
8. Cambria, E., Gastaldo, P., Bisio, F., Zunino, R.: An ELM-based model for affective analogical reasoning. Neurocomputing **149**, 443–455 (2015)
9. Cambria, E.: Affective computing and sentiment analysis. IEEE Intelligent Systems **31**(2), (2016)
10. Cambria, E., Hussain, A.: Sentic computing: a common-sense-based framework for concept-level sentiment analysis. Springer, Cham, Switzerland (2015)
11. Cambria, E., Hussain, A., Havasi, C., Eckl, C.: SenticSpace: visualizing opinions and sentiments in a multi-dimensional vector space. In: Setchi, R., Jordanov, I., Howlett, R., Jain, L. (eds.) Knowledge-Based and Intelligent Information and Engineering Systems. Lecture Notes in Artificial Intelligence, vol. 6279, pp. 385–393. Springer, Berlin (2010)
12. Cambria, E., Livingstone, A., Hussain, A.: The hourglass of emotions. In: Esposito, A., Vinciarelli, A., Hoffmann, R., Muller, V. (eds.) Cognitive Behavioral Systems. Lecture Notes in Computer Science, vol. 7403, pp. 144–157. Springer, Berlin Heidelberg (2012)
13. Cambria, E., Olsher, D., Kwok, K.: Sentic activation: a two-level affective common sense reasoning framework. In: AAAI. pp. 186–192. Toronto (2012)
14. Cambria, E., Olsher, D., Rajagopal, D.: SenticNet 3: a common and common-sense knowledge base for cognition-driven sentiment analysis. In: AAAI. pp. 1515–1521. Quebec City (2014)
15. Cambria, E., Poria, S., Bisio, F., Bajpai, R., Chaturvedi, I.: The clsa model: a novel framework for concept-level sentiment analysis. In: Computational Linguistics and Intelligent Text Processing, pp. 3–22. Springer (2015)
16. Cambria, E., Rajagopal, D., Kwok, K., Sepulveda, J.: GECKA: game engine for commonsense knowledge acquisition. In: FLAIRS, pp. 282–287 (2015)
17. Cambria, E., Schuller, B., Liu, B., Wang, H., Havasi, C.: Knowledge-based approaches to concept-level sentiment analysis. IEEE Intell. Syst. **28**(2), 12–14 (2013)
18. Cambria, E., Schuller, B., Liu, B., Wang, H., Havasi, C.: Statistical approaches to concept-level sentiment analysis. IEEE Intell. Syst. **28**(3), 6–9 (2013)
19. Cambria, E., Xia, Y., Hussain, A.: Affective common sense knowledge acquisition for sentiment analysis. In: LREC, pp. 3580–3585. Istanbul (2012)
20. Chikersal, P., Poria, S., Cambria, E.: Sentu: Sentiment analysis of tweets by combining a rule-based classifier with supervised learning. In: Proceedings of the International Workshop on Semantic Evaluation (SemEval 2015) (2015)
21. Chinthala, S., Mande, R., Manne, S., Vemuri, S.: Sentiment analysis on twitter streaming data. In: Emerging ICT for Bridging the Future-Proceedings of the 49th Annual Convention of the Computer Society of India (CSI), vol. 1, pp. 161–168. Springer (2015)

22. van Cranenburgh, A., Huygens, I., Koolen, C.: Identifying literary texts with bigrams. In: Computational Linguistics for Literature, p. 58 (2015)
23. Davidov, D., Tsur, O., Rappoport, A.: Enhanced sentiment learning using twitter hashtags and smileys. In: Proceedings of the 23rd International Conference on Computational Linguistics: Posters, pp. 241–249. Association for Computational Linguistics (2010)
24. Di Fabbrizio, G., Aker, A., Gaizauskas, R.: Summarizing on-line product and service reviews using aspect rating distributions and language modeling. IEEE Intell. Syst. **28**(3), 28–37 (2013)
25. Dyer, M.G.: Connectionist natural language processing: a status report. In: Computational Architectures Integrating Neural and Symbolic Processes, pp. 389–429. Springer (1995)
26. Elliott, C.D.: The affective reasoner: a process model of emotions in a multi-agent system. Northwestern University (1992)
27. García-Moya, L., Anaya-Sanchez, H., Berlanga-Llavori, R.: A language model approach for retrieving product features and opinions from customer reviews. IEEE Intell. Syst. **28**(3), 19–27 (2013)
28. Hatzivassiloglou, V., McKeown, K.R.: Predicting the semantic orientation of adjectives. In: Proceedings of the 35th Annual Meeting of the Association for Computational Linguistics and Eighth Conference of the European Chapter of the Association for Computational Linguistics, pp. 174–181. Association for Computational Linguistics (1997)
29. Honkela, T., Korhonen, J., Lagus, K., Saarinen, E.: Five-dimensional sentiment analysis of corpora, documents and words. In: Advances in Self-Organizing Maps and Learning Vector Quantization, pp. 209–218. Springer (2014)
30. Hung, C., Lin, H.K.: Using objective words in SentiWordNet to improve sentiment classification for word of mouth. IEEE Intell. Syst. **28**(2), 47–54 (2013)
31. Kamps, J., Marx, M., Mokken, R.J., De Rijke, M.: Using wordnet to measure semantic orientations of adjectives. In: LREC. vol. 4, pp. 1115–1118. Citeseer (2004)
32. Kim, S.M., Hovy, E.: Automatic detection of opinion bearing words and sentences. In: Companion Volume to the Proceedings of the International Joint Conference on Natural Language Processing (IJCNLP), pp. 61–66 (2005)
33. Kim, S.M., Hovy, E.: Extracting opinions, opinion holders, and topics expressed in online news media text. In: Proceedings of the Workshop on Sentiment and Subjectivity in Text, pp. 1–8. Association for Computational Linguistics (2006)
34. Kouloumpis, E., Wilson, T., Moore, J.: Twitter sentiment analysis: the good the bad and the omg!. Icwsm **11**, 538–541 (2011)
35. Meda, C., Bisio, F., Gastaldo, P., Zunino, R., Surlinelli, R., Scillia, E., Ottaviano, A.V.: Content-adaptive analysis and filtering of microblogs traffic for event-monitoring applications. In: Proceedings of the 18th Asia Pacific Symposium on Intelligent and Evolutionary Systems, vol. 1, pp. 155–170. Springer (2015)
36. Melville, P., Gryc, W., Lawrence, R.D.: Sentiment analysis of blogs by combining lexical knowledge with text classification. In: Proceedings of the 15th ACM SIGKDD International Conference on Knowledge Discovery and Data Mining, pp. 1275–1284. ACM (2009)
37. Mohammad, S.M., Kiritchenko, S., Zhu, X.: Nrc-canada: building the state-of-the-art in sentiment analysis of tweets. In: Second Joint Conference on Lexical and Computational Semantics (*SEM), vol. 2, pp. 321–327 (2013)
38. Murtagh, F., Ganz, A.: Pattern recognition in narrative: analysis of narratives of emotion (2014). arXiv preprint arXiv:1405.3539
39. Ortigosa, A., Martín, J.M., Carro, R.M.: Sentiment analysis in facebook and its application to e-learning. Comput. Hum. Behav. **31**, 527–541 (2014)
40. Ortony, A., Clore, G., Collins, A.: Cogn. Struct. Emotions. Cambridge University Press, Cambridge (1988)
41. Pang, B., Lee, L.: A sentimental education: sentiment analysis using subjectivity summarization based on minimum cuts. In: Proceedings of the 42nd Annual Meeting on Association for Computational Linguistics, p. 271. Association for Computational Linguistics (2004)
42. Pang, B., Lee, L.: Seeing stars: exploiting class relationships for sentiment categorization with respect to rating scales. In: Proceedings of the 43rd Annual Meeting on Association for Computational Linguistics, pp. 115–124. Association for Computational Linguistics (2005)

43. Pang, B., Lee, L.: Opinion mining and sentiment analysis. Found. Trends Inf. Retrieval **2**(1–2), 1–135 (2008)
44. Pang, B., Lee, L., Vaithyanathan, S.: Thumbs up? Sentiment classification using machine learning techniques. In: EMNLP, pp. 79–86. Philadelphia (2002)
45. Plutchik, R.: The nature of emotions. Am. Sci. **89**(4), 344–350 (2001)
46. Popescu, A.M., Etzioni, O.: Extracting product features and opinions from reviews. In: Natural Language Processing and Text Mining, pp. 9–28. Springer (2007)
47. Poria, S., Gelbukh, A., Hussain, A., Howard, N., Das, D., Bandyopadhyay, S.: Enhanced SenticNet with affective labels for concept-based opinion mining. Intell. Sys. IEEE **28**(2), 31–38 (2013)
48. Rao, D., Ravichandran, D.: Semi-supervised polarity lexicon induction. In: EACL, pp. 675–682. Athens (2009)
49. Riloff, E., Wiebe, J.: Learning extraction patterns for subjective expressions. In: Proceedings of the 2003 Conference on Empirical Methods in Natural Language Processing, pp. 105–112. Association for Computational Linguistics (2003)
50. Rosenthal, S., Nakov, P., Kiritchenko, S., Mohammad, S.M., Ritter, A., Stoyanov, V.: Semeval-2015 task 10: Sentiment analysis in twitter. In: Proceedings of the 9th International Workshop on Semantic Evaluation, SemEval (2015)
51. Sangiacomo, F., Leoncini, A., Decherchi, S., Gastaldo, P., Zunino, R.: Sealab advanced information retrieval. In: IEEE Fourth International Conference on Semantic Computing (ICSC), pp. 444–445. IEEE (2010)
52. dos Santos, C.N., Gatti, M.: Deep convolutional neural networks for sentiment analysis of short texts. In: Proceedings of the 25th International Conference on Computational Linguistics (COLING), Dublin, Ireland (2014)
53. Snyder, B., Barzilay, R.: Multiple aspect ranking using the good grief algorithm. In: HLT-NAACL, pp. 300–307 (2007)
54. Socher, R., Perelygin, A., Wu, J.Y., Chuang, J., Manning, C.D., Ng, A.Y., Potts, C.: Recursive deep models for semantic compositionality over a sentiment treebank. In: Proceedings of the Conference on Empirical Methods in Natural Language Processing (EMNLP), vol. 1631, p. 1642. Citeseer (2013)
55. Somasundaran, S., Wiebe, J., Ruppenhofer, J.: Discourse level opinion interpretation. In: COLING, pp. 801–808. Manchester (2008)
56. Speer, R., Havasi, C.: ConceptNet 5: a large semantic network for relational knowledge. In: Hovy, E., Johnson, M., Hirst, G. (eds.) Theory and Applications of Natural Language Processing, chap. 6. Springer (2012)
57. Stevenson, R., Mikels, J., James, T.: Characterization of the affective norms for english words by discrete emotional categories. Behav. Res. Methods **39**, 1020–1024 (2007)
58. Strapparava, C., Valitutti, A.: WordNet-Affect: An affective extension of WordNet. In: LREC, pp. 1083–1086. Lisbon (2004)
59. Tang, D., Wei, F., Qin, B., Liu, T., Zhou, M.: Coooolll: a deep learning system for twitter sentiment classification. In: Proceedings of the 8th International Workshop on Semantic Evaluation (SemEval 2014), pp. 208–212 (2014)
60. Tang, D., Wei, F., Yang, N., Zhou, M., Liu, T., Qin, B.: Learning sentiment-specific word embedding for twitter sentiment classification. In: Proceedings of the 52nd Annual Meeting of the Association for Computational Linguistics, vol. 1, pp. 1555–1565 (2014)
61. Tsai, A., Tsai, R., Hsu, J.: Building a concept-level sentiment dictionary based on commonsense knowledge. IEEE Intell. Syst. **28**(2), 22–30 (2013)
62. Turney, P.D.: Thumbs up or thumbs down?: semantic orientation applied to unsupervised classification of reviews. In: Proceedings of the 40th Annual Meeting on Association for Computational linguistics, pp. 417–424. Association for Computational Linguistics (2002)
63. Vahdat, M., Oneto, L., Anguita, D., Funk, M., Rauterberg, M.: Can machine learning explain human learning? Neurocomputing (In Press)
64. Wang, Q., Cambria, E., Liu, C., Hussain, A.: Common sense knowledge for handwritten chinese recognition. Cogn. Comput. **5**(2), 234–242 (2013)

65. Weichselbraun, A., Gindl, S., Scharl, A.: Extracting and grounding context-aware sentiment lexicons. IEEE Intell. Syst. **28**(2), 39–46 (2013)
66. Wiebe, J., Wilson, T., Cardie, C.: Annotating expressions of opinions and emotions in language. Lang. Resour. Eval. **39**(2), 165–210 (2005)
67. Wu, W., Li, H., Wang, H., Zhu, K.: Probase: a probabilistic taxonomy for text understanding. In: SIGMOD, pp. 481–492. Scottsdale (2012)
68. Yu, H., Hatzivassiloglou, V.: Towards answering opinion questions: separating facts from opinions and identifying the polarity of opinion sentences. In: Proceedings of the 2003 Conference on Empirical Methods in Natural Language Processing, pp. 129–136. Association for Computational Linguistics (2003)
69. Zirn, C., Niepert, M., Stuckenschmidt, H., Strube, M.: Fine-grained sentiment analysis with structural features. In: IJCNLP, pp. 336–344 (2011)

Interpretability of Computational Models for Sentiment Analysis

Han Liu, Mihaela Cocea and Alexander Gegov

Abstract Sentiment analysis, which is also known as opinion mining, has been an increasingly popular research area focusing on sentiment classification/regression. In many studies, computational models have been considered as effective and efficient tools for sentiment analysis. Computational models could be built by using expert knowledge or learning from data. From this viewpoint, the design of computational models could be categorized into expert based design and data based design. Due to the vast and rapid increase in data, the latter approach of design has become increasingly more popular for building computational models. A data based design typically follows machine learning approaches, each of which involves a particular strategy of learning. Therefore, the resulting computational models are usually represented in different forms. For example, neural network learning results in models in the form of multi-layer perceptron network whereas decision tree learning results in a rule set in the form of decision tree. On the basis of above description, interpretability has become a main problem that arises with computational models. This chapter explores the significance of interpretability for computational models as well as analyzes the factors that impact on interpretability. This chapter also introduces several ways to evaluate and improve the interpretability for computational models which are used as sentiment analysis systems. In particular, rule based systems, a special type of computational models, are used as an example for illustration with respects to evaluation and improvements through the use of computational intelligence methodologies.

Keywords Sentiment prediction · Computational intelligence · Machine learning · Rule based systems · Interpretability analysis · Interpretability evaluation · Fuzzy computational models · Rule based networks

H. Liu (✉) · M. Cocea · A. Gegov
School of Computing, University of Portsmouth, Buckingham Building,
Lion Terrace, Portsmouth PO1 3HE, UK
e-mail: Han.Liu@port.ac.uk

M. Cocea
e-mail: Mihaela.Cocea@port.ac.uk

A. Gegov
e-mail: Alexander.Gegov@port.ac.uk

© Springer International Publishing Switzerland 2016
W. Pedrycz and S.-M. Chen (eds.), *Sentiment Analysis and Ontology Engineering*,
Studies in Computational Intelligence 639, DOI 10.1007/978-3-319-30319-2_9

1 Introduction

Sentiment analysis, which is also known as opinion mining, generally aims to identify the attitude of a speaker/writer using natural language processing, text analysis and computational linguistics. In recent years, research has been increasingly focusing on sentiment classification/regression. In other words, sentiment analysis can be regarded as a classification task if the sentiment is analyzed in qualitative terms, e.g. positive/negative and rating, and as a regression task if the sentiment is analyzed in quantitative terms, e.g. numerical values.

Many studies have shown that computational models are effective and efficient tools for sentiment classification/regression. Such models can be designed through adopting computational intelligence approaches. In particular, these can be done by using expert knowledge or learning from real data. From this point of view, the design of computational models can be divided into expert based design and data based design. The former follows traditional system engineering approaches [1, 2] whereas the latter typically follows machine learning approaches [3].

The expert based approach is in general domain dependent. It is necessary to have knowledge or requirements acquired from experts at first and then to identify the relationships between inputs and outputs. Modelling, which is the most important step, is to be executed in order to build a system. Once the modelling is complete, then simulation is started to check the model towards fulfillment of systematic complexity such as predictive accuracy and computational efficiency. Finally, statistical analysis is undertaken in order to validate whether the computational model is reliable and efficient in practical application.

On the other hand, the data based approach is in general domain independent. As mentioned earlier in this section, it typically follows machine learning approaches. Machine learning is a branch of artificial intelligence and involves two stages: training and testing [3]. Training aims to learn something from known properties by using learning algorithms and testing aims to make predictions on unknown properties by using the knowledge learned in the training stage. From this point of view, training and testing are also known as learning and prediction respectively. In practice, a machine learning task aims to build a model that is further used to make predictions through the use of learning algorithms. Therefore, this task is usually referred to as predictive modelling. In the context of sentiment analysis, supervised modelling approaches, which can be involved in classification or regression tasks, are popularly used for solving particular issues.

As justified in [4], the data based approach is more effective and efficient than expert based approach in modern development. The main reason is explained with the following arguments: expert knowledge may be incomplete or inaccurate; some of experts' viewpoints may be biased; engineers may misunderstand requirements or have technical designs with defects. In other words, with regards to solving problems with high complexity, it is difficult for both domain experts and engineers to have all possible cases considered or to have perfect technical designs. Once a failure arises, domain experts and engineers may have to find the problem and fix it through reanalysis or redesign. However, the real world has had big data available. Some previously unknown information or knowledge may be discovered from data. Data may potentially be used as supporting evidence to reflect some useful and important pattern by using modeling techniques. More importantly, the model could be modified automatically along with the update of database in real time if the data based modeling technique is used. On the basis of above description, the data based approach is recommended instead of the expert based approach for design of computational models.

In machine learning research, computational models can be evaluated in three main dimensions, namely accuracy, efficiency and interpretability. This chapter focuses on the analysis of interpretability for computational models since interpretable models have been increasingly required. In particular, computational models can be used not only to make predictions but also to extract knowledge which needs to be communicated to people. For example, a computational model is built as a sentiment analysis system in order to provide people with the judged rating score of a review. People may not trust the judgment made by the sentiment analysis system unless they can fully understand the reasons of the judgment making. Consequently, interpretability of computational models has become a significant aspect which needs further research.

As different machine learning algorithms involve different learning strategies, the computational models built by using these algorithms are usually represented in different forms and thus are characterized by different levels of interpretability. Some machine learning algorithms lead to computational models that operate in a black box manner. In other words, the models can make highly accurate predictions as outputs but it is difficult to interpret the reason why the outputs are derived from the models. On this basis, the interpretability of a computational model has become increasingly significant for knowledge usage.

As introduced in [4], most computational models can work based on different types of logic such as deterministic logic, probabilistic logic and fuzzy logic. Ross stated in [5] that logic is a small part of human capability for reasoning, which is used to assist people in making decisions or judgments. As mentioned in [6], in the context of Boolean logic, each variable is only assigned a binary truth value: 0 (false) or 1 (true). It indicates that reasoning and judgment are made under certainty resulting in a deterministic outcome. From this point of view, this type of logic is also referred to as deterministic logic. However, in reality, people usually can only reason and make decisions and judgments under uncertainty. Therefore, the other two types of logic, namely probabilistic logic and fuzzy logic, are used more widely, both of which can be seen as an extension of deterministic logic. The main difference is that the truth value is not binary but continuous between 0 and 1. The truth value implies a probability of truth between true and false in probabilistic logic and a degree of that in fuzzy logic, both of which need to be interpretable through the use of computational modes in order to indicate explicitly the probability or degree.

The rest of this paper is organized as follows: Sect. 2 explores the significance of interpretability. Section 3 identifies a list of impact factors that may affect the interpretability of a computational model. Section 4 introduces some criteria for evaluation against interpretability. Some recommendations are given with respect to improvements on interpretability in Sect. 5. Section 6 summarizes the completed work and suggests some further directions on the development of computational modelling with respect to interpretability.

2 Significance of Interpretability

As mentioned in Sect. 1, interpretability of computational models has been increasingly significant in practice for knowledge usage. This section justifies why interpretability of a computational model is so significant.

In practice, machine learning methods can be used for two main purposes. One is to build a predictive model that is used to make predictions. The other one is to discover some meaningful and useful knowledge from data [7]. For the latter purpose, the knowledge discovered is used to provide insights for a knowledge domain. From this point of view, it is required to have a computational model which works in a white box manner. This is in order to make the model transparent so that people can understand the reasons why the output is derived from the model.

However, as mentioned in Sect. 1, different learning algorithms involve different strategies of learning that usually result in different ways of representing knowledge. In terms of knowledge representation, Higgins justified in [8] that an interpretable model needs to be able to provide the explanation with regard to the reason of an output and that a rule based knowledge representation makes computational models more interpretable with the following arguments:

- A network was conceived in [9], which needs a number of nodes exponential in the number of attributes in order to restore the information on conditional probabilities of any combination of inputs. It is argued in [8] that the network restores a large amount of information that is mostly less valuable.
- Another type of networks known as Bayesian Networks introduced in [10] needs a number of nodes which is the same as the number of attributes. However, the network only restores the information on joint probabilities based on the assumption that each of the input attributes is totally independent of the others. Therefore, it is argued in [8] that this network is unlikely to predict more complex relationships between attributes due to the lack of information on correlational probabilities between attributes.
- There are some other methods that fill the gaps that exist in Bayesian Networks by deciding to only choose some higher-order conjunctive probabilities, such as the first neural networks [11] and a method based on correlation/dependency measure [12]. However, it is argued in [8] that these methods still need to be based on the assumption that all attributes are independent of each other.

On the basis of above arguments, Higgins motivated the use of rule based knowledge representation due mainly to the advantage that rules specify relationships between attributes, which can provide explanations with regard to a decision from a computational model [8]. Therefore, Higgins argues the significance of interpretability, i.e. the need to explain the output of a computational model based on the reasoning of that model.

In the machine learning context, due to the presence of massively large data, it is very likely to have a complex model built, which makes the knowledge extracted from such a model cumbersome and less readable for people. In this case, it is necessary to represent the model in a way that has a high level of interpretability.

On the other hand, different people would usually have different level of cognitive capability. In other words, the same message may take different meanings to different people due to their different levels of capability of reading and understanding. In addition, different people would also have different levels of expertise and different preferences with regard to the way of receiving information. All these issues raised above make it necessary that knowledge extracted from a computational model needs to be represented in a way that suits people to read and understand. In other words, high interpretability of a computational model is required in order to make the knowledge extracted from the model more readable and understandable to different groups of people.

In sentiment analysis, a great number of studies focus on sentiment classification/regression. In particular, machine learning methods, such as Naïve Bayes and Support Vector Machine, are popularly used to improve the accuracy of sentiment prediction [13, 14]. However, as mentioned in Sect. 1, sentiment analysis is also known as opinion mining, which means to discover opinions from texts. In other words, sentiment analysis is considered as a data mining task and the opinions discovered from texts need to be communicated to people, which again indicates the significance of interpretability of computational models. Those popular machine learning methods used in sentiment analysis such as Naïve Bayes and Support Vector Machine are useful to build accurate sentiment prediction systems but are difficult to provide interpretable models due to the nature of their learning strategies.

In terms of Naïve Bayes, the interpretability problem is similar to that of Bayesian Networks. As mentioned earlier in this section, this type of learning algorithms need to work based on the assumption that all input attributes are totally independent of each other. This assumption makes generated models unable to represent the correlations among input attributes and thus results in insufficient interpretability.

In terms of Support Vector Machine, the interpretability problem is present due to the limitations in transparency and depth of learning. With regard to transparency, support vector machine is not working in a black box manner but models built by using this method are still less transparent to a general audience. This is because the model built by using this algorithm appears to be function lines as decision boundaries in geometric form or a piecewise function in algebraic form. As mentioned earlier, due to different levels of cognitive capability from different people, this type of model representation is usually less interpretable to a non-technical audience who does not know mathematics well. With regard to the depth of learning, this method involves a sound theory in learning strategy but does not aim to learn in depth. In particular, this method does not go through all data instances but just takes a look at a few instances that are selected as support vectors, in order to make predictions. In this case, there could be a great amount of useful information failed to be discovered from data.

On the basis of the above discussion, interpretability is considered as a significant issue in sentiment analysis, and thus it is required to be improved through advancing computational intelligence approaches. However, interpretability is in general a domain-independent problem. Therefore, the rest of this chapter provides a theoretical analysis of issues relating to interpretability as well as an empirical investigation on ways of improving interpretability of computational models in a domain-independent manner.

3 Impact Factors for Interpretability

As described in Sect. 2, the interpretability of computational models is significant due to the presence of some problems from machine learning methods, size of data, model representation and human expertise and preferences. This section discusses how each of these factors has an influence on interpretability.

3.1 Learning Strategy

As mentioned in Sect. 2, different machine learning algorithms usually involve different strategies of learning. This would usually result in differences in two aspects namely, transparency and model complexity.

In terms of transparency, a typical example would be the natural difference between neural networks and rule based learning methods in terms of learning strategy. Neural network learning aims to construct a network topology that consists of a number of layers and that has a number of nodes, each of which represents a perceptron. As a neural network is working in a black box manner with regard to providing an output, the transparency is poor, i.e. people cannot see in an explicit way the reason why the output is given. On the basis of the above description, neural networks have been judged poorly interpretable in [15]. On the other hand, a rule based method aims to generate a rule set typically in the form of either a decision tree or if-then rules. As also mentioned in Sect. 2, rule based knowledge representation is able to explain the reason explicitly with regard to providing an output and, thus, it is well transparent. In addition, another popular machine learning algorithm, which is known as k nearest neighbor, involves lazy learning. In other words, the learning algorithm does not aim to learn in depth to gain some pattern from data but just to make as many correct predictions as possible. In the training stage, there is no actual learning but just some data loaded into computer memory. In this sense, there is no model built in the training stage so there is nothing to be visible for people to gain some useful patterns. On the basis of the above description relating to transparency, the strategies of learning involved in learning algorithms are an impact factor that affects interpretability.

In terms of model complexity, the learning strategy is a significant impact factor. For example, as mentioned earlier, k nearest neighbor does not build a model in the training stage whereas a rule based method would build a rule set. For the former algorithm, the model complexity is 0 as there is no model built. For the latter algorithm, the model complexity would be determined by the overall number of rule terms that is dependent on the number of rules and the average number of terms per rule, i.e. a large number of complex rules indicates the rule set has a high complexity whereas a small number of general rules indicates the rule set has a low complexity. For rule based methods, the strategy of rule induction would usually significantly affect the model complexity. As mentioned in [16, 17], the induction of classification rules could be divided into two categories: 'Divide and Conquer' [18] and 'Separate and Conquer' [19]. The two approaches are also referred to as Top-Down Induction of Decision Trees (TDIDT) and covering approach respectively. As introduced in [20, 21], Prism, which is a rule induction method that follows the 'Separate and Conquer' approach, is likely to generate fewer and more general rules than decision tree, which is another rule induction method that follows the 'Divide and Conquer' approach. The above phenomenon is due mainly to the strategy of rule learning. As mentioned in [20], the rule set generated by TDIDT needs to have at least one common attribute to be in the form of decision trees. The same also applies to each of the subtrees of a decision tree, which requires to have at least one common attribute represented as the root of the subtree. Due to this requirement, TDIDT is likely to generate a large number of complex rules with many redundant terms such as the replicated subtree problem [20] illustrated in Fig. 1 and thus results in a model of high complexity.

The above description relating to model complexity also indicates that learning strategies involved in learning algorithms can be an impact factor that may affect the interpretability.

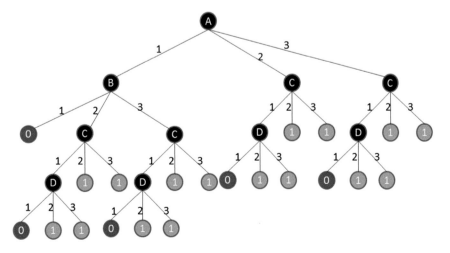

Fig. 1 Cendrowska's replicated subtree example [22]

3.2   Data Size

As mentioned in Sect. 3.1, different learning algorithms involve different strategies of learning and thus generate models with different level of complexity. In this sense, when the same data set is used, different learning algorithms would usually lead to different model complexities. However, for the same algorithm, data of different size would also usually result in the generation of models with different levels of complexity. The rest of this subsection justifies the potential correlation between data size and model complexity using rule based methods and Naïve Bayes as examples.

As mentioned earlier, rule based methods involve the generation of rule sets. The complexity of a rule set is determined by the total number of rule terms, which is dependent upon the number of rules and the average number of terms per rule. However, the total number of rule terms is also affected by the data size in terms of both dimensionality (number of attributes) and sample size (number of instances). For example, a data set has n attributes, each of which has t values, and its sample contains m instances and covers all possible values for each of the attributes. In this example, the model complexity would be equal to $\sum t^i$, while $i = 0, 1, 2, \ldots, n$, but no greater than $m \times n$ in the worst case. This indicates that a rule set consists of a default rule, which is also referred to as 'else' rule, and t^i rules, each of which has i terms, for $i = 0, 1, 2, \ldots, n$ respectively. However, each rule usually covers more than one instance and the rule set is expected to cover all instances. Therefore, the number of rules from a rule set is usually less than the number of instances from a data set. As also justified above, each rule would have up to n (the number of attributes) terms due to the requirement that each attribute can only appear once comprising one of its possible values in any of the rules. On the basis of above description, the complexity of a rule set is up to the product of dimensionality and sample size of a data set.

On the other hand, Naïve Bayes, which is another machine learning algorithm, aims to identify a list of probabilistic correlations, such as $P(class = 1|x_1 = 1) = 0.5$, between an input attribute and the class attribute [23]. The model complexity would be dependent on the number of correlation pairs. For example, a data set contains n input attributes plus one class attribute and each of the attributes (including the class attribute) has t values included in the data set. In this example, the complexity of the model would be equal to $n \times t^2$. This is because Naïve Bayes aims to identify the probabilistic correlation of each attribute-value pair to each class label. In this sense, each attribute has t values and the class attribute has t labels so the number of correlation pairs is t^2 for the attribute and thus the overall number is the same as n times the above number (t^2).

The above description relating to Naïve Bayes also indicates that a large number of attributes with many possible values is likely to make a data set in a large size, and thus data size is also one of the impact factors that may affect the interpretability.

3.3   Model Representation

As mentioned in Sect. 2, different types of machine learning algorithms may generate models represented in different forms. For example, the 'divide and conquer' approach generates a rule set in the form of a decision tree as illustrated in Fig. 1 whereas the 'separate and conquer' approach would generate if-then rules represented in a linear list. In addition, neural network learning algorithm would generate a multi-layer network with a number of interconnected nodes, each of which represents a perceptron. As also described in Sect. 2, models generated by rule based learning methods are in white box and thus well transparent whereas models constructed by neural network learning methods are in black box and thus poorly transparent. As justified in Sect. 3.1, the level of transparency can affect the level of interpretability. However, models that demonstrate the same level of transparency may also have different levels of interpretability due to their differences in terms of representation such as rule based models. The rest of this subsection justifies why and how the nature of model representation can affect the level of interpretability using rule based models as a special example.

As mentioned earlier in this subsection, a rule set can be represented in two different forms namely, decision tree and linear list.

Decision tree representation is criticized by Cendrowska and identified as a major cause of overfitting in [20] due to the replicated sub-tree problem as illustrated in Fig. 1. It can be seen from the Fig. 1 that the four sub-trees which have node C as their roots are identical. This is due to the requirement that all rules must have at least one common attribute involved as mentioned in Sect. 3.1. This kind of problems mentioned above could be referred to as redundancy, which would lower the interpretability as argued in [19] that decision trees are often quite complex and difficult to understand and thus inscrutable to provide insight into a domain for knowledge usage [18, 19].

In comparison with decision trees, linear lists do not have the constraint that all rules must have common attributes and thus reduces the presence of redundant terms in a rule set. However, redundancy may still arise with this representation. This is because the same attribute may repetitively appear in different rules as illustrated by the example below:

Rule 1: If $x = 0$ and $y = 0$ Then class = 0;
Rule 2: If $x = 0$ and $y = 1$ Then class = 1;
Rule 3: If $x = 1$ and $y = 0$ Then class = 1;
Rule 2: If $x = 1$ and $y = 1$ Then class = 0;

When a training set is large, there would be a large number of complex rules generated. In this case, the presence of redundancy would make the rule set (represented in a linear list) very cumbersome and difficult to interpret for knowledge usage. In other words, a large number of complex rules represented in this way is quite like a large number of long paragraphs in an article that would be very difficult for people to read and understand. Instead, people prefer to look at diagrams to gain information.

In this sense, a graphical representation of rules would be expected to improve the interpretability of a model. More details about ways to improve interpretability will be presented in Sect. 5.

3.4 Human Characteristics

As mentioned in Sect. 2, different people may have different levels of expertise and preferences and thus different levels of cognitive capability to understand the knowledge extracted from a particular computational model. The rest of this subsection justifies why and how human expertise and personality may affect the interpretability of computational models.

In terms of expertise, due to the fact that an expert system is used to act as a domain expert to extract knowledge or make predictions, people need to have the relevant expertise in order to be able to understand the context. From this point of view, the exactly same model may demonstrate different levels of interpretability for different people due to their different levels of expertise in this domain.

In terms of preferences, due to the fact that different people may have different preferences with respect to the way of reading, the exactly same model may also demonstrate different levels of interpretability for different people due to their different preferences. From this point of view, human characteristics are also an impact factor that may affect the interpretability of a model.

As mentioned in Sect. 3.3, model representation can affect the interpretability with respect to the level of redundancy. In other words, the same model can have different levels of interpretability depending on its representation. However, due to the difference in expertise and preferences, a particular representation may be understandable to some people but not to others. For example, some people in nature sciences/engineering would prefer to read diagrams/mathematical formulas whereas others may dislike them. From this point of view, model representation, human expertise and characteristics may jointly determine the cognitive capability for people to read and understand the knowledge extracted from a model.

4 Criteria for Evaluation of Interpretability

On the basis of description in Sect. 3, the list of identified impact factors would have causal relationships to the interpretability of a model as illustrated in Fig. 2.

Figure 2 indicates that the evaluation of a model's interpretability could be based on several criteria, namely model transparency, model complexity, model redundancy and cognitive capability, due to their direct relationships to interpretability.

In terms of model transparency, as mentioned in Sect. 3.1, the evaluation is based on information visualization. In other words, to what extent the information is visible or hidden to people. For example, a neural network learning method generates a

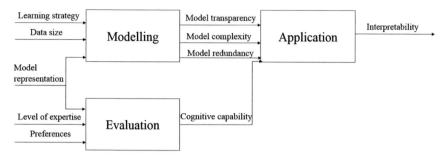

Fig. 2 Causal relationship between impact factors and interpretability [24]

model in black box, which means that the information in the model is mostly hidden to people and thus poorly transparent. In contrast, a rule based learning method generates a model in white box, which means the information in the model is totally visible to people and thus well transparent.

In terms of model complexity, the evaluation is subject to the type of learning algorithms to some extent. For example, with regard to rule based methods, the model complexity could be measured by checking the total number of rule terms in a rule set, which is referred to as rule set complexity. For the rule set given below, the complexity would be 8.

Rule 1: If x = 0 and y = 0 Then class = 0;
Rule 2: If x = 0 and y = 1 Then class = 0;
Rule 3: If x = 1 and y = 0 Then class = 0;
Rule 4: If x = 1 and y = 1 Then class = 1;

In addition, with regard to Naïve Bayes method, the complexity could be measured by checking the number of probabilistic correlation pairs such as $P(class = 1 | x_1 = 1)$ = 0.5. For the example shown in Table 1 and in the list of probabilistic correlations, the complexity would be 8. Besides, the model complexity also partially depends on data size as mentioned in Sect. 3.2.

Pair 1: $P(y = 0 | x_1 = 0) = 0.5$
Pair 2: $P(y = 1 | x_1 = 0) = 0.5$
Pair 3: $P(y = 0 | x_1 = 1) = 0.5$

Table 1 Probabilistic correlations

x_1	x_2	y
0	0	1
0	1	0
1	0	0
1	1	1

Pair 4: $P(y = 1|x_1 = 1) = 0.5$
Pair 5: $P(y = 0|x_2 = 0) = 0.5$
Pair 6: $P(y = 1|x_2 = 0) = 0.5$
Pair 7: $P(y = 0|x_2 = 1) = 0.5$
Pair 8: $P(y = 1|x_2 = 1) = 0.5$

In terms of model redundancy, the evaluation could be based on the extent of information duplication. For example, a rule set may be represented in different forms namely, decision tree, linear list and rule based network. As mentioned in Sect. 3.3, both of the first two representations may include duplicated information. For decision tree, the replicated subtree problem is a typical example to indicate that redundancy is a principal problem that arises with this representation. As can be seen from Fig. 1, there are four identical subtrees. For a linear list, as can be seen from the rule set given earlier in this section, all of the four rules have two common attributes, namely 'x' and 'y', which are repeated. The authors have recently developed two types of network topologies [22] in order to reduce redundancy. One is attribute-value-oriented and the other one is attribute oriented. More details on the improvements of interpretability by using network topologies will be described in Sect. 5.

In terms of cognitive capability, the evaluation would be based on empirical analysis following machine learning approaches. This is in order to analyze to what extent the model representation is understandable to particular people. In detail, this could be designed as a classification task to predict the cognitive capability in qualitative aspects or as a regression task to predict in quantitative aspects. Briefly speaking, the analysis could be done by collecting the data records on expertise and preferences from previous people who have high similarities to the current people and then taking the majority voting for a classification task or averaging for a regression task. With respect to the cognitive capability. The above task can be done effectively using k nearest neighbor, which is a popular machine learning method.

5   Improvement of Interpretability

Section 3 listed some impact factors for interpretability and Sect. 4 introduced some criteria with regard to evaluation of interpretability. In particular, transparency is dependent on learning algorithms. Model complexity is subject to both learning algorithms and data size. Model redundancy is subject to the form of model representation. Cognitive capability is determined jointly by model representation, human expertise and characteristics. This section demonstrates in detail some ways recommended in [24] towards improvements of interpretability through advancing and using computational intelligence approaches.

Table 2 Data example from [21]

Object	Height	Hair	Eyes	Class
O1	Short	Blond	Blue	C1
O2	Short	Blond	Brown	C2
O3	Tall	Red	Blue	C1
O4	Tall	Dark	Blue	C2
O5	Tall	Dark	Blue	C2
O6	Tall	Blond	Blue	C1
O7	Tall	Dark	Brown	C2
O8	Short	Blond	Brown	C2

5.1 Scaling up Algorithms

As mentioned in Sect. 3, the strategy of a learning algorithm may affect the model transparency. In this case, the transparency could be improved by scaling up algorithms in terms of depth of learning. For example, rule based methods usually generate models with good transparency because this type of learning is in great depth and on an inductive basis. As also mentioned in Sect. 2, models represented by rules can provide straightforward explanations with regard to the outputs of a computational model.

On the other hand, the performance of a learning algorithm would also affect the model complexity as mentioned in Sect. 3. In this case, the model complexity could be reduced by scaling up algorithms. In the context of rule based models, complexity could be reduced through proper selection of rule generation approaches. As mentioned in Sect. 3, there are typically two generation approaches namely, 'divide and conquer' and 'separate and conquer'. As mentioned in [19], the latter is usually likely to generate less complex rule sets than the former. The following example is given for illustration (Table 2):

According to [21], both ID3 and Prism generate four rules as follows:
Rule set by ID3:

- if Hair = red then Class = C1;
- if Hair = blond and Eyes = blue then Class = C1;
- if Hair = blond and Eyes = brown then Class = C2;
- if Hair = dark then Class = C2;

Rule set by Prism:

- if Hair = red then Class = C1;
- if Hair = blond and Eyes = blue then Class = C1;
- if Eyes = brown then Class = C2;
- if Hair = dark then Class = C2;

As can be seen from the above example, ID3, which follows the divide and conquer rule learning, generates a redundant term (Hair = blond) in the third rule whereas

Table 3 Number of rules and average number of rule terms

Data set	C4.5		Prism	
	Count (rules)	Avg (terms)	Count (rules)	Avg (terms)
Anneal	35	3.02	**18**	**2.28**
Breast-w	14	4.33	**12**	**1.42**
cmc	157	6.23	**36**	**4.67**
Credit-a	30	3.9	**12**	**1.0**
Credit-g	103	7.3	**20**	**1.8**

Prism, which follows the separate and conquer rule learning, successfully manages to remove the redundancy. The above example also indicates Prism manages to reduce the complexity of rule set through the removal of the redundant term from the third rule. In addition, Cendrowska compared ID3 with Prism in [20] on the training set, which has 7 attributes and 647 instances and is provided by the King-Knight-King-Rook chess end game [25]. As reported in [20], the rule set generated by ID3 contains 52 rules and 337 terms whereas that generated by Prism only has 15 rules and 48 terms.

On the other hand, the following empirical investigation using the data sets retrieved from UCI repository [26], which is illustrated in Table 3, also indicates that separate and conquer rule learning algorithms are likely to generate fewer and more general rules than algorithms that follow divide and conquer rule learning even when small training data is used.

In addition, it is also helpful to employ pruning algorithms, such as reduced error pruning [27] and Jmid-pruning [16], to simplify rule sets [18–20]. In this way, some redundant or irrelevant information is removed and thus the interpretability is improved. Empirical results, which are reported in [24] and illustrated in Table 4, indicate that applying Jmid-pruning to Prism usually reduces the number of rules and the average number of terms per rule and thus helps improve the interpretability of a rule set.

In sentiment analysis, Support Vector Machines and Naïve Bayes are two popular methods used for prediction of sentiments. However, as mentioned in Sect. 2, computational models generated by using these two methods are generally less transparent, which again indicates the necessity of improvement of model transparency through scaling up algorithms.

Reduction of model complexity is also necessary in sentiment analysis. This is because sentiment data is typically in the form of text which is unstructured unlike the data sets retrieved from machine learning repositories such as UCI [26]. As introduced in [28, 29], features in sentiment data usually include unigrams, bigrams and trigrams or various combinations between them, which indicates that such data is generally of high complexity. This again shows that it is highly necessary to involve effective management of model complexity through scaling up algorithms. For example, if rule learning methods are used to build models for sentiment analysis,

Table 4 Number of rules and average number of rule terms by Prism [24]

Dataset	Prism without pruning		Prism with Jmid-pruning	
	Count (rules)	Avg (terms)	Count (rules)	Avg (terms)
cmc	36	4.67	**25**	**4.48**
Vote	25	6.28	**15**	**5.13**
kr-vs-kp	63	5.84	**21**	**5.52**
Ecoli	24	**1.88**	**17**	1.94
Anneal.ORIG	16	**1.56**	**12**	3.67
Audiology	48	3.60	**38**	**2.79**
Car	**2**	**1.0**	3	2.0
Optdigits	431	7.46	**197**	**6.53**
Glass	26	**2.85**	**24**	3.29
Lymph	**10**	1.3	**10**	**1.11**
Yeast	37	1.68	**20**	**1.5**
Shuttle	30	3.87	**12**	**1.0**
Analcatdataasbestos	**5**	1.6	**5**	**1.4**
Irish	**10**	1.5	11	**1.27**
Breast-cancer	**11**	1.09	**11**	**1.0**

proper selection of rule generation approaches and employing pruning algorithms are generally helpful for reducing model complexity.

As mentioned in [28], general preprocessing techniques partially aim to remove redundancy that exists in data. However, it is currently still difficult to guarantee that such redundancy can be effectively eliminated from data. When the preprocessing techniques fail to detect and remove some redundancy from data, it would be very likely to result in the occurrence of model redundancy. On the basis of the above description, it is worth to manage the effective reduction of model redundancy on algorithms side through scaling up algorithms rather than to rely only on data preprocessing techniques. The proper selection of rule generation approaches and employing pruning algorithms mentioned above are also helpful here.

5.2   Scaling down Data

As mentioned in Sect. 3, the size of data may also affect the model complexity. In other words, if a data set has a large number of attributes with various values and instances, the generated model is very likely to be complex.

The dimensionality issue can be resolved by using feature selection techniques such as entropy [30] and information gain [3], both of which are based on information theory pre-measuring uncertainty present in the data. In other words, the aim is to remove those irrelevant attributes and thus make a model simpler. In addition, the issue can also be resolved through feature extraction methods, such as Principal Component Analysis (PCA) [31] and Linear Discriminant Analysis (LDA) [32].

When a dataset contains a large number of instances, it is usually required to take advantage of sampling methods to choose the most representative instances. Some popular methods comprise simple random sampling [33], probabilistic sampling [34] and cluster sampling [35].

Besides, it is also necessary to remove attribute values due to the presence of irrelevant attribute values. For example, in a rule based method, an attribute-value pair may be never involved in any rules as a rule term. In this case, the value of this attribute can be judged irrelevant and thus removed. In some cases, it is also necessary to merge some values for an attribute in order to reduce the attribute complexity especially when the attribute is continuous with a large interval. There are some ways to deal with continuous attributes such as ChiMerge [36] and use of fuzzy linguistic terms [5].

In the context of rule based systems, as analyzed in Sect. 3.2, dimensionality reduction can effectively reduce the average number of rule terms per rule. This is because each single rule can have only up to n rule terms, where n is the data dimensionality. As also analyzed in Sect. 3.2, reduction of the number of values for each input attribute can effectively reduce the number of rules. For example, three attribute a, b, c have 2, 3 and 4 values respectively. In this case, the number of first order rules (with one rule term) is $2 + 3 + 4 = 9$; the number of second order rules (with two rule terms) is $2 \times 3 + 2 \times 4 + 3 \times 4 = 26$; the number of third order rules (with three rule terms) is $2 \times 3 \times 4 = 24$. In addition, effective sampling of data can reduce the number of instances and may also reduce the number of values for some or all input attributes. Therefore, dimensionality reduction, data sampling and reduction of the number of attribute values are all generally effective towards reduction of model complexity.

On the basis of above description, the complexity could be reduced through dimensionality reduction, data sampling and removal or merging of attribute values. In sentiment analysis, such scaling down data is even more necessary due to the variety of features in sentiment data as mentioned in Sect. 5.1. In addition, as reported in [37], rule learning methods have been popularly used as sentiment classification techniques [38–40], which indicates to a larger extent the necessity of scaling down data through the ways introduced above.

5.3 Selection of Model Representation

As mentioned in Sect. 3, a change of model representation would usually result in the change of model interpretability. As also introduced, rule based models could be represented in different forms such as decision tree and linear list. These two representations usually have redundancy issues. For example, a decision tree may have the replicated subtree problem and a linear list may have the attribute appear

Fig. 3 Deterministic rule based network [22]

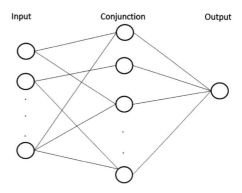

in different rules on a repetitive basis. This kind of problem may be resolved by converting to a rule based network representation as illustrated in Fig. 3.

In general, this is a three layer network. In the first layer, each node represents an input attribute and this layer is referred to as input layer. In the middle layer, each node represents a rule to make the conjunction among inputs and provide outputs for the node in the last layer and thus the middle layer is referred to as conjunction layer. The only node in the last layer represents the class output and thus this layer is referred to as output layer. In addition, the nodes in the input layer usually have connections to other nodes in the conjunction layer. Each of the connections represents a condition judgment which is explained further using specific examples. However, a node in the input layer may not necessarily have connections to other nodes in the conjunction layer. This is due to a special case that an attribute may be totally irrelevant to making a classification. In other words, this attribute is not involved in any rules in the form of rule terms.

In terms of model redundancy, this network based representation solves this problem. As illustrated in Fig. 3, each of the attributes is located in the input layer and only appears once with some branches connected to the nodes, each of which represents a rule, in the conjunction layer. This totally removes the repetitive appearance of the same attribute in different rules, which is likely to arise with decision tree and linear list representations. In the context of programming, a selection structure could be represented in the form of 'if-else' statement or 'switch-case'. The network based representation is philosophically similar to the 'switch-case' whereas the linear list is similar to the 'if-else' statement. Due to the requirement that the program needs to have a good readability, 'switch-case' is used instead of 'if-else' statement in many cases. In 'switch-case', a variable only appears once as an input parameter of a switch function and each of the possible values of the variable is involved in a particular case respectively. This removes the repetitive judgment with regard to the value of the same variable in comparison with the use of the 'if-else' statement.

In addition, the network based representation has another advantage that the network could be represented in colors to highlight important information. For example, there is a rule set given below:

Fig. 4 Deterministic rule based network example [22]

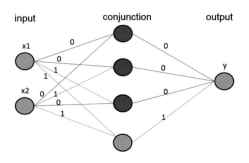

Rule 1: If $x_1 = 0$ and $x_2 = 0$ Then class = 0;
Rule 2: If $x_1 = 0$ and $x_2 = 1$ Then class = 0;
Rule 3: If $x_1 = 1$ and $x_2 = 0$ Then class = 0;
Rule 2: If $x_1 = 1$ and $x_2 = 1$ Then class = 1;

The corresponding representation in network form is illustrated in Fig. 4. In this diagram, the rules that are firing and the conditions that are satisfied for each of the rules can be explicitly highlighted by using the two colors namely, green and red. The former color generally means the condition is met and it is permitted to pass through whereas the latter color means the opposite case.

On the other hand, the network based representation can also be generalized to different logics such as deterministic, probabilistic and fuzzy logic, see Fig. 5.

In this network topology, the modifications are made to the one illustrated in Fig. 3 by adding two new layers called input values and disjunction respectively, and

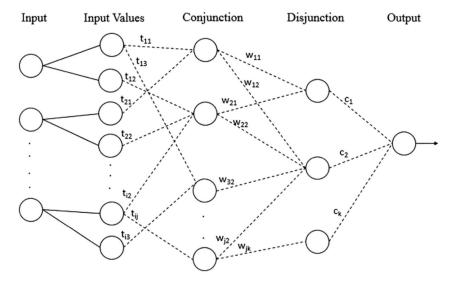

Fig. 5 Unified rule based network

assigning a weight to each of the connections between nodes. In the input values layer, each node represents one of the values of an input attribute. For those values of the same input attribute, they cannot commonly appear in the same rule as rule terms. Therefore, for these nodes in the input values layer, if they are connected to the same node in the input layer, then they cannot be commonly connected to the same node in the conjunction layer. In the disjunction layer, each node represents a class label. The topology allows representing inconsistent rules, which means that the same rule antecedent could be mapped to different classes (consequents). For example, the first node in the conjunction layer is mapped to both the first and the second node in the disjunction layer as illustrated in Fig. 5. With regard to the weights assigned to the connections between nodes, they would represent the truth values if the computation is based on deterministic or fuzzy logic. The truth value would be crisp (0 or 1) for deterministic logic whereas it would be continuous (between 0 and 1) for probabilistic and fuzzy logic.

For fuzzy computational models, it is required not only to interpret the way that the inputs determine the outputs but also the degree of likelihood. The nature of the generalized network representation would fulfil the above requirement by simply assigning a weight to each of the connections between nodes as illustrated in Fig. 5. This is because each of the connections is only involved in one rule in this representation. In contrast, in a decision tree representation, a connection may be involved in more rules with the need that different weights are assigned for the involvements in different rules. In addition, for a linear list representation, if each of the rule terms is assigned a weight, it is likely to make the rules less readable.

On the other hand, the network topology illustrated in Fig. 5 can interpret well the fuzzy truth values derived in different stages of fuzzy simulation namely, fuzzification, application, implication, aggregation and defuzzification.

In particular, each of the weights assigned to each of the connections between the second and third layers indicates the fuzzy membership degree of each input value derived in the fuzzification stage.

In the application stage, the firing strength of a fuzzy rule is computed through conjunction of all the fuzzy membership degrees derived in the first stage. In other words, the minimum among these fuzzy membership degrees is taken as the firing strength of the fuzzy rule.

Each of the weights assigned to each of the connections between the third and fourth layers indicates the fuzzy membership degrees of given rule consequents, each of which is derived in the implication stage through the conjunction of the rule firing strength and the fuzzy membership degree of the corresponding output value.

Each of the weights assigned to each of the connections between the last two layers indicates the overall fuzzy membership degree of each output value, which is derived through the disjunction of the fuzzy membership degrees for each output value from a particular fuzzy rule.

The final output illustrated in Fig. 5 is derived in the defuzzification stage through weighted majority voting for fuzzy classification or weighted averaging for fuzzy regression.

On the basis of above description in this section, the model interpretability can be improved by selecting a model representation with high transparency and low redundancy. In sentiment analysis, both probabilistic and fuzzy logic have become increasingly popular for uncertainty handling in prediction of sentiments [41–43]. From this point of view, advancing interpretability of such computational models has become highly significant, which again shows the necessity of proper selection of a model representation.

5.4 Assessment of Cognitive Capability

As mentioned in Sect. 3, due to the difference in level of expertise and personal preferences, the same model representation may show different levels of comprehensibility for different people. For example, people who do not have a good background in mathematics may not like to read information in mathematical notations. In addition, people in social sciences may not understand technical diagrams used in engineering fields. On the basis of above description, cognitive capability needs to be assessed to make the knowledge extracted from computational models more interpretable to people in different domains, due to the fact that sentiment analysis is actually involved in different domains. This can be resolved by using expert knowledge in cognitive psychology and human-machine engineering, or following machine learning approaches to predict the capability as mentioned in Sect. 4.

6 Conclusion

As mentioned in Sect. 2, many studies focused on the improvement of accuracy of sentiment predictions using computational intelligence approaches. However, there are very few studies focused on addressing the issues of interpretability of sentiment analysis systems. This paper justifies the significance of interpretability of computational models, which are used as sentiment analysis systems, and identifies some significant impact factors for their interpretability. According to each of the impact factors, some criteria are proposed for the evaluation of interpretability using computational intelligence methodologies. This paper also provides recommendations towards improvements in terms of interpretability through the use of computational intelligence approaches. In particular, scaling up algorithms and scaling down data, which can be effectively used for reduction of model overfitting, are considered also capable of improvement of interpretability through reduction of model complexity. In addition, a generalized rule based network topology is proposed for the improvement of interpretability through reduction of model redundancy. These proposed solutions have also been investigated theoretically and empirically with expected outcomes. The aspects relating to human characteristics introduced in the paper will be further investigated in the future research. In addition, in sentiment analysis, data preprocessing is an important stage in order to have the data structured and provided with relevant features [13, 28]. Therefore, in what way to provide the structural data with an acceptable number of relevant features would be a significant impact factor for interpretability of sentiment analysis systems. In other words, it is significant to determine what elements need to be kept or removed as well as ways to use n-grams [28] through the selection of features such as unigrams, bigrams and trigrams or combinations between them. The above description indicates the needs of further research in computational intelligence towards improvement of model interpretability without loss of accuracy for sentiment analysis.

References

1. Schlager, J.: Systems engineering: key to modern development. IRE Trans. EM **3**(3), 64–66 (1956)
2. Hall, A.D.: A Methodology for Systems Engineering. Van Nostrand Reinhold (1962)
3. Mitchell, T.M.: Machine Learning, McGraw Hill (1997)
4. Liu, H., Gegov, A., Stahl, F.: Categorization and construction of rule based systems. In: 15th International Conference on Engineering Applications of Neural Networks, Sofia, Bulgaria (2014)
5. Ross, T.J.: Fuzzy Logic with Engineering Applications. John Wiley & Sons Ltd, West Sussex (2004)
6. Simpson, S.G.: Mathematical Logic, PA (2013)
7. Tan, P.-N., Steinbach, M., Kumar, V.: Introduction to Data Mining. Pearson Education, New Jersey (2006)
8. Higgins, C.M.: Classification and Approximation with Rule Based Networks. Pasadena, California (1993)
9. Uttley, A.M.: The design of conditional probability computers. Inf. Control **2**, 1–24 (1959)
10. Kononenko, I.: Bayesain Neual Networks. Biol. Cybern. **61**, 361–370 (1989)
11. Rosenblatt, F.: Principles of Neurodynamics: Perceptron and the Theory of Brain Mechanisms. Spartan Books, Washington, DC (1962)
12. Ekeberg, O., Lansner, A.: Automatic generation of internal representations in a probabilistic artificial neural network. In: Proceedings of the First European Conference on Neural Networks (1988)
13. Altrabsheh, N., Cocea, M., Fallahkhair, S.: Sentiment analysis: towards a tool for analysing real-time students feedback. In: 2014 IEEE 26th International Conference on Tools with Artificial Intelligence, Limassol, Cyprus (2014)
14. Liu, B.: Sentiment Analysis and Opinion Mining. Morgan & Claypool Publishers (2012)
15. Stahl, F., Jordanov, I.: An overview of use of neural networks for data mining tasks. In: WIREs: Data Mining and Knowledge Discovery, pp. 193–208 (2012)
16. Liu, H., Gegov, A., Stahl, F.: J-measure based hybrid pruning for complexity reduction in classification rules. WSEAS Trans. Syst. **12**(9), 433–446 (2013)
17. Liu, H., Gegov, A., Stahl, F.: Unified framework for construction of rule based classification systems. In: Pedrycz, W., Chen, S.M. (eds.) Inforamtion Granularity, Big Data and Computational Intelligence, vol. 8, pp. 209–230, Springer (2015)
18. Quinlan, R.: C4.5: Programs for Machine Learning, Morgan Kaufman (1993)
19. Fürnkranz, J.: Separate-and-conquer rule learning. Artif. Intell. Rev. **13**, 3–54 (1999)
20. Cendrowska, J.: PRISM: an algorithm for inducing modular rules. Int. J. Man-Machine Stud. **27**, 349–370 (1987)
21. Deng, X.: A Covering-Based Algorithm for Classification: PRISM. Regina (2012)

22. Liu, H., Gegov, A., Cocea, M.: Network Based Rule Representation for Knowledge Discovery and Predictive Modelling. In: IEEE International Conference on Fuzzy Systems, Istanbul (2015)

23. Rish, I.: An empirical study of the Naïve Bayes classifier. In: IJCAI 2001 Workshop on Empirical Methods in Artificial Intelligence, vol. 3, no. 22, pp. 41–46 (2001)

24. Liu, H., Gegov, A., Cocea, M.: Rule Based Systems for Big Data: A Machine Learning Approach, vol. 13. Springer, Berlin (2016)

25. Quinlan, R.: Discovering rules from large collections of examples: a case study. In: Michie, D. (ed.) Expert Systems in the Micro-Electronic Age, pp. 168–201. Edinburgh, Edinburgh (1979)

26. Lichman, M.: Machine Learning Repository, University of California, School of Information and Computer Science (2013). http://archive.ics.uci.edu/ml

27. Elomaa, T., Kääriäinen, M.: An analysis of reduced error pruning. J. Artif. Intell. Res. **15**(1), 163–187 (2001)

28. Altrabsheh, N., Cocea, M., Fallahkhair, S.: Learning sentiment from students' feedback for real-time interventions in classrooms. In: Adaptive and intelligent systems: Third International Conference, ICAIS 2014, Bournemouth (2014)

29. Kummer, O., Savoy, J.: Feature Selection in Sentiment Analysis. In: CORIA 2012, Bordeaux (2012)

30. Shanno, C.E.: A mathematical theory of communication. Bell Syst. Tech. J. **27**(3), 379–423 (1948)

31. Jolliffe, I.T.: Principal Component Analysis. Springer, New York (2002)

32. Yu, H., Yang, J.: A direct LDA algorithm for high-dimensional data—with application to face recognition. Patt. Recogn. **34**(10), 2067–2069 (2001)

33. Yates, D.S., Moore, D.S., Starnes, D.S.: The Practice of Statistics, 3rd ed. Freeman (2008)

34. Deming, W.E.: On probability as a basis for action. Am. Stat. **29**(4), 146–152 (1975)

35. Kerry, S.M., Bland, J.M.: The intracluster correlation coefficient in cluster randomisation. Br. Med. J. **316**(7142), 1455–1460 (1998)

36. Kerber, R.: ChiMerge: discretization of numeric attributes. In: Proceedings of the 10th National Conference on Artificial Intelligence, California (1992)

37. Medhat, W., Hassan, A., Korashy, H.: Sentiment analysis algorithms and applications: a survey. Ain Shams Eng. J. **5**(4), 1093–1113 (2014)

38. Yi, H., Li, W.: Document sentiment classification by exploring description model of topical terms. Comput. Speech Lang. **50**, 386–403 (2011)

39. Lewis, D.D., Ringuette, M.: A comparison of two learning algorithms for text categorization. In: Third Annual Symposium on Document Analysis and Information Retrieval, Las Vegas (1994)

40. Chakrabarti, S., Roy, S., Soundalgekar, M.V.: Fast and accurate text classification via multiple linear discriminant projections. Int. J. Very Large Data Bases **12**(2), 170–185 (2003)

41. Zhao, C., Wang, S., Li, D.: Fuzzy sentiment membership determining for sentiment classification. In: IEEE International Conference on Data Mining Workshop (ICDMW), Shenzhen (2014)

42. Nadali, S., Murad, M., Kadir, R.: Sentiment classification of customer reviews based on fuzzy logic. In: International Symposium in Information Technology (ITSim), Kuala Lumpur (2010)

43. Colace, F., Santo, M.D., Greco, L.: A probabilistic approach to Tweets' sentiment classification. In: 2013 Humaine Association Conference on Affective Computing and Intelligent Interaction, Geneva (2013)

Chinese Micro-Blog Emotion Classification by Exploiting Linguistic Features and SVM^{perf}

Hua Xu, Fan Zhang, Jiushuo Wang and Weiwei Yang

Abstract These years, micro-blog emotion mining becomes one of the research hotspots in social network data mining. Different from state of the art study, this paper presents a novel method for emotion classification, which is SVM^{perf} based method combined with syntactic structure of Chinese micro-blogs. The classified emotion type includes Happiness, Anger, Disgust, Fear, Sadness and Surprise. For the proposed method, an emotional lexicon is constructed and linguistic features are extracted from micro-blog corpus firstly. Secondly, for the current feature space dimension is higher, Chi-square test is used to extract the high-frequency and high-class relevance keywords. At the same time, Pointwise Mutual Information (PMI) is used to pick the effective low frequency words in feature dimension reduction, which can reduce the computational complexity. Finally, SVM^{perf} is applied for the emotion classification. In order to illustrate the effectiveness of the algorithm, LIBSVM and SVM-Light are used as the baseline. The data from Sina Micro-blog (weibo.com) have been used as the experiment data. The experiment results demonstrate that all the above features contribute to emotion classification in micro-blogs, and the results validate the feasibility of the proposed approach. It also shows that SVM^{perf} is an appropriate choice of classifier for emotion classification.

Keywords Micro-blog · Text mining · Emotion classification · SVM^{perf}

1 Introduction

In the era of information explosion, social network applications present a platform for people to share various news and information. Millions of people record and share their daily lives in micro-blog. As a result, they produce a large number of social network data, containing almost all kinds of opinions and sentiments. These

H. Xu (✉) · F. Zhang · J. Wang · W. Yang
State Key Laboratory of Intelligent Technology and Systems, Tsinghua National
Laboratory for Information Science and Technology, Department of Computer Science
and Technology, Tsinghua University, Beijing 100084, China
e-mail: xuhua@tsinghua.edu.cn

© Springer International Publishing Switzerland 2016
W. Pedrycz and S.-M. Chen (eds.), *Sentiment Analysis and Ontology Engineering*,
Studies in Computational Intelligence 639, DOI 10.1007/978-3-319-30319-2_10

kinds of data usually include the description of emergencies, incidents, disasters and some other hot events, attaching people's emotions and sentiments as well.

Currently, emotion processing in text attracts more and more researchers' attention in the field of natural language processing. In early studies, they mainly tend to determine the polarity of the given text for positive or negative orientation [1]. But these orientations cannot adequately express complex emotions of human. The study of fine-grained emotions becomes important for mining the further information on emotions. Wen and Wan [2] proposed an approach based on class sequential rules to classify the given texts into seven emotion types. And more researchers usually use rule-based approaches, the probability distribution model and machine learning (e.g., decision tree, SVM and Naive Bayes classifier) for emotion analysis [3, 4]. Making full use of micro-blog features (e.g. subject characteristics, expression features, emotion labels) can identify emotions effectively [5]. But existing methods based on rule or machine learning usually cannot consider the syntactic features in Chinese micro-blogs comprehensively.

In this paper, we adopt Ekman and Friesen's list of six basic emotions (i.e., "Happiness", "Anger", "Disgust", "Fear", "Sadness", and "Surprise") [6]. Then we present a novel method for emotion classification based on SVM^{perf}. The innovation of this paper seeks to promote the capacity of feature extraction based on the syntactic and grammar structure in Chinese micro-blogs. This novel method not only overcomes the difficulty of Chinese's complexity but also improves the performance of emotion analysis in Chinese blogs, an unchartered territory yet. χ^2 (Chi-square) test and PMI are applied to feature dimension reduction for reducing the computational complexity. Finally, this paper uses the SVM^{perf} for classification training and testing. Within the micro-blog domain, the proposed approaches have scientific values on short-text emotion analysis, and the experimental results show the effectiveness of the approach.

The remaining sections are organized as follows. Section 2 discusses the related works on emotion classification. Section 3 focuses on linguistic feature extraction. Section 4 introduces the method on feature dimension reduction. Section 5 presents the classification algorithm. Section 6 shows the experimental results. In the end, we discuss the remaining challenges and possibilities for the future research.

2 Related Work

In the aspect of feature extraction, Li et al. [7] used the technique of emotion cause extraction for feature selection. Liu et al. [8] proposed the emotional dictionaries including internet slang and emoticons for emotion analysis based on the phrase path and multiple characteristics. Pei et al. [9] used emoticons to construct the emoticon networks model by FP-growth algorithm for micro-blog sentiment analysis. To obtain the feature of the emotional word, Bai et al. [10] proposed an emotional dictionary by using word2vec according to the Semantic Orientation Pointwise Similarity Distance

(SO-SD) model. Some research focused on the lexical and semantic features and so on [11].

Meanwhile, mining the syntactic information is another way to extract features. Some work focused on the emotion of a sentence by mining the motivated features in context emotion [12]. Kim et al. [13] proposed a method for lyrics-based emotion classification using four kinds of syntactic analysis rules (i.e., negative word combination, time of emotions, emotion condition change and interrogative sentence). Agrawal et al. [14] proposed an unsupervised approach to detect emotions by extracting the noun, adjective, verb, adverb and the syntactic grammatical relation (such as nominal subject, negation and adjectival modifier etc.), and computed the emotional vector for each affected word. Ho et al. [15] proposed a high-order Hidden Markov Model for emotion analysis by taking the psychological characteristic of the emotion and linguistic information of the input text into consideration.

Generally speaking, there are two common approaches for emotion classification, i.e., the lexicon methods and the machine learning methods. Li et al. [16] proposed a lexicon-based multi-class semantic orientation analysis for micro-blog. Liu's group [17] proposed a multi-label classification based approach for emotion analysis, including text segmentation, feature extraction and multi-label classification. A multi-task multi-label classification model was proposed to perform classification based on both emotions and topics concurrently [18]. An unsupervised domain independent method for sentiment classification is proposed in order to build one emotion vocabulary list for text emotion classification [19]. In He's work [20], the researcher compared the performance of the three methods (Naive Bayesian, SVM, and SMO) in micro-blog emotion classification.

Unlike the above work, this paper presents a novel emotion classification based on SVM^{perf}. The innovation of this paper lies in mining the effective features by using the method of feature extraction based on the syntactic and grammar structure. Secondly, the proposed approach reduces the dimension of the feature space by combining χ^2 test with PMI. Finally, SVM^{perf} is used for classification training and testing and we take LIBSVM [21] and SVM-Light [22] as the comparison algorithms of the baseline.

3 Linguistic Feature Extraction

3.1 Emotional Lexicon Construction

Emotional words indicate people's attitudes and opinions. They are indispensable features in the process of emotion recognition. As different words can express different emotions, how to construct an effective emotional lexicon becomes a key task to improve the accuracy of emotion classification. Generally, the emotional lexicon can be constructed manually or automatically from the corpus [23]. This paper presents our method in the following two aspects.

Table 1 Closest words of "poet" extracting by word2vec

Word	Cosine distance
Novelist	0.6225367
Shakespeare	0.5734583
Litterateur	0.5559585
Composer	0.5498531
Authoress	0.5474696
Dramatist	0.5451957
Ideologist	0.5451323
Tagore	0.5419989
Poetry	0.5286906
Poems	0.5231616

Firstly, the words are selected manually as the standard words by several professional people. Meanwhile, we classify them into 6 different emotion types. Those words are from HowNet Dictionary,[1] National Taiwan University Sentiment Dictionary,[2] and Dalian University of Technology Sentiment Dictionary [24], respectively. Secondly, to obtain the larger capacity, we utilize word2vec[3] to expand the lexicon by acquiring words from the corpus automatically. Word2vec provides an efficient implementation for computing vector representation of words using the continuous bag-of-words and skip-gram architectures [25]. On the basis of word2vec, we can do the word clustering and choose the synonyms. To begin with, a large number of micro-blogs are randomly crawled from the Sina Micro-blog website (weibo.com) to constitute a 1.5G micro-blog corpus, which is used as the input file ("train.txt"). Then the corpus is transformed to vector representation of words by the following command:

/word2vec -train train.txt -output wordvectors.bin -cbow 0 -size 200 -window 5 -negative 0 -hs 1 -sample 1e-3 -threads 12 -binary 1

This paper uses the skip-gram model and the hierarchical soft ax in this model, each word vector dimensionality is set as 200. The number of the contextual window is 5. Finally, we put the result in the output file ("wordvectors.bin"). For example, if we input the word "poet", then we can get the closest words of it (see Table 1).

Finally, the closest words of the standard emotion words are chosen as the candidate words. It is necessary to compute the similarity between the candidate words and the standard words. We append the candidate words with the maximum similarity into the lexicon. The emotional lexicon based on 6 types of emotions is shown as follows (see Table 2) and the distribution of words in the prior-polarity lexicon is shown in Table 3.

---

[1] http://www.keenage.com/.

[2] http://nlg18.csie.ntu.edu.tw:8080/opinion/index.html.

[3] http://nlg18.csie.ntu.edu.tw:8080/opinion/index.html.

Table 2 Details of the emotional lexicon

Happiness	Sadness	Surprise	Fear	Disgust	Anger	Total
13043	6358	976	13765	10293	6796	51231

Table 3 Distribution of words in the prior-polarity lexicon

Positive	Negative
25.46%	74.54%

Table 4 Emotions set

Emotion	Examples of the emotions
Happiness	
Anger	
Disgust	
Sadness	
Surprise	

3.2 Emoticon Features

Bloggers often use various emoticons to express their current emotions. Emoticon features can express some more complex emotions, so mining them is important. As for the social domain, according to Twitter statistics, the emoticon " " accounts for 10 % in the top 100 emoticons. In previous research works, researchers tend to classify emoticons into three types: positive, negative and neutral. We think some more specific emoticons should be taken into account. In this paper, we establish a relationship between the emoticons and the six emotions. As for the fear emotion, we often use the keyword, such as "Jing Kong[4] (means Fear in English)", to represent this emotion. As for the other emotions, we can show them in Table 4. For example, the emoticon " " represents "Happiness" emotion, " " presents "Anger" emotion and so on. For example "We can't go out to play! ", it is reasonable that this presents a "Sad" emotion.

---

[4]It is the original Chinese character and the tone, and the corresponding literal English translation is shown in the bracket.

Table 5 The details of patterns of POS tags

POS tags group	The number of phrases
(adjective, nouns)	350
(Adverb, verbs)	2507
(Adverb, adjective)	1023
(nouns, adjective)	298
(adjective, adjective)	175
(Adverb, adverb)	555
(verbs, adverb)	608

3.3 Part-Of-Speech Features

Part-Of-Speech (POS) features are indispensable factors in textual emotion processing, particularly in syntactic analysis. They usually play different roles in emotion analysis. By mining the POS features, we find many nouns, adjectives, verbs and adverbs usually contain emotions. On the basis of the patterns of POS tags for extracting two-word phrases in [26], 1000 sentences with emotions are segmented by using the Chinese segmentation, and 49747 segmented results are 2 It is the original Chinese character and the tone, and the corresponding literal English translation is shown in the bracket. 6 Authors Suppressed Due to Excessive Length obtained. Table 5 describes the group of patterns of POS tags for extracting two-word phrases. From Table 5, it is clear that the group of (adverb and verbs), together with (adverb and adjective) can strongly express some emotions. Meanwhile, some nouns or verbs may express weak or less emotion in micro-blog domain.

3.4 Syntactic Structure Features

Mining the features in micro-blog mainly reflects the syntactic structure in one sentence and the dependence relationship in context. Syntactic structure analysis is a basic technology in the field of natural language processing. In accordance with dependency parsing of Chinese sentences, various types of relationships among the sentence's syntactic units are captured, such as the relationships between "subject", "predicate", "object", "attributive", "adverbial" and so on.

On the base of LTP[5] platform, the work in this paper extracts 24 interdependent relationships. We do some research works in syntactic structure analysis. However, not all of the dependencies are considered. For example, the "MT(voice-tense)" structure is only a simple grammatical feature, which is difficult to express the emotion well. With regard to the "SIM (similarity)" structure, it is one of the rhetorical

---

[5]http://www.ltp-cloud.com/demo/.

devices, and it gives some characteristics of one thing to another. Only if people know the features of another thing, can they estimate the characteristics.

In order to mine more specific features based on the syntactic structure of the sentence, this paper presents the dependencies from the following three aspects.

3.5 Interdependent Relationship Between the Conjunctions and the Emotional Words

In some case, conjunctions may change the emotional meaning of the two adjacent words. Some words may keep their original emotions, while others perhaps turn into their opposite emotions. There exist different kinds of conjunctions such as coordinating conjunctions, adversative conjunctions, and causal conjunctions and so on.

In this paper, we summarize the relationship between the conjunctions and the emotional words which are shown in Table 6. For example, we can use "He2 (and)" is one of the coordinating conjunctions. The emotional words which are before and after the coordinating conjunction have the same polarities. Meanwhile, the interdependent relationship between the conjunctions and the emotional words can also be obtained by dependency parsing, such as COO (coordinate), CNJ (conjunctive) etc.

Table 6 Relationship between the conjunctions and the emotional words

Coordinating conjunctions	He2 (and), Yu3 (with), Kuang4 Qie2 (in addition), etc.	These emotional words have the same polarities with the prior emotional words
Continuing conjunctions	Ze2 (then), Nai3 (thus), Yu2 Shi4 (hence), etc.	The intensities of the emotional words are stronger than the prior emotional words
Adversative conjunctions	Que4 (however), Dan4 Shi4 (but), Ran2 Er3 (nevertheless), etc.	These emotional words have opposite polarities with the prior emotional words
Concessive conjunctions	Sui1 Ran2 (although), Jin3 Guan3 (though), Ji2 Shi3 (even though), etc.	If the main clause is in the front of the conjunction, the intensities of the prior emotional words are stronger than the center emotional words; otherwise the intensities of the emotional words are stronger than the prior emotional words
Progressive conjunctions	Bu4 Dan4...Hai2 (not only...but also), Shen Zhi (even), etc.	The intensities of the emotional words are stronger than the prior emotional words

Fig. 1 An example of dependency parsing

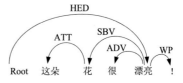

(Zhe4 Duo3 Hua1 Hen3 Piao4 Liang4 !)

(This flower is very beautiful.)

3.6 Interdependent Relationship Between the Modifiers and the Emotional Words

Modifiers in sentences have different effects on the emotional words. Some can increase the intensity of the emotion or weaken its intensity. This paper uses the intensifier lexicon including 219 degree adverbs, which are separated into five levels: "Ji2 Qi2 | extreme"; "Hen3 | very"; "Jiao4 | more"; "Shao1 | a little" and "Qian4 | insufficiently" [27]. For example, the word "extreme" which can strongly modify the corresponding emotional intensity, and the level of emotion is changing from one polarity to the opposite. The interdependent relationship between the modifiers and the emotional words can also be obtained by dependency parsing. For example, "This flower is very beautiful!", the dependency between the modifier "very" and the emotional word "beautiful" is ADV (adverbial), see Fig. 1.

3.7 Interdependent Relationship Between the Negation Words and the Emotional Words

Sometimes negation words can affect the negative transformation of the corresponding emotion or their intensities. Double negation and the position modification of the negation word are highlights in particular. As for double negation, usually there are some fixed patterns such as two negative adverbs appearing in succession or the combination of one negative adverb with one rhetorical question and so on. Most of them can express an affirmative meaning, and as a result, their emotion intensities are also strengthened. If the location of the negation word changes, the emotion and the corresponding emotion intensity will also change. For instance, the difference between "not very bitter" and "not bitter" is that the former expresses an affirmative meaning (only the emotion intensity of the word is weakening), but the orientation of the latter changes into negative. In accordance with the above situations, this paper summarizes them from the following four aspects (see Table 7). Here, "N" means the negation word, "D" means the modifier, and "E" means the emotional word.

During the process of feature extraction, the relationship between the keywords and the dimensions can be determined, and the feature space can be established. After processing the micro-blog post one by one, we can modify the value of the cor-

Table 7 Description of negation word

Pattern	Explanation
N+N+E	Expressing an affirmative meaning
N+D+E	Having the same type of emotion with the emotional word, but weaker than its intensity
N+E	Having the reversed type of emotion with the emotional word, or it is neutral
D+N+E	Having the reversed type of emotion with the emotional word, and its intensity can be modified by the modifier

responding dimension on the basis of the keywords in each micro-blog. Eventually, the corresponding VSM space vector of each micro-blog can be generated.

4 Feature Dimension Reduction

The higher feature space dimension is, the higher computational complexity is. Meanwhile, there exists a few noises and it is necessary to filter them in these dimensions. It is necessary to choose the effective features and reduce the computational complexity.

The purpose of feature dimension reduction is to choose the high-class relevance keywords. These features may appear in the corpus few times. This paper does this work in two aspects: one is picking the effective high frequency words, and the other is picking the effective low frequency words.

χ^2 (Chi-square) test can extract the high-frequency and high-class relevance keywords perfectly. It can be calculated by formula (1) below. Here, N_i is the observed frequency of class i, n is the total frequency p_i, is the expected frequency of class i. But it is not perfect because some words with higher relevance computation results may have few relations with the class. In this paper, we use the word frequency statistics as the auxiliary algorithm. This paper counts the word frequency ratio of the keywords on the top of χ^2 test, and confirms the selective number and threshold through circulation. Then we can obtain the keywords which exceed the threshold to form the high frequency words lexicon.

$$\chi^2 = \sum_{i=1}^{k} \frac{(N_i - np_i)^2}{np_i} \tag{1}$$

The basis of picking the effective low frequency words comes from PMI (i.e., Pointwise Mutual Information), which can compute the relevancy between the feature word w and the emotion class c. The formula of PMI is shown as follows. Here, P (w, c) stands for the probability of the post that includes the feature word w and belong to the emotion class c, P(w) is the probability that the post includes the feature word w, and P(c) is the probability that the post belongs to the emotion class c. This method can counteract the effect of the word frequency and do well in picking the effective low frequency words. Finally, we confirm the threshold iteratively, and obtain the keywords which exceed the threshold so as to form the low frequency words lexicon.

$$PMI(w, c) = \log \frac{P(w, c)}{P(w) P(c)} \tag{2}$$

Finally, the high frequency words lexicon and the low frequency words lexicon will be integrated to form the new feature space after feature dimension reduction.

5 Classification

In this section, we present SVM^{perf} as the core classification algorithm. It is a Support Vector Method for optimizing multivariate nonlinear performance measures described in [28]. Its basis lies on the alternative formulation of the SVM optimization problem, which shows a different form of sparsity compared to the conventional formulation [29]. Through the Cutting-Plane Subspace Pursuit (CPSP) algorithm [30], it is reasonable to train the large datasets with sparse feature vectors and get fast predictions on testing set. As SVM^{perf} is a binary classifier, we can use more classifiers to achieve the goal based on the idea of hierarchical classification. Firstly, we use one classifier to classify the sentence as the emotional one and the neutral one. As for the sentence with emotion, we use another classifier to classify it as positive emotion and negative emotions. Then the "Happiness" can be seen as the former class. The "Anger", "Disgust", "Fear", "Sadness", and "Surprise" belong to the latter. We can get the classification results through more than one classifier. In the process of training, we need to adjust the parameter C which have been given in [29] and the threshold to confirm the best combination and get the perfect classification results.

6 Experiment Results and Analysis

6.1 The Micro-Blog Emotion Dataset

As Sina Micro-blog website (weibo.com) is one of the biggest websites in China, it is reasonable to choose it as the source of the experimental dataset. As for obtaining the dataset, we have chosen 10,893 original Chinese micro-blog posts from Sina Weibo

Table 8 Details of the micro-blog emotion dataset

Emotions	The number of posts	Ratios
Happiness	1727	0.1585
Anger	689	0.0633
Disgust	873	0.0801
Fear	384	0.0353
Sadness	1072	0.0984
Surprise	348	0.0319
Neutral	5800	0.5325
Total	10893	1.0000

randomly and crawled them as dataset for keeping the authenticity and practicality of the posts [31]. Meanwhile, these posts are published between 15:00 and 16:30 on March 22, 2013. After removing duplicates, filtering irrelevant results and doing some conversion, the micro-blogs with obvious emotions are marked the emotion types, such as "*Happiness*", "*Anger*", "*Disgust*", "*Fear*", "*Sadness*" and "*Surprise*".

In the process of marking, we comply with the following rules. If the micro-blog does not belong to any emotion type, it will not be marked. If both the markers cannot distinguish which emotion the micro-blog belongs to, then the micro-blog will be marked with neutral; If there's a conflict between the two markers, the third marker will determine the final result.

The marked result of the experiment corpus is shown in Table 8. For reflecting the realistic situation in micro-blogs, we use the imbalanced dataset. The ratios of all classes are also presented in Table 8. More than 50 % of posts are neutral. Among the emotional posts, the ratio of different emotions also differs, e.g. happiness (0.1585) and sadness (0.0984).

We remove four kinds of elements, which are explained as follows, as they definitely does not contain emotions. Firstly, if user A wants to share something with user B, he will include B's username by adding one @ symbol before it in his post. As this part is surely non-emotional, so we take it away by detecting the @ symbol and remove it together with the username. Secondly, every user can take part in discussions under a certain topic. To participate, users only need to include the topic in posts denoted by two #, e.g. # Emotion Analysis #. This part, of course, does not contain any emotion and is removed in preprocessing procedure. Thirdly, users can include links in their posts. The links will be converted into short links by the micro-blog platform to reduce occupied spaces. Fourthly, micro-blog platforms allow users to add position information at the end of posts, which will not help in emotion classification. Position information has fixed prefix, such as "I am here": or "I am at":, which makes it easier for preprocessing program to identify.

Table 9 Details of the experiments

Experiments	Details
Baseline group 1	We only extract **the linguistic features** as the set of features, applying **LIBSVM** within this group
Baseline group 2	We extract **the linguistic features and reduce the feature dimension** and then take them as the set of features. Finally, we apply **LIBSVM** within this group
Baseline group 3	We only extract **the linguistic features** as the set of features, applying **SVM**^{perf} within this group
Baseline group 4	We only extract **the linguistic features** as the set of features, applying **SVM-Light** within this group
Baseline group 5	We extract **the linguistic features and reduce the feature dimension** and then take them as the set of features. We apply **SVM-Light** within this group
Experimental group	We extract **the linguistic features and reduce the feature dimension** and then take them as the set of features. Finally, we apply **SVM**^{perf} within this group

6.2 Experiment Results of Emotion Classification

In order to illustrate the effectiveness of the algorithm, three baseline systems are developed to evaluate our method. We take LIBSVM as the classification algorithm of the baselines. The details of our experiments are shown in Table 9. Meanwhile, 67 % micro-blogs in our corpus are seen as the training set, and others are as the testing set. The classification results are shown in Table 10.

Table 9 shows the performances of six groups in emotion classification. In details, our method has higher precisions than the baselines for most emotions (*"Happiness"*, *Sadness"*, *"Fear"*, *and "Disgust"*), especially for the classes of *"Fear"*, the precision increases to 91.67 %. As for *"Surprise"* and *"Anger"*, in spite of the precisions of our method do not exceed the precisions of the baseline group 2, their recalls are higher. Meanwhile, in terms of F-score, our method outperforms the baseline group 1, the baseline group 3, the baseline group 4 and the baseline group 5 in all classes, but the F-scores of "Fear" and "Anger" are lower than the baseline group 2. We can infer that as the imbalanced dataset has been used, the number of micro-blogs with *"Surprise"*, *"Anger"* and *"Fear"* are relatively less than others and the training sets will decline as well, which may lead to the results. From the perspective of micro-blog users, there may be less things in daily life that make them feel strong emotions like *"Fear"*, so they tend to describe other things to express other emotions.

Table 10 Details of the experiment results

		Precision	Recall	F-score
Baseline group 1	Happiness	0.7570	0.3794	0.5055
	Sadness	0.6919	0.6263	0.6575
	Surprise	0.4595	0.7683	0.5751
	Fear	0.5124	0.9241	0.6593
	Anger	0.5246	0.6906	0.5963
	Disgust	0.7188	0.5902	0.6482
Baseline group 2	Happiness	0.5803	0.8633	0.6941
	Sadness	0.5886	0.7939	0.6760
	Surprise	0.7905	0.6455	0.7107
	Fear	0.7560	0.7330	0.7443
	Anger	0.8679	0.7302	0.7931
	Disgust	0.6306	0.8456	0.6941
Baseline group 3	Happiness	0.6364	0.5497	0.5899
	Sadness	0.7345	0.6806	0.7065
	Surprise	0.8182	0.2812	0.4186
	Fear	0.7632	0.3580	0.4874
	Anger	0.7348	0.6584	0.6945
	Disgust	0.6739	0.2422	0.3563
Baseline group 4	Happiness	0.6789	0.3874	0.4933
	Sadness	0.7971	0.5759	0.6686
	Surprise	1.0000	0.3125	0.4761
	Fear	0.8387	0.3210	0.4642
	Anger	0.7669	0.6188	0.6849
	Disgust	0.6364	0.1641	0.2609
Baseline group 5	Happiness	0.6937	0.4031	0.5099
	Sadness	0.8000	0.4817	0.6013
	Surprise	1.0000	0.4062	0.5777
	Fear	0.6304	0.3580	0.4566
	Anger	0.7669	0.6188	0.6849
	Disgust	0.6500	0.2031	0.3094
Experimental group	Happiness	0.7978	0.8906	0.8417
	Sadness	0.8486	0.8296	0.8390
	Surprise	0.7714	0.8438	0.8060
	Fear	0.9167	0.5593	0.6947
	Anger	0.8182	0.7397	0.7770
	Disgust	0.8056	0.6976	0.7477

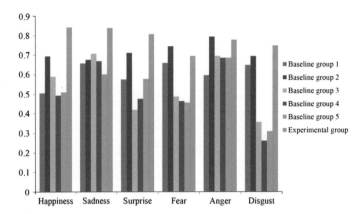

Fig. 2 Comparison of all the baseline groups and experimental group

Figure 2 presents the comparison among the corresponding baseline experiments and the proposed method in terms of F-score. We note obvious improvement. The results of the experimental group using our set of features significantly outperform both the baselines. Particularly, the F-score adds up to 84.17 % for happiness, 83.90 % for sadness. It indicates that with the consistency of the classification algorithm (LIBSVM, SVM-Light and SVM^{perf}), reducing the feature dimension during feature extraction is necessary. Meanwhile, we also compare those results with the baseline group 2, the baseline group 5 and the experimental group. It shows that with the consistency of features selection, SVM^{perf} is an appropriate choice of classifier for emotion classification.

In addition, using the word2vec to expand the emotional lexicon can contribute to the multi-label emotion classification of micro-blogs. The new emotional lexicon can serve as a good reference for researchers in the field of emotion processing. Meanwhile, mining different linguistic structures and dependency structures in micro-blogs can identify useful features effectively, which improves the classification performances.

There are also some great practice significations in the aspect of emotion classification. First, we will apply our algorithm to consumer behavior analysis. For example, by analyzing the online review information of consumer commodity and the emotion characteristics of consumers, we can get the characteristics of the consumer behavior and the change trend of consumer preferences. Meanwhile, we can also get the psychological state of consumers for the review of commodity or service. It can realize the function of recommendations specifically. Second, we can also apply our algorithm to the effect analysis of commercial promotion. By analyzing the emotion of consumers for the effect of promotions, we can provide a basis of decision-making for improving the marketing behavior. Third, it is possible for us to track the emotion changes characteristics of micro-blog users, and we can track their happiness and the happiness index of certain areas and so on.

7 Conclusions and Future Works

Emotion analysis for micro-blog has great research values and many practical applications, such as monitoring internet public emotion, commercial advertising recommendation and so on. There are also some insightful future research directions. Firstly, an emotional word may have different polarity in different application domains. If one emotional word which describes one agent or object is positive, then it may be negative when the agent or the object changes to the other. The agent /object mainly represents the person or the object, which can be extracted with the help of the HowNet corpus, such as mom, apple and tree. The emotional word mainly refers to the emotional word in the emotional lexicon, such as Gao1 Xing4 De (*happy*), Nan2 Guo4 (*sadness*) word in this situation in the future. Secondly, there also exist some complicated linguistic patterns, such as the internet language with a variety of forms. The complicated linguistic patterns may be constituted by several simple linguistic patterns. Thirdly, mining the correlations among emotions for some time is another research direction. For example, in one time, the emotion is happy, while in another time, the emotion changes to sadness. They can help researchers find the rules of emotion changes. How to deal with these issues is our future research work.

Acknowledgments Supported by National Basic Research Program of China (973 Program) (Grant No:2012CB316301), National Natural Science Foundation of China (Grant No: 61175110), and Chinese National Programs for High Technology Research and Development (863 Program) (Grant No. 2013AA013702).

References

1. Su, Z., Zhou, B., Li, A., Han, Y.: Analysis on chinese microblog sentiment based on syntax parsing and support vector machine. Web Technol. Appl. **8710**, 104–114 (2014)
2. Wen, S., Wan, X.: Emotion classification in microblog texts using class sequential rules. In: Twenty-Eighth AAAI Conference on Artificial Intelligence, pp. 187–193. Association for the Advancement of Artificial Intelligence (2014)
3. Wei, W., Gulla, J.A.: Sentiment analysis in a hybrid hierarchical classification process. In: Proceedings of 7th International Conference on Digital Information Management, pp. 47–55. IEEE (2012)
4. Yang, D.H., Yu, G.: A method of feature selection and sentiment similarity for chinese micro-blogs. J. Inform. Sci. **39**(1), 429–441 (2013)
5. Yuan, Z., Purver, M.: Predicting emotion labels for chinese microblog texts. In: Proceedings of the First International Workshop on Sentiment Discovery from Affective Data (SDAD 2012), pp. 40–47. CEUR (2012)
6. Ekman, P., Friesen, W.V.: Constants across cultures in the face and emotion. J. Pers. Soc. Psychol. **17**(2), 124–129 (1971)
7. Li, W., Xu, H.: Text-based emotion classification using emotion cause extraction. Expert Syst. Appl. **41**(4), 1742–1749 (2014)
8. Liu, Q., Feng, C., Huang, H.: Emotional tendency identification for micro-blog topics based on multiple characteristics. In: Proceedings of the 26th Pacific Asia Conference on Language, Information and Computation, pp. 280–288. Universitas Indonesia, Faculty of Computer Science (2012)

9. Pei, S., Zhang, L., Li, A.: Microblog sentiment analysis model based on emoticons. Web Technol. Appl. **8710**, 127–135 (2014)

10. Bai, X., Chen, F., Zhan, S.: A study on sentiment computing and classification of sina weibo with word2vec. In: Proceedings of 2014 IEEE International Congress on Big Data (BigData Congress), pp. 358–363. IEEE (2014)

11. Desmet, B., Hoste, V.: Emotion detection in suicide notes. Expert Syst. Appl. **40**(16), 6351–6358 (2013)

12. Ghazi, D., Inkpen, D., Szpakowicz, S.: Prior and contextual emotion of words in sentential context. Comput. Speech Lang. **28**, 76–92 (2014)

13. Kim, M., Kwon, H.C.: Lyrics-based emotion classification using feature selection by partial syntactic analysis. In: Proceedings of the 23rd IEEE International Conference on Tools with Artificial Intelligence (ICTAI), pp. 960–964. IEEE (2011)

14. Agrawal, A., An, A.: Unsupervised emotion detection from text using semantic and syntactic relations. In: Proceedings of 2012 IEEE/WIC/ACM International Conferences on Web Intelligence and Intelligent Agent Technology (WI-IAT), pp. 346–353. IEEE (2012)

15. Ho, D.T., Cao, T.H.: A high-order hidden markov model for emotion detection from textual data. Knowl. Manage. Acquis. Intell. Syst. **7457**, 94–105 (2012)

16. Li, Y., Li, X., Li, F., Zhang, X.: A lexicon-based multi-class semantic orientation analysis for microblogs. Web Technol. Appl. **8709**, 81–92 (2014)

17. Liu, S.M., Chen, J.H.: A multi-label classification based approach for sentiment classification. Expert Syst. Appl. **42**(3), 1083–1093 (2015)

18. Huang, S., Peng, W., Li, J., Lee, D.: Sentiment and topic analysis on social media: a multi-task multi-label classification approach. In: Proceedings of the 5th Annual ACM Web Science Conference, pp. 172–181. ACM (2013)

19. Li, T., Xiao, X., Xue, Q.: An unsupervised approach for sentiment classification. In: Proceedings of 2012 IEEE Symposium on Robotics and Applications, pp. 638–640. IEEE (2012)

20. He, H.: Sentiment analysis of sina weibo based on semantic sentiment space model. In: Proceedings of management science and engineering, pp. 206–211. IEEE (2013)

21. Chang, C.C., Lin, C.J.: LIBSVM: A library for support vector machines. ACM Trans. Intell. Syst. Technol. **2**, 1–27 (2011)

22. Joachims, T.: Making large-scale SVM learning practical. In: Schölkopf, B., Burges, C., Smola, A. (eds.) Advances in Kernel Methods—Support Vector Learning, chap. 11, pp. 169–184. MIT Press, Cambridge, MA (1999)

23. Jiang, F., Cui, A., Liu, Y., Zhang, M., Ma, S.: Every term has sentiment: learning from emoticon evidences for chinese microblog sentiment analysis. In: Natural Language Processing and Chinese Computing, vol. 400, pp. 224–235 (2013)

24. Xu, L., Liu, H., Pan, Y., Ren, H., Chen, J.: Constructing the affective lexicon ontology. J. China Soc. Sci. Tech. Inf. **27**(2), 180–185 (2008)

25. Mikolov, T., Chen, K., Corrado, G., Dean, J.: Efficient estimation of word representations in vector space. In: Proceedings of the 1st International Conference on Learning Representations (2013)

26. Liu, B.: Sentiment analysis and opinion mining. Synth. Lect. Hum. Lang. Technol. **5**, 1–167 (2012)

27. Zhang, P., He, Z.: A weakly supervised approach to chinese sentiment classification using partitioned self-training. J. Inform. Sci. **39**(6), 815–831 (2013)

28. Joachims, T.: A support vector method for multivariate performance measures. In: Proceedings of the 22th International Conference on Machine Learning (ICML), pp. 377–384. ACM (2005)

29. Joachims, T.: Training linear svms in linear time. In: Proceedings of the 12th ACM SIGKDD International Conference on Knowledge Discovery and Data Mining, pp. 217–226. ACM (2006)

30. Joachims, T., Yu, C.N.J.: Sparse kernel svms via cutting-plane training. Mach. Learn. **76**(2–3), 179–193 (2009)

31. Gao, K., Zhou, E.L., Grover, S.: Applied methods and techniques for modeling and control on micro-blog data crawler. In: Applied Methods and Techniques for Mechatronic Systems, vol. 452, pp. 171–188 (2014)

Social Media and News Sentiment Analysis for Advanced Investment Strategies

Steve Y. Yang and Sheung Yin Kevin Mo

Abstract The motivation of this chapter hinges on the growing popularity in the use of news and social media information and their increasing influence on the financial investment community. This chapter investigates the interplay between news/social sentiment and financial market movement in the form of empirical impact. The underlying belief is that news and social media influence investor sentiment, which in turn drives financial decisions and predicates the upward or downward movement of the financial markets. This book chapter contributes to the existing literature of sentiment analysis in the following three areas: (a) It provides a review of existing findings about influence of social media and news sentiment to asset prices and documents the persistent correlation between media sentiment and market movement. (b) It shows that abnormal news sentiment can be a predictive proxy for financial market returns and volatility, based on the intuition that extreme investor sentiment changes tend to have long and last effects to market movement. (c) It presents a number of approaches to formulate investment strategies based on the sentiment trend, shocks and feedback strength. The results show that the sentiment-based strategies yield superior risk-adjusted returns over other benchmark strategies. Altogether, this chapter provides a framework of existing empirical knowledge on the impact of sentiment on financial markets and further prescribes advanced investment strategies based on sentiment analytics.

Keywords Sentiment analysis · Financial investment community · Genetic algorithms · Investment strategies

S.Y. Yang (✉)
Faculty of Financial Engineering, School of Systems & Enterprises,
Stevens Institute of Technology, Hoboken, USA
e-mail: steve.yang@stevens.edu

S.Y.K. Mo
Faculty of Financial Engineering, Stevens Institute of Technology, Hoboken, USA
e-mail: smo@stevens.edu

© Springer International Publishing Switzerland 2016
W. Pedrycz and S.-M. Chen (eds.), *Sentiment Analysis and Ontology Engineering*,
Studies in Computational Intelligence 639, DOI 10.1007/978-3-319-30319-2_11

1 Introduction

With increasing digitization of textual information and computation capability, large-scale and sophisticated sentiment analysis has become a feasible alternative for leveraging the use of computational intelligence technologies to advance understanding of the financial markets. The field of behavioral finance, in particular, studies how psychology and cognition influence decision-making of real-world investors in an irrational manner. Past studies have relied heavily on the impact of textual form of financial documents such as news and press releases. Psychological evidence suggests that sentiment, emotion and mood play a key role in affecting investors when making financial decisions [9, 12, 17, 35]. Barberis et al. developed a theory of investor sentiment to illustrate the impact of investor overreaction and underreaction to public information on generating on post-earnings announcement drift, momentum, long-term reversals and predictive power or scaled-price ratio [7]. Daniel et al. further enriched the idea of investor sentiment with private information leading to overconfidence [14, 15]. On the empirical front, a number of studies found different measures of investor sentiment significant in explaining asset price and volatility movements. Chopra et al. showed that prior losing portfolios significantly outperform prior winning portfolio by 5–10% annually during the next 5 years, validating the overreaction effect [11]. La Porta et al. also displayed evidence that the correction of the extreme investor sentiment tends to revert during earnings announcements when investors realize their initial beliefs were too extreme [27, 36, 43]. These studies are instrumental in demonstrating the existence of investor sentiment along with its impact on the financial markets. The motivation of this book chapter rests on the thesis that news and social media sentiment reflect the societal states that affect individual investors to react and therefore predicates upward or downward movement of the financial markets. In addition, this chapter aims to investigate the interplay between social media/news sentiment and financial market movement, and to demonstrate how the existing findings can be leveraged with computational intelligence techniques to develop advanced investment strategies. This book chapter contributes to the existing literature of sentiment analysis in the following three areas:

1. It provides a review of existing findings about influence of social media and news sentiment to asset prices and documents the persistent correlation between media sentiment and market movement. Furthermore, it shows the presence of a financial community on Twitter whose primary interests are consistently aligned with financial market-related knowledge and information. By harnessing the sentiment expressed by the influential Twitter users within the community, one can construct a better proxy for quantifying social sentiment.
2. It shows that abnormal news sentiment can be a predictive proxy for financial market returns and volatility, based on the intuition that extreme investor sentiment changes tend to have long and last effects to market movement.
3. It presents a number of approaches to formulate investment strategies based on the sentiment trend, shocks and feedback strength. The results show that the sentiment-based strategies yield superior risk-adjusted returns over other bench-

mark strategies. Altogether, this chapter provides a framework of existing empirical knowledge on the impact of sentiment on financial markets and further prescribes advanced investment strategies based on sentiment analytics.

2 Market Sentiment Analysis

This section provides a review of existing literature about common sentiment analysis techniques among financial studies. Market sentiment algorithms can be categorized into two major groups in lexicon-based approach and machine learning techniques. At the end of the literature review, two additional innovative approaches are presented for extracting market sentiment in using social media financial community and media sentiment feedback.

2.1 *Lexicon Based Sentiment Approach*

The lexicon-based approach refers to the use of specific sets of vocabulary to identify semantic orientation of a textual source, which has been widely used in recent research studies [30]. Moreo et al. proposed a lexicon-based news sentiment analyzer that incorporates non-standard language and generates sentiment measures based on specific topics of interest [34]. A study conducted by Schumaker et al. investigated the effectiveness of the Arizona Financial Text system, which leverages the use of OpinionFinder in identifying the tone and polarity of the underlying text [37]. In addition, Li et al. evaluated financial news articles with a lexicon-based approach using the Harvard psychological dictionary and Loughran-McDonald financial sentiment dictionary for sentiment generation [29]. Other related studies emphasis the use of emotional words such as "soar" and "fall" to enhance the sentiment measure of news articles [48].

The lexicon-based approach is popular among academic and industry studies because of its direct usage of specialized word dictionaries to generate relevant sentiment scores. A common source of reference is the SentiWordNet dictionary which is a lexical resource with words linked to sentimental scores [5]. Our studies performed in later sections also rely on the application of the lexicon-based sentiment approach. Through a four-step process, the objective of the sentiment algorithm is to convert raw text data into daily sentiment score for the empirical study. With the complex textual structure, the raw text is initially decomposed into individual words with the removal of stop words such as "a", "but", "how" and "to". Lemmatization techniques are then applied to convert different inflicted forms of a word into a uniform entity. For instances, the words "rising", "risen" and "rises" are regarded as the entity "rise". For each word entity in the textual data, the algorithm extracts the associated score from the sentiment dictionary and finally, generates the sentiment score for each news text by averaging all individual word scores.

2.2 Machine Learning Based Sentiment Approach

Related to financial news mining, machine learning has become more popular as a feasible approach to extract text sentiment. In a recent study, Tan et al. showcased a sentiment mining analyzer based on machine learning techniques using polarity lexicon [21]. Xie et al. illustrated a novel approach in quantifying news document in semantic tree structures and then using tree kernel support vector machines to predict stock market movement [44]. In addition, the five-step procedure was illustrated related to the computerized techniques for handling news content [23] (see Fig. 1).

2.3 Social Media Financial Community Based Sentiment Approach

Apart from the mainstream sentiment analysis approaches, financial sentiment can be extracted from the social media messages of key influencers among the social media financial community. This novel approach has been developed and implemented in our recent publications [33, 46]. It performs sentiment analysis on social media

1. News Filtering
- To ensure the relevant stories are filtered for traders and investors
- Clustering to avoid displaying the same news for the same event

2. News Association
- Search for associated news that might affect the underlying entity
- For example: GOOG being affected by competitors and sectors

3. News Analysis
- Interpretation of sentiment (good or bad to evaluate its impact on market)
- Key to extract the exact context of any key words or phrases.

4. News Interpretation
- To estimate how much the information should change the asset's price.
- To identify and extract key information from news for NLP-based approach
- The new target prices generated can then be used to make decisions

5. News Mining
- Combines the analysis of historical data with information from news archives
- Use Pattern recognition techniques to infer the market impact of news
- Analyzing short-term impact of news is easier than mid and long-term effects

Fig. 1 Computerized news handling techniques

messages according to their relevance to key financial topics. The approach can be further described in the following components:

(a) Financial Entity Matching

A natural processing step of examining Twitter data is to pinpoint those messages that are related to the financial market topics. The rationale behind this critical step is to ensure that the level of noise is minimized in the study. We propose an approach to first form a list of financial entities by extracting commonly used languages from the top financial news broadcasters and traders' Twitter accounts, and then matching them against the individual words and phrases in the tweet message. Furthermore, each financial entity is quantitatively assigned a score to reflect its proximity to the financial related topics.

- *Entity Extraction from top financial Twitter users*: Sentiment analysis does not perform well if members of the financial community tweet messages that are unrelated to financial market. We conducted an entity matching process to extract messages with financial interests. A financial keyword corpus was then created using a sample dataset from 10/05/2013 to 02/05/2014 from the top financial news broadcasters and traders' Twitter accounts. The sample dataset was categorized according to the type of message initiator. Messages from the top news agencies and traders were labeled as "key messages" while messages from less important community users were "noise messages". By running text parsing and text filtering processes for the two groups, two respective keyword lists are obtained with frequency of occurrence reported. Through text parsing, stop words were dropped and only nouns and noun phrases were extracted from tweet messages. Entities that demonstrated higher occurrence in the "key messages" group were preserved to form the entity corpus while others are discarded. This step effectively screens for financial entities that are more commonly mentioned through established financial Twitter accounts. In addition, the corpus pairs each entity with a weight based on the frequency of occurrence. Higher weight can be interpreted as closer relevance to financial market topics.
- *Company name and ticker*: Another source of financial entity identification is the name and ticker symbol of major U.S. domestic companies. If a specific company or its ticker is mentioned, there is a high likelihood that the message contains financial-related information. As a result, the financial entity dictionary contains the names and their ticker symbols for over 6,000 companies listed on the three largest U.S. exchanges: the New York Stock Exchange (NYSE), the NASDAQ stock market (NASDAQ) and the American Stock Exchange (AMEX).
- *Financial entity score computation*: All words and phrases from the tweet message are matched against the financial entity dictionary. Each message contains a score based on the degree of relevance to financial topics. The higher the financial entity score is, the heavier its weight counting towards its sentiment is. For instance, a tweet message containing the ticker symbol for General Electric 'GE' has a financial entity score of $+1$. If multiple financial entities are matched in one Twitter

message, the financial entity score weight is equal to the highest weight among the matched financial entities (1).

$$S^i_{entity} = \max \left(\omega \left(W^i_{message} \cap W_{fe} \right) \right) \quad i = 1, 2, \ldots, N \tag{1}$$

where S^i_{entity} is the financial entity score of message i, $W^i_{message}$ is the word set split from message i, W_{fe} is the financial entity word set, $\omega(W)$ is the financial entity weights set of word set W, N is the number of message and $n_{matched}$ is the number of matched financial entity.

(b) Message Sentiment Computation

With the number of word occurrence and negative flag detection, the sentiment of a single tweet message can be generated. The message sentiment score ranges from -1 to 1, with -1 indicating the most negative sentiment, 0 being neutral, and $+1$ the most positive sentiment. The formula for computing sentiment score of message I can be found in (2). Among all the messages initiated by the top 2,500 critical nodes, the sentiment distribution approximately follows a normal distribution.

$$S^i_{sentiment} = \frac{\sum_j n^j_i \times s(j)}{\sum_j n^j_i} \times S^i_{entity} \times sgn(i)$$

$$sgn(i) = \begin{cases} -1 & if \; W^i_{message} \cap W_{neg} \neq \emptyset \\ 1 & others \end{cases} \tag{2}$$

where $S^i_{sentiment}$ is the sentiment score of message S^i_{entity} is the financial entity score of message i, $W^i_{message}$ is the word set split from message i, W_{neg} is the negative connotation word set, n^j_i is the number of occurrence of SentiWordNet word j in message i, $s(j)$ if the sentiment score of word j (Fig. 2).

(c) Sentiment Daily Score Computation

The last procedure in the sentiment analysis algorithm is to compute the sentiment score for the day of observation for a given user. Three components are factored into the computation process: the score of the financial entity, the message sentiment score and the centrality score for the message initiator (3). The algorithm then takes the average of the active user sentiment generated for each day to generate the daily sentiment score for regression studies.

$$S(t) = \frac{1}{N} \sum_{j=1}^{N} W^j_c \times \frac{\sum_{k=1}^{n_j(t)} S^k_{sentiment}(t)}{\sum_{k=1}^{n_j(t)} S^k_{entity}(t)} \tag{3}$$

Data: Twitter Messages
Result: Message Sentiment Score

while *not at the end of document* **do**
  $Message_{raw}$ ← read current message;
  Step 1: Data-Preprocessing;
  $Message_{splitted}$ ← Word-splitting(Message);
  $Message_{splitted}$ ← Remove-stop-words($Message_{splitted}$);
  $Message_{splitted}$ ← Word-stemming($Message_{splitted}$);
  Step 2: Financial Entity List Processing;
  $WordList_{FinancialEntity}$ ← Financial Entity List created from
  SAS;
  $MatchedWords_{FinancialEntity}$ ←
  $Message_{splitted} \cap WordList_{FinancialEntity}$;
  $Score_{FinancialEntity}$ ← max$\{MatchedWords_{FinancialEntity}\}$;
  if $Score_{FinancialEntity} > 0$ **then**
    Step 3: Sentiment Word List Processing;
    $WordList_{Sentiment}$ ← $SentiWordNetList$;
    $MatchedWords_{Sentiment}$ ←
    $Message_{splitted} \cap WordList_{Sentiment}$;
    $Score_{SentiWordNet}$ ← $\frac{1}{N}\sum\{MatchedWords_{Sentiment}\}$;
    Step 4: Inverse Factor;
    if *negative words/phrases (not n't) detected* **then**
    | $InverseFactor$ ← −1
    else
    | $InverseFactor$ ← 1
    end
    Step 5: Final Message Sentiment Score [-1,1];
    $Score_{Sentiment}$ ←
    $Score_{FinancialEntity} \times Score_{SentiWordNet} \times InverseFactor$
  else
  | Drop $Message_{raw}$; continue;
  end
end

Fig. 2 Sentiment analysis algorithm

where $S(t)$ is the daily sentiment score of day t, ω_c^j is the centrality weight of user j, $n_j(t)$ is the number of message by user j on day t, $S_{entity}^k(t)$ and $S_{sentiment}^k(t)$ are entity score and sentiment score of message k computed by Eq. (10) and Eq. (11) respectively.

2.4 Media Sentiment Feedback Effects

Feedback mechanisms have been explored in the field of finance, mainly through the examination of its effects on price and volatility. Sentiment can be quantified in the form of its feedback effects. Hirshleifer et al. presented a theoretical framework that justifies irrational investors to earn abnormal profits based on a feedback mechanism from stock prices to cash flows [18]. Crude oil prices were found to contain feedback effects along with an inverse leverage impact with its implied volatility [1]. Khanna and Sonti showed the feedback effect of stock prices on firm value through a herding equilibrium model and investigated into the incentive for traders to conduct price manipulation [25]. The volatility feedback effect is an empirical observation of feedback effect between squared volatility and stock price. Inkaya and Okur estimated the volatility feedback effect rate using Malliavin calculus and suggested its predictability of large price declines [22]. They showed that large feedback effect rate is a useful indicator for measuring market stability [22].

There is also empirical evidence that feedback trading, a self-perpetuating pattern of investor's behavior, is present in G7 stock markets, and other international markets [2, 36]. The effect of feedback trading was found to vary across business cycle [10] and the strongest influence was observed during periods of financial crisis with declining futures prices [36]. Hou and Li developed a regression model of feedback trading to analyze CSI300 stock returns and demonstrated that lagged index returns can predict market index return and conditional volatility [20]. In addition, feedback trading was found to significantly influence exchange rate movements [28]. Using a theoretical framework, Arnold and Brunner showed that positive feedback trading causes price overreaction and the impacts of feedback trading would be dampened if news is incorporated into price in time [4].

3 Market Sentiment Based Investment Strategies

This section presents the latest research findings on using market sentiment as a source to develop financial investment strategies. It covers the economic intuition behind the study, the methodology along with the data sources, the key findings and their implications. The first study centers on the use of one of the most popular social media platforms Twitter and how its messages can be used to generate sentiment for predictive signals. Using news data, the second study focuses on firm-specific sentiment and how it is correlated with the financial market returns and volatility. The last study examines the interaction effect between social media and news sentiment and showcases that the feedback effect can be quantified for developing investment strategies using a genetic programming framework. These studies revolve around the central theme of market sentiment based investment strategies and display the respective theoretical basis along with empirical evidence for their practical applications.

3.1 Twitter Financial Community and Its Predictive Relation to Stock Market

Twitter, one of the several major social media platforms, has been identified as an influential factor to financial markets by multiple academic and professional publications in recent years. The motivation of this method hinges on the growing popularity of the use of Twitter and the increasing prevalence of its influence among the financial investment community. We present an empirical evidence of the existence of a financial community on Twitter in which users' interests align with the financial market related topics. We establish a methodology to identify relevant Twitter users who form the financial community, and we also present the empirical findings of network characteristics of the financial community. We observe that this financial community behaves similarly to a small-world network, and we further identify groups of critical nodes and analyze their influence within the financial community based on several network centrality measures. Using a novel sentiment analysis algorithm, we construct a weighted sentiment measure using tweet messages from these critical nodes, and we discover that it is significantly correlated with the returns of the major financial market indices. By forming a financial community within the Twitter universe, we argue that the influential Twitter users within the financial community provide a better proxy between social sentiment and financial market movement. Hence, we conclude that the weighted sentiment constructed from these critical nodes within the financial community provides a more robust predictor of financial markets than the general social sentiment.

Our hypothesis is that Twitter sentiment reflects the market participants' beliefs and behaviors toward future outcomes and the aggregate of the societal mood can present itself as a reliable predictor of financial market movement. However, not all users are equally influential in the social media, and those influential social media users will certainly have higher impact to the societal mood or sentiment. Reported evidence demonstrates that there exists a community on Twitter whose primary concern is about financial investment. Those users who are harvesting information from these influential sources on the social media for their daily trading decisions forms the robust linkage between the social mood and financial market asset price movement. Hence this community would be more representative to market participant's beliefs, and consequently the sentiment extracted from this financial community would serve as a better predictor to the market movement.

We seek to identify the corresponding investment community and pinpoint its major influencers in the social networks context. The primary research question is whether the beliefs and behaviors of major key players in such community reveal better signals to financial market movement. From a large-scale data crawling effort, we define a financial community as a group of relevant Twitter users with interests aligned with the financial market. We first identify 50 well-recognized investment experts' accounts in Twitter and use their common keywords to create the interests of the financial investment community. By constructing the two layers of the experts' followers, we apply a multitude of rigorous filtering criteria to establish a financial community boundary based on their persistent interests in the topic of financial investment (Fig. 3 and Table 1).

Fig. 3 Financial community
from Twitter universe

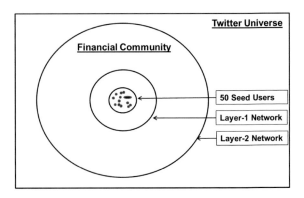

Table 1 Financial community network summary statistics

Number of nodes	154,327
Number of links	4,846,805
Average out-degree centrality	35.71
Average betweenness centrality	420.2
Clustering coefficient	0.15
Network diameter	6
Average path length	2.72
Connected component	1

3.1.1 Critical Node Analysis in the Financial Community

In the financial community, there exist users who play a central role in the connectedness of the network. These users, known as critical nodes, situate at the most critical locations of the community network and therefore bear a large weight in the network dynamic properties such as connectedness and message propagation pattern. Analyzing these nodes is essential for understanding the financial community because they represent the most influential users in the community in terms of facilitating the message propagation process and stabilizing the network structure.

Through social network analysis, we identify these critical nodes by applying centrality measures: out-degree centrality, betweenness centrality and closeness centrality (4), (5) and (6). Our data captures the direction of the friend-follower relationship which contributes to the formation of a directed network. The three centrality measures incorporate direction as information propagates from the sender to their followers, and, more importantly, capture unique aspects regarding the node's relative centrality level among the community. We track the profile information and tweet messages of the top 2,500 users ranked by each of the three centrality measures. The description of each centrality measure is defined as:

1. **Out-degree centrality** measures the number of followers a node has in the network. A higher out-degree centrality value indicates that the specific node is well-connected to many nodes in the network.

$$C_D(v_i) = \sum_{j=1}^{n} a_{ij} \qquad (4)$$

where A denotes the adjacency matrix, a_{ij} is a binary term that values 1 if node i out-ties with j and values 0 otherwise, and n is the number of nodes in network [8].

2. **Betweenness centrality** captures the number of shortest paths from all vertices to other nodes in the network that passes through the specific node. With a higher number of shortest paths passing through a specific node in the network, its betweenness centrality measure will be higher.

$$C_B(v_i) = \sum_{j \neq i} \sum_{k \neq i} \frac{g_{jik}}{g_{jk}} \qquad (5)$$

where g_{jk} denotes the number of geodesic paths from node j to node k and g_{jik} denotes the number of geodesic paths from node j to node k that pass through node i [8].

3. **Closeness centrality** is calculated based on the aggregation of geodesic distances from each node to all of other nodes in the network (also known as the farness). If a specific node is located at a more central location relative to another node, the farness will be lower because of its shorter distance from all other nodes in the network.

$$C_C(v_i) = \sum_{j=1}^{n} d_{ij} \qquad (6)$$

where C is the geodesic distance matrix, d_{ij} is the geodesic distance between node i and node j, and n is the number of nodes in network [8].

Each group of these critical nodes shares certain common attributes. The profile data consists of the nodes' location, username, description, the number of messages they have posted, and the number of followers and friends. We examine the three important attributes: the number of tweeted messages, the number of followers and the number of friends. Samples of these critical nodes are provided in Table 2.

When we analyzed critical nodes with the highest out-degree centrality, we observed that they were followed by a large portion of the users in the community. A tweet message initiated by this group can be spread to a large domain in the network. Furthermore, nodes with the highest out-degree centrality can be much more influential than other nodes as their large number of followers can gain significant interest from the financial community and therefore attract more users who follow them. From the profile dataset, these top nodes with the highest out-degree centrality are broadcasters who have posted more than 10,000 tweets over the

Table 2 Critical node sample users

Critical nodes	Sample users
Out-degree centrality	@TheEconomist, @BreakingNews, @FinancialTimes, @FortuneMagazine, @CMEGroup
Betweenness centrality	@themotleyfool, @Vanguard_Group, @ReformedBroker, @TheStreet, @NYSEEuronext
Closeness centrality	@YESBANK, @currency4trades, @QNBGroup, @Bizzun, @FFinancialGroup, @sobertrader

lifetime of the account, a much higher tweeting rate than that of normal broadcasters and the community. They also appeared to have a longer account history compared to the other two groups of critical nodes.

As an illustration of the importance of centrality measures related to the community structure, we are interested in exploring how the current financial community compares against sample community structures. We select four sample communities to showcase extreme scenarios with distinct characteristics. We then calculate the normalized centrality score for every community node and compute the average among all the node scores to gauge the connectedness of the community (see Fig. 4). Figure 4a shows a symmetric structure with two nodes as the central hub. The peripheral nodes are directly attached to the center of the community and therefore the overall structure has a relatively high average closeness centrality score. Figure 4b features a similar structure with one central node with more connections. This biased phenomenon results in a lower average centrality scores across all three measures. Figure 4c is another symmetric structure featuring three nodes as the central hub and higher average centrality scores. In contrast, Fig. 4d is a fully connected community with each node linked to all other nodes. This structure is an ideal framework for message propagation as the distance between any two nodes in the community is 1. Therefore, the betweenness centrality and the closeness centrality for all nodes is 0 and 1, respectively. In summary, these sample communities feature distinct network structures in terms of key centrality measures. For the financial community, the average scores of out-degree centrality, betweenness centrality and closeness centrality are 0.720, 0.005 and 0.360, respectively. These scores are normalized from 0 to 1 to facilitate the comparison with the sample structures. The high average out-degree centrality score illustrates that the financial community contains more direct linkages among nodes, while the low average betweenness centrality score shows that the central hub among the network is widely spread among many nodes. In addition, the average closeness centrality score reflects that a substantial portion of the community users are close to the center of the network. These observations show that the financial community is more densely connected at the local level. In addition,

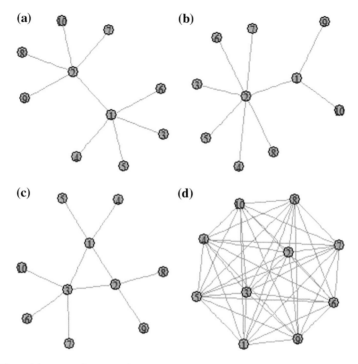

Fig. 4 Financial community network topology. **a** avg_dc = 0.36 avg_bc = 0.20 avg_cc = 0.70.
b avg_dc = 0.26 avg_bc = 0.15 avg_cc = 0.62. **c** avg_dc = 0.40 avg_bc = 0.24 avg_cc = 0.71.
d avg_dc = 1.00 avg_bc = 0.00 avg_cc = 1.00

the connectivity is not strong at the global level but there is evidence of clustering
around the center of the network.

Based on the location data, we extract the top 2,500 users with the highest betweenness centrality from the financial community in the U.S. and map their population
density (see Fig. 5). New York and California are identified as the top two states
containing the largest number of critical nodes, a fact reflective of their wealth distribution and status as financial and media centers. The next level consists of the states
of Massachusetts, Texas, Illinois and Florida. These are among the U.S. states with
the largest and wealthiest population. It is not surprising that some of their cities,
such as Boston and Chicago, serve as major hubs of the U.S. financial system. Lastly,
the final level comprises mostly states on the east coast like Pennsylvania, Virginia,
Georgia, New Jersey, Maryland and District of Columbia. They tend to be situated
with close proximity to New York City and have significant business ties with regard
to the number of financial corporation headquarters. The population map of the critical nodes has two significant interpretations: First, it reflects to a certain degree
where the most influential nodes in the financial community are most likely located.
Second, their tweeting activities might have direct implications on the location of

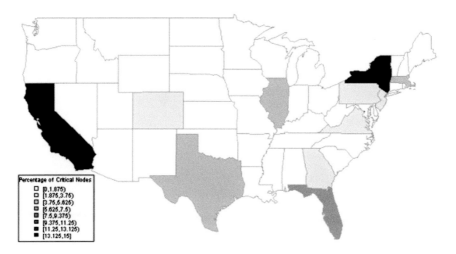

Fig. 5 Critical nodes location in financial community

the events. Knowing the location of the source provides a competitive advantage in tracking the scope of the events.

After settling on a definition, we examine how messages from key influencers in the community interact with social mood or sentiment that tend to signal an impending upward or downward swing in the market price movement. We use key network metrics such as out-degree centrality and betweenness centrality to identify the financial community influencers and we conjecture that these key influencers along with their weight of their influence in the financial community will provide better predictors of financial market movement measures.

This section demonstrates the value of extracting sentiment based on the social structure of the financial community. A key hypothesis of this approach is that Twitter sentiment extracted from the network's critical nodes serves as a reliable predictor of financial market movement. We believe that not all messages carry equivalent weight of information and therefore the message initiated from a more credible source should have larger influence on the financial community. Through regression analysis, we test whether specific critical nodes in the financial community have predictive power to key financial market measures. From the previous section, we identified three groups of 2,500 critical nodes based on key centrality measures from the financial community: betweenness centrality, out-degree centrality and closeness centrality. Through their tweet messages, we extracted their sentiments using the sentiment analysis algorithm and determined the statistical relationship with the historical daily return of major market returns and volatility indices.

3.1.2   Data Sources

We applied the regression analysis on 1,606,104 tweet messages for the period between 02/15/2014 and 06/15/2014. The daily sentiment series for the three major critical node groups were generated from their messages and then weighed with respect to the normalized centrality measures. In addition, we adopted the returns series for 6 exchange-traded funds which served as proxies for the historical daily market returns and volatility. The daily return is computed by taking the log return of market price from 02/15/2014 to 06/15/2014 (7).

$$r_t = \log\left(\frac{p_t}{p_{t-1}}\right) \tag{7}$$

where r_t denotes the return of day t and p_t is the price of day t.

3.1.3   Linear Regression Model of Sentiment

A linear regression model is applied to examine the relationship between the daily index return and the lag-1 Twitter message sentiment from the critical nodes. In particular, we investigated whether the lagged series of message sentiment have significant statistical relations with market returns and volatility. The dependent variable in the regression model is the daily return of respective market index. In this analysis, we used SPY, DIA, QQQ, and IWV as the market returns and VIX as the volatility returns. These market indices represent major distinct components of the financial market by the characteristics of the underlying stock such as market capitalization and industry sector (see Table 3). The independent variable is the lagged time series of the message sentiment from three critical node groups. In the experiment, we test the lag-1 sentiment for their respective significance to the returns series (8). If a significant relationship is observed, it suggests that returns lag behind the sentiment movement and therefore sentiment has predictive property over returns.

$$r_t = \beta_0 + \beta_1 S_{t-1} + \varepsilon \tag{8}$$

Table 3 Exchange-traded funds ticker and corresponding index

Market return	
SPY	S&P 500
DIA	Dow Jones industrial average
QQQ	NASDAQ
IWV	Russell 2000
Market volatility	
VIX	CBOE volatility index

where r_t denotes the daily return of day t and S_{t-1} denotes the daily sentiment score of day $t - 1$.

3.1.4 Comparison Among Centrality Groups

It is important to examine whether there is a fundamental difference of the regression result across the three different centrality measures in relation to the financial market returns and volatility. Varying the number of critical nodes in each group, we found that the betweenness centrality (BC) group consistently outperformed the degree centrality (DC) and closeness centrality groups (CC) (see Appendix 4B). The sentiment regression model of the BC group has shown significance across all market returns at the level of 95 %. In addition, the positive coefficients of the model demonstrate the predictive capability that the more positive the message sentiment is, the higher market returns it leads to. For volatility, the betweenness centrality group is also more significant than the two other groups in terms of its significance level. The result shows that more positive sentiment leads to lower volatility level, vice versa. It is consistent with the observation that negative sentiment can cause a higher volatility spike, suggesting that bad news on Twitter increases the volatility of price return in the stock market.

3.1.5 Comparison Among Number of Top Critical Nodes

Along the same intuition that not all messages carry equivalence of information, we investigated whether there is an optimal number of a critical node for each centrality group in explaining financial market movement. For the extreme scenarios, too many critical node users may introduce unnecessary noise but too few users may omit key contributing sentiment for the regression study. Varying the number of critical nodes for each centrality group might yield results that reveal the emerging critical point and, therefore, lead to an enhanced indicator for explaining market movements. In this analysis, we investigate all three centrality measures starting from the top 100 users to 2,500 users in each group at an increment of 100 additional users. For instances, we first examine the top 100 users in the betweenness centrality group and then the top 200 users in the same group. We observe that an optimal point exists when the number of critical nodes in the group is 200 (see Fig. 6). The coefficient for the 200-user group with the highest centrality measure is the most significant and consistent among all market indices. With the incorporation of more critical nodes in the regression model, we find that the p-value remains stabilized under the 0.05 level (except VIX volatility measure) for the models against market returns. This illustrates that our regression result is robust across different number of top critical nodes (Fig. 7).

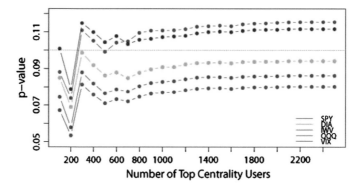

Fig. 6 Different market indices comparison (*p*-value)

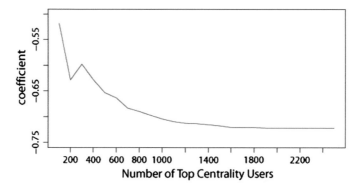

Fig. 7 Different market indices comparison (coefficient)

3.1.6 Discussion

The market sentiment regression highlights the significant influence of critical nodes towards movements in the financial market. Through performance comparison, the sentiment expressed by the betweeness centrality group is more significant and impactful than those by other community members. Critical nodes ranked by betweeness centrality degree yield the highest parameter significance close to the 99 % level. In addition, the BC group also yields the highest positive sentiment coefficients against market return at 0.15**, 0.16**, 0.18** and 0.19** for the top 200, 500, 1000 and 2000 user group respectively. In Table 4, the volatility regression exhibits similar observations of sentiment coefficients at −0.94*, −1.01*, −1.11*, −1.15* with better significance level by the critical nodes ranked by betweenness centrality. This is consistent with the definition of critical nodes because of their central contribution to network connectedness. Grouping users in financial community by centrality analysis transforms the regression result to be more precise and accurate in explaining market movements (Table 5).

Table 4 Lag-1 sentiment linear regression statistics (n = 200)

Market indices	BC score[a]		DC score[b]		CC score[c]	
	Coeff	p-value	Coeff	p-value	Coeff	p-value
SPY	0.15	0.01**	0.15	0.07*	0.11	0.35
DIA	0.12	0.03**	0.13	0.11	0.09	0.41
QQQ	0.17	0.05**	0.16	0.20	0.13	0.46
IWV	0.16	0.01**	0.17	0.07*	0.11	0.38
VIX	−0.94	0.10*	−0.81	0.29	−1.14	0.29

*$p < 0.10$; **$p < 0.05$
[a]Sentiment score weighted by betweenness centrality
[b]Sentiment score weighted by degree centrality
[c]Sentiment score weighted by closeness centrality

Table 5 Lag-1 sentiment linear regression statistics (n = 500)

Market indices	BC score[a]		DC score[b]		CC Score[c]	
	Coeff	p-value	Coeff	p-value	Coeff	p-value
SPY	0.16	0.01**	0.18	0.12	−0.10	0.54
DIA	0.14	0.03**	0.15	0.18	−0.06	0.69
QQQ	0.19	0.05**	0.17	0.30	−0.22	0.35
IWV	0.18	0.01**	0.19	0.11	−0.10	0.55
VIX	−1.01	0.09*	−0.95	0.36	0.91	0.53

Another significant finding is the effect of using a smaller subset of the critical node groups. The tradeoff analysis of varying the number of critical nodes reveals the optimal point in yielding significant signals to financial market returns and volatility. We found that selecting the top 200 users in the betweenness centrality group provide the most significant signal. A group with more than 200 critical node users dilutes the significance of the model but the result remains robust at a high significance level. The systematic search for the optimal number among the key centrality groups reinforces the main principle that not all messages should be treated with equal weight. Lastly, our comparison among different market indices signifies the value of extracting message sentiment based on social structure of the investment financial community. The sentiment series has potential applications on pairs trading strategy across multiple market indices.

We show that the behavior of critical nodes can be used to yield a more reliable indicator for the financial asset's price movement. It is worth noting that the current critical node analysis does not factor in the effect of tweets being read by unregistered users who may be active investors. This channel of information propagation is achieved by searching the Twitter website for specific keywords and the associated tweets would appear regardless whether the users are followers or not. This would facilitate the propagation mechanism of tweets to a wider community, including those tweets broadcasted by the critical nodes. Measuring the impact of this unobservable user group to the financial market will be a challenging problem, and we plan to address this issue in future studies. This empirical study of the financial community has three major contributions to the current literature of financial market and Twitter sentiment. First, it addresses the hypothesis that Twitter sentiment reflects the market participants' beliefs and behaviors toward future outcomes and the aggregate of the societal mood can present itself as a reliable predictor of financial market movement. Second, the concepts of leveraging critical nodes in the financial community generate a robust linkage between the social mood and financial market asset price movement. Moreover, the findings of critical nodes serve as an important guidance for regulatory authorities in paying attention to avoid manipulative or malicious actions such as the 2013 Associated Press hacking incident. Lastly, the empirical study provides insightful observations about the demographics and network structure of the financial community. By decomposing the community into unique user types, beliefs and behaviors of market participants can be better understood. With the continual growth of the Twitter universe, the dynamic characteristics of the financial community can be better understood in terms of its network structure and the message sentiment influence towards the financial market movement.

3.1.7   Conclusion

Based on the intuition that "not everyone on Twitter has the same influence on market sentiment", we document that there exists a financial community on Twitter, and the weighted sentiment of its key influencers has significant predictive power to market movement. In proving this hypothesis, we first document a methodology to identify the financial community, and then we illustrate the key properties of the financial community networks. We also demonstrate that the betweenness centrality measure of the network nodes is a better measure of influence of the nodes of the financial community networks than the other popular influential measures, i.e. the degree centrality and the closeness centrality. We show that different groups of critical nodes exert different degree of impacts on financial asset prices and volatility movement. In conclusion, we document that there is a robust correlation between the weighted Twitter financial community sentiment and financial market movement measured by lagged daily prices. This study covers the major financial market indices such as Dow Jones (DIA), S&P 500 (SPY), NASDAQ (QQQ) and Russell 3000 ETF (IWV), and the model significant levels are all less than 0.05 (p-value). A key finding of this study is that Twitter sentiment generated from critical nodes in the Twitter financial community provides a robust surrogate to predict financial market movement: the weighted sentiment of the critical nodes has significant predictive power over major market returns, and it consistently predicates market volatility (VIX) as well [45].

3.2 News Sentiment as Predictive Proxy to Financial Market

News sentiment has been empirically observed to have significant impact on financial market returns. In this study, we investigate firm-specific news from the Thomson Reuters News Analytics data from 2003 to 2014 and propose an optimal trading strategy based on a sentiment shock score and a sentiment trend score which measure extreme positive and negative sentiment levels for individual stocks. The intuition behind this approach is that the impact of events that generate extreme investor sentiment changes tends to have long and lasting effects to market movement and hence provides better prediction to market returns. We document that there exists an optimal signal region for both indicators. In addition, we demonstrate that extreme positive sentiment provides a better signal than extreme negative sentiment, which presents an asymmetric market behavior in terms of news sentiment impact. The back-test results show that extreme positive sentiment yields superior trading signals across market conditions, and its risk-adjusted returns significantly outperform the S&P 500 index over the same time period.

Many studies have demonstrated that news media can affect financial markets and often becomes drivers of market activities [3, 6, 7, 31, 32, 38, 41]. Analyzing news contents and translating them into trading signals have become an attractive research topic in both academia and industry. There have been a number of studies that further document the value of using media sentiment to make trading decisions [13, 24, 40, 49]. The motivation of this study is based on recent findings that news content affects investor sentiment and market volatility [6, 16, 39, 42]. We propose a trading strategy based on extreme news sentiment levels on individual stocks, and we further explore the effect of a long and short strategy based on extreme positive and negative sentiment on these stocks.

3.2.1 Data Source

This study utilizes the Thomson Reuters News Analytics package as the sole financial news data source. The package is used to quantify individual news events into sentiment, and its numerical form is then supplied for the trading system. The duration of the data ranges from January 2003 to December 2014. With more than 80 metadata fields in the Thomson Reuters News Analytics package, the corresponding fields are used in this study below.

- **Date/Time**: The date and time of the news article.
- **Stock RIC**: Reuters Instrument Code (RIC) of the stock for which the sentiment scores apply.
- **Sentiment Classification**: An integer number indicate the predominant sentiment value for news with respect to a stock identified by the RIC. Possible values are 1 for positive sentiment, 0 for neutral and -1 for negative sentiment.
- **Sent_POS**: Positive Sentiment Probability, the probability that the sentiment of the news article is positive for the stock. The possible value ranges from 0 to 1.

- **Sent_NEUT**: Neutral Sentiment Probability, the probability that the sentiment of the news article is neutral for the stock. The possible value ranges from 0 to 1.
- **Sent_NEG**: Negative Sentiment Probability, the probability that the sentiment of the news article is negative for the stock. The possible value ranges from 0 to 1. The sum of the three probabilities (Sent_POS, Sent_NEUT, Sent_NEG) equals 1.
- **Relevance**: A real-valued number between 0 and 1 indicating the relevance of the news item to a stock. A single news article may refer to multiple stocks, by comparing the number of occurrences within the text, the stock with the most mentions will be assigned with the highest relevance, and a stock with a lower number of mentions will have a lower relevance value.

In order to calculate a sentiment score for each stock mentioned in one news item, we first calculate the expected value of the sentiment score, and then generate the weighted expected value using its relevance value. Finally, the weighted weekly average sentiment score is calculated as follows:

$$Avg\_Sent = \frac{1}{N} \sum_N \begin{array}{l} ((Sent\_POS \times 1 \\ +Sent\_NEUT \times 0 \\ +Sent\_NEG \times (-1)) \\ \times Relevance), \end{array} \tag{9}$$

where N is the total number of new articles for a stock within one week. The weighted weekly average sentiment is later used as the input for computing the other two sentiment scores.

The summary statistics of the news sentiment data, including the mean, standard deviation, maximum/minimum, and 5th/50th/95th percentile of each variable, is displayed below (see Table 8). We also plot the monthly aggregated average news sentiment for all 596 stocks, the total number of news articles (hereinafter "number of news") for each month, and the S&P 500 index monthly return (see Fig. 8). The data indicates that the average news sentiment is positively correlated with market return with a correlation coefficient of 0.21, while the total number of news is negatively correlated with market return with a correlation coefficient of -0.14. Through conducting a lead-lag analysis, the news sentiment is shown to lead the market return, while there is no opposite effect from market return on future news sentiment (Table 6).

Table 6 Statistics of calculated average news sentiment

	Mean	STD.	Max	Min	5%	50%	95%
Average sentiment	0.09	0.24	0.83	−0.78	−0.20	0.00	0.60
Number of news items	5.18	11.69	830	0	0	2	22

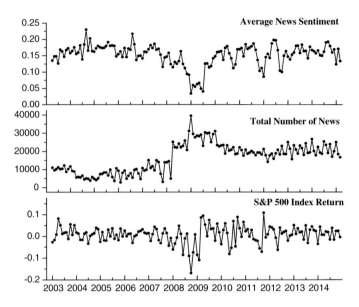

Fig. 8 Monthly aggregated news data comparison with market returns

3.2.2   Sentiment Shock and Trend Scores

With the preliminary observations on correlation, we explore the abnormal levels of
the news sentiment data as a way to create more insightful and unique indicators
for the trading system. We propose two sentiment scores to characterize shocks (i.e.
spike up or down) and trends in the sentiment time series. The calculation is based
on the average weekly news sentiment scores for each stock. In order to reduce the
number of parameters in the trading strategy and avoid over-fitting, we optimize the
calculation parameters for each GICS sector, so that all stocks within the same sector
use the same parameter. The trading strategy is designed to monitor the calculated
sentiment scores and generate buy-and-sell signals for each stock.

Sentiment shocks are the abnormal spikes observed from the time series. These
sentiment shocks are often caused by the release of unexpected macroeconomic data,
financial report results, and corporate actions. The sentiment shock score is calculated
as below:

$$(S_{t0} - \mu)/\sigma, \tag{10}$$

where S_{t0} is sentiment value on week $t0$, $t0$ represents the current week, μ is the
mean of sentiment values from week $t0 - N$ to $t0 - 1$ S_{t0-N} to S_{t0-1}, and σ is the
standard deviation of sentiment values from week $t0 - N$ to $t0 - 1$. N is the total
number of look-back weeks.

Sentiment trend score measures the aggregated change of sentiment over a historical period. The sentiment trend measure reveals more information than the sentiment

shock approach when comparing abnormal changes of sentiment over a long period
of time.

$$\sum_{i=t0-N}^{t0} \Delta S_i \tag{11}$$

where ΔS_i is the change of sentiment in week i, and represents the current week. N
is the moving window size, summing the change of sentiment within it.

3.2.3   Parameters Optimization

Each of the sentiment shock or trend score has a parameter N (the look-back window)
to choose. To find the best parameter, we use Spearman rank correlation as the
objective value. In order to reduce the number of parameters and avoid over-fitting,
N is optimized for each GISC sector, and stocks in the same sector use the same
value. The method we use to optimize these parameters is to maximize the Spearman
rank correlation between the sentiment scores and the next week's stock return. The
Spearman rank correlation is a measure of rank dependence between two variables.
For a sample of size n, the two variables X_i, Y_i are converted to ranks x_i, y_i, the
correlation coefficient is computed as:

$$\rho = 1 - \frac{6 \sum d_i^2}{n(n^2 - 1)} \tag{12}$$

where $d_i = x_i - y_i$. By maximizing the rank correlation, the calculated sentiment
scores are most informative for future stock return.

3.2.4   Trading Strategy Construction

Using the optimized look-back windows for each sector, the time series of sentiment
shock and sentiment trend scores are constructed for each stock from 2003 to 2014.
The design of our trading strategy is based on the hypothesis that extreme sentiment
has a persistent effect on subsequent stock returns. Therefore, the trading strategy
involves establishing long positions on stocks with unusually high positive sentiment
scores and vice versa.

   In terms of the trading strategy parameters, the cutoff percentile between extreme
sentiment score versus normal score is defined according to the empirical probability
distribution of sentiment scores during the training period (see Fig. 9). A threshold
of bottom 5 % means the sentiment score in the 5 % bottom percentile in the training
data is the break point of extreme negative sentiment, and scores lower than that
threshold are considered as short signals. As suggested in Fig. 9, the majority of the
sentiment scores are centered on zero. With the selection of larger thresholds, the
sentiment value diminishes quickly and only the extreme values are of significance
in providing signals. This effect is particularly pronounced in the sentiment trend

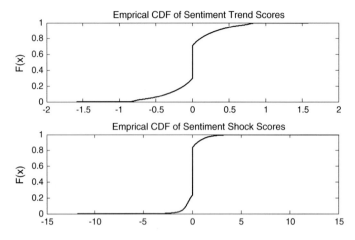

Fig. 9 Empirical CDFs of sentiment trend and sentiment shock scores

score, as it has a broader non-zero region in the cumulative distribution function than that in the sentiment shock score. We will further discuss the optimal selection criterion for this threshold next (Table 7).

3.2.5 Trading Strategy Implementation

We design a dynamic trading framework with an evaluation period of 4 weeks (see Fig. 10). We determine the threshold of extreme sentiment with 90 % as positive threshold and 10 % as negative threshold. Each firm-specific sentiment score is then evaluated and compared with the threshold to make trading decision. If the firm's sentiment exceeds the positive threshold, we establish a long position or vice versa. The portfolio is rebalanced every 4 weeks, with the risk control of a 10 % stop loss limit order in place. In the trading system, the strategy return is recorded weekly. During the training process, we use 4 years of data from 2003 to 2006 as the training period so that sufficient out-of-sample data is available including the 2008 financial crisis. Table 8 summarizes the optimized look-back windows of sentiment indicators for each sector (Table 9).

3.2.6 Strategy Performance and Discussion

The proposed trading strategy using sentiment shock and trend scores was back-tested from 2007 to 2014. For the long strategy, both shock indicator and trend indicator yield higher Sharpe ratios than the S&P 500 index (see Table 10). Interestingly, higher cutoff percentile led to the phenomenon that Sharpe ratio rises to a peak and then gradually flattens out. This can be explained from the two perspectives. (1) When the

Table 7 Optimized number of weeks for sentiment scores by sector

Sector name	Sentiment shock	Sentiment trend
Consumer discretionary	15	14
Information technology	11	30
Consumer staples	18	19
Materials	15	16
Industrials	21	18
Utilities	16	28
Health care	10	15
Energy	25	20
Financials	11	25
Telecommunication services	19	24

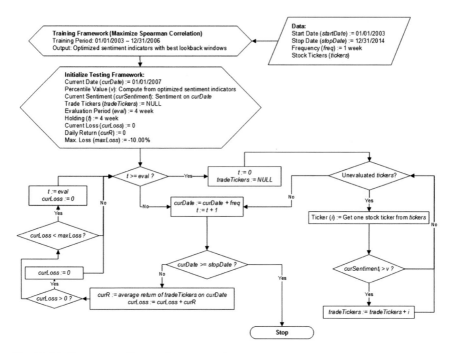

Fig. 10 Trading strategy diagram (long strategy)

cutoff percentile rises, more stocks are added into the trading portfolio. The Sharpe ratio increases in the initial stage because more companies with superior returns are included for better diversification. (2) The subsequent decline of Sharpe ratio is due to the diminishing effect of news influence. The other key result is that the trend indicator strategy performs better than the shock indicator strategy in terms of higher Sharpe ratio consistently across all cutoff percentiles. The distinctively

Table 8 Backtest statistics for long strategies

Strategy	Max. drawdown (%)	No. of trades	No. of wining trades	No. of losing trades	Avg. holding period (weeks)
Sentiment trend	49.09	673	449	224	9.51
Sentiment shock	56.37	896	576	320	6.96

different results from the long and short strategies demonstrate the asymmetric market response to extreme positive and negative sentiment. In order to test the robustness of the trading strategies, we recorded the trading activities for each strategy (Table 8). For both sentiment indicators, the number of winning trades almost doubles the number of losing trades, which suggests the consistency of the strategy in generating positive return (Fig. 11).

The top chart of Fig. 12 shows the cumulative returns of our trading strategies, with benchmark of the buy-and-hold strategy return of the S&P 500. The bottom chart of Fig. 12 illustrates the market volatility at corresponding period. The long strategies outperform the buy-and-hold strategy for the entire test period, confirming that the sentiment trend trading strategy has superior performance over the sentiment shock strategy. To validate the performance of the trading strategies in different market conditions, we split the back-test period into high and low volatility regime using 6-month realized market volatility. The high volatility regime during 2003 to

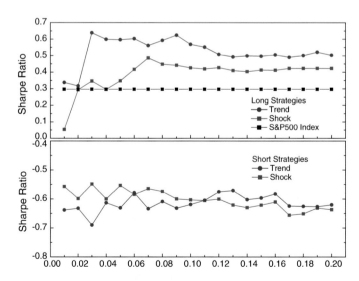

Fig. 11 Strategy sharpe ratios by changing extreme sentiment selection percentile. *Top chart* shows long strategy with threshold from *top* 1 to 20 % and *bottom chart* shows short strategy with threshold from *bottom* 1 to 20 %

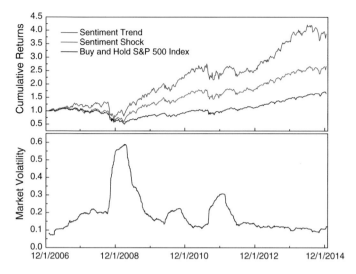

Fig. 12 Cumulative returns of sentiment trend and shock strategy benckmarked with Buy and Hold S&P 500 Index. *Top chart* shows the long strategy, *bottom chart* shows the market volatilty for the same time peroid

2014 was from 10/2008 to 05/2009. Both sentiment indicator strategies show higher profitability than the benchmark strategy in high volatility regime. In the low volatility regime that was bull market period, the trend indicator outperforms the benchmark in terms of higher return and Sharpe ratio. The shock indicator exhibits the same level of performance compared with the buy-and-hold strategy with a slightly lower Sharpe ratio (see Table 9). This result demonstrates that both sentiment indicators have good performance in predicting subsequent market returns in the long run, and the sentiment trend indicator provides more robust trading signals than the sentiment shock indicator. As shown in Table 9, the long strategies using the extreme positive sentiment outperform the S&P 500 index in both high and low market volatility regimes.

3.2.7   Conclusion

We demonstrate the value of using news sentiment data in formulating potential trading signals. Specifically, we use firm-specific news data from Thomson Reuters News Analytics, and we propose a sentiment shock score and a sentiment trend score for individual stocks to identify extreme sentiment levels and used them as trading signals. A previous study has shown that abnormal news sentiment, like sentiment shocks and trends, are predictive for future market return and volatility [47]. For individual stock level, the same intuition still applies that a big jump of the sentiment or a trend of sentiment change in the same direction will trigger persistent

Table 9 Backtest results in different market conditions

Strategy	Annualized performance measures		
	Mean return (%)	Volatility (%)	Sharpe ratio
Total backtest period			
Sentiment trend	16.90	26.50	0.64
Sentiment shock	12.24	25.22	0.49
Buy and Hold S&P 500	6.30	21.18	0.30
High volatility regime			
Sentiment trend	56.40	52.60	1.07
Sentiment shock	49.21	54.28	0.91
Buy and Hold S&P 500	−19.94	47.92	−0.42
Low volatility regime			
Sentiment trend	13.61	23.11	0.59
Sentiment shock	9.15	21.18	0.43
Buy and Hold S&P 500	8.49	17.29	0.49

impact on stock price movement. The back-test results of the trading strategy support the intuition that extreme positive sentiment has an impact on market returns and volatility. Our results show that the extreme positive sentiment for individual stocks generates more reliable trading signals than the extreme negative sentiment, which suggests the asymmetric response of the market to positive and negative sentiment.

3.3 Sentiment Genetic Programming Based Trading Strategies

This approach is motivated by the empirical findings that news and social media Twitter messages (tweets) exhibit persistent and predictive power on financial market movement. Based on the evidence that tweets is faster than news in revealing new market information, whereas news is regarded broadly a more reliable source of information than tweets, we propose a superior trading strategy based on the sentiment feedback strength between the news and tweets using generic programming optimization method. The key intuition behind the feedback strength based approach is that the joint momentum of the two sentiment series leads to significant market signals, which can be exploited to generate above-average trading profits. With the trade-off between information speed and its authenticity, we aim to develop a trading strategy with the objective to maximize the Sterling ratio. We find that the sentiment feedback based strategy yields superior market returns with small standard deviation over the two years from 2012 to 2014. When compared across the Sterling ratio and other risk measures, the proposed sentiment feedback based strategy generates better results over both the technical indicator-based strategy and the basic buy-and-hold strategy.

Using both news and tweets sentiments, we present a novel framework for developing an optimal trading strategy using genetic programming. It leverages existing empirical findings on the relationships observed among new sentiment, tweets sentiment and market returns. The key intuition behind the sentiment indicator is that the joint momentum of the two sentiment series leads to significant market anomalies which can be exploited in the form of above-average trading profits. For instance, if both news and tweets sentiments show strong momentum in trending in one direction, the market return is likely to follow in the same direction. An investor can therefore establish a long position when the sentiment indicator generates such signal and exits when the reversal appears. In addition, the two information sources also display key distinguishable characteristics that the trading rules can be constructed by choosing the optimal tradeoff between the speed of information release and the authenticity of the publishers. Our previous studies demonstrate that news sentiment has a more delayed impact than tweets sentiment on financial market returns, and we extend the findings by formulating an optimization problem to maximize risk-adjusted returns with the sentiment indicator [33, 45]. We prefer the use of genetic programming because of its flexibility in handling character strings and capability to search in a large population. In the study, we find that the sentiment-based genetic programming approach yields an above-average trading profit over the two years from 2012 to 2014. The out-performance suggests that the sentiment-based indicator can be regarded as a valuable source of information along with technical indicators.

3.3.1   Genetic Programming as an Optimization Approach

Genetic programming is a special class of genetic algorithm, which was first developed by John Holland in 1992. Genetic algorithm was built on the premise of the natural selection process that individual action with condition is evaluated with a pre-specified fitness function until the optimal combination is reached. Holland illustrated that "a population of fixed length character strings can be genetically bred using the Darwinian operation of fitness proportionate reproduction and the genetic operation of recombination" [26]. The central goal of using genetic algorithm is to exploit a vast region in the search space and at the same time to manipulate variations of strings [19]. The difference between genetic programming and genetic algorithm lies on the representation of the varying string length in the search space. Genetic programming allows solutions to be represented by a flexible string length with the Boolean connectors connecting the sentiment indicator with other technical indicators. For example, we can construct solutions with different combinations of indicators and parameters in contrast to the fixed set of indicators that we have to use for each search. Moreover, GP requires input solutions to be represented in a tree structure to accommodate the flexibility. Three major genetic operators are applied to a given problem during the optimization process: mutation, crossover and encoding. These

1: Randomly create an initial population of programs from the available primitives;
repeat
    2: Execute each program and ascertain its fitness;
    3: Select one or two program(s) from the population with a probability based on gitness to participate in genetic operations;
    4: Create new individual program(s) by applying genetic operations with specified probabilities;
until *an acceptable solution is found or some other stopping condition is met*;
return the best-so-far individual;

Fig. 13 Genetic programming algorithm

operators are crucial in the effectiveness of the genetic programming framework to converge towards the optimal solution within the search space.

We apply the standard framework of genetic programming to locate the optimal trading strategy with the proposed sentiment indicator (see Fig. 13). The framework allows the comparison of the large number of combinations among variations of indicators and parameters. The goal of the algorithm is to locate the optimal trading strategy with the highest Sterling ratio, which is set to be the fitness function. In the genetic programming (GP) framework, we first initialize the population of programs constructed from the sentiment feedback strength indicator and the two technical indicators. Through the search process, we incorporate the Boolean operators, "AND" and "OR", for allowing different combinations of the indicators. For each strategy, the algorithm generates trading signals in the form of "TRUE/FALSE" signal at each time period. Since there is no short position allowed, the "FALSE" signal is effective only when an existing position is open. For a "TRUE" signal, the system records the cumulative returns over the holding period until a reversal of the trading signal appears. For example, if a position is established on day 1 and closed on day 10, the trading return is calculated as the cumulative returns over the 10-day period. With the trading record of the strategy, the algorithm ascertains its fitness with the Sterling ratio and then performs genetic operations in crossover and mutation with specified probabilities. We select the crossover rate and mutation rate as 0.50 and 0.90 respectively to extend the search population and increase the likelihood of solution achieving global convergence. The GP algorithm is implemented through multiple iterations to generate the optimal combinations of the indicators connected by Boolean operators and numeric parameters in each indicator.

3.3.2 Trading Strategy Performance Comparison

This section presents the performance comparison of the two sentiment feedback strength based trading strategies against two benchmark strategies. The first benchmark strategy utilizes the genetic programming framework to generate trading signals based on entirely technical indicators only. The rationale behind this strategy is that GP can generate useful technical trading rules with optimal set of parameters. The

Table 10 GP Algorithm parameter set

Parameter	Value
Population size	100
Number of iterations	5,000
Selection method	Roulette wheel
Fitness function	Stirling ratio
Boolean operators	"AND", "OR"
Numeric parameters	$U(0,1)$
Crossover rate	0.5
Mutation rate	0.9
Max initial tree depth	5
Max following tree depth	5

second benchmark is the traditional buy-and-hold strategy that is commonly utilized by small investors and mutual funds.

The results show that the combined trading strategy with both sentiment feedback strength and technical indicators provides a clear edge over the other three trading strategies in terms of higher Sterling ratio and total profit/loss. The optimal strategy generates over 21.3 % Sterling ratio compared to 15.5 %, 12.1 % and 8.6 % from the feedback strength-only strategy, technical indicators-only strategy, and the buy-and-hold strategy respectively (see Table 11). For the comparison of the total cumulative returns over the evaluation period, the combination of sentiment feedback strength and technical indicators yields the best performance at 21.7 % (see Fig. 14). On the other hand, the result related to the trading strategy based entirely on sentiment feedback strength only suggests that sentiment provides support in controlling loss indicated by the significantly lower maximum dropdown at -0.6% in contrast to -2.1% for the three strategies. We find that the percentage of winning trades is higher at 73.61 % despite the lower number of trades at 72. From a standpoint of evaluating the strategy risk, the standard deviation of the daily returns is also lower at 0.30 % compared to 0.47 % and 0.58 % respectively.

Through the genetic programming optimization, we find that the lag-1 news sentiment and lag-2 tweets sentiment are the most dominant factors in the formulation of the optimal trading strategy. In other words, the trading signals based on the feedback strength indicator rely significantly on business news articles published one day ago and tweet messages generated by the Twitter financial community two days ago. The results suggest that the combination of the two factors yields the optimal performance in terms of Sterling ratio and the percentage of winning trades. Furthermore, the lag-2 tweets sentiment exhibits a stronger effect on triggering trading signals over the lag-1 news sentiment, demonstrated by the higher optimal weight determined by the algorithm. On the contrary, the lag-1 news sentiment displays a greater sensitivity in affecting market returns, reflected by the lower summation threshold in the sentiment feedback strength indicator.

Table 11 GP optimization trading strategy performance

	Sentiment + Technical indicators	Sentiment indicator	Technical indicators	Buy-and-hold
Number of observations	132	132	132	132
Number of trades	102	72	92	132
Trading %	77.3	54.5	69.7	100
Number of winning trades	74	53	61	81
Percentage of winning trades (%)	72.5	73.6	66.3	61.4
Total profit/loss (%)	21.7	15.6	12.3	8.8
Average profit/loss per trade (%)	0.21	0.22	0.13	0.07
Standard deviation (%)	0.44	0.30	0.47	0.58
Maximum drawdown (%)	−2.1	−0.6	−2.1	−2.1
Stirling ratio (%)	21.3	15.5	12.1	8.6

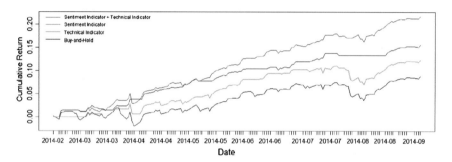

Fig. 14 Cumulative returns of sentiment-based trading strategy

3.3.3 Conclusion

We introduce a genetic programming approach to develop an optimal trading strategy with news and tweet sentiments. The proposed feedback strength indicator, a measure of the joint momentum between the news and tweet sentiments, was found to provide a significant improvement in trading performance over the S&P 500 financial market index ETF. By quantifying the joint momentum of the sentiment series, we can detect significant market anomalies that can be exploited in the form of modest trading returns. We find that the sentiment-based genetic programming approach yields positive market returns with small standard deviation over the two years from 2012 to 2014. When comparing the Sterling ratio and other risk measures, the proposed strategy is superior to the technical indicators and the traditional buy-and-hold strategy. The out-performance suggests that news and tweet sentiments can be regarded as valuable sources of information in constructing meaningful trading system along with technical indicators [45].

4   Summary and Conclusion

This chapter focuses on showcasing the capabilities of financial market sentiment analysis. It explores the impact of news and social media on investor sentiment and financial markets. Through their empirical relations, it further introduces how existing findings can be combined with artificial intelligence techniques to develop advanced investment strategies. The main contribution of the book chapter rests on surveying existing literature findings related to sentiment analysis and financial market and addressing research questions within the field of behavioral finance whether investor sentiment can be quantified through news and tweet sentiment. Research has shown that financial market sentiment can be leveraged as a source to develop practical financial solutions in the form of a modestly profitable trading strategy. It reinforces the empirical evidence in the literature of behavioral finance that there exists opportunities in the area of investor sentiment for generating above-average returns. Using advanced techniques in processing and analyzing textual sources of information, the specific contributions of this book chapter are summarized as follows:

- The concept of community construction using social network analysis is a novel approach in the field of analyzing social media impact on financial market. By harnessing the sentiment expressed by the influential Twitter users within the community, the empirical study provides a better proxy over existing studies between social sentiment and financial market movements
- Using news sentiment data, a firm-specific trading strategy is developed based on the detection of abnormal sentiment levels. The study shows that news sentiment can be a proxy for future market returns and volatility. The above-normal positive sentiment for individual stocks generates more reliable trading signals than the extreme negative sentiment. This is also an indication of the asymmetric response of the market to positive and negative sentiment.
- Based on the finding of the time scale difference between news and social sentiment, we present a novel approach to formulate an optimization problem for identifying a trading strategy based on the sentiment feedback strength using genetic programming. The proposed sentiment feedback based strategy shows better performance over the technical indicator-based strategy and the basic buy-and-hold strategy and further validates the value of both news and tweets sentiment in exploiting trading opportunities.

Acknowledgments The authors would like to thank the Financial Engineering Division at Stevens Institute of Technology for providing a state-of-the-art research environment with data access and hardware support. They would also like to acknowledge the support from the Civil Group of Northrop Grumman Corporation.

References

1. Aboura, S., Chevallier, J.: Leverage vs. feedback: which effect drives the oil market? Finan. Res. Lett. **10**(3), 131–141 (2013)
2. Antoniou, A., Koutmos, G., Pericli, A.: Index futures and positive feedback trading: evidence from major stock exchanges. J. Empirical Finan. **12**(2), 219–238 (2005)
3. Antweiler, W., Frank, M.Z.: Is all that talk just noise? The information content of internet stock message boards. J. Finan. **59**(3), 1259–1294 (2004). http://doi.org/10.1111/j.1540-6261.2004.00662.x
4. Arnold, L.G., Brunner, S.: The economics of rational speculation in the presence of positive feedback trading. Q. Rev. Econ. Finan. (2014)
5. Baccianella, S., Esuli, A., Sebastiani, F.: SentiWordNet 3.0: an enhanced lexical resource for sentiment analysis and opinion mining. In: Calzolari, N., Choukri, K., Maegaard, B., Mariani, J., Odijk, J., Piperidis, S., ... Tapias, D. (eds.) Proceedings of the Seventh International Conference on Language Resources and Evaluation (LREC'10), pp. 2200–2204. European Language Resources Association (ELRA) (2010)
6. Barber, B.M., Odean, T.: All that glitters: the effect of attention and news on the buying behavior of individual and institutional investors. Rev. Finan. Stud. **21**(2), 785–818 (2007)
7. Barberis, N., Shleifer, A., Vishny, R.: A model of investor sentiment. J. Finan. Econ. **49**, 307–343 (1998)
8. Borgatti, S.P., Everett, M.G.: A Graph-theoretic perspective on centrality. Soc. Netw. **28**(4), 466–484 (2006)
9. Brown, G.W., Cliff, M.T.: Investor sentiment and the near-term stock market. J. Empirical Finan. **11**(1), 1–27 (2004)
10. Chau, F., Deesomsak, R.: Business cycle variation in positive feedback trading: evidence from the G-7 economies. J. Int. Finan. Mark. Inst. Money (2014)
11. Chopra, N., Lakonishok, J., Ritter, J.R.: Measuring abnormal performance. J. Finan. Econ. **31**, 235–268 (1992)
12. Cohen, G., Kudryavtsev, A.: Investor rationality and financial decisions. J. Behav. Finan. **13**(1), 11–16 (2012)
13. Crone, S.F., Koeppel, C.: Predicting exchange rates with sentiment indicators: an empirical evaluation using text mining and multilayer perceptrons. In: 2014 IEEE Conference on Computational Intelligence for Financial Engineering and Economics (CIFEr), pp. 114–121 (2014)
14. Daniel, K.D., Hirshleifer, D., Subrahmanyam, A.: Overconfidence, arbitrage, and equilibrium asset pricing. J. Finan. **LVI**, 921–965 (2001)
15. Daniel, K., Hirshleifer, D., Subrahmanyam, A.: Investor psychology and security market under- and overreactions. J. Finan. **53**, 1839–1885 (1998)
16. DiBartolomeo, D., Warrick, S.: Linear Factor Models in Finance. Elsevier (2005)
17. Hilton, D.J.: The psychology of financial decision-making: applications to trading, dealing, and investment analysis. J. Psychol. Finan. Mark. **2**(1), 37–53 (2001)
18. Hirshleifer, D., Subrahmanyam, A., Titman, S.: Feedback and the success of irrational investors. J. Finan. Econ. **81**(2), 311–338 (2006)
19. Holland, J.: Genetic algorithms. Sci. Am. 66–72 (1992)
20. Hou, Y., Li, S.: The impact of the CSI 300 stock index futures: positive feedback trading and autocorrelation of stock returns. Int. Rev. Econ. Finan. **33**, 319–337 (2014)

21. Im, T.L., San, P.W., On, C.K., Alfred, R., Anthony, P.: Impact of financial news headline and content to market sentiment. Int. J. Mach. Learn. Comput. **4**(3), 237–242 (2014)
22. İnkaya, A., Yolcu Okur, Y.: Analysis of volatility feedback and leverage effects on the ISE30 index using high frequency data. J. Comput. Appl. Math. **259**, 377–384 (2014)
23. Johnson, B.: Algorithmic Trading & DMA: An Introduction to Direct Access Trading Strategies, p. 574. 4Myeloma Press (2010)
24. Kaya, M.: Stock price prediction using financial news articles. In: 2nd IEEE International Conference on Information and Financial Engineering ICIFE 2010, pp. 478–482 (2010)
25. Khanna, N., Sonti, R.: Value creating stock manipulation: feedback effect of stock prices on firm value. J. Finan. Mark. **7**(3), 237–270 (2004)
26. Koza, J.R.: Evolution of a Subsumption Architecture that Performs a Wall Following Task for an Autonomous Mobile Robot via Genetic Programming, pp. 1–39 (1992)
27. La Porta, R., Lakonishok, J., Shleifer, A., Vishny, R.W.: Good news for value stocks: further evidence on market efficiency. J. Finan. **52**, 859–874 (1997)
28. Laopodis, N.T.: Feedback trading and autocorrelation interactions in the foreign exchange market: further evidence. Econ. Modell. **22**(5), 811–827 (2005)
29. Li, X., Xie, H., Chen, L., Wang, J., Deng, X.: News impact on stock price return via sentiment analysis. Knowl-Based Syst. **69**, 14–23 (2014)
30. Medhat, W., Hassan, A., Korashy, H.: Sentiment analysis algorithms and applications: a survey. Ain Shams Eng. J. **5**(4), 1093–1113 (2014)
31. Mitchell, M., Mulherin, H.: The impact of public information on the stock market. J. Finan. **49**(3), 923–950 (1994)
32. Mitra, L., Mitra, G.: The Handbook of News Analytics in Finance. Wiley, Chichester, West Sussex, UK (2011)
33. Mo, S.Y.K., Liu, A., Yang, S.: News sentiment to market impact and its feedback. SSRN Working Paper (Submitted) (2015)
34. Moreo, A., Romero, M., Castro, J.L., Zurita, J.M.: Lexicon-based comments-oriented news sentiment analyzer system. Expert Syst. Appl. **39**(10), 9166–9180 (2012)
35. Nofsinger, J.R.: Social mood and financial economics. J. Behav. Finan. **6**(3), 144–160 (2005)
36. Salm, C.A., Schuppli, M.: Positive feedback trading in stock index futures: international evidence. Int. Rev. Finan. Anal. **19**(5), 313–322 (2010)
37. Schumaker, R.P., Zhang, Y., Huang, C.-N., Chen, H.: Evaluating sentiment in financial news articles. Decis. Support Syst. **53**(3), 458–464 (2012)
38. Scott, J., Stumpp, M., Xu, P.: News, not trading volume, builds momentum. Finan. Anal. J. **59**(2), 45–54 (2003)
39. Smales, L.A.: Asymmetric volatility response to news sentiment in gold futures. J. Int. Finan. Mark. Inst. Money **34**, 161–172 (2015)
40. Song, Q., Liu, A., Yang, S.Y., Datta, K., Deane, A.: An extreme firm-specific news sentiment asymmetry based trading strategy. In: 2015 IEEE Conference on Computational Intelligence for Financial Engineering and Economics (CIFEr 2015) (2015)
41. Tetlock:. Giving content to investor sentiment: the role of media in the stock market. J. Finan. **62**(3), 1139–1168 (2007)
42. Tetlock, P.C., Saar-Tsechansky, M., MacSkassy, S.: More than words: quantifying language to measure firms' fundamentals. J. Finan. **63**, 1437–1467 (2008). http://doi.org/10.1111/j.1540-6261.2008.01362.x
43. Thaler, R.H.: Advances in Behavioral Finance, vol. 2. Princeton University Press, Princeton, NJ (2005)
44. Xie, B., Wang, D., Passonneau, R.J.: Semantic feature representation to capture news impact. In: Twenty-Seventh International Florida Artificial Intelligence Research Society Conference (2014)
45. Yang, S., Mo, S.Y.K., Liu, A., Kirilenko, A.: Genetic programming optimization for a sentiment feedback strength based trading strategy. SSRN Working Paper (2015)
46. Yang, S.Y., Mo, S.Y.K., Liu, A.: Twitter financial community sentiment and its predictive relationship to stock market movement. Quant. Finan. **15**(10), 1637–1656 (2015)

47. Yang, S.Y., Song, Q., Yin, S., Mo, K., Datta, K., Deane, A.: The impact of abnormal news sentiment on financial markets. J. Bus. Econ. **6**(10), 1682–1694 (2015)
48. Yu, L.-C., Wu, J.-L., Chang, P.-C., Chu, H.-S.: Using a contextual entropy model to expand emotion words and their intensity for the sentiment classification of stock market news. Knowl.-Based Syst. **41**, 89–97 (2013)
49. Zhang, W., Skiena, S.: Trading Strategies To Exploit Blog and News Sentiment (2010). Chan (2003)

Context Aware Customer Experience Management: A Development Framework Based on Ontologies and Computational Intelligence

Hafedh Mili, Imen Benzarti, Marie-Jean Meurs, Abdellatif Obaid, Javier Gonzalez-Huerta, Narjes Haj-Salem and Anis Boubaker

Abstract Customer experience management (CEM) denotes a set of practices, processes, and tools, that aim at personalizing customer's interactions with a company according to customer's needs and desires (Weijters et al., J Serv Res 10(1):3–21, 2007 [29]). E-business specialists have long realized the potential of ubiquitous computing to develop context-aware CEM applications (CA-CEM), and have been imagining CA-CEM scenarios that exploit a rich combination of sensor data, customer profile data, and historical data about the customer's interactions with his environment. However, to realize this potential, e-commerce tool vendors need to figure out which software functionalities to incorporate into their products that their customers (e.g. retailers) could use/configure to build CA-CEM solutions. **We propose** to provide such functionalities in the form of **an application framework within which CA-CEM functionalities can be specified, designed, and implemented**. **Our framework relies on, (1) a cognitive modeling of the purchasing process**, identifying potential touchpoints between sellers and buyers, and relevant influence factors, **(2) an ontology to represent relevant information about consumer categories, property *types*, products, and promotional material, (3) computational intelligence techniques to compute consumer- or category-specific property *values***, and **(4) approximate reasoning algorithms to implement** some of **the CEM functionalities**. In this paper, we present the principles underlying our framework, and outline steps for using the framework for particular purchase scenarios. We conclude by discussing directions for future research.

Keywords Customer experience management · Customer behavior modeling · Service design · Customer profile ontology · Product ontology · Recommendation systems · Context awareness · Computational Intelligence · Data mining

H. Mili (✉) · I. Benzarti · M.-J. Meurs · A. Obaid · J. Gonzalez-Huerta ·
N. Haj-Salem · A. Boubaker
LATECE Laboratory, Université du Québec à Montréal, Montréal, Canada
e-mail: mili.hafedh@uqam.ca

© Springer International Publishing Switzerland 2016
W. Pedrycz and S.-M. Chen (eds.), *Sentiment Analysis and Ontology Engineering*,
Studies in Computational Intelligence 639, DOI 10.1007/978-3-319-30319-2_12

1 Introduction

Customer experience management (CEM) denotes a set of practices, processes, and tools, that aim at personalizing customer's interactions with a company according to customer's needs and desires [30]. Ubiquitous computing is a computing paradigm where a computation of interest to a stakeholder is collaboratively performed by a variety of often specialized devices with limited capacities that interact "spontaneously" on behalf of the stakeholder [20, 28]. E-business specialists have long realized the potential of ubiquitous computing to develop context-aware CEM applications (CA-CEM). Many e-business visionaries fantasize about what our shopping experience could be like at our favorite food store, entertainment place, or clothing store if the "system" knew about our lifestyle, our age, our heartbeat, what we ate this week, the weather outside, the people we hang out with, in real life and social networks, the places we visit, on foot or by mouse, what we like, what we wrote in our blogs, etc. What would be possible if we shared the same identity across devices, media, and applications, and if all the "things" within our environment were connected [30], including our cell phone, the store we walked into, the steak slab we put in our shopping cart.

This raises a multitude of questions, both for the e-commerce software vendors who want to offer CA-CEM functionalities, and for customers of such applications (e.g. retailers), who wish to take advantage of such functionalities.

Examples of these questions include:

- *What CEM functionalities do we want to provide?* Different actions can be taken, depending on the phase of the consumer experience. For example, during pre-purchase, we can try to anticipate a customer need ("your car is due for an oil change"), or respond to a product search on the company website by promoting a particular product. During purchase, we can develop functionalities for cross-selling ("would you like to buy a tie with your shirt?") or up-selling ("did you check [the better] brand X?"). After the purchase, we can conduct customer satisfaction surveys, or monitor customer posts on product forums, blogs, or social networks.
- *What information about the customer, the product, the promotional material, and the environment do we need in order to support various CEM functionalities?* For example, to anticipate a customer's need, what kind of information should we know about the customer? This depends on the need. For some basic needs (e.g. food), knowing little information is enough. For *lifestyle* purchases, a lot more is necessary. The same can be said about products. For some commodities, price may be all that consumers care about. However, health- or socially-conscious consumers may care about the *production process* of what they are buying, including environmental considerations, fairness, sustainability, or labor practices.
- *Once identified, how to capture customer's data supporting the CEM functionalities?* This involves a number of issues. Consumers are—understandably—loathe to supply personal/demographic information, and marketers have to be creative to get that information from actual or potential customers. Then, there is the concern of *subjective* information about the consumers, such as their beliefs, values,

and sentiments about products, processes, and issues. Finally, there are *legal* and *ethical* issues related to the capture and usage of such data.

We propose to address these challenges within the context of *CA-CEM development framework* that enables us to:

(1) specify the CA-CEM functionalities that we wish to support within the context of a purchasing process/experience;
(2) translate these CA-CEM functional requirements into *software specifications*, in terms of required data structures and algorithms to support the CA-CEM functionalities;
(3) translate those software specifications into actual code using predefined software artefacts (libraries, templates, generators, etc.).

Traditional so-called *application frameworks* embody an architecture/a design, with predefined *variation points* that can be instantiated for the application at hand. As such, they support only the third step of our framework. By comparison, the *development framework* that we propose would cover all the steps from business/user requirements to code.

To support the requirements phase, we need to understand/delimit the *problem space*, i.e. the *requirements space for CEM*. Toward this end, we rely on **a cognitive modeling of the purchasing process to identify** the various **decision points**, and the decision criteria/**influence factors** relevant to each decision point (see Sect. 3). This modeling helps us identify the touchpoints between sellers and customers that are needed to manage the customer experience. In other words, this cognitive model enables us to *script the interactions between sellers and customers in a way that allows us to **read** or **change** the mind of customers at pivotal moments of the purchasing experience* (see Sect. 4).

The cognitive/functional design of interactions between sellers and customers relies on:

(1) the identification of the relevant information about: (a) the customer (*customer profile*), (b) the products, and (c) the communication content between the customer and the seller (e.g. promotional material);
(2) the selection or design of algorithms and tools to get that information from available sources;
(3) the selection and design of algorithms to customize the interactions between the seller and the customer based on the available information.

The **relevant information is presented in the form of ontologies** in Sect. 5 **to be specialized or instantiated for specific purchasing scenarios** described in Sect. 7.3. We rely on **computational intelligence techniques to: (1) fill out some of the property values including subjective information from, and about the customer** (e.g. *preferences*, *sentiments*, or *beliefs*, see Sect. 6.1), and (2) **customize the interactions between seller and customer** through product recommendation, targeted marketing, and the like as described in Sect. 6.2.

This paper presents the principles supporting our approach. Section 2 introduces the process view of CEM, using a retail example. Section 3 describes the purchasing process, from an operational and a cognitive point of view. It identifies the different steps of the purchasing process, and the various factors that can influence customers in their decision making. Section 4 lays the foundations of our framework by going from the cognitive model of the purchasing process to, (a) a generic CEM pattern, and (b) sample instantiations of the pattern for step-specific of the purchasing process. These instantiations help design the ontologies needed to implement CEM functionalities, which are presented in Sect. 5. Section 6 shows how computational intelligence techniques can be used to fill out consumer profiles, and to customize the interactions with them. Section 7 shows how this framework can be instantiated to handle a particular purchasing scenario. We conclude in Sect. 8 with a discussion about future work.

2   A Process View of CEM

In this section, we use an example of a purchasing scenario that illustrates some of the possibilities and requirements of a CA-CEM framework. We start by presenting the scenario, and then present our process view of CEM.

2.1   A CA-CEM Scenario

Chris is a young urban professional in her 30's who walks into her favourite grocery store where she usually shops. As she drops items in her shopping cart, the food labels are automatically displayed on her (latest) iPhone. As she drops a box of crackers, she gets a warning because of its high-level of sodium—given her family history of blood pressure. She walks through the produce section, and gets notices about latest arrivals of fair trade certified products. She is pleased, of course, being an active member of Équiterre,[1] but wonders if every shopper gets notified. Walking into the meat section, her attention is drawn to a special on lamb chops that her significant other enjoys immensely. She picks a rack and drops it into the shopping cart. She gets wine suggestions: two thumbs up for a Syrah from Northern Rhône, and one thumb up for a Shiraz.[2] The Shiraz wins with the ongoing special on Australian wines. Cheese with that? A French baguette? In the seafood section, Chris picks up a slab of Tuna steak. She gets a warning to the effect of having consumed big fish already

---

[1] http://www.equiterre.org, whose mission statement includes "Équiterre helps build a social movement by encouraging individuals, organizations and governments to make ecological and equitable choices, in a spirit of solidarity. We see the everyday choices we all make - food, transportation, housing, gardening, shopping - as an opportunity to change the world, one step at a time…".

[2] Gracieuseté of an entry in the INTOWINE web site, http://www.intowine.com/best-wine-pair-lamb-chops, accessed 1/12/2014.

four times already this week.[3] While getting toothpaste, she gets a notification about size 4 diapers on special. Considering that she has been buying size 3 diapers for the past 6 months, it is about time she switched to size 4!

Some might find this "idyllic" shopping experience frightening, and rightfully so. We do not presume that this scenario is desirable: we will simply explore the capabilities that make it possible. In particular, we explore the kind of data/knowledge needed about the products and the customer to make such a scenario possible. We will do this step by step:

(1) "As she drops items in her shopping cart, the food labels are automatically displayed on her iPhone". This, of course, is easy enough, and can be done in many different ways, either by having short range RFIDs, which are probed by emitters in the cart, or simple bar codes scanned by devices upon being dropped in the cart. The technology for this already exists in many different forms.

(2) "As she drops a box of crackers, she gets a warning because of its high-level of sodium—given her family history of blood pressure". This requires the system to have access to a medical profile of the customer: what conditions they already suffer from, or they are predisposed to.

(3) "She walks through the produce section, and gets notices about latest arrivals of fair trade certified products". This notification depends on the system having access to the customer's social values. These values could have been entered directly by the customer when signing up for a particular service or social network. Alternatively, we can infer the customer's interests, based on their likes (Facebook profile), on the groups they are following (LinkedIn), on their membership to various advocacy groups that promote specific values (e.g. Équiterre), or on posts in various social medias.[4]

(4) "Walking into the meat section, her attention is drawn to a special on lamb chops that her significant other enjoys immensely". Knowing that someone likes a particular food product is easy enough. However, underlying this recommendation is a deeper understanding of human relationships, and what they entail: (1) people *live* with their spouses, and (2) as such, they eat together. If Chris' office buddy, or sibling like lamb chops, the system should make no such recommendation.

(5) "She gets wine suggestions: two thumbs up for a Syrah from northern Rhône, and one thumb up for a Shiraz. The Shiraz wins with the ongoing special on Australian wines. Cheese with that? A French baguette?". This is standard market basket analysis, combined with recommender functionality: it does not rely on any data specific to the customer.

(6) "Chris picks up a slab of Tuna steak. She gets a warning to the effect of having consumed big fish already four times already this week". Such a functionality depends on the system embedding a number of health advisories, and a history of the customer's purchases.

---

[3]Big fish, because they are higher up in the food chain, contain more toxic substances such as mercury.

[4]Analyzed using big data and text mining techniques such as sentiment analysis for instance.

(7) "She gets a notification about size 4 diapers on special. Considering that she has been buying size 3 diapers for the past 6 months, it is about time she switched to size 4". This is an extreme/fine-grained case of *family life cycle marketing*. The concept of *family lifecycle marketing* recognizes that families go through a predictable set of stages (*family lifecycle*) during which their needs, means, and decision patterns, evolve [31]. Marketers take advantage of this lifecycle to better identify the target market segment, an adapt their marketing message accordingly. In our scenario, not only does our system recognize that Chris is in the so-called "full nest I" stage[5]—which can be inferred from her track record of repeated diaper purchases—but it is also predicting her *child's progression* through her/his lifecycle stages.

Hence, to make this idyllic (or frightening) purchasing scenario possible, the system needs:

- electronically accessible detailed product information, including nutritional information (1), production mode, toxicity (6), and value-based assessments and certifications (3),
- product associations (5),
- a detailed customer profile, including medical history (2), demographic data and lifecycle stage (7), relationships (4), tastes (4, spouse), beliefs and values (3),
- a history of purchases (6 and 7).

Note the history of purchases can be used to infer—or more accurately to guess—other customer profile data, such as tastes or beliefs and values. For the time being, we will not consider how the different pieces of information can be obtained. That concern will be addressed later.

In the next section, we analyze these different pieces of information within the context of a process-oriented view of the interaction between an organization offering a product, and its customers.

2.2   A Process View of CEM

CEM aims at managing the interactions between a company and its customers. An actual or potential customer interacts with a company to fulfill a particular need, in the form of a product or service provided by the company. A company interacts with its customers, actual or potential, because that is its raison d'être: fulfilling the needs of its customers. A for-profit organization gets paid in return for the fulfillment of that needs at a price that is higher than its production cost. A public service organization (e.g. a government department) draws its value from the fulfillment of the needs of the citizenry.

---

[5]Identified as "young married couple with dependent children".

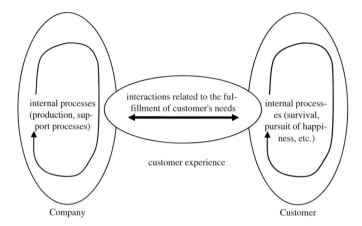

Fig. 1 Customer experience as the set of interactions between a company and a customer around the fulfillment of customer's needs

A *customer* has a 'life of his/her own' pursuing his/her objectives of survival and happiness. In the process of pursuing those objectives, they have needs that can or need to be fulfilled by a company. Customer experience deals with the interactions between customers and companies around the act of need fulfillment: the product or service sale. Figure 1 illustrates this interaction. The family lifecycle theory recognizes that the needs of people evolve during their life, as they enter different stages of the cycle. Those needs depend on many factors, including the processes underlying those stages (e.g. raising children), as well as the means that are typically available to customers at those various stages.

In the next section, we will look into the mechanisms of interactions covered by the customer experience by identifying the different stages of a purchasing process, and the various factors that influence the purchasing decision.

3 Understanding the Purchasing Process: A Cognitive Approach

Consumer behavior has been studied thoroughly by marketers and social psychologists trying to understand its mechanisms. They use the term "consumer", in a broad sense where the object of consumption can be a product/service (food, jeans, a cell phone service package, a car, a house), a behavior (exercising, dieting), or a belief (social values, political affiliation). At a fairly basic level, consumption is a conscious purposeful behavior, whose goal is to address a need or a desire. A number of psychological models have been proposed, including the *theory of reasoned action* [14], the *theory of planned behaviour* [1] the *MODE model* [13], the *theory of trying* [9], the *theory of self-regulation* [6], and subsequent variations thereof.

The theory of reasoned action (TRA) perceives all actions as a purposeful behavior that starts by forming an *intention* to act, followed by the performance of the action itself [14]. TRA recognizes that *intentions* are influenced by two factors, namely, the *attitude*(s) towards the action, and the so-called *subjective norm*. The attitude towards the action is defined as the perceived likelihood of some outcome occurring. The subjective norm refers to the actor/consumer's perceived belief that (certain) people expect him/her to behave/act or not. To use a simple example, I intend to buy Nike shoes, because I think I will look cool (expected outcome for me), and I think my peers expect me to wear sneakers ("peer pressure"). Researchers and exper-iments have poked holes in this theory, which was amended to include other *salient* influences, such as actors' belief in their ability to complete the action (*perceived behavioral control*), leading to the theory of planned behaviour [1], and to account for *impulsive* action/consumption, leading to the MODE (*Motivation and Oppor-tunity as DEterminants*) [13]. Other amendments took into account the *complexity* of the actions needed to consume, where the actions towards the consumption goal become themselves goals, leading to the *theory of trying* [9].

For the sake of clarity, we recall in this section the synthesis presented by Bagozzi et al. [8], which integrates all of the influences that have been identified by researchers. The model is shown in Fig. 2. We briefly explain the various steps, and the influencing factors. The significance of 's model for CEM will be explored in Sect. 4.

First, we start with general observations/principles. First, there are different paths through the consumption process, depending on the type and complexity of con-sumption decisions. For example, habitual or low-risk consumption activities (e.g. grabbing a carton of milk) involve deliberation or planning. Second, social psycholo-gists make the distinction between desirable **goals** and desirable **behaviors**, with the former often preceding the latter. For example, my goal could be to lead a healthier life, which represents a desirable end state. This goal may entail alternate (set of) behavior(s), such as dieting and exercising, where the objective is a *behavior*.[6] For the sake of simplicity, we represent them as two independent process paths, ignor-ing the precedence relationship that can exist between goals and behaviors. Third, this consumption model makes a distinction between *desires*, *intentions*, *plans*, and actual *actions*:

- A consumer may have several desires (**goal desire** or **behavior desire** in Fig. 2), but intends to pursue only one (**goal intention** or **behavior intention** in Fig. 2). I would love to have a car to take my kids to school, and a motorcycle to take leisurely rides during the week-end (two goal desires), but I am going to stick with buying a car (goal intention). The same applies to behaviors—otherwise known as New Year's resolutions.
- Having decided to pursue a particular goal (goal intention) or behavior (behavior intention), the consumer needs to figure out *how* to achieve that goal, i.e. s/he

---

[6]A good number of the studies that helped build these models concern *behaviors* like smoking, drinking, dieting, recycling, exercising, etc. Thus, the consumption process under study actually starts with the behavior desire.

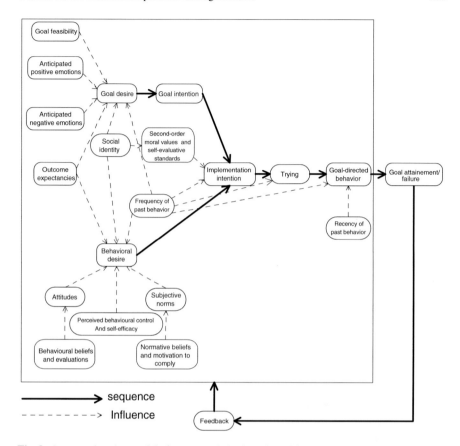

Fig. 2 A comprehensive model of consumer behavior adapted from [8]

needs a *plan of action*. For example, having decided to purchase a car, I need to figure out how to do it. This is as ***implementation intention*** in Fig. 2.

- Having devised a plan, the consumer then executes the plan by performing the actions of the plan. This is covered by the two steps, ***trying*** and ***goal-directed behavior*** in Fig. 2.

We will comment on some of the finer distinctions below. In the following, we describe each step of the process depicted in Fig. 2, along with their relevant influence factors:

(1) ***Goal desire***: corresponds to identifying the various needs and desires of the customer. Several factors influence the setting of a goal:

  (a) *Goal feasibility*: I may fantasize about owning an executive jet, but I will not consider it as a potential goal because it is unattainable.
  (b) *Anticipated positive emotions*: this represents the perceived reward of attaining a particular goal, and it is a combination of positive emotions resulting

from success (how good I will feel), and the expectation of success (how likely I am to succeed).

(c) *Anticipated negative emotions*: this represents the perceived penalty of failing to attain the goal, which is a combination of negative emotions resulting from failing (how bad I will feel if I fail), and the expectation of failure (how likely I am to fail).

(d) *Outcome expectancies*: the expected outcome of the pursuit of the goal (success vs. failure)

(e) *Social identity*: an individual's membership to a particular group, and the "emotional and evaluative [positive or negative] significance of this membership" [7].

(f) *Frequency of past behavior*: this was not included in 's framework,[7] but is an important aspect of the *theory of trying*, which is one of the foundations of this framework. The basic idea is that customers may undergo the complex cognitive processes involved in goal setting once, or the first few times, but after that, they make a cognitive shortcut: "I have thought this through many times, and found it is worth considering or pursuing this goal".

(2) **Goal intention**: Goal desire is concerned with the desirability of goals, but not with the decision to pursue them. This is step is where the decision is made.

(3) **Behavioral desire**: Researchers have identified six influences: (1) social identity, (2) outcome expectancies, (3) frequency of past behavior, (4) attitudes, (5) subjective norms, and (6) perceived behavioral control and self-efficacy. The first three were discussed above. We discuss the remaining others hereafter:

(a) *Attitudes*: were identified by the *theory of reasoned action* (TRA, [14]) as influencers of the pursuit, or not, of some actions. A *positive attitude* about an action leads the consumer to act, whereas a *negative attitude* inhibits him/her. The attitude towards an action is a combination of my belief about what would result from the action (what is the outcome), and my evaluation of that result. For example, if I buy jeans, I *believe* that I will look fashionable (the outcome), and I *like* being fashionable.

(b) *Subjective norms*: were defined by Ajzen and Fishbein as an individual's *perception* of the social pressure to perform, or not perform, an action [2]. That pressure is a combination of how confident I am that my spouse/colleague/peer group expects me to perform the action (a probability), and the perceived reward (or penalty) from complying or not with that expectation. For example, I am *90% confident* that my colleagues expect me to wear jeans, and if I do not comply, I *believe* that I will be eating alone.

(c) *Perceived behavioral control and self-efficacy*: Perceived behavioral control was introduced by the *theory of planned behavior* as a determinant in the decision to undertake or not, some actions/behaviors. It is defined as an individual's perception of how easy or difficult it is to perform a particular behavior [3]. It can be thought of as combination of my belief in me

---

[7]Figure 4.5, p. 97, in [7].

possessing a factor I think is needed to perform the behavior, and my belief in how important that factor is. For example, to stop smoking, I think that will power is everything, and I concede that I do not have strong will power. For the purposes of this paper, we will consider perceived behavioral control and self-efficacy as synonymous.[8]

(4) *Implementation intention*: This is the planning phase. Once I have chosen a goal or a goal-directed behavior to perform, I plan for it. This is particularly relevant when the purpose of my consumption is an *end state*, i.e. when we are dealing with a *goal desire* as opposed to a *behavior desire*. In such a case, implementation intention consists of identifying the steps/individual actions I need to perform to reach my goal. These steps can themselves become goals, in their own right, if their performance is problematic [7]. The planning phase is influenced by the frequency of past behavior: if I solved the problem (of goal attainment) several times before, I can reuse the same solution (implementation plan). According to [7], a second influencing factor relates to *second-order moral values and self-evaluative standards*. Values are defined as the criteria or frame of reference by which people justify actions and judge actions and other people [25]. *Self-evaluative standards* represent how consumers see themselves, or what they want to become [7]. Our shopper from Sect. 2.1 is a socially responsible consumer, and that is why the grocery store pitched the latest arrival of politically correct products. Figure 2 shows the moral values and self-evaluative standards are influenced by the consumer's *social identity*.

(5) *Trying*: Within the context of the original *theory of trying*, the act of *trying* was thought of as "a singular subjective state summarizing the extent to which a person believes s/he has tried or will try to act" [9]. The definition was later extended to the actual execution of the plan established in the previous step, where each action in the plan can be problematic, and be forestalled [7]. Trying involves monitoring progress towards the objective, and making adjustments to the plan, as appropriate. Bagozzi uses the example of showing appreciation to a friend at their birthday (a *goal intention*) by buying her/him a gift (a *behavior desire*), which involves a plan of going to a shopping mall, selecting stores to visit, and browsing through merchandise to select a gift within the selected budget (*implementation intention*), to actually executing the plan (*trying*), which involves going to the mall, etc. [8]. The consumer is not guaranteed to be able to do any of the steps required to make the purchase, including: (1) getting to the mall (transportation problem), (2) finding a set of stores of interest (not the right kind of merchandise), and (3) finding merchandise that fits the taste of the gift receiver, and the budget of the gift giver. The *trying* phase is influenced by the frequency of past behavior, discussed above, and the recency of past behavior. Within the context of the *theory of trying* [9], the recency of past behavior can

---

[8]There is some debate as to whether perceived behavioral control and self-efficacy are the same thing. Ajzen thinks so (e.g. http://people.umass.edu/aizen/faqtxt.html). Armitage and Conner do not [4].

influence current behavior in the same way that the frequency of past behavior does, because the behavior is still fresh in the consumer's mind.

(6) ***Goal-directed behavior***: This refers to the final act in the consumption process, for example making a purchase.

(7) ***Goal attainment/failure***: This is the step where consumers assess the extent to which they have reached their goal.

(8) ***Feedback***: Based on the previous assessment, the consumers can adjust any of the choices or actions made in the consumption process, including the choice of goals to pursue.

This comprehensive model of consumer behavior, proposed by [8] with very minor adaptations of our own, is more or less the result of merging consumer behavior models pertaining to different kinds of consumptions for which some of the steps, or the influencing factors, may be irrelevant. Our framework for CA-CEM should provide guidance (or tools) for customizing this process for specific consumption or CEM scenarios. That is discussed in Sect. 7.1.

4   A Cognitive Approach to CA-CEM Design

4.1   A CEM Pattern

Section 2.2 showed a view of CEM as managing the interactions between two processes executed on behalf of two entities, the enterprise/seller and the customer, each with their own objectives. Those interactions are centered on the act of consuming. Section 3 presented the different steps of the consumption process, and identified the factors that influence each step, without considering interaction points between enterprise and consumer. In this section, we put it all together.

Figure 3 shows the idea. Consumers are active systems (living organisms) whose internal processes (to stay alive, pursue happiness, etc.) require a number of resources (food, clothing, transportation means) or conditions (fulfillment, happiness, etc.), triggering consumption processes to replenish the resources ("we are out of cereal") or to attain those conditions ("I need to go/dine out"). Those consumption processes involve a number of steps.

Decisions and choices taken in those steps are influenced by a number of factors, which have been identified by psycho-sociological studies of the consumer, and discussed in Sect. 3. Some of the steps of the consumption process involve interactions between enterprise and consumer. The interaction can be triggered by the consumer, for example by accessing the enterprise's web site, or triggered by the enterprise, by communicating with the customer, nominally or through advertising.

To the extent that some of the steps of the consumption process involve choices and decisions that are influenced by a number of psycho-sociological factors, the enterprise gains from knowing the values of those factors. Some of these influence factors are malleable, such as the *anticipated positive emotions*, and the enterprise

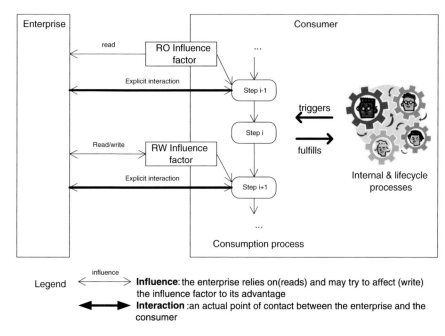

Legend
influence

Influence: the enterprise relies on(reads) and may try to affect (write) the influence factor to its advantage

Interaction : an actual point of contact between the enterprise and the consumer

Fig. 3 Towards a systematic view of consumer experience management

may gain from modifying their values to its advantage, for example, by showing potential consumers how good life would be if they would purchased its products, or how easy it is to get approved for credit to make the purchase (e.g. perceived behavior control). Such factors are referred to as *read-write influence factors* in Fig. 3, as opposed to *read-only* (RO) influence factors, that enterprises cannot modify or act upon. Furthermore, while all the *read-write influence factors* are *potentially actionable* from a CEM point of view, some may not be worth pursuing.

4.2 Towards an Integrated View of Customer Experiences

Figure 4 shows a first cut naive application of the pattern shown in Fig. 3 to all the steps of the consumption process, incorporating the full roster of influence factors identified by the literature on social psychology of consumer behavior [8]. However, such a view is reductive in the following ways:

(1) A seller, acting through its e-commerce system, does not necessarily accompany a (potential) buyer through all the steps of the purchasing process. This is particularly true for the early steps of the process, such as the goal desire stage. This is not to say that sellers have no influence on goal desire but quite the contrary. For example, the tobacco industry, the beer industry, and the fashion industry,

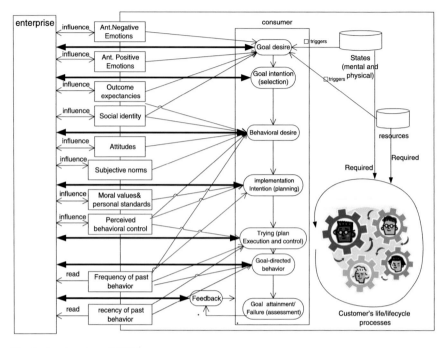

Fig. 4 A naive view of CEM

to name a few, all act upon the anticipated (positive) emotions, and to a lesser extent, social identity associated with the consumption of their products. However, this influence does not happen within the context of a one-to-one interaction between the customer and the seller's e-commerce system.

(2) As a corollary of the first observation, different channels will typically accompany different subsets of the purchasing process, often corresponding to different entry points into the purchasing process. For example, I can get into an e-commerce site through a web search for a product, where the returned page will display the product specs, and propose to add it to a shopping cart, thereby skipping "goal/behavior desire", "goal intention", "implementation intention", and part of "trying". However, a walk into a particular store will instantiate/trigger the full cycle, starting with a welcoming display featuring and promoting the desirability of a specific product.

(3) A second corollary is related to the fact that a consumer may interact with several sellers for some steps of the process. This may be the case early in the process before the customer commits to a product and a seller. For example, you walk into a shopping mall and the location-based services on your phone send you advertisements from the different restaurants or shoe stores in the mall. Knowing that you are in competition with other contenders will impact the way you interact with the (potential) customer.

(4) A lot happen in the trying step (executing a consumption plan) and the goal directed behavior step, that may include information search, comparison shopping, consulting product reviews, and so forth, i.e. lots of sub-steps that do involve interactions between seller and consumer. The pattern of Fig. 3 will need to be applied to those sub-steps.

With this is mind, in the next Section we study the various steps of the purchasing process, and see *whether* and *how* we can apply the pattern of Fig. 3.

4.3   CEM as Service Design

The design of E-commerce applications in general, and CEM applications in particular, can be studied within the context of *service design*. We all have an intuitive understanding of what services are (as opposed to goods). They include things such as health care, education, transportation, hospitality, phone or telecommunications, etc. Wikipedia defines a *service* as an *intangible commodity*, where a *commodity* is a "marketable item produced to satisfy wants and needs".[9] Service delivery typically involves a process that orchestrates people, components, structures, and automated systems—typically IT—to satisfy the needs/wants of the customer. *Service design* is concerned with the selection of the people, components, structures and automated systems needed to deliver the service and the design of the interactions between them in a way that is economically viable for the service provider and that maximizes customer satisfaction. Service design theory involves a number of disciplines, including marketing, operations management, organizational design, and technology (see e.g. [12, 16, 19]). The outputs of service design (good and bad) include the work organization at your local bank branch, your Department of Motors and Vehicles bureau, or your favorite department store.

There has been a growing interest in the research community to look at experience management within the context of service design (see e.g. [12, 27]), with the intent of incorporating *customer experience requirements* along with the other requirements that drive the service design. Teixeira et al. introduced *customer experience modeling* as a "model-based method [...] to represent and systematize customer experiences for service design." [27]. Customer experience modeling draws on Constantine's *Human Activity Modeling* (HAM) [11] on *customer experience requirements,* themselves from Mylopoulos' et al. *goal-oriented requirements analysis* [18], and on Patricio's et al. *multi-level service design* [21]. Figure 5 shows the concepts of CEM and their relationships. Note that customer experience requirements apply to both *activities* and the *context* within which these activities take place, including other actors (e.g. a sales clerk), artifacts (e.g. a passive display), and technology enabled systems, including computing devices, sensors, and the like.

---

[9]See https://en.wikipedia.org/wiki/Commodity, accessed on 18/8/2015.

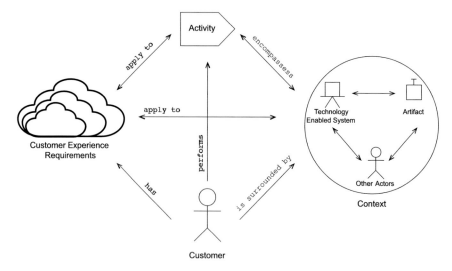

Fig. 5 Incorporating *customer experience requirements* in service design. From [27]

An important aspect of service design is *customer scripting* (see e.g. [12]), which is the specification/design of the interaction process between a customer and a service provider delivering the service. This involves the careful design of both, (a) the individual interactions and interaction touchpoints, and (b) the orchestration of such interactions. As an example of (a), we explore different ways for the consumers to identify themselves, and select the easiest ID method. An example of (b) is whether to request identification from the beginning of a service script (which would result into more appropriate/customized service interactions), or adopt a light process, and wait to request identification until the customer proceeds to checkout. For the most part, the *customer experience modeling* method is *descriptive*, providing modeling concepts and notations. However, it offers little guidance as how to design a service in a way that it conforms to the customer experience requirements. Cook et al. [12] identified some of the *human* and *subjective* issues surrounding customers' appreciation of their service encounters/experiences, independently of the intrinsic qualities of the product/service rendered, and of the customer specifics. For example, one of the script design guidelines, based on psychological studies, suggests that we design a customer script that puts unpleasant interactions/steps first, and ends with a high note [12].

Within the context of service design, the description of the purchasing process shown in Fig. 2 provides a *high-level solution space for consumer scripting*. As mentioned in Sect. 3, our e-commerce (or experience management) application will have to implement a sub-script (sub-process) of the process described in Fig. 2, depending on the type of purchase (carton of milk, versus car) and on the channel (brick or click). The pattern of Fig. 3 shows us what to look for to design the individual interactions/steps of the script. We see next how to turn the psychological influences described in Fig. 2 into specific actions/interactions to incorporate into a consumer script.

4.4   Turning Psychological Influences into Actionable Consumer Experiences

In this section, we apply the pattern of Fig. 3 to the steps of the purchasing process depicted in Fig. 2. We start by outlining the principles (Sect. 4.4.1), and then we look at specific examples (Sect. 4.4.2).

4.4.1   Principles

We start by categorizing the purchasing process steps, and their influence. Then, we present the different ways we can exploit those influences.

Roughly speaking, the purchasing process involves two kinds of activities, (1) *internal* cognitive activities, taking place within the consumer's head, and (2) *external,* possibly *physical*, activities, some of which involving explicit interactions between the customer and the seller/service provider. Examples of cognitive activities include *goal/behavioral desire*, *goal intention* (selecting one worthwhile goal among many) and *goal implementation* (planning a way to achieve the goal). Examples of external activities include *trying* (plan execution) and *goal directed behavior*. An example of external activity that involves an interaction with the seller is doing an on-line product search on the seller's portal, asking an in-store sales clerk for advice, or checkout (online or in store).

We categorize the influence factors based on two dimensions:

(1) *Subject/scope*: customer-specific factors (socio-demographic data, social iden-tity, moral values and personal standards), product/service-specific factors (prod-uct specifications), and factors related to the <customer, product> relationship (anticipated positive emotions, anticipated negative emotions, history of past behavior),
(2) *Type*: objective/factual factors (customer socio-demographic data, history of past behavior, product specifications) versus subjective/emotional/perception-based factors (attitudes, subjective norms, perceived behavioral control).

From a service script/interaction design point of view, for each activity of the purchasing process, we can do different things, depending on the kind of activity, and the kind of factors influencing that activity:

(1) Purely cognitive activities. These activities happen inside the consumer's head (e.g. goal intention), and do not require interaction with the seller/service provider. Thus, we need to *provoke* an interaction, as part of our service script design. This interaction can take place:

    (a) *prior to the activity* to modify the subjective/emotional/perception-based factors. For example, prior to goal desire/goal intention, I can strengthen the anticipated positive (negative) emotions associated with a goal, to make the customer desire it *and* select it over *other* competing goals.

(b) *after the activity*. In this case, we cannot influence the activity, but we can at least hope to get a *reading* on the choices made so that we can better prepare the response of the service provider to the subsequent activities. For example, as a retailer I may not be able to influence goal intention (selecting one goal to pursue, among many desired, e.g. which item of clothing you came into the store to purchase, among all the ones we offer). But if I can tell (or ask!) which item you came in for, I can direct you.

(2) External activities that *normally* involve an interaction with the seller/service provider. In this case, the activity is part of the normal script, and the customization concerns the information content, and how it is delivered, both adapted to what we know about the customer. An example of such activities includes searching the seller's online catalogue. If we know who the customer is, we can refine the search query, and customize the presentation of the results.

(3) External activities that do not normally involve an interaction with the service provider. Here too, we need to *provoke* an interaction with the customer, either before the activity takes place, to influence its outcome, or, failing that, after it has taken place, to get a reading on what the customer did, to plan the subsequent steps of the process/service script. An example of such an activity is searching for product reviews (pre-purchase), or entering a product review (post-purchase), *in a third party site*.

The difficulty with the activities that do not (normally) involve an interaction with the seller/service provider (via their e-commerce software) is to *guess* where the customer stands along the purchasing process. A seller may adopt a default strategy of assuming that you are at the (pre) goal desire stage, and send you advertisements until/unless you do something specific. However, when we get into the intermediate stages (e.g. goal intention, goal implementation, and the various substeps/sub-stages of trying), things get tricky. Not knowing where the customer stands in the purchasing process can be challenging even when the customer initiates an interaction. When you walk into a dealership, the salesperson first asks you whether you are looking for a model in particular, or just looking, and if you answer the latter, they may go back to playing Solitaire. Similarly, when you do a product search on an online catalog, you may be looking for a particular model whose model number you forgot, or looking for any product that matches your specifications. Depending on the context, there are different strategies for completing a sale!

4.4.2 Examples

In this section, we show how to use the concepts and principles described above to design interactions for specific purchasing process steps. The purpose here is to illustrate the kind of analyses a customer experience designer needs to make. The final answer would rely/depend on marketing knowledge. This analysis will also help us start determining how the identified influence factors translate into data that an e-commerce application can manipulate. Section 4.4.1 will propose a first-cut ontology/metamodel to support the experience management functionalities illustrated throughout the paper. We start by a thorough analysis of the goal desire stage, and then present pointers related to other stages.

At a basic level, goal desire is triggered by the lack or deficit of physical resources ("we are out of milk") and physical or mental states("I am out of shape","I need some fun") (see Fig. 4). With the exception of biological needs, for which we need no prompting, companies/marketers can *create* a need[10] or strengthen a need through advertising. According to 's framework (see Fig. 2), the *strength* of such desire is influenced by anticipated emotions (positive and negative), outcome expectancies, and social identity. This raises two kinds of questions:

(1) What is the appropriate advertising message *content* to strengthen the desirability of a goal? In particular, we need to think of the extent to which a (better) knowledge of the customer profile can help customize/select the appropriate message
(2) How to deliver that message within the context of a CEM application, *considering the stage of the purchasing cycle we are at*?

At a basic level, we can strengthen the desirability of a product/service or behavior by projecting images that: (a) strengthen the anticipated positive emotions ("how good it will feel to own this product or adopt this behavior"), (b) strengthen the anticipated negative emotions ("how bad I will feel by *not* owning the product or not adopting the behavior"), and (c) show how easy it is to succeed (in acquiring the product or adopting the behavior).

In what ways does our knowledge of the customer make such images more effective?

(1) By depicting someone the customer can identify with. If we want to "sell" the latest Android smart phone to a teenager or professional, we present an advertisement that features a teenager, or professional, respectively. Idem with gender, ethnicity, geography, etc.[11]
(2) By sending images of more specific goals, appropriate for customers' category/social identity. While owning the latest and greatest in smart phones may be a broadly shared desire, the advertising message could focus on social media and multimedia capabilities when targeted towards teenagers, or focus on office productivity tools, when targeted towards professionals. Similarly, if a product is too expensive for my social category, the marketing message can stress the possibility or ease with which I can finance the purchase, playing on outcome expectancy.
(3) By sending images of goals *we know the customer has not yet achieved*. My cell phone provider, or the dealer who sold me my last car, know the features of the current product I own, and know which feature(s) to highlight or stress in the marketing message.

---

[10]Marketers would disagree with the manipulative "create": they prefer the term "recognize".

[11]Typically, marketers produce different variants of the same marketing theme, aimed at different populations.

Having determined/selected the advertising material, we need to determine ways to deliver it. This depends on the combination of two things: whether the customer is known to the seller, and the channel. For example, if the seller already knows the customer, then we have several possibilities:

- We can push the advertising material through an unsolicited e-mailing campaign (**on-line**).
- We can present the material to the customer when they log in (**on-line**).
- When the customer enters the store, and s/he is positively IDed, the message is displayed on an appropriately located monitor (**in-store**).

If we do not know the customer, then we broadcast the message variants, both through the company portal and in-stores.

In summary, for goal desire:

- We can act upon the desirability of goals through advertising.
- Knowing that anticipated positive emotions, anticipated negative emotions, and outcome expectancies influence the desirability of a goal helps determine the *orientation* or *content category* of the message carried by the advertising material.
- Knowing the customer (social identity, history of consumption) can help customize the message.
- The channel determines the advertising material delivery method.

Similar analyzes can be performed on the subsequent steps of the purchasing process. We provide some highlights for some of those steps:

- *Goal intention*: Here, the issue is in selecting a goal to pursue among many that are deemed desirable. This is another internal cognitive activity, that is practically indistinguishable from the goal desire. The priority of goals depends on means, lifecycle stage, values, social identity, etc. I can influence goal selection through advertising by matching advertising content to consumer profile. For instance, if I am a specialty outdoor equipment manufacturer, and I am talking to a young person who can afford *either* a new car or a (top of the line) mountain bike, I can show her/him an add that features healthy and handsome young people taking their fancy bikes to an extreme mountain trail in a cheap ride, or even hitchhiking there!
- *Goal implementation*: This is the planning stage. Planning makes sense in a multichannel shopping experience. For example, I have already made the decision to buy a new fridge, and I am planning my shopping experience. How can a retailer influence the planning stage to his advantage? Several general strategies include offering a simpler shopping experience (e.g. order from a catalog, or a better online ordering system), and making sure that any shopping plan includes an interaction with the retailer ("we offer the best price guaranteed", or "we will best the lowest price you will find by 5 %"). Retailers can also take advantage of knowledge about customers like their location, by suggesting stores to visit to shop or pick up their merchandise, or their shopping style, by activating/proposing appropriate shopping scripts following an on-line search.

- *Trying*: This is the execution of the consumption plan (*goal implementation*). If the plan includes in-store browsing, then we can think of the CA-CEM scenario presented in Sect. 2.1, and various product recommendation strategies especially if we know who the customer is. Imagine walking into a department store and having your profile displayed on the PDAs of nearby salespeople who hence call you by name, and direct you to your style section.

Through these analyses, we do not pretend to provide perfect answers from a marketing point of view. However, we hope to provide the *design vocabulary* that service designers, including marketing specialists, can use to design CEM-enabled services.

5   Ontologies for CA-CEM

Section 4 showed how the combination of cognitive modeling of the purchasing process and service design theory can help us in:

(1) *identifying the steps* (cognitive or physical) of the purchasing process *that, either naturally lend themselves to an interaction between seller and customer, or would benefit from provoking such an interaction.*
(2) *selecting the modalities for such interactions*, depending on the stage of the purchasing process, and the purchasing channel.
(3) *selecting the contents of such interactions*, depending on our knowledge of the customer and the product.

In this section, we focus on the last aspect, **the modeling of the knowledge we need to have about the customer and the products**.

We present the required knowledge models in the form of *ontologies*. We use the term *ontology* in two complementary ways:

(1) In the computer science/information systems sense, as specification of a set of representational primitives modeling a domain of knowledge [15]. If we think in terms of UML and MOF, the above definition corresponds to the *UML metamodel* (M0), i.e. the UML modeling language, a subset of UML itself. An ontology can also be understood in terms of description logics, where an *ontology language* can be seen as the realization of an underlying description logic.
(2) In the sense of reflecting a *shared conceptualization of a domain*.

Indeed, within the context of our CA-CEM development framework, we want to specify the modeling ingredients that analysts can use to represent user's and product profiles required by purchase scenarios. We also use the term *ontology* in the sense of representing shared knowledge about domains, namely marketing knowledge in the context of CEM. Such marketing knowledge may include things such as known customer categories, their likes, tastes, and spending habits, known product categories, and their appeal, etc. However, to *specify* such knowledge, we need representational primitives—hence the complementarity of the two perspectives.

In this section, we focus on the *representational primitives* needed to represent the data/knowledge about customers, products, promotional material that may be needed, within the range of possible purchasing scenarios, and CA-CEM functionalities.

5.1 Consumer Data

Knowing the consumer is critical to a successful CEM. As shown in Sects. 2 and 3, there is a lot of relevant data. The different kinds of data are discussed in the following.

As mentioned in Sect. 4.1, the influence factors that are relevant to the consumption process steps represent information that consumers use to select their choices and make their decisions. Thus, companies benefit from knowing what that information is, and should incorporate it in the consumer's *profile*. Let us take a couple of examples from the scenario presented in Sect. 2.1:

(1) The customer's like for lamb chops.[12] This is an example of *anticipated positive emotion* (enjoyment) resulting from eating/getting lamb chops.
(2) The customer's sustainable/equitable development values. This is part of the *moral values and personal standards* factor, which depends on the consumer's *social identity*[13] [5].
(3) The same-week purchases of big fish. This is part of the *frequency/recency of past behavior* factors. In this case, it consists of customer's transaction history.

Part of this data is specific to a given customer, for example her/his transaction history or her/his liking of lamb chops, while other parts pertains to a category that the customer belongs to, for example fair traders. Also, parts of the data are specific to a product (e.g. liking lamb chops), while some are more generic (e.g. liking any fair trade product). We discuss product variations in Sect. 5.2.

Bagozzi's framework does not explicitly mention the consumer's socio-demographic category, although part of this information is implicitly contained in the consumer's social identity. We see such data, especially the one relating to lifecycle processes, as the *main driver of needs*, particularly, basic ones (food, clothing, shelter, healthcare, education). Our imaginary scenario of Sect. 2.1 also highlights the role of *relationships* in anticipating consumers' needs. Some of these relationships are implicit in the lifecycle concept, namely in the nesting stages where the consumer makes purchases (or purchasing decisions) for the benefit, and on behalf of dependents.

---

[12]Ignoring, for the time being, the fact that it is the customer's significant other who likes lamb chops.

[13]Which may be *characterized*, if not *defined*, by the customer's membership to, or militancy within, various organizations and associations. In our example, our shopper Chris is member of http://www.equiterre.org.

At any given point in time, consumers may be in a particular lifecycle stage. They can go from one stage to the next after some time has elapsed, and/or after some event has occurred. For example, the typical family lifecycle has a stage for young adults with preschool-age children, followed by a stage for school-age children. For a given child, the duration of the "young adults with preschool-age children" stage is 4–5 years. Similarly, the family lifecycle has an "unattached young adult" stage, followed by a "newly married adult, no children" stage. Marriage or otherwise attachment indicates the transition from one stage to the next. To the extent that a CEM system will typically *not* have access to court marriage certificates, some events will have to be *inferred* from *property value changes*. For example, when teenage or young adult children move out, they start having basic household needs (furniture, telecom services, etc.). Such an event can be detected by the change of address, from their parents' home address to an outside address. The reverse move back home may also be of interest to marketers, as the disposable income of the returning children increases. Figure 6 shows a customer profile metamodel. Starting with the upper half, it shows that **Consumer**s have properties (**Property**) and values for those properties (**PropertyValue**). The association between a **Consumer** and a **PropertyValue** is characterized by a *confidence level*, reflecting our confidence in

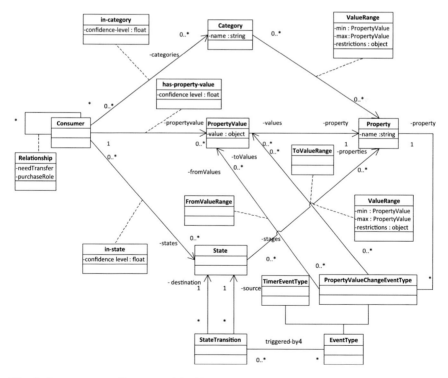

Fig. 6 A customer profile metamodel taking into account decision factors, demographics, and lifecycle data

the value. For example, a value that was entered by the consumer when signing up for a service is more reliable than a value inferred by data mining techniques.

Consumers belong to **Category**(ies), with some confidence level. For example, we may not know for sure that Chris is a socially responsible consumer (a **Category**), but we know that she constantly accesses the web site of equiterre.org, and posts comments on their articles. From which we can infer with 80 % confidence level that she belongs to the category of socially responsible consumers.

Category(ies) are also characterized by instances of **Property**. The relationship between a **Category** and a **Property** (the association class **ValueRange**) shows which property values are admissible, or expected, for this category. For example, for the property **PreferredTransportationMode**, socially responsible people may be known to prefer public transportation or self-propelled vehicles. Value ranges for properties of categories provide *default property values for members of that category*. For example, we may have no value for the **PreferredTransportationMode** property for Chris, though she is a member of the socially responsible consumer category, hence we can infer that she prefers public transportation.

Similar to the concept of **Category**, we have the concept of **State**. A consumer can be in different *states*, which correspond to qualitatively different consumption behavior in terms of needs, attitudes, etc. This accounts for the family lifecycle theory, where states represent lifecycle stages. It can also be used to represent finer concepts such as diaper stages, discussed in the scenario of Sect. 2.1 **StateTransition**s are triggered by events (**EventType**), which can be property change events (**PropertyValueChangeEventType**, as in getting married), or timer events (**TimerEventType**), as in going from size 3 to size 4 diapers after a fixed number of months. A property value may change with a worthwhile event in general, or if the initial value of the property equals a specific value (e.g. moving out of the parents' home, signalling need for furniture), or if the destination value equals a specific value (changing marital status, or moving to a significant other's house, or back to the parents house), hence the association classes **FromValueRange** and **ToValueRange** from **PropertyValueChangeEventType** and **PropertyValue**.

The **Relationship** association class from/to **Consumer** represents personal relationships of consumers, to the extent that they are relevant to consumption behavior. In our fictional shopping scenario, Chris gets lamb chops because her mate likes them. She was also told about size 4 diapers because the system suspected that she cared for a toddler, who had been size 3 for X months. Some relationships *transfer* needs: if I care for a baby, the baby needs diapers, then I need diapers, and I will be attentive to diaper adds, for example. The message sent to me may be adapted to my role in the purchasing process: whether I am an advisor, a decision maker, or a doer. We use the attributes needTransfer, and purchaseRole to represent these distinctions.

The model in Fig. 6 is more of *a metamodel* or a *high-level ontology* than a customer profile data model per se. An actual customer profile model can be constructed by *instantiating* the classes shown in Fig. 6. For example, we need to figure out which properties are worth representing about a customer depending on the product or service being sold, which consumer categories are worth considering, which lifecycle stages are relevant to the kind of product we sell, etc. This will be part of the framework instantiation discussed in Sect. 7.

5.2 Product Data

We showed in the previous subsections the kind of data that a company needs about its customers to assess their needs and desires (e.g. demographic and lifecycle data), and understand their attitudes, biases, emotions, and values that come into in the process of satisfying those needs. This information is used in identifying those, among its products, that best help address the needs and desires of those customers, and positioning those products in a way that appeals to customers, considering their (known) attitudes, biases, emotions, and values. To take full advantage of our knowledge about consumers, we need a comparably rich representation of the products and services sold by the company.

In our grocery shopping scenario, we knew the function of our products (nourishment), their composition (sodium content for crackers, but also mercury content for big fish), their assortments (a Shiraz, to go with lamb chops), and their production process (faire trade or not). Different kinds of products have different facets. Figure 6 partially illustrates what our data could look like. We will comment on the most important aspects of the model.

Products are represented using five facets: (1) *function*, which represents the objective utility of the product (e.g. nourishment, transportation, cleaning product), (2) *form*, which represents things related to packaging (e.g. bulk, by unit, six-pack) and presentation (visual aspects), (3) *structure*, which can be used to represent the ingredients or nutritional information of a food item, or the major functional components of a device (e.g. a car having a V6 turbo engine, all-wheel drive, 17 in. wheels), (4) *emotions*, which represents, in a shorthand form, the emotional function of the product (the emotions it elicits), and (5) *manufacture*, which provides information about the manufacturing/construction process of the product. As a first-cut model, we include the process itself, the jurisdiction where the product was produced, and the manufacturer, all of which being information that environmentalists, fair traders, and various social activists might care about. A conscientious consumer may refuse to purchase the product of an environmentally unfriendly process (e.g. fish captured by bottom trawling), or manufactured in a jurisdiction known for poor environmental or labor regulations, or by a manufacturer operating in such a jurisdiction, or headquartered in a jurisdiction known as a tax haven. Finally, the association class **Association** represents positive ("goes with") or negative ("does not go with") associations between products, and the strength of the association ("goes really well with"), as in "the *Bleu de Brest* blue cheese goes really well with the 2011 Australian Shiraz".

In the same way that consumers are members of categories (market segments), products belong to categories. Each product category is characterized by the same five facets (*function*, *form*, *structure*, *emotion*, and *manufacture*). Categories hold *default information* known about classes/sets of products, which can be overridden for specific products. The categories are organized along specialization hierarchies using the subcategory relationship. We can have different categorizations, based on function, country of origin, packaging, etc. Finally, similar to the associations between products (goes with/does not go with), we have associations between categories: red wines go well with meats, white wines go well with fish, and polka dot shirts do not go well with pinstripe suits.

5.3   Other Kinds of Data

Different types of data may be relevant to interactions between seller and customer, with the purpose of capturing the needs of customers, their appreciation of a particular product or product category, information about products, such as technical product reviews, product comparisons, as well as promotional material, as described in Sect. 4.4.2 Good experience management depends on soliciting (customer → seller) or presenting (seller → customer) the right information at the right time. In this section, we illustrate the kind of representations of some of this data that would support fine-grained matches between products, customers, and context.

Let us take the example of promotional material. We showed in Sect. 4.4.2 how the message conveyed by promotional material can be matched to the customer and the situation at hand for maximum impact. To be able to realize this match automatically, we need to index the promotional material with a precise description of the message.

Figure 8 shows a sample model for representing promotional material. The model is broken in two parts separated by the dashed line. The upper part reproduces a subset of the models shown in Fig. 6 and Fig. 7 that embodies our marketing knowledge, namely:

(1) Certain consumer categories (**ConsumerCategory**) have a need for specific functions (**Function**).
(2) Those functions are supported by certain products (**Product**) and product categories (**Category**).
(3) Those products or product categories induce **Emotion**s.

Recall from Sect. 4.4.2 that in order to re-enforce/strengthen customers' desire of a particular product or function (the so-called *anticipated positive emotions*), one can show them a person they can identify with, who is using the product or the function, and who is experiencing those *positive emotions*. Hence it involves an **Actor** (as in comedian in a video), who is identified as belonging to a particular category. A video clip (or some other promotional material) would be indexed with the action taking place, which is a **ProductUsage** of the target **Function** or the target **Product**, and should be shown to experience the target **Emotion**. Thus, for promoting the latest smart phone, I could have two different clips: one featuring college students

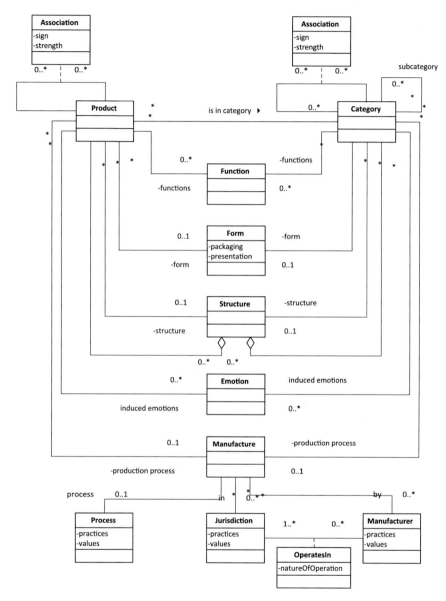

Fig. 7 An example of product representation for rich CEM functionality

exchanging using their phones social media, and another clip showing business peo-
ple/professionals using the office productivity tools of the phone. I would then send
the appropriate clip to the appropriate customer by matching on customer category.

General knowledge

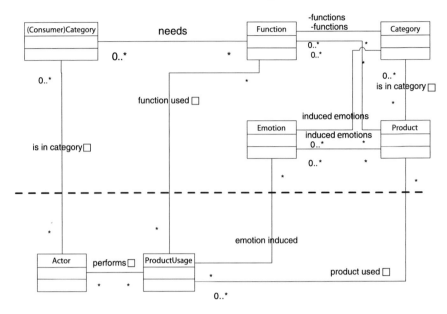

Fig. 8 A representation of promotional material

6 Computational Intelligence for CEM

6.1 Data Mining Techniques for E-Commerce

Data mining techniques have long been used in electronic commerce, for a variety
of usages. Some of the first usages included the use of navigation traces for various
purposes, including knowing more about incoming traffic (geographic location of
users/potential customers, referring pages), optimizing the design of websites, and
recommender functionalities, suggesting pages to look at from the pattern of traces,
or suggesting products, based on customer profiles, on purchase history, or basket
market-like analyses, of the kind we find on amazon.com (see e.g. for a survey of
the early uses [24]).

   As B2C customers started leaving *textual* traces on the internet, *text mining*
techniques found many additional customer *relationship* management applications,
including intelligent routing of users queries left on service portals, *sentiment analysis* to discover how customers *feel* about a company's products or actions in terms of
a unidimensional polarity, and *opinion mining* for finer analyses of customers' views

towards products and actions (see e.g. [17] for a thorough treatment of the various mining flavours).

The advent of social media has opened up a whole new set of possibilities and challenges:

(1) On the positive side, whereas the textual traces left on customer service portals were typically limited to *actual* customers *after* they had made purchasing decisions, companies can now find out what is said about them and their products on other media, by non-customers, or not-yet customers *before* they make a decision. Companies are then capable to act upon, and influence some of the *earlier* steps of the purchasing process (see e.g. [22]).

(2) On the negative side, the proliferation of social media poses many operational challenges:

   a. CRM/CEM specialists now need to collect the information from *many* different sources (specialty blogs, portals, Facebook posts, Twitter tweets, etc.).
   b. We are no longer able (or less able) to trace the opinions or sentiments that are expressed on the web to individuals that we can reach out to address their needs or change their perceptions.[14]
   c. We are not always able to manage identities across different social media. For example, if we find three similar complaints posted on three different media, under three different pseudonyms, do we have three different service failures, or is it the same person blasting us everywhere.

In the next section, we look at the different potential usages of data mining techniques within the context of CA-CEM.

6.2 Data Mining for CEM

Data mining techniques can be used in at least three areas:

(1) Categorization refers to both the *identification/specification* of categories, as in identifying or characterizing a particular consumer group (e.g. DINKs), and the assignment of individuals (a particular consumer) to categories, e.g. recognizing that the customer in front of me is a DINK.

(2) Opinion/belief mining tries to figure out what the customer thinks or feels about things or issues.

(3) Feedback assessment refers to the ability of sellers to automatically assess, and act upon, the feedback left by customers on their product ownership experience, or on their *customer* experience, i.e. their service interactions with the seller.

We discuss below the main issues raised by these three applications, and their operational implications on our CA-CEM development framework.

---

[14]If an unhappy customer leaves a complaint on the company's portal, the company is able to connect with that customer and remedy the situation.

6.2.1 Categorization

Categorization is a very important aspect of CEM. The gist of CEM is knowing the customer, i.e. associating them with a consumer group that has identifiable needs, desires, and consumption patterns. As we saw in Fig. 6, a lot of information about the consumer will be inferred from the knowledge of her/his membership in specific consumer categories, as opposed as being specifically given/known for that consumer. Categorization raises two related issues: how to discover/codify categories, and how to categorize/classify new instances of individuals. The two issues are related, and will be discussed jointly. We discuss below two types of categorization, depending on how much prior marketing knowledge was have about the market we serve.

Categorization Through an Explicit Codification of Marketing Knowledge

We codify marketing knowledge in the form of identifiable consumer groups along with socio-demographic properties/property value ranges. Those properties can be defining (for example for an age group, age is defining) or characteristic (for example for residents of a particular neighborhood, income level or ethnic origin are characteristics). Knowledge representation theory may dispute the distinction.

Although definitional properties may be difficult to obtain from customers (age, income), we can predict them from characteristic properties directly observed. These predictions can be made by machine learning based algorithms trained on customer data sets for which we have categorization and characteristic properties. We can also imagine a tool that enables us to tag the properties of a model as defining versus characteristic, thus triggering real time analyzes as data is inserted.

New instances get categorized based on whichever combination of properties are available. The system can then predict the value of other properties from the definition of the category with different confidence levels, which would determine the strength of the inferences that we draw from those inferred property values.

Categorization Through Unsupervised Learning

The idea here is to use unsupervised learning techniques to discover new consumer categories. Clustering algorithms are fed customer's data. The quality of the categories will depend on the richness of the data. This is somewhat related to the earlier discussion about *definitional* versus *characteristic* properties. Among the relevant issues is whether the data that is being captured is discriminating for the classification of new instances.

Once instance data (consumer or product) has been clustered, we can compute for each cluster the value distribution of each property for the instances to get a better characterization of the cluster. Thus, if a customer is found to belong to a certain category, and is missing some property values (e.g. yearly income), we can infer values for those properties based on the property value distribution for that category.[15] For example, if I know that a customer is a DINK, and I know the family

---

[15]In fact, the property value *is/can be mapped* to the property value distribution of the category.

yearly income distribution for DINKs, I can translate that knowledge into income level probabilities, or fuzzy income values. This, in turn, will determine the strength of the inferences I draw from those income values.

Clustering algorithms tend to be computationally intensive. If customer categorization functionality is used by checkout/cash register clerks for cross-sells,[16] then we need to make the categorization algorithm fairly quick. This could mean many things including using incremental clustering algorithms, or making an approximate and temporary category assignment, to be refined later on in batch mode during off or low traffic hours.

One of the major problems with automatic clustering algorithms is that the resulting clusters/categories are not labeled in a way that is meaningful to a marketer. Within the context of a CA-CEM development framework, we need to provide analysts/CA-CEM application designers with a set of tools that can perform such computations through minimal configuration.

6.2.2 Mining Beliefs and Values

Section 3 showed that *social identity*, and its corollary, *second-order moral values and self-evaluative standards* influence key purchasing decision steps, namely, *goal desire*, *behavior desire*, and *implementation intention*. In other words, who we are (or we think we are), and our values/what we believe in, influence both *what* we desire, and *how* we go about fulfilling that desire. Chris, our mystery shopper of Sect. 2.1, is a "fair trader", and as such, was told about the latest fair trade certified arrivals. However, how can we tell if a customer is environmentally conscious, or a fair trader, or someone who cares not only about the price of products they purchase, but also the labor practices of the companies, or jurisdictions that produce them?

There are many strategies for inferring such opinions or beliefs. For example, a shopper whose basket regularly includes items that are certified organic, or simply picked up from the organic/natural foods section, can be safely assumed to be health conscious, and *probably* relatively well. Idem for fair trade coffee. As more and more food store chains carry kosher or halal sections, we can even guess consumers' religions. Thus, basket market analysis techniques can be used if we have a history of purchases, and appropriate product labeling.

Alternatively, we can mine consumers' opinions from the traces they leave on the web. Again, it is difficult to devise a general strategy. It depends on the kind of opinion that is sought, and on the traces that are available. Roughly speaking, the search for opinions can be decomposed into two sub problems: finding out if a customer *cares* about a particular issue (*topic*), and finding out *how* they feel about that issue. To answer the first question (*topic*), we can check issue-specific social media. For example, any article or comment you enter on http://www.climatecentral.org shows that you care about climate change. The same can be said if the site is listed

---

[16]For example, the cashier just scanned cheese, and they know that the consumer is a DINK, they may suggest them a matching bottle of wine *in the mid-to-expensive* price range.

on your Facebook page, or if you subscribe to #ClimateChange hashtag through your Twitter account. Idem if the phrase 'climate change' occurs in your posts *anywhere* on the web. Generally speaking, the more *diffuse* the topic of the medium, the more sophisticated the opinion mining technique is needed.

Once we know that you care about a particular issue, discovering *how* you feel about it becomes slighter easier, thanks to *sentiment analysis* techniques. For example, to find out whether you believe in human-induced climate change or not, I can look-up your posts on http://www.climatecentral.org and do a simple polarity analysis.[17] However, there are cases where simpler cues can go a long way towards mining opinions. For example, if you use Monsanto and Genetically Modified Food (GMF) in the same sentence, then your position on the subject of GMF is easy to guess.

From an operational point of view, one of the issues we need to address is whether our CA-CEM application should keep updating users' profiles incrementally, and if so, what should be the triggering event for the incremental re-evaluation. Should it be a new Facebook post, a nightly web crawl for my customers, a purchase, or should I run a batch process on my customer database to compute their environmental opinions?

Despite advances in semantic opinion mining (see e.g. [10, 17, 22, 23]), opinion mining remains an art, because the best (or the only) available strategy to use in each case depends on or greatly benefits from the selection of *sources* from which to mine such opinions (e.g. *available* media specialized in the topic of interest), and the exploitation of information or constraints specific to that opinion (e.g. the *public image* of Monsanto, a *specific company*, in the agricultural food industry). Within the context of our framework, we can only provide a tool-box that opinion miners can use. It is probably something that should be done off-line, e.g. disconnected from the transactional databases/systems.

6.2.3 Assessing Consumer Feedback

The model of the purchasing process shown in Fig. 2 of Sect. 3 showed that the purchasing process ends with a post-purchase feedback whereby the consumers express, in one way or another, whether they have attained the desired goal that triggered the whole process in the first place. Naturally, companies are very interested in what their customers have to say about the products they purchased, and the quality of the *service* they got during the purchasing process, where the latter has been steadily gaining in importance in building customer satisfaction and loyalty.

There are many ways of getting this feedback. Post-service customer surveys are fairly common nowadays, where customers are given incentives to fill out customer surveys. While such surveys are undoubtedly very valuable because of their granularity and focus, they suffer from a number of problems:

---

[17]Knowing the *orientation* of the site alone does not tell us whether a frequent commentator shares the editorial line of the site: a regular commentator may express systematically contrarian opinions.

(1) A low return rate, despite the incentives.
(2) They do not address image and perception issues in the general public to the extent that they focus on actual customers.
(3) The artificially accommodating/agreeable reviews that all but the most irate consumers end up entering at the insistent requests of customer service personnel.

Thanks to the internet and social media, consumers are now leaving unstructured, unscripted, and uninhibited product and service reviews all over the place. This is a very active research area (see e.g. [10, 17, 22, 23, 29]). Thanks to knowledge base combining lexical, semantic, and sentiment/affective knowledge, recent opinion mining techniques are able to go well beyond *polarity* to explore *emotions* and *intensity,* and even identify *fake* product reviews (see e.g. [17, 22]). As we saw in Sect. 3, the *anticipated positive emotions* and *anticipated negative emotions* are important deciding factors in goal/behavior desirability. Emotions are not only important indicators of product appreciations, but they are very valuable measures of the success or failures of *service encounters* (see e.g. Cook et al. [12]).

Within the context of our framework, we need to provide resources and tools to CA-CEM applications designers that they can use to mine textual/unstructured product or service reviews left on the seller's portal. Anything beyond the data left on the seller's own web site offers the same opportunities and challenges described in Sect. 6.2.2.

6.3 Approximate Reasoning to Deal with Incomplete Data

As mentioned earlier, the personalisation embodied in experience management centers around selecting the products and services appropriate for the consumer's needs and desires, and presenting them to the consumer in an appealing way. Both CEM functions rely on the company's knowledge of its products and its customers. The models of Fig. 6 and Fig. 7 show fairly elaborate representations of customers and products, respectively. It is very likely that the company will not have this level of information about its products or customers. Furthermore, for many products and purchase types, much of this information will not be relevant.

We outline four strategies for dealing with incomplete data:

Instance data versus class data: For both products and consumers, we made a distinction between individuals (**Consumer** and **Product**) and classes/categories (**Category**, in both Figs. 6 and 7, and State in Fig. 6). If we do not have information about an individual, we look it up in the category. Thus, if I do not know that lamb chops go with the 2011 Shiraz, at least I know that meat goes with red wine, and lamb chops being meat, we know that we can recommend red wine, but not the 2011 Shiraz specifically. Thus, we have graceful degradation in the recommendation quality.

(1) Category inheritance: If some information is missing from a category, we can look it up in its super-categories. Marketers may know something about transportation needs of DINKs in general, but not specifically about under thirty DINKs.
(2) Because of the above, our confidence in the information about a consumer or a product depends on the length of the inference path (from instance to category, and up in the category hierarchy).
(3) Use a scoring approach, whereby the different facets contribute positively or negatively to a <product, consumer> match. Consider two customers C_1(needs: detergent, values: environment) and C_2(needs: detergent, values: none), and three detergents D_1(function: detergent, manufacturing.process: contains phosphate), D_2(function: detergent, manufacturing.process: does not contain phosphate), and D_3(function: detergent, manufacturing.process: NA). Assume that a match on function counts for 100, and a match on manufacturing process counts for 20 (or -20). The additive scoring approach would yield the following:

   a. For customer C_1, match$(C_1, D_1) = 100 - 20 = 80$, match$(C_1, D_2) = 100 + 20 = 120$, and match$(C_1, D_3) = 100 + 0 = 100$. Thus, D_2 is the best match.
   b. For customer C_2, match$(C_2, D_1) =$ match$(C_2, D_2) =$ match$(C_2, D_3) = 100 + 0 = 100$. In other words, customer C_2 does not care. They may use other criteria to select (e.g. cost).

Note that, when the customer does not care, the company's *own values* could come into play, to position one product higher than another.

7 Instantiating the Framework for a CEM Scenario

Section 4 laid the foundation for CEM by outlining the *range* of data and functions that *may be used* to support CEM functionality within the context of a generic e-business system. In this section, we outline steps for selecting and customizing the generic functionality presented in Sect. 4 to handle a particular CEM scenario.

7.1 Specifying the Relevant Purchase Scenario

In Sect. 3, we presented the generic structure of a purchasing process as a goal-directed behavior, and identified the various factors that influence or are relevant to the purchasing process. The process described in Fig. 2 may be an overkill for grabbing a carton of milk, or may not reflect the reality of impulsive purchases where one goes window-shopping and ends up making purchases, or walks into an electronic store to buy a USB key and comes out with a laptop.

Thus, the first step in using our CA-CEM framework is to specify the purchasing scenario(s) appropriate for the business by answering the following questions:

(1) Which steps of the process in Fig. 2 are relevant? This depends on the type of purchase (see e.g. [26]).
(2) Which factors are relevant to those steps? This depends on the type of need (e.g. food versus entertainment), and the type of product (basic versus luxury).
(3) When and where do the various steps occur? This is important for many reasons: when, and how much time does the consumer have to make a decision, and what is the best way to position the products to the customer. This also depends on the kind of store/products (e.g. Sharper Image versus Home Depot), and the sale channel used (in-store vs on-line).

A given business can enact several purchase scenarios, depending on the product and the channel, and a CA-CEM system needs to support all the scenarios. Furthermore, the system may *need* to guess which type of purchasing process is in progress, e.g. by using the browsing pattern (physical or on-line) in the store.

The outcome of this step will be a set of processes, each one of which representing a subset of the process of Fig. 2.

7.2 Specifying the Desired Level of Functionality

The model of Fig. 3 shows that all the steps of the purchasing process and all the influence factors are amenable to CEM intervention/functionality. Based on the process specified in Sect. 7.1, and when and where each step takes place, a retail business needs to figure out for each process step:

(1) Whether intervention at a particular process step brings value: a grocery store may not need to intervene at the **Goal desire** step, as a consumer needs no enticements to buy a carton of milk.
(2) Whether it has the means to intervene. This is an issue of opportunity and physical means. If you drop lamb chops in the shopping cart, and I have no way of finding that out (physical limitation) until you reach the cashier, it is probably too late (opportunity) to suggest an appropriate wine.
(3) What is the best modality. Intervening at various steps of the purchasing process can easily become intrusive. Should I "push" unsolicited specials to the consumer (the Australian wine special), or wait until s/he asks for wine recommendations, or suggest a wine when s/he pick a product that goes with wine?

The answers to these questions can help a brick-and-mortar retailer decide which and where to put sensors to get the best cost-effective CEM solution. Again, this approach contrasts with "stick an RFID tag to everything that moves, and see what you can do with it".

7.3   Build the Relevant Data Schemas

CEM functionalities rely on our knowledge of purchasing processes, products, and consumers. We saw in Sects. 5.1 and 6 the *range* of information we can manipulate about consumers and products to support CEM, and how default information about products and consumers can be gleaned from the categories to which they belong.

The desired level of functionality specified in the previous sections enables us to figure out what/how much data about consumers and products we need. We now have to specify the structure of the data, and populate it for categories (for consumers and products) and states (for consumers).

The models shown in Figs. 6 and 7 represent ontologies, i.e. *metamodels* of the actual data models needed by our CA-CEM functionality. In this step, we need to *instantiate* those ontologies to specify the actual data schemas needed by our system. For example, the model in Fig. 6 shows that a **Consumer** can have a bunch of **Property**(ies), and that each **Category** is characterized by a range of *values* for those properties. In this step, we need to specify the properties that we are interested in. For a **Consumer**, such properties include your typical CRM data such as age (or birth date/birth year), address, payment methods, profession, perhaps income bracket, marital status, and transaction history. To this, we add significant relationships, social identity(ies), and values. Consumer likes for specific products or product categories can be represented by ternary relationships.

Most of these **Property**(ies) may be used to characterize **Category**(ies) (e.g. DINKs, generation Y) and **State**s (married people, various family lifecycle stages), but some are exclusive to instances (e.g. name, transaction history). When a property applies to a **Category** (or **State**), it is represented by a *range* of values. For example, DINKs, may have an income bracket of [100 k, ∞[, like sports cars, while generation Y have an age range of [18–35]. Entering value ranges for these properties is crucial, since learning of a consumer's membership to a particular **Category** or **State** can help us infer a lot of information about their needs and desires. ***Those values embody the marketing knowledge that the business has about its customers or potential customers***.

A similar process needs to take place for the product data. We need to first specify which properties are worth representing. Then, we populate information about product *categories* (see the model of Fig. 6). That is where we specify that red wines go with meats, that big fish tend to have high mercury contents, or that Absurdistan has poor labor laws and lax environmental regulations.

7.4   Populate Instance Data

The product instance data is a one-time deal for the products sold by the business. Idem for promotional material (see Sect. 5.3). The consumer instance data, however, is a work in progress. We show below the lifecycle of a consumer record:

(1) Creation. Businesses learn about new or potential customers in many ways: (a) when customers go through the cash register the first time around, (b) when they access anonymously the web site of the company—in which case they may be known by IP address, (c) when they create accounts on the company's web site, or post comments with an ID from another portal, and (d) by purchasing consumer lists. Depending on the channel, the business will have more or less information.

(2) Category/state assignment. This is the process through which a consumer is identified with a category (a generation Y), or with a state/lifecycle state (young married couple in early childbearing phase). This information can be entered explicitly (filling out a questionnaire when applying for a store credit card), or inferred. For example, someone who buys baby formula or diapers regularly is assumed with a high-level of certainty to be in the early childbearing phase. This assignment allows to infer other data.

(3) Property value updates. Property values, or our confidence in their values, may be regularly updated through explicit entries or accumulation of evidence. For example, if someone has bought baby formula three times in the past 2 weeks, s/he is *probably* in childbearing stage (confidence level of 70 %)—although s/he could just be staying with someone who is. If s/he keeps at it week after week, our confidence level goes up. Generally speaking, consumer classification rules need to kick-in each time information about the customer is updated.

(4) State change. Recall, from the model in Fig. 6, that state changes can be triggered by two kinds of events, timer events, and property-value change events. A pre-school toddler will remain in this state for a maximum of 5 years. A teenager who changes home address, from his/her parents' home to an outside address has probably moved into young adulthood. Thus, *each time a consumer property value is updated, we run statement assignment rules*.

We mention the possibility that category data be updated from consumer instance data. For example, our marketing department came up with a category (market segment) based on age an income. We later realize that all of the consumers of that category live in a particular area. The zip code could become a *characteristic feature* of this category, i.e. its value used as a default for consumers of the category, but it would not be used as a criterion for membership (i.e. it is not *definitional*).

8 Discussion

CEM aims at personalizing a customer's interactions with a company around the customer's needs and desires [30]. Researchers and e-business visionaries have been fantasizing about the kind of personalization afforded by ubiquitous computing (e.g. the Internet of Things). While all the technical ingredients for such fantastic scenarios exist today (sensor technology, middleware, mobile computing, social network analysis, data mining techniques), there are no guidelines to help either e-business software vendors, to figure out which functionalities to provide in their software to support CEM functionalities, or e-business software clients to figure out which CEM functionalities to implement, let alone *how* to implement them, given the type of products they sell, the channels through which they sell those products, and their software and hardware capabilities.

Consumers interact with companies to acquire products or services that satisfy their needs and desires: a *purchase* [7]. To personalize such interactions, we need to understand them. Toward this end, we relied on studies of *consumer behavior* from marketing and social psychology. Such studies identified the steps undertaken by a consumer in a purchase process, and the factors that influence their decision making in those steps (Sect. 3). For the purposes of CEM, these steps are potential interaction opportunities between the company and the consumer, and the influence factors represent relevant data that the company could use to personalize those interaction (Sect. 4.1). This enabled us to identify an *experience management pattern* that could be applied to various steps of the purchasing process. However, we showed that a blind application of the pattern to all the steps of the purchasing process described in Sect. 3 would not be appropriate (Sect. 4.2). We argued that the application of this pattern is best considered within the context of *service design* (Sect. 4.3), and showed an example of the analyzes a service designer could make to instantiate this pattern (Sect. 4.4). While Sect. 4 explored the design vocabulary for *customer experiences*, Sect. 5 presented the kind of representation that is required for consumers and products to be able to match products to consumer's needs/desires, preferences, biases, and attitudes. Section 6 discussed how to deal with incomplete customer or product data. Given such an infrastructure, we proposed the first elements of a methodology enabling e-business software users to specify the CEM functionalities they wish to support (Sects. 7.1 and 7.2), set-up the data infrastructure (Sect. 7.3), and populate the consumer data (Sect. 7.4).

This work is at an early stage. In this paper, we focused on the foundations for *a methodology, for designing and implementing CA-CEM functionalities*. The proposed approach gives us a frame of reference to study the multitude of issues raised by CEM functionalities within the context of e-commerce software. We are currently testing the theory on a specific purchasing scenario provided by an e-commerce suite vendor. It will enable us to validate the theory, and implement a first cut of some of the software artifacts of our framework.

References

1. Ajzen, I.: From intentions to actions: a theory of planned behavior. In: Action Control, pp. 11–39. Springer (1985)
2. Ajzen, I., Fishbein, M.: Understanding Attitudes and Predicting Social Behavior. Prentice-Hall (1980)
3. Ajzen, I., Madden, T.: Prediction of goal-directed behavior: attitudes, intentions, and perceived behavioral control. J. Exp. Soc. Psychol. (1986)

4. Armitage, C., Conner, M.: The theory of planned behaviour: assessment of predictive validity and'perceived control. Br. J. Soc. Psychol. (1999)
5. Bagozzi, R., et al.: Network analyses of hierarchical cognitive connections between concrete and abstract goals: an application to consumer recycling attitudes and behaviors (1996)
6. Bagozzi, R.: The self-regulation of attitudes, intentions, and behavior. Soc. Psychol. Q. **55**(2), 178–204 (1992)
7. Bagozzi, R., et al.: The Social Psychology of Consumer Behaviour. Open University Press (2007)
8. Bagozzi, R., Dholakia, U.: Antecedents and purchase consequences of customer participation in small group brand communities. Int. J. Res. Mark. (2006)
9. Bagozzi, R., Warshaw, P.: Trying to consume. J. Consum. Res. **17**(2), 127–140 (1990)
10. Cambria, E., et al.: SenticNet 2: a semantic and affective resource for opinion mining and sentiment analysis. In: FLAIRS Conference (2012)
11. Constantine, L.: Human activity modeling: toward a pragmatic integration of activity theory and usage-centered design. Human-Centered Softw. Eng. 27–51 (2009)
12. Cook, L.S., et al.: Human issues in service design. J. Oper. Manag. **20**(2), 159–174 (2002)
13. Fazio, R.: How do attitudes guide behavior. In: Handbook of Motivation and Cognition: Foundations of Social Behavior. The Guildford Press (1986)
14. Fishbein, M., Ajzen, I.: Belief, attitude, intention and behavior: an introduction to theory and research (1975)
15. Gruber, T.: Ontology. Encyclopedia of Database Systems. pp. 1963–1965. (2009). doi:10.1007/978-0-387-39940-9_1318. http://dx.doi.org/10.1007/978-0-387-39940-9_1318
16. Hill, A.V., et al.: Research opportunities in service process design. J. Oper. Manag. **20**(2), 189–202 (2002)
17. Liu, B.: Sentiment Analysis Mining Opinions, Sentiments, and Emotions. Cambridge University Press (2015)
18. Mylopoulos, J., et al.: From object-oriented to goal-oriented requirements analysis. Commun. ACM. **42**(1), 31–37 (1999)
19. Ostrom, A.L., et al.: Moving forward and making a difference: research priorities for the science of service. J. Serv. Res. **13**(1), 4–36 (2010)
20. Palmer, A.: Customer experience management: a critical review of an emerging idea. J. Serv. Mark. (2010)
21. Patricio, L., et al.: Multilevel service design: from customer value constellation to service experience blueprinting. J. Serv. Res. **14**(2), 180–200 (2011)
22. Poria, S., et al.: Enhanced SenticNet with affective labels for concept-based opinion mining. IEEE Intell. (2013)
23. Poria, S., et al.: Towards an intelligent framework for multimodal affective data analysis. Neural Netw. (2015)
24. Sarwar, B., et al.: Analysis of recommendation algorithms for e-commerce. In: 2nd ACM Conference on Electronic Commerce, pp. 158–167, Minneapolis, MN, USA (2000)
25. Schwartz, S.: Universals in the content and structure of values: theoretical advances and empirical tests in 20 countries. Adv. Exp. Soc. Psychol. **25**(1), 1–65 (1992)
26. Solomon, M., et al.: Consumer Behavior: Buying, Having, and Being. Pearson (2009)
27. Teixeira, J., et al.: Customer experience modeling: from customer experience to service design. J. Serv. Manag. **23**(3), 362–376 (2012)
28. Verhoef, P.C., et al.: Customer experience creation: determinants, dynamics and management strategies. J. Retail. **85**(1), 31–41 (2009)
29. Vinodhini, G., Chandrasekaran, R.: Sentiment analysis and opinion mining: a survey. Int. J. Adv. Res. Comput. Sci. Softw. Eng. **2**, 6 (2012)
30. Weijters, B., et al.: Determinants and outcomes of customers' use of self-service technology in a retail setting. J. Serv. Res. **10**(1), 3–21 (2007)
31. Wells, W., Gubar, G.: Life cycle concept in marketing research. J. Mark. Res. (1966)

An Overview of Sentiment Analysis in Social Media and Its Applications in Disaster Relief

Ghazaleh Beigi, Xia Hu, Ross Maciejewski and Huan Liu

Abstract Sentiment analysis refers to the class of computational and natural language processing based techniques used to identify, extract or characterize subjective information, such as opinions, expressed in a given piece of text. The main purpose of sentiment analysis is to classify a writer's attitude towards various topics into positive, negative or neutral categories. Sentiment analysis has many applications in different domains including, but not limited to, business intelligence, politics, sociology, etc. Recent years, on the other hand, have witnessed the advent of social networking websites, microblogs, wikis and Web applications and consequently, an unprecedented growth in user-generated data is poised for sentiment mining. Data such as web-postings, Tweets, videos, etc., all express opinions on various topics and events, offer immense opportunities to study and analyze human opinions and sentiment. In this chapter, we study the information published by individuals in social media in cases of natural disasters and emergencies and investigate if such information could be used by first responders to improve situational awareness and crisis management. In particular, we explore applications of sentiment analysis and demonstrate how sentiment mining in social media can be exploited to determine how local crowds react during a disaster, and how such information can be used to improve disaster management. Such information can also be used to help assess the extent of the devastation and find people who are in specific need during an emergency situation. We first provide the formal definition of sentiment analysis in social media and cover traditional and the state-of-the-art approaches while highlighting contributions, shortcomings, and pitfalls due to the composition of online

G. Beigi (✉) · R. Maciejewski · H. Liu
Computer Science and Engineering, Arizona State University, Tempe, USA
e-mail: gbeigi@asu.edu

R. Maciejewski
e-mail: rmacieje@asu.edu

H. Liu
e-mail: huan.liu@asu.edu

X. Hu
Department of Computer Science and Engineering, Texas A&M University,
College Station, USA
e-mail: hu@cse.tamu.edu

© Springer International Publishing Switzerland 2016   313
W. Pedrycz and S.-M. Chen (eds.), *Sentiment Analysis and Ontology Engineering*,
Studies in Computational Intelligence 639, DOI 10.1007/978-3-319-30319-2_13

media streams. Next we discuss the relationship among social media, disaster relief and situational awareness and explain how social media is used in these contexts with the focus on sentiment analysis. In order to enable quick analysis of real-time geo-distributed data, we will detail applications of visual analytics with an emphasis on sentiment visualization. Finally, we conclude the chapter with a discussion of research challenges in sentiment analysis and its application in disaster relief.

Keywords Sentiment analysis · Disaster relief · Visualization · Social media

1 Introduction

With the explosive growth of social media (e.g. blogs, micro-blogs, forum discussions and reviews) in the last decade, the web has drastically changed to the extent that nowadays billions of people all around the globe are freely allowed to conduct many activities such as interacting, sharing, posting and manipulating contents. This enables us to be connected and interact with each other anytime without geographical boundaries, as opposed to the traditional structured data available in databases. The resulted unstructured user-generated data mandates new computational techniques from social media mining, while it provides us opportunities to study and understand individuals at unprecedented scales [1–7]. Sentiment analysis (a.k.a opinion mining) is one class of computational techniques which automatically extracts and summarizes the opinions of such immense volume of data which the average human reader is unable to process. This ocean of opinionated postings in social media is central to the individuals' activities as they impact our behaviors and help reshape businesses. Nowadays, not only individuals are no longer limited to asking friends and family about products but also businesses, organizations and companies do not require to conduct surveys or polls for opinions about products, as there are tons of user reviews and discussions in public forums on the Web. There are thus numerous immediate and practical applications and industrial interests of collecting and studying such opinions by using computational sentiment analysis techniques, spreading from consumer products, services, healthcare, and financial services to social events, political elections and more recently crisis management and natural disasters.

Social media has pervasively played an increasing role and they have become an important alternative information channel to traditional media in the last five years during emergencies and disasters, where they rank as the fourth most popular sources to access necessary information during emergencies [8, 9]. In particular, individuals and communities have used social media for many tasks from warning others of unsafe areas to fund raising for disaster relief [8]. The days of one-way communication where only official sources used to provide bulletins during emergencies are actually gone. In 2005 for instance, when Hurricane Katrina slammed the U.S. gulf coast, there was no Twitter for news update while Facebook was not that much famous yet. Compare, for example, Hurricane Katrina to the Haiti earthquake on January 2010. During latter, people used Twitter, Facebook, Flicker, blogs and YouTube to post their experience in form of texts, photos and videos during the earthquake which resulted in donating 8 million U.S. dollars to the Red Cross which vividly demonstrates the power of social media in propagating information during emergencies [10]. Hurricane Sandy on 2012, is another example to show the positive impact of social media during disasters. By that time, using social media had become an important part of disaster response. There are numerous similar examples that show how social media have come to the rescue in disaster situations including Hurricane Irene, California gas explosion on 2010, Japan earthquakes, Genoa flooding and more recently Ebola. Social media could be actually leveraged to keep the problem informed, help locate loved ones, and express support or notify authorities during emergencies and disasters. Sentiment analysis of disaster related posts in social media in could help to detect posts that contribute to the situational awareness and better understand the dynamics of the network including users' feelings, panics and concerns by identifying the polarity of sentiments expressed by users during disaster events to improve decision making. Sentiment information could also be used to project the information regarding the devastation and recovery situation and donation requests to the crowd in better ways.

Interactive tools such as visual analytic methods could help us to make a large amount of complex information more readable and interpretable, if integrated by computational approaches, as the effectiveness of most computational techniques is limited due to several factors [11]. Interactive visual analytics provide intuitive ways of making sense of large amount of posts available in social media. These techniques are now widely used in social media data and contribute in many areas of exploratory data analysis. Despite most social media visualization approaches which rely solely on geographical and temporal features, there are some systems which are able to exploit the sentiments of the data such which help improving visualization. Besides disaster related data management in social media, the ability to drawing out important features could be used for better and quick understanding of situation which leads to rapid decision making in critical situations. Moreover, the data produced by social media during disasters and events, is staggering and hard for an individual to process. Therefore, visualization is needed for facilitating pattern discovery.

The goal of this chapter is to give the reader a concrete overview of sentiment analysis in social media and how it could be leveraged for disaster relief during emergencies and disasters. In particular, we cover state-of-the-art sentiment analysis approaches and highlight their contributions and shortcomings and then discuss the application of social media and sentiment analysis in disaster relief and situational awareness. We conclude the chapter by detailing applications of visual analytics with an emphasis on sentiment analysis and then discussing the research challenges in sentiment analysis and disaster relief. By the end of this chapter, the reader is expected to learn about the sentiment analysis and disaster management concepts as well as the state-of-the-art approaches and the applications of visual analytics in these contexts.

2 Sentiment Analysis

Sentiment analysis (a.k.a sentiment classification, opinion mining, subjectivity analysis, polarity classification, affect analysis, etc.) is the multidisciplinary field of study that deals with analyzing people's sentiments, attitudes, emotions and opinions about different entities such as products, services, individuals, companies, organizations, events and topics and includes multiple fields such as natural language processing (NLP), computational linguistics, information retrieval, machine learning and artificial intelligence. It is set of computational and NLP based techniques which could be leveraged in order to extract subjective information in a given text unlike factual information, opinions and sentiments are subjective [12].

Despite the recent surge of interest in sentiment analysis since the term was coined by Nasukawa et al. [13] in 2003, the demand for information on sentiment and opinion during decision-making situations dates back to long before the widespread use of the World Wide Web. Opinions are central to almost all human activities as they could influence our behaviors specially when making a decision. For example, many of us may have asked their friends to recommend a dishwasher or to explain who they might vote for during elections, or even requested reference letters from colleagues regarding job applications. Now, opinions and experiences of numerous people that are neither our acquaintances nor professional critics are readily available thanks the Internet and the Web [14]. This is not limited to individuals only; businesses, organizations and companies are also eager to know consumers' opinions about their products and services. In the past, when a business needed consumer opinions, it conducted surveys and opinion polls. Nowadays, one is no longer limited to asking friends and family or conducting surveys for opinions about products; instead one can use volumes of user reviews and discussions in public forums on the Web [12]. Indeed, the Web has dramatically changed the way that people express their opinions about products, services, companies, individuals and social events. There are now many Internet forums, discussion groups, blogs and even micro-blogs that are well suited for the users to freely post reviews about products and express their views on almost anything online. These users-generated contents and word-of-mouth behavior are sources of information with many immediate and practical applications.

The research on sentiment analysis appeared even earlier than 2003 [15–20], while there were also some other earlier work [21, 22] on beliefs as frontiers or later work [23–30] on interpretation of metaphors, sentiment adjectives, subjectivity, view points, affects and related areas [12, 14]. In contrast to the long history of linguistics and NLP, the area of analyzing people's opinions and sentiments has been virtually untrodden before the year 2000. However, since then, the literature witnessed literally hundreds of studies [31–39] due to several factors, including: (1) the rise of machine learning techniques in natural language processing and information retrieval; (2) access to datasets for training machine learning techniques because of the World Wide Web and specifically review-aggregation Websites, and; (3) realizing the huge applications in industry that the area started to offer [14]. This rapid growth of sentiment analysis and, more importantly its coincidence with the explosive popularity of the social media, have made the sentiment analysis the central point in the social media research [12].

Research on sentiment analysis has been investigated from different perspectives. Perhaps the most popular perspective is to categorize these studies into three levels, document level, sentence level, and entity and aspect level [12] described as follows:

- **Document level**: The aim here is to determine the overall sentiment of an entire document. For example given a product review, the task is to determine whether it expresses positive or negative opinions about the product. This level looks at the document as a single entity, thus it is not extensible to multiple documents.
- **Sentence level**: This level of analysis is very close to subjectivity classification and the task at this level is limited to the sentences and their expressed opinions. Specifically, this level determines whether each sentence expresses a positive, negative or neutral opinion.
- **Entity and aspect level**: Instead of solely analyzing language constructs (e.g. documents, paragraphs, sentences), this level (a.k.a feature level) provides finer-grained analysis for each aspect(or feature) i.e., it directly looks at the opinions for different aspects itself. The aspect-level is more challenging than both document and sentence levels and consists of several sub-problems. It finds different available sentiment

Sentiment analysis methods could be categorized into two groups, language processing based and application oriented methods. We describe the state-of-the-art approaches in each category and highlight their contributions. Then we conclude this section with a brief overview on visual analytics approaches in sentiment analysis.

2.1 Language Processing Based Methods

Meanwhile, some other works have addressed sentiment analysis from two different aspects, namely, lexicon-based, and linguistic analysis. The most obvious yet important indicators of sentiments are sentiment or opinion words such as *good, amazing, poor, bad* as well as some phrases and idioms which are used to express positive or negative opinions. A sentiment lexicon (a.k.a opinion lexicon) is the list of such words and phrases and is necessary but not sufficient for sentiment analysis. In addition to exploiting lexicons, linguistic based approaches also use the grammatical structure of the text for sentiment classification. There are two kinds of lexicon generation methods, namely, dictionary based [32, 33, 40] and corpus-based [31] approaches. The first category starts with a small set of opinion words and expands the lexicon through bootstrapping a certain dictionary while the second category generates the opinion lexicon through learning the dataset. For example, authors et al. [32] propose to employ predefined syntactic templates to capture links between opinion words and their targets. The links can then be used to infer more opinion words. Wu et al. [41] further extended the link structure to syntactic trees and expand the opinion lexicon accordingly. Kouloumpis et al. [31] investigate the utility of linguistic features including lexicon and part of speech for sentiment analysis of microblogging posts by deploying AdaBoost.MH model [42]. The authors then concluded that sentiment lexicon features demonstrate better performance. Another lexicon based method is proposed by Thelwall et al. [33] that authors study the relation between importance of events and sentiment intensity of messages posted in Twitter. Time series based analysis of sentiment can be used to predict offline phenomena of online topics or investigating the role of sentiment in important online events.Thelwall et al. [33] incorporate SentiStrength algorithm [43], which simultaneously assigns positive and negative scores to the token in the text, to extract the intensity of posts sentiments. The algorithm assigns scores to the tokens in a dictionary which includes common emoticons while modifier words or symbols can boost the score. The final score is the maximum score among all tokens' scores. The authors conclude with the observation that negative sentiment intensity of microblogging posts increases with the importance of the event.

2.2   Application Oriented Methods

Pervasive real-life applications of sentiment analysis and opinion mining are one reason that sentiment analysis is now a popular research problem. Due to these offered applications by sentiment analysis, many activities specially those of industries have risen in recent years which have spread to almost every possible domain including products, services, healthcare, financial services, social events and political elections. Sentiment analysis also helps to understand opinions of people regarding different online topics. It has many applications in social media and many works have been done recently in applying sentiment analysis methods to social media data. These applications include, but are not limited to, movie reviews [17], product reviews [12, 44], App reviews [45], stock market predictions [46] and trend detection [47]. For example, Pang et al. [17] discuss the performance of machine learning classifiers including Naïve Bayes [48], maximum entropy [49] and support vector machines [50] in the specific domain of movie reviews. They use various feature extractors based on unigrams and bigrams. Star ratings are also used as polarity signals in training data. Another application of sentiment analysis is in stock market prediction [46], investigating whether the stock market could be predicted using public sentiment expressed in daily Twitter posts. The authors use OpinionFinder [51] and Google-Profile of Mood States(GPOMS) tools to measure users sentiments and mood to further analyze resulted time series and the effect of public mood states on prediction of stock market. In spite of these applications and the abundance of public information available for sentiment analysis, finding opinion Websites on the Internet and processing the information contained in them is still a tedious task for the average human reader. Therefore, automated sentiment analysis systems are critical. One good example of such a system is the Stanford CoreNLP toolkit [52] which is an extensible and annotation-based NLP processing pipeline that provides several core natural language analysis steps, from tokenization to coreference resolution.

There are also some other studies that leverage network information into sentiment analysis applications. For example, some works exploit social network information [34, 35] to improve sentiment analysis performance. Authors in [34] incorporate user-user relations including follower/followee network corresponding into the principle of homophily [53] and "@"-network which shows the attention of users to each other, i.e. mentions network. They propose a semi-supervsied model using a factor-graph model to leverage social network information for user-level sentiment polarity classification. Tan et al. show considering both homophily and attention at the same time could result in a significant improve even when representative information is very sparse, as long as demonstrating a strong correlation between users shared opinions. Another work [35] confirms the existence of two social theories *sentiment consistency* and *emotional contagion* in microblogging data. Sentiment consistency [54] refers to sentiment consistency of messages posted by the same user in comparison to two different random users' posts. Emotional contagion [55] suggests that messages posted by two friends have more similar sentiment than those of random users in the network. The proposed method, then, integrates the social theories using sentiment relations between tweets as a regularization parameter into the definition of traditional supervised classification method and finally uses optimization techniques to solve the problem. In order to handle noisy and short texts in microblogging data, Hu et al. [35] utilizes sparse learning method lasso [56] for the classification feature space.

Different from these works, some efforts have been made to exploit emoticons [36, 37, 57]. For example, Go et al. [57] apply machine learning classifiers using distant supervision to consider microblogging post with emoticons as training data. Distant supervision is introduced by Read [58] which shows that emoticons could be used as sentiment labels to reduce dependencies in machine learning techniques. In a similar attempt, Zhao et al. [36] train a Naïve Bayes classifier on a dataset from Weibo[1] using emoticons as noisy label information. Another work by Hu et al. [37] further investigate using emotional signals in the sentiment analysis problem. Emotional signals refer to any information in the data which reveals the sentiment polarity of a post such as emoticons, product ratings or some words which carry clear semantic meaning. Emotional signals can be grouped into two categories: emotion indication and emotion correlation. The former refers to the signals that strongly demonstrate the sentiment of a post or word, like emoticons or product ratings, and the latter includes posts that demonstrate the correlation between post or words such as synonym correlation or sentiment consistency theory. This theory states that two co-occurred words are more likely to be sentiment consistent rather than two random words. The proposed method, first verifies the existence of two representative emotional signals, emoticons and sentiment consistency. Then it models them with regularization factors considering both post- and word-level aspects of them and exploits them in the orthogonal nonnegative matrix tri-factroization model (ONMTF) [59] for sentiment classification.

---

[1]http://Weibo.com.

There exist some other studies for sentiment classification that are different from the previous methods. For instance, the impact of human factors such as textual, topical, demographical, spatial and temporal features on the attitude and sentimentality of posts is studied by Kucuktunc et al. [40]. This study proposes to use posts on Yahoo! Answers,[2] a large online question answering system and uses [43] as a sentiment extraction tool to extract sentiments and further deploys gradient boosted decision trees [60, 61] to predict the attitude a question will provoke in answers. As an another example, Hu et al. [62] propose a method to identify sentiments of segments and topics of event related tweets which have received praise or criticism. Their framework decomposes the tweet-vocabulary matrix into four factors to demonstrate the relation of tweets with different segments, topics and sentiments so that it could characterize the segment and topics of events via aggregated Twitter sentiment. To compensate the learning process, three types of prior knowledges have been exploited in their method including sentiment lexicon, manually labeled tweets and tweet/event alignment using [63]. Balahur et al. [64] apply sentiment analysis methods for news opinion mining since it is different from that of other text types in clarifying target, separation of good and bad news from good and bad sentiment, and analysis of explicitly marked opinion. Recent work of [38], presents three neural networks to exploit and learn sentiment-specific word embedding (SSWE) for sentiment classification on Twitter without any manual annotations. In more details, they apply sentiment-specific word embedding for Twitter sentiment analysis under a supervised learning framework. In this study, instead of hand-crafting features, authors incorporate the continues representation of word and phrases as the feature of a tweet. Another study [39], compares three different approaches including *replacement*, *augmentation* and *interpolation* for exploiting semantic features for Twitter sentiment analysis and concludes that the best results are achieved by interpolating the generative model of words given semantic concepts into the unigram language of model of the Naïve Bayes classifier.

In Table 1, we summarize some of the recent studies with their datasets along with their proposed approaches and their best reported accuracies.

2.3   Sentiment Visualization

Visual analytics focuses on providing an intuitive way of making sense of large amount of posts available in social media. It is widely used in social media data and contributes in many areas of exploratory data analysis, such as geographical analysis [65], information diffusion [66] and business prediction [67]. While most social media visualization approaches rely on geographical and temporal features, some systems exploit the semantic of the data such as sentiments to improve visualization. For example, there exist some Websites that provide mashup applications to visualize

---

[2]http://answers.yahoo.com.

Table 1 Best results reported by different studies using different approaches on different data sets

Paper	Dataset	Approach	Evaluation
[17]	IMDb Movie Review	Support vector machine	82.9% (Accuracy)
[31]	iSieve Twitter dataset	AdaBoost.MH	75% (Accuracy)
[32]	Customer Review Dataset	Others	70% (FScore)
[33]	Twitter dataset	SentiStrength	
[34]	Twittes dataset on Different Topics	Graphical models incorporating users relation network	80% (Accuracy)
[35]	Stanford Twitter Sentiment; Obama-McCain Debate	Incorporating sociological sciences into multiple sentiment classification	79.6% (Accuracy)
[36]	Weibo Tweets	Incorporating emoticons into Naive Bayes classifier	58.3% (FScore)
[37]	Stanford Twitter Sentiment; Obama-McCain Debate	Leveraging emotional signals in ONTMF [59]	72.6% (Accuracy)
[38]	Twitter sentiment classification benchmark dataset in SemEval	Neural networks	86.48% (FScore)
[39]	Stanford Twitter Sentiment; Obama-McCain Debate; Health Care Reform	Incorporating semantics features into Naive Bayes	84.25% (Accuracy)
[40]	Yahoo answers	SentiStrength	0.4939 (RMSE)
[57]	Self crawled Twitter dataset	Incorporating emoticons in support vector machine	83% (Accuracy)
[62]	Denver Debate Twitter dataset; ME Speech Twitter dataset	Matrix factorization by leveraging prior knowledge of lexicon, sentiment and alignment of tweets to the events	76.8% (Accuracy)
[64]	News articles quotes	SentiWordNet	82.0% (Accuracy)

Some studies may have reported different accuracies in their original papers

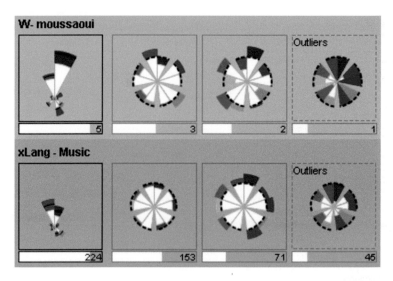

Fig. 1 Sentiment distribution of two blog topics using IN-SPIRE. The sentiment of lexicons categorized into different groups visualize as a *rose plot* with different colors. Reprinted with permission from [68]

and analyze tweets, including TrendsMap,[3] Twitalyzer,[4] and Geotwitterous,[5] some of which provide sentiment analysis as well. In the remainder of this section, we explore some of the existing systems and tools that are able to visualize sentiment.

IN-SPIRE [68] is a visual analytic tool for blog analysis which helps users to harvest blogs and classify them with respect to their contents. IN-SPIRE supports sentiment visualization of blogs and streaming contents. The sentiment of lexicons are categorized into positive/negative, virtue/vice, power coop/conflict and pleasure/pain according to Gregory et al. [69] and is visualize as a rose plot. Each pair of category is shown with a different shade of a same color. Moreover, the size of petals are different in terms of the amount of sentiment. For example, Fig. 1 depicts sentiment distribution in two event-related and music-related blogs regarding the September 11th attacks.

As another example of sentiment visualization prototype, Pulse [70], simultaneously detects topic and sentiment of a document in the sentence-level in the following way. It first constructs a tree-map [71] to visualize the clusters and their associated sentiments. Each cluster is depicted by a box whose size indicates the number of sentences in it and also its red or green color translates into the positive or negative sentiment respectively. Naive Bayes classifier with Expectation Maximization (EM) and bootstrapping [72] are finally applied to find the sentiment of the sentence. Figure 2 shows a screenshot of Pulse user interface system. Oelke et al [73] develops

---

[3]http://trendsmap.com/.

[4]http://www.twitalyzer.com/.

[5]http://ouseful.open.ac.uk/geotwitterous/.

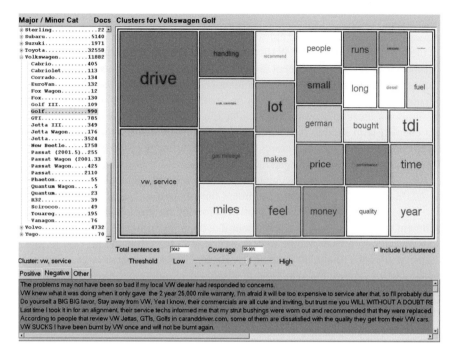

Fig. 2 Screenshot of Pulse user interface. Each cluster of corespondent tree-map is shown with a box whose size indicates the number of sentences in it. Positive and negative sentiments are shown with *red* and *green* color respectively. Reprinted with permission from [70]

a tool that analyzes and visualizes customer feedbacks using a 2-D matrix to show the sentiment overview of customer feedbacks. Then it clusters the reviews and uses special thumbnails to provide their relationship. It uses a circular correlation map to analyze the correlations among different aspects of the dataset such as numerical ratings, sentiment orientations and product features of the reviews.

Similarly, VISA [74], is an extension to the generic text visualization tool TIARA [75] with the further sentiment analysis abilities which helps average human reader to understand sentiments of a huge amount of textual data. The novel concept of this tool is sentiment tuple which is the core of the data model and acts as an interface between backend sentiment analysis system and the frontend visualization. This tuple of feature, aspect, opinion, and polarity helps the system to incorporate any external sentiment analysis method results which could be easily transformed into this representation. VISA is also a mashup visualization which shows the temporal sentiment dynamics, comparison of sentiment among different topics, document snippet to provide details about the context of sentiment and chart visualization for different structured features for more accurate sentiment perception. It deploys three domain-specific *entity*, *word*, and *opposite* dictionaries to determine features, aspects and polarity of the document.

Eventscape tool [76] combines time, visual media, mood, and controversy by performing topic modeling and clustering to determine events and topics and arranging them in time. It uses ANEW dictionary to find emotion valence and arousal of document text. It shows sentiment diversity of documents by focusing on interactive colour-coded timeline, lists of tweets, and mood maps. TwitInfo [77] is also a system for summarizing and visualizing event on Twitter and has a timeline based display which allows users to brows a large collection of tweets. It plots the geographical data coupled with sample tweets, color-coded for sentiment. TwitInfo uses Naive Bayes classifier on unigram features following the strategy of Go et al. [57] to classify tweets as positive and negative. It also introduces a normalization procedure for sentiment summarization to guarantee that sentiment methods produce correct results. MediaWatch [78] creates sentiment flow diagrams to analyze the evolution of sentiment over time with an innovative color-coding display. It uses aggregated sentiment polarity method as a sentiment detection algorithm. A screenshot of the MediaWatch on climate change data is shown in Fig. 3. Shook et al. [79] propose a geospatial visual analytics method, TwitterBeat, which analyze large volume of textual data to capture the sentiment to display it in emotional heatmaps. Figure 4 depicts a sample of TwitterBeat on US sentiment feed.

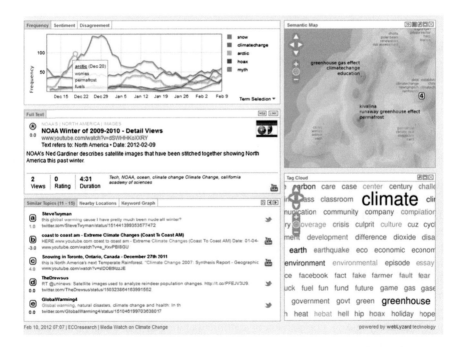

Fig. 3 Screenshot of the MediaWatch on climate change data. Sentiment flow diagram is shown to analyze evolution of sentiment over time. Reprinted with permission from [78]

Fig. 4 Heat Maps of US Sentiment on Twitter. Reprinted with permission from [79]

3 Sentiment Analysis for Disaster Relief

Recent years have witnessed an increase in severity and frequency of natural disasters [80]. Many natural disasters around the globe, have killed people and impacted human lives worldwide. Consequently there is a critical need to use a mechanism to best allocate investment for prevention, preparedness, response, and recovery to enhance safety and reduce the cost and social effects of emergencies and disasters [81]. In the following subsections, first we describe the application of social media during disasters and emergencies as one of such mechanisms and then explain how sentiment analysis can be deployed for disaster relief and finally review some of the state-of-the-art visual analytics techniques used to accommodate the task of disaster management.

3.1 Social Media in Disaster Relief

Social media have become an important alternative information channel to traditional media during emergencies and disasters. They have been used for many tasks including warning others of unsafe areas and fund raising for disaster relief [8]. Social media help to keep informed, locate loved ones, express support or notify authorities during emergencies and disasters. Due to popularity and diverse areas of discussion of social media, it has been used recently as a tool by first responders-those who provide first hand aids- for disaster relief and crisis management. Microblogging systems are used by millions around the world to broadcast just about anything. Therefore they can be used in disaster management as they provide an information platform with easy accessibility. For example Twitter was used to provide real time situation updates during major disasters [82–85]. [86] used 2008 Sichuan Earthquake in 2008 as a case study to show that individuals use social media to gather and disperse useful information regarding the disasters. [87] also provides a study on how microblogging systems could be used to facilitate disaster response.

The applications of social media in disaster management can be roughly categorized into two groups, situational awareness and information sharing [88]. Situational awareness means identifying, processing, and comprehending critical elements of an incident or situation to provide useful insight into time- and safety-critical situations [8, 89]. It assists first responders in assessing the amount of damage, victims' locations and needs. Information sharing also shows how people behave and share information in social media regarding the disasters and could be used for directing needed resources into public. Both situational awareness and information sharing help to accelerate disaster response and alleviate both property and human losses in crisis managements.

Analyzing and identifying posts related to the disaster [9, 90], detecting posts originated from within a crisis region [91], studying automatic methods for extracting information [85, 92–99] and behavioral analysis methods during disasters [87, 100, 101] are among different methods used to improve situational awareness. There are also some applications and analysis tools which help humanitarian and disaster relief organizations to track, analyze, and monitor microblog event related posts such as [102–108].

3.2 Applications of Sentiment Analysis in Disaster Relief

Sentiment analysis of disaster related posts in social media is one of the techniques that could gear up detecting posts for situational awareness. In particular, it is useful to better understand the dynamics of the network including users' feelings, panics and concerns as it is used to identify polarity of sentiments expressed by users during disaster events to improve decision making. It helps authorities to find answers to their questions and make better decisions regarding the event assistance without paying the cost as the traditional public surveys. Sentiment information could also be used to project the information regarding the devastation and recovery situation and donation requests to the crowd in better ways. Using the results obtained from sentiment analysis, authorities can figure out where they should look for particular information regarding the disaster such as the most affected areas, types of emergency needs [109].

Many methods have been proposed to analyze the role of sentiment analysis in disaster management where most of them deploy variant machine learning techniques [89, 109–121] and other techniques such as swarm intelligence [122]. The datasets used for evaluations includes the social media posts related to events such as Hurricane Sandy, Hurricane Irene, Red River flood in 2009 and 2010, Haiti earthquake, California gas explosion, and Terrorist attack in Mumbai.

One of the earlier machine learning based works for disaster management using sentiment analysis is proposed by Verma et al. [89]. In particular, they build a classifier to automatically detect situational awareness tweets by categorizing them to several dimensions including subjectivity, personal or impersonal style and linguistic style (formal or informal) using combination of hand-annotated and automatically linguistic features. Subjectivity features could be used to assess the amount of emotion a user expressed in her tweets for situational awareness and information extraction. This study, uses OpinionFinder [123, 124] to extract subjectivity of tweets as a linguistic features for their developed classifier. Similarly, authors in [109] use Bayesian Networks to detect sentiment of tweets of California gas explosion based on combination of sentiment ontology, emoticons,frequent lists of sentiment words SentiWordNet [125] and AFINN [126]. In addition to classifying sentiments of tweets into positive and negative, study of Nagy et al. [109] could be used to detect information-based tweets as a way to increase credibility of the post. Using different features such as bag of words, pruning features and lexicon based ones, authors in [110] train a sentiment classifier to categorize messages based on level of concern. Then they leverage trained classifier for understanding public perception towards disasters by investigating relation of sentiment and demographic features including location, gender and time. The authors use Hurricane Irene dataset for evaluation.

Dong et al. [111] develop a prototype to automatically collect, analyze and visualize live disaster related social media data from Twitter and also studies how individuals on Twitter react to disaster. Finally they utilize Granger causality analysis [127] to correlate positive/negative sentiments of users on Twitter with distance of Hurricane Sandy approaching the disaster location. Another study on Hurricane Sandy is [112] where proposes a fine-grained sentiment analysis using machine learning classifiers, SVM, Naive Bayes Binary and Multinomial model to distinguish crisis-related micro-posts from irrelevant information. In particular, all of the tweets have been preprocessed to remove irrelevant terms and features and extract general categories from microposts using OpenCalais API[6] and also to handle abbreviations based on noslang dictionary.[7] Then, several features are extracted including bag of words, part of speech tags, n-grams, emoticons, and sentiment features which are obtained from AFINN and SentiWordNet word list. The proposed method identifies seven emotion classes anger, disgust, fear, happiness, sadness, and surprise. Likewise, after collecting Hurricane Sandy relevant tweets, authors in [113] annotate the tweets with one of four sentiment labels, positive, fear, anger and other using SVM and Naïve Bayes. As opposed to [109, 112], Caragea et al. [114] classify tweets sentiments into three classes of positive, negative and neutral associating tweet geo-location to demonstrate general mood on the ground. Authors identify the sentiments of tweets by applying machine learning techniques on the combination of bag of words and sentiment features such as emoticons, acronyms and polarity clues as the feature representation provided to the classifier. Then they assigns locations

---

[6]http://www.opencalais.com.

[7]http://noslang.com.

of tweets to their sentiments to show how sentiments of users change with respect to relative distance from the disaster. Mapping disaster information helps response organizations to have a real time map which shows population's response to disaster using sentiment analysis as a measure.

Using Japan earthquakes as a case study, Vo et al. [117] track crowd emotions during earthquake in microblog posts. The authors consider 6 different emotion classes namely calm, unpleasantness, sadness, anxiety, fear, and relief. Then it deploys two machine learning classifier for identifying earthquake related tweets and also classification of them based on the expressed emotions. By tracking crowd sentiment, Vo et al. [117] conclude that fear and anxiety are two emotions that users express right after the events while calm and unpleasantness are not exposed clearly during small earthquakes but in the large tremor. As an another case study of sentiment analysis for disaster management, authors in [116] apply sentiment analysis methods on Kenya Westgate Mall attack dataset to assess and understand how emergency organizations use social media to improve their responses. They study the difference between sentiment of posts published by emergency organizations and managers and deploy document-level sentiment analysis of TwitterMate which utilizes Alchemy API[8] [39]. Authors also apply statistical t-test to confirm that managers posts have more positive sentiment than organizations' as they are considered more approachable by the public due to the positive language of their posts. Similar to other machine learning techniques, considering eight types of emotions including accusation, anger, disgust, fear, happiness, sadness, surprise, and no emotion, Torkildson et al. [118] classify Gulf Oil Spills event related tweets using ALOE [128] which deployed SVM.

Buscaldi et al. [119] use subjectivity, polarity and irony detection tool named as IRADABE proposed in [129] (which is based on SVM) to identify event related tweets. They assumed that subjective tweets are more likely from a person who is involved in the event, ironic tweets are posted after the event to criticize or blame and event-related tweets have more negative sentiments rather than day-to-day ones. The authors extracted tweets related to 2014 Genoa Floodings to analyze how sentiment analysis and natural language processing methods affect extracting useful information. Another study [120] investigates the impact of disasters on the underlying sentiment of social media streams. In more details, it explores the underlying trends in positive and negative sentiment with respect to the disasters and geographically related sentiments. In particular, it first proposes an uncertainty measure to evaluate the disagreement among multiple sentiment classifiers SentiwordNet [130], SentriStrength [131] and CoreNLP [132] using vote entropy and then uses a committee vote scheme to decide on a Tweet's sentiment class [133]. Due to the vast amount of information published in social media during an event, problem reports and corresponding aid messages were not successfully exchanged most of the time. Therefore, many resources would be wasted and victims would not received necessary aids. So discovering matches between problem reports and aid agencies facilitate communication between victims and humanitarian organizations. Varga et al. [121] try to solve problem of aid and problem report messages recognition and matching

---

[8]http://www.alchemyapi.com/.

using machine learning techniques by considering different features such as lexicon, semantic and sentiment features.

Furthermore, information flow during catastrophic events such as hurricanes is a critical part of disaster management.In an attempt, authors in [115] analyze sentiments of over 50 million tweets before, during and after Hurricane Sandy to study people's behavior changes in Twitter based on the times hurricane reached different cities in terms of number of published posts and expressed sentiments in their tweets. They observe that sentiment of tweets deviate from those of normal ones and conclude that analyzing sentiments from tweets, along with other metrics enables use of sentiment sensing for detecting and locating disasters. Authors use three different methods, Topsy [134], Linguistic Inquiry and Word Count(LIWC) [135] and SentiStrength as sentiment detection algorithms. Other than machine learning techniques, there exist some other methods which are developed for sentiment analysis in disaster management. For example, Chakraborty et al. [122] model the evolution of social dynamics, sentiment, opinion and views of people after a major event. They use swarm intelligence technique to identify the dynamic and time bound property of social events.

3.3   Visual Analytics

Besides disaster related data management in social media, the ability to drawing out important features is essential for better and quick understanding of situation which leads to rapid decision making in critical situations. Moreover, the data produced by social media during disasters and events, is staggering and hard for an individual to process. Therefore, visualization is needed for facilitating pattern discovery [136]. Using visualization, people can find their answers regarding the disasters more quickly and they will figure out where they should look for to find their answers more easily.

Some of the methods that have been introduced previously such as [111, 113, 114, 118, 120], develop visualization tools to increase meaningfulness of analysis. Dong et al. [111] develop a prototype to visualize automatically collected and analyzed live disaster data from Twitter using word clouds, spatial maps and dynamic online activity graphs. Figure 5 is an example of how the proposed visualization tool utilized on Hurricane Sandy dataset. Authors in [113] simply visualize proportion of messages classified with different labels and [114] maps sentiment of disaster-related tweets to their locations and visualize it on a geographical map to have a real time analysis of population's response to disaster. Figure 6 shows maps of tweets sentiment analyzed by [114] in global and regional scale. Another study [118] displays the frequency of emotion labels during different hours of each event while example tweets of each label are also shown to make sense about the event. Authors in [120] develop a visual analytics framework for analyzing and modeling sentiment on disaster related Twitter data. They particularly uses Ebola Twitter dataset in order to evaluate their visual analytic framework. The proposed framework is particularly useful for exploring

Fig. 5 Visualization tool applied on Hurricane Sandy corpus. Reprinted with permission from [111]

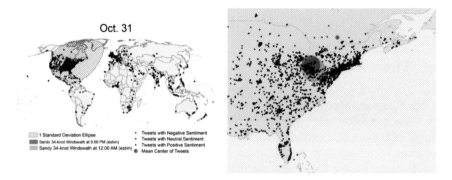

Fig. 6 Maps of positive, neutral, and negative tweets at global and regional scale during Hurricane Sandy. Reprinted with permission from [114]

the patterns of sentiment distributions and also comparing between the distributions. Figure 7 depicts results of sentiment analysis and visualization on Ebola dataset where blurred circles show high sentiment uncertainty.

The main challenge in crisis-related social media data visualization, is the immediate analysis of data for emergency management. For example, Calderon et al. [105] design a visual analytics prototype to support real-time analysis of sentiment in social media in emergency management. The proposed model addresses three domain-specific issues as follows: dealing with high volume of generated social media data during disaster, real-time extraction of relevant features and analysis and visualization of social media streams in the absence of critical attributes e.g. geo-location when real-time analysis is needed. It uses animation to represent time update considering the dynamic nature of stream data. SentiStrength [132] is applied as an algorithm for

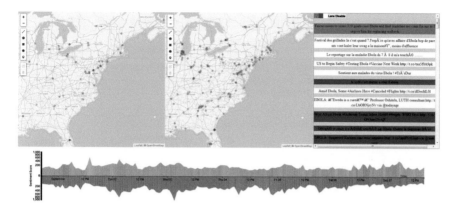

Fig. 7 Sentiment analysis and visualization overview on Ebola Twitter dataset. The two maps are used for geo-comparison view. The list on the right depicts ordered tweets by retweet count. The bottom view is entropy sentiment river. Reprinted with permission from [120]

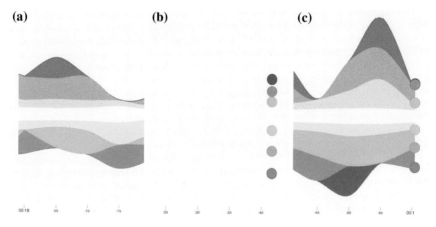

Fig. 8 Screenshots from 3 animations used to represent a stream of emergency-related tweets classified by sentiment analysis. Reprinted with permission from [105]

sentiment analysis. Figure 8 shows screenshots generated by the proposed dynamic visualization prototype.

We summarize the recent works in application of sentiment analysis in disaster relief with highlighting their datasets, approaches and reported accuracies in Table 2. We distinguish between studies with and without visualization abilities.

Table 2 Recent approaches in the application of sentiment analysis in disaster relief along with their datasets and applied methods

Paper	Dataset	Approach	Evaluation	Visualization
[89]	OK Fire; Red River flood 2009–2010; Haiti earthquake	Maximum Entropy	88 % (Accuracy)	
[105]	Hurricane Sandy	SentiStrength	20 % (Accuracy for animation visualization)	✓
[109]	California gas explosion and resulting fires	Bayesian Networks	94.9 % (FScore)	
[110]	Hurricane Irene	Maximum Entropy	84.27 % (Accuracy)	
[111]	Hurricane Sandy	Granger Causality Analysis	–	✓
[112]	Hurricane Sandy	Binary Naïve Bayes	65.8 % (Accuracy)	
[113]	Hurricane Sandy	Support Vector Machine	75.3 % (Accuracy)	✓
[114]	Hurricane Sandy	Support Vector Machine	75.91 % (Accuracy)	✓
[115]	Hurricane Sandy	Topsy, LIWC, SentiStrength	–	
[116]	Kenya Westgate Mall attack	Alchemy API	–	
[117]	Japan Earthquakes	Multinomial Naïve Bayes	87.8 % (FScore)	
[118]	Gulf Oil Spill	Support Vector Machine	91 % (Accuracy)	✓
[119]	2014 Genoa Floodings	IRADABE Tool	66 % (Accuracy)	
[120]	Ebola	SentiWordNet, SentiStrength, CoreNLP	–	✓
[122]	Terrorist attack in Mumbai	Swarm Intelligence Technique	–	

Methods applying visualization are check marked. Some studies may have reported different accuracies in their original papers

4 Discussions and Future Directions

This chapter presented an overview of sentiment analysis in social media and how it could be leveraged for disaster relief during emergencies and disasters. We covered state-of-the-art sentiment analysis approaches and highlight their contributions and then discussed the application of social media and sentiment analysis in disaster relief and situational awareness, while we also detailed applications of visual analytics with an emphasis on sentiment analysis. In this section we discuss some of the challenges facing the studies in sentiment analysis and its application in disaster relief, as well as visual analytics, as potential research directions for further considerations.

Despite few works that have exploited network information such as followee-follower networks and @-networks for more accurate sentiment analysis, majority of the works have not incorporated such useful information; they have considered the problem of sentiment analysis with solely taking into account the lexicon or linguistic based features. One potential avenue of future work is to deploy extra information such as emotional correlation information including spatial-temporal patterns and homophily effect as a measurement of sentiment of social media posts. For example, during winters, people in Florida are expected to be happier than people in Wisconsin. Also, homophily effect suggest that similar people might behave in a more similar way than other people regarding a specific event. Moreover, as the task of collecting sentiment labels in social media is extremely time consuming and tedious, unsupervised or semi-supervised approaches are required to reduce the cost of sentiment classification. Although some studies have addressed this issue, the literature is premature in this realm and still lacks strong contributions. For example, future studies could explore the contributions of other available emotion indications in social media such as product ratings and reviews.

Most studies in disaster relief have used plain machine learning techniques with simple lexicon features. Therefore, more complex machine learning based approaches along with stronger features are required. Furthermore, leveraging the findings of psychological and sociological studies on individuals' behaviors (e.g. hope, fear) during disasters, could be another interesting research direction. This additional information could help better understand people's behaviors and feelings during disasters and also help decision makers know how they can handle this situation. Investigating how this information could be preprocessed to be immediately usable by corresponding authorities, is another interesting future research direction. Furthermore, visualization techniques need to be improved to allow for real time visual analytics of disasters related posts in order to help first responders easily track the changes during disasters and make proper and quick decisions.

Acknowledgments This material is based upon the work supported by, or in part by, Office of Naval Research (ONR) under grant number N000141410095, the NSF under Grant Number 1350573, and the U.S. Department of Homeland Security's VACCINE Center under Award Number 2009-ST-061-CI0001.

References

1. Zafarani, R., Abbasi, M.A., Liu, H.: Social Media Mining: An Introduction. Cambridge University Press, Cambridge (2014)
2. Leskovec, J., Huttenlocher, D., Kleinberg, J.: Predicting positive and negative links in online social networks. In: Proceedings of the 19th International Conference on World Wide Web, pp. 641–650. ACM (2010)
3. Beigi, G., Jalili, M., Alvari, H., Sukthankar, G.: Leveraging community detection for accurate trust prediction. In: ASE International Conference on Social Computing, June 2014
4. Tang, J., Xia, H., Liu, H.: Social recommendation: a review. Soc. Netw. Anal. Min. **3**(4), 1113–1133 (2013)
5. Alvari, H., Hajibagheri, A., Sukthankar, G.: Community detection in dynamic social networks: a game-theoretic approach. In: 2014 IEEE/ACM International Conference on Advances in Social Networks Analysis and Mining (ASONAM), pp. 101–107. IEEE (2014)
6. Goetz, M., Leskovec, J., McGlohon, M., Faloutsos, C.: Modeling blog dynamics (2009)
7. Alvari, H., Hashemi, S., Hamzeh, A.: Discovering overlapping communities in social networks: a novel game-theoretic approach. AI Commun. **26**(2), 161–177 (2013)
8. Lindsay, B.R.: Social Media and Disasters: Current Uses, Future Options, and Policy Considerations. Technical report, Congressional Research Service, September
9. Cobo, A., Parra, D., Navón, J.: Identifying relevant messages in a twitter-based citizen channel for natural disaster situations. In: Proceedings of the 24th International Conference on World Wide Web Companion, pp. 1189–1194 (2015)
10. Gao, H., Barbier, G., Goolsby, R.: Harnessing the crowdsourcing power of social media for disaster relief. IEEE Intell. Syst. **3**, 10–14 (2011)
11. Kumar, S., Morstatter, F., Liu, H.: Twitter Data Analytics. Springer, New York (2014)
12. Liu, B.: Sentiment Analysis and Opinion Mining. Synth. Lect. Hum. Lang. Technol. **5**(1), 1–167. Morgan & Claypool Publishers (2012)
13. Nasukawa, T., Yi, J.: Sentiment analysis: capturing favorability using natural language processing. In: Proceedings of the 2nd International Conference on Knowledge Capture, K-CAP'03, pp. 70–77. ACM, New York (2003)
14. Pang, B., Lee, L.: Opinion mining and sentiment analysis. Found. Trends Inf. Retr. **2**(1–2), 1–135 (2008)
15. Das, S., Chen, M.: Yahoo! for amazon: extracting market sentiment from stock message boards. In: Asia Pacific Finance Association Annual Conference (APFA) (2001)
16. Morinaga, S., Yamanishi, K., Tateishi, K., Fukushima, T.: Mining product reputations on the web. In: Proceedings of the ACM SIGKDD Conference on Knowledge Discovery and Data Mining (KDD), pp. 341–349 (2002). Industry track
17. Pang, B., Lee, L., Vaithyanathan, S.: Thumbs up? sentiment classification using machine learning techniques. In: Proceedings of the ACL-02 Conference on Empirical Methods in Natural Language Processing, vol. 10, pp. 79–86. Association for Computational Linguistics (2002)
18. Tong, R.M.: An operational system for detecting and tracking opinions in on-line discussion. In: Proceedings of the Workshop on Operational Text Classification (OTC) (2001)
19. Turney, P.D.: Thumbs up or thumbs down? Semantic orientation applied to unsupervised classification of reviews (2002). arXiv:cs.LG/0212032
20. Wiebe, J.: Learning subjective adjectives from corpora. In: Kautz, H.A., Porter, B.W. (eds.) AAAI/IAAI, pp. 735–740. AAAI Press/The MIT Press (2000)
21. Carlson, G.: Review of "subjective understanding, computer models of belief systems Jaime G. Carbonell". UMI Research Press, Ann Arbor Michigan Copyright 1981. SIGART Bull. **80**, 12 (1982)
22. Wilks, Y., Bien, J.: Beliefs, points of view and multiple environments. In: Proceedings of the International NATO Symposium on Artificial and Human Intelligence, pp. 147–171. Elsevier North-Holland Inc., New York (1984)

23. Hearst, M.: Direction-based text interpretation as an information access refinement. In: Jacobs, P. (ed.) Text-Based Intelligent Systems, pp. 257–274. Lawrence Erlbaum Associates (1992)

24. Huettner, A., Subasic, P.: Fuzzy typing for document management. In: ACL 2000 Companion Volume: Tutorial Abstracts and Demonstration Notes, pp. 26–27 (2000)

25. Sack, W.: On the computation of point of view. In: Proceedings of the Twelfth National Conference on Artificial Intelligence, AAAI'94. American Association for Artificial Intelligence, Menlo Park, CA, USA, vol. 2, p. 1488 (1994)

26. Wiebe, J., Bruce, R.: Probabilistic classifiers for tracking point of view. In: Working Notes of the AAAI Spring Symposium on Empirical Methods in Discourse Interpretation (1995)

27. Wiebe, J.M.: Identifying subjective characters in narrative. In: Proceedings of the 13th Conference on Computational Linguistics, COLING'90, vol. 2, pp. 401–406. Association for Computational Linguistics, Stroudsburg (1990)

28. Wiebe, J.M.: Tracking point of view in narrative. Comput. Linguist. **20**(2), 233–287 (1994)

29. Wiebe, J.M., Bruce, R.F., O'Hara, T.P.: Development and use of a gold-standard data set for subjectivity classifications. In: Proceedings of the 37th Annual Meeting of the Association for Computational Linguistics on Computational Linguistics, ACL'99, pp. 246–253. Association for Computational Linguistics, Stroudsburg (1999)

30. Wiebe, J., Rapaport, W.J.: A computational theory of perspective and reference in narrative. In: Hobbs, J.R. (ed.) ACL, pp. 131–138 (1988)

31. Kouloumpis, E., Wilson, T., Moore, J.: Twitter sentiment analysis: the good the bad and the omg!. ICWSM **11**, 538–541 (2011)

32. Qiu, G., Liu, B., Bu, J., Chen, C.: Expanding domain sentiment lexicon through double propagation. In: Boutilier, C. (ed.) IJCAI, pp. 1199–1204 (2009)

33. Thelwall, M., Buckley, K., Paltoglou, G.: Sentiment in twitter events. J. Am. Soc. Inf. Sci. Technol. **62**(2), 406–418 (2011)

34. Tan, C., Lee, L., Tang, J., Jiang, L., Zhou, M., Li, P.: User-level sentiment analysis incorporating social networks. In: Proceedings of the 17th ACM SIGKDD International Conference on Knowledge Discovery and Data Mining, pp. 1397–1405. ACM (2011)

35. Hu, X., Tang, L., Tang, J., Liu, H.: Exploiting social relations for sentiment analysis in microblogging. In: Proceedings of the Sixth ACM International Conference on Web Search and Data Mining, pp. 537–546. ACM (2013)

36. Zhao, J., Dong, L., Wu, J., Xu, K.: Moodlens: an emoticon-based sentiment analysis system for Chinese tweets. In: Proceedings of the 18th ACM SIGKDD International Conference on Knowledge Discovery and Data Mining, pp. 1528–1531. ACM (2012)

37. Hu, X., Tang, J., Gao, H., Liu, H.: Unsupervised sentiment analysis with emotional signals. In: Proceedings of the 22nd International Conference on World Wide Web, pp. 607–618. International World Wide Web Conferences Steering Committee (2013)

38. Tang, D., Wei, F., Yang, N., Zhou, M., Liu, T., Qin, B.: Learning sentiment-specific word embedding for twitter sentiment classification. In: Proceedings of the 52nd Annual Meeting of the Association for Computational Linguistics, ACL 2014, June 22–27, 2014, Baltimore, MD, USA, Long Papers, vol. 1, pp. 1555–1565 (2014)

39. Saif, H., He, Y., Alani, H.: Semantic sentiment analysis of twitter. In: Cudré-Mauroux, P., et al. (eds.) The Semantic Web-ISWC 2012, pp. 508–524. Springer, Berlin (2012)

40. Kucuktunc, O., Cambazoglu, B.B., Weber, I., Ferhatosmanoglu, H.: A large-scale sentiment analysis for yahoo! answers. In: Proceedings of the fifth ACM International Conference on Web Search and Data Mining, pp. 633–642. ACM (2012)

41. Wu, L., Zhou, Y., Tan, F., Yang, F., Li, J.: Generating syntactic tree templates for feature-based opinion mining. In: Tang, J., King, I., Chen, L., Wang, J. (eds.) Advanced Data Mining and Applications, pp. 1–12. Springer, Berlin (2011)

42. Schapire, E.R., Singer, Y.: Boostexter: a boosting-based system for text categorization. Mach. Learn. **39**(2), 135–168 (2000)

43. Thelwall, M., Buckley, K., Paltoglou, G., Cai, D., Kappas, A.: Sentiment strength detection in short informal text. J. Am. Soc. Inf. Sci. Technol. **61**(12), 2544–2558 (2010)

44. Hu, M., Liu, B.: Mining and summarizing customer reviews. In: Proceedings of the Tenth ACM SIGKDD International Conference on Knowledge Discovery and Data Mining, pp. 168–177. ACM (2004)
45. Fu, B., Lin, J., Li, L., Faloutsos, C., Hong, J., Sadeh, N.: Why people hate your app: making sense of user feedback in a mobile app store. In: Proceedings of the 19th ACM SIGKDD International Conference on Knowledge Discovery and Data Mining, pp. 1276–1284. ACM (2013)
46. Bollen, J., Mao, H., Zeng, X.: Twitter mood predicts the stock market. J. Comput. Sci. **2**(1), 1–8 (2011)
47. Hennig, P., Berger, P., Lehmann, C., Mascher, A., Meinel, C.: Accelerate the detection of trends by using sentiment analysis within the blogosphere. In: 2014 IEEE/ACM International Conference on Advances in Social Networks Analysis and Mining (ASONAM), pp. 503–508, Aug 2014
48. Manning, C.D., Schütze, H.: Foundations of Statistical Natural Language Processing. MIT Press, Cambridge (1999)
49. Nigam, K., Lafferty, J., McCallum, A.: Using maximum entropy for text classification. In: IJCAI-99 Workshop on Machine Learning for Information Filtering, vol. 1, pp. 61–67 (1999)
50. Cristianini, N., Shawe-Taylor, J.: An introduction to support vector Machines and Other Kernel-Based Learning Methods. Cambridge University Press, Cambridge (2000)
51. Wilson, T., Hoffmann, P., Somasundaran, S., Kessler, J., Wiebe, J., Choi, Y., Cardie, C., Riloff, E., Patwardhan, S.: Opinionfinder: a system for subjectivity analysis. In: Proceedings of HLT/EMNLP on Interactive Demonstrations, pp. 34–35. Association for Computational Linguistics (2005)
52. Manning, C.D., Surdeanu, M., Bauer, J., Finkel, J., Bethard, S.J., McClosky, D.: The stanford corenlp natural language processing toolkit. In: Proceedings of 52nd Annual Meeting of the Association for Computational Linguistics: System Demonstrations, pp. 55–60 (2014)
53. McPherson, M., Smith-Lovin, L., Cook, J.M.: Birds of a feather: Homophily in Social Networks. Annual Review of Sociology, pp. 415–444 (2001)
54. Abelson, R.P.: Whatever became of consistency theory?. Pers. Soc. Psychol. Bull. Sage Publications (1983)
55. Hatfield, E., Cacioppo, J.T., Rapson, R.L.: Emotional Contagion. Cambridge University Press, Cambridge (1994)
56. Friedman, J., Hastie, T., Tibshirani, R.: The Elements of Statistical Learning. Springer Series in Statistics, vol. 1. Springer, Berlin (2001)
57. Go, A., Bhayani, R., Huang, L.: Twitter sentiment classification using distant supervision. CS224N Project report, Stanford, pp. 1–12 (2009)
58. Read, J.: Using emoticons to reduce dependency in machine learning techniques for sentiment classification. In: Proceedings of the ACL Student Research Workshop, pp. 43–48. Association for Computational Linguistics (2005)
59. Ding, C., Li, T., Peng, W., Park, H.: Orthogonal nonnegative matrix t-factorizations for clustering. In: Proceedings of the 12th ACM SIGKDD International Conference on Knowledge Discovery and Data Mining, pp. 126–135. ACM (2006)
60. Friedman, J.H.: Greedy function approximation: a gradient boosting machine. Ann. Stat. 1189–1232 (2001)
61. Ye, J., Chow, J.-H., Chen, J., Zheng, Z.: Stochastic gradient boosted distributed decision trees. In: Proceedings of the 18th ACM Conference on Information and Knowledge Management, pp. 2061–2064. ACM (2009)
62. Hu, Y., Wang, F., Kambhampati, S.: Listening to the crowd: automated analysis of events via aggregated twitter sentiment. In: Proceedings of the Twenty-Third International Joint Conference on Artificial Intelligence, pp. 2640–2646. AAAI Press (2013)
63. Yuheng, H., John, A., Wang, F., Kambhampati, S.: Et-lda: joint topic modeling for aligning events and their twitter feedback. AAAI **12**, 59–65 (2012)
64. Balahur, A., Steinberger, R., Kabadjov, M., Zavarella, V., Van Der Goot, E., Halkia, M., Pouliquen, B., Belyaeva, J.: Sentiment analysis in the news (2013). arXiv:1309.6202

65. Chae, J., Thom, D., Bosch, H., Jang, Y., Maciejewski, R., Ebert, D.S., Ertl, T.: Spatiotemporal social media analytics for abnormal event detection and examination using seasonal-trend decomposition. In: 2012 IEEE Conference on Visual Analytics Science and Technology (VAST), pp. 143–152. IEEE (2012)
66. Zhao, J., Cao, N., Wen, Z., Song, Y., Lin, Y.-R., Collins, C.M.: # fluxflow: visual analysis of anomalous information spreading on social media. IEEE Trans.Vis. Comput. Graph. **20**(12), 1773–1782 (2014)
67. Yafeng, L., Wang, F., Maciejewski, R.: Business intelligence from social media: a study from the vast box office challenge. IEEE Comput. Graph. Appl. **34**(5), 58–69 (2014)
68. Gregory, M.L., Payne, D., McColgin, D., Cramer, N., Love, D.: Visual analysis of weblog content. In: ICWSM (2007)
69. Gregory, M.L., Chinchor, N., Whitney, P., Carter, R., Hetzler, E., Turner, A.: User-directed sentiment analysis: visualizing the affective content of documents. In: Proceedings of the Workshop on Sentiment and Subjectivity in Text, pp. 23–30. Association for Computational Linguistics (2006)
70. Gamon, M., Aue, A., Corston-Oliver, S., Ringger, E.: Pulse: mining customer opinions from free text. In: Famili, A.F., Kok, J.N., Peña, J.M., Siebes, A., Feelders, A. (eds.) Advances in Intelligent Data Analysis VI, pp. 121–132. Springer, Berlin (2005)
71. Smith, M.A., Fiore, A.T.: Visualization components for persistent conversations. In: Proceedings of the SIGCHI Conference on Human Factors in Computing Systems, pp. 136–143. ACM (2001)
72. Nigam, K., McCallum, A.K., Thrun, S., Mitchell, T.: Text classification from labeled and unlabeled documents using em. Mach. Learn. **39**(2–3), 103–134 (2000)
73. Oelke, D., Hao, M., Rohrdantz, C., Keim, D., Dayal, U., Haug, L.-E., Janetzko, H. et al.: Visual opinion analysis of customer feedback data. In: IEEE Symposium on Visual Analytics Science and Technology, 2009. VAST 2009, pp. 187–194. IEEE (2009)
74. Duan, D., Qian, W., Pan, S., Shi, L., Lin, C.: Visa: a visual sentiment analysis system. In: Proceedings of the 5th International Symposium on Visual Information Communication and Interaction, pp. 22–28. ACM (2012)
75. Wei, F., Liu, S., Song, Y., Pan, S., Zhou, M.X., Qian, W., Shi, L., Tan, L., Zhang, Q.: Tiara: a visual exploratory text analytic system. In: Proceedings of the 16th ACM SIGKDD International Conference on Knowledge Discovery and Data Mining, pp. 153–162. ACM (2010)
76. Adams, B., Phung, D., Venkatesh, S.: Eventscapes: visualizing events over time with emotive facets. In: Proceedings of the 19th ACM International Conference on Multimedia, pp. 1477–1480. ACM (2011)
77. Marcus, A., Bernstein, M.S., Badar, O., Karger, D.R., Madden, S., Miller, R.C.: Twitinfo: aggregating and visualizing microblogs for event exploration. In: Proceedings of the SIGCHI Conference on Human Factors in Computing Systems, pp. 227–236. ACM (2011)
78. Hubmann-Haidvogel, A., Brasoveanu, A.M.P., Scharl, A., Sabou, M., Gindl, S.: Visualizing contextual and dynamic features of micropost streams (2012)
79. Shook, E., Leetaru, K., Cao, G., Padmanabhan, A., Wang, S.: Happy or not: generating topic-based emotional heatmaps for culturomics using cybergis. In: 2012 IEEE 8th International Conference on E-Science (e-Science), pp. 1–6, Oct 2012
80. Guha-Sapir, D., Vos, F., Below, R., Ponserre, S.: Annual disaster statistical review 2010. Centre for Research on the Epidemiology of Disasters (2011)
81. Mejova, Y., Weber, I., Macy, M.W.: Twitter: A Digital Socioscope. Cambridge University Press, Cambridge (2015)
82. Glaser, M.: California wildfire coverage by local media, blogs, twitter, maps and more. PBS MediaShift (2007)
83. Starbird, K., Palen, L., Hughes, A.L., Vieweg, S.: Chatter on the red: what hazards threat reveals about the social life of microblogged information. In: Proceedings of the 2010 ACM Conference on Computer Supported Cooperative Work, pp. 241–250. ACM (2010)
84. Sutton, J., Palen, L., Shklovski, I.: Backchannels on the front lines: emergent uses of social media in the 2007 southern california wildfires. In: Proceedings of the 5th International ISCRAM Conference, Washington, DC, pp. 624–632 (2008)

85. Vieweg, S., Hughes, A.L., Starbird, K., Palen, L.: Microblogging during two natural hazards events: what twitter may contribute to situational awareness. In: Proceedings of the SIGCHI Conference on Human Factors in Computing Systems, pp. 1079–1088. ACM (2010)
86. Qu, Y., Wu, P.F., Wang, X.: Online community response to major disaster: a study of tianya forum in the 2008 sichuan earthquake. In: 42nd Hawaii International Conference on System Sciences, 2009. HICSS'09, pp. 1–11. IEEE (2009)
87. Qu, Y., Huang, C., Zhang, P., Zhang, J.: Microblogging after a major disaster in china: a case study of the 2010 yushu earthquake. In: Proceedings of the ACM 2011 Conference on Computer Supported Cooperative Work, pp. 25–34. ACM (2011)
88. Sakaki, T., Toriumi, F., Uchiyama, K., Matsuo, Y., Shinoda, K., Kazama, K., Kurihara, S., Noda, I.: The possibility of social media analysis for disaster management. In: Humanitarian Technology Conference (R10-HTC), 2013 IEEE Region 10, pp. 238–243. IEEE (2013)
89. Verma, S., Vieweg, S., Corvey, W.J., Palen, L., Martin, J.H., Palmer, M., Schram, A., Anderson, K.M.: Natural language processing to the rescue? Extracting "situational awareness" tweets during mass emergency. In: ICWSM (2011)
90. Terpstra, T., de Vries, A., Stronkman, R., Paradies, G.L.: Towards a realtime twitter analysis during crises for operational crisis management. In: Proceedings of the 9th International ISCRAM Conference, Vancouver, Canada (2012)
91. Morstatter, F., Lubold, N., Pon-Barry, H., Pfeffer, J., Liu, H.: Finding eyewitness tweets during crises (2014). arXiv:1403.1773
92. Imran, M., Elbassuoni, S., Castillo, C., Diaz, F., Meier, P.: Practical extraction of disaster-relevant information from social media. In: Proceedings of the 22nd International Conference on World Wide Web Companion, pp. 1021–1024. International World Wide Web Conferences Steering Committee (2013)
93. Sakaki, T., Okazaki, M., Matsuo, Y.: Earthquake shakes twitter users: real-time event detection by social sensors. In: Proceedings of the 19th International Conference on World Wide Web, pp. 851–860. ACM (2010)
94. Li, H., Guevara, N., Herndon, N., Caragea, D., Neppalli, K., Caragea, C., Squicciarini, A., Tapia, A.H.: Twitter mining for disaster response: a domain adaptation approach
95. Imran, M., Elbassuoni, S.M., Castillo, C., Diaz, F., Meier, P.: Extracting information nuggets from disaster-related messages in social media. In: Proceedings of ISCRAM, Baden-Baden, Germany (2013)
96. Chowdhury, S.R., Imran, M., Asghar, M.R., Amer-Yahia, S., Castillo, C.: Tweet4act: using incident-specific profiles for classifying crisis-related messages. In: 10th International ISCRAM Conference (2013)
97. Truong, B., Caragea, C., Squicciarini, A., Tapia, A.H.: Identifying valuable information from twitter during natural disasters. Proc. Am. Soc. Inf. Sci. Technol. **51**(1), 1–4 (2014)
98. Ashktorab, Z., Brown, C., Nandi, M., Culotta, A.: Tweedr: Mining twitter to inform disaster response. In: Proceedings of ISCRAM (2014)
99. Sen, A., Rudrat, K., Ghosh, S.: Extracting situational awareness from microblogs during disaster events
100. Mendoza, M., Poblete, B., Castillo, C.: Twitter under crisis: can we trust what we rt? In: Proceedings of the First Workshop on Social Media Analytics, pp. 71–79. ACM (2010)
101. Athanasia, N., Stavros, P.T.: Twitter as an instrument for crisis response: the typhoon haiyan case study
102. MacEachren, A.M., Robinson, A.C., Jaiswal, A., Pezanowski, S., Savelyev, A., Blanford, J., Mitra, P.: Geo-twitter analytics: applications in crisis management. In: 25th International Cartographic Conference, pp. 3–8 (2011)
103. Kumar, S., Barbier, G., Abbasi, M.A., Liu, H.: Tweettracker: an analysis tool for humanitarian and disaster relief. In: ICWSM (2011)
104. Morstatter, F., Kumar, S., Liu, H., Maciejewski, R.: Understanding twitter data with tweetxplorer. In: Proceedings of the 19th ACM SIGKDD International Conference on Knowledge Discovery and Data Mining, pp. 1482–1485. ACM (2013)

105. Calderon, N., Arias-Hernandez, R., Fisher, B., et al.: Studying animation for real-time visual analytics: a design study of social media analytics in emergency management. In: 2014 47th Hawaii International Conference on System Sciences (HICSS), pp. 1364–1373. IEEE (2014)
106. MacEachren, A.M., Jaiswal, A., Robinson, A.C., Pezanowski, S., Savelyev, A., Mitra, P., Zhang, X., Blanford, J.: Senseplace2: geotwitter analytics support for situational awareness. In: 2011 IEEE Conference on Visual Analytics Science and Technology (VAST), pp. 181–190. IEEE (2011)
107. Ren, D., Zhang, X., Wang, Z., Li, J., Yuan, X.: Weiboevents: a crowd sourcing weibo visual analytic system. In: Pacific Visualization Symposium (PacificVis), 2014 IEEE, pp. 330–334. IEEE (2014)
108. Abel, F., Hauff, C., Houben, G.-J., Stronkman, R., Tao, K.: Semantics + filtering + search = twitcident. exploring information in social web streams. In: Proceedings of the 23rd ACM Conference on Hypertext and Social Media, HT'12, pp. 285–294. ACM, New York (2012)
109. Nagy, A., Stamberger, J.: Crowd sentiment detection during disasters and crises. In Proceedings of the 9th International ISCRAM Conference, pp. 1–9 (2012)
110. Mandel, B., Culotta, A., Boulahanis, J., Stark, D., Lewis, B., Rodrigue, J.: A demographic analysis of online sentiment during hurricane irene. In: Proceedings of the Second Workshop on Language in Social Media, pp. 27–36. Association for Computational Linguistics (2012)
111. Dong, H., Halem, M., Zhou, S.: Social media data analytics applied to hurricane sandy. In: 2013 International Conference on Social Computing (SocialCom), pp. 963–966. IEEE (2013)
112. Schulz, A., Thanh, T., Paulheim, H., Schweizer, I.: A fine-grained sentiment analysis approach for detecting crisis related microposts. In: ISCRAM 2013 (2013)
113. Brynielsson, J., Johansson, F., Westling, A.: Learning to classify emotional content in crisis-related tweets. In: 2013 IEEE International Conference on Intelligence and Security Informatics (ISI), pp. 33–38. IEEE (2013)
114. Caragea, C., Squicciarini, A., Stehle, S., Neppalli, K., Tapia, A.: Mapping moods: geo-mapped sentiment analysis during hurricane sandy. In: Proceedings of ISCRAM (2014)
115. Kryvasheyeu, Y., Chen, H., Moro, E., Van Hentenryck, P., Cebrian, M.: Performance of social network sensors during hurricane sandy. PLoS one **10**, e0117288–e0117288 (2015)
116. Simon, T., Goldberg, A., Aharonson-Daniel, L., Leykin, D., Adini, B.: Twitter in the cross firethe use of social media in the westgate mall terror attack in Kenya (2014)
117. Vo, B.K.H., Collier, N.: Twitter emotion analysis in earthquake situations. Int. J. Comput. Linguist. Appl. **4**(1), 159–173 (2013)
118. Torkildson, M.K., Starbird, K., Aragon, C.: Analysis and visualization of sentiment and emotion on crisis tweets. In: Luo, Y. (ed.) Cooperative Design, Visualization, and Engineering, pp. 64–67. Springer, New York (2014)
119. Buscaldi, D., Hernandez-Farias, I.: Sentiment analysis on microblogs for natural disasters management: a study on the 2014 genoa floodings. In: Proceedings of the 24th International Conference on World Wide Web Companion. International World Wide Web Conferences Steering Committee, pp. 1185–1188 (2015)
120. Lu, Y., Hu, X., Wang, F., Kumar, S., Liu, H., Maciejewski, R.: Visualizing social media sentiment in disaster scenarios. In: Proceedings of the 24th International Conference on World Wide Web Companion. International World Wide Web Conferences Steering Committee, pp. 1211–1215 (2015)
121. Varga, I., Sano, M., Torisawa, K., Hashimoto, C., Ohtake, K., Kawai, T., Jong-Hoon, O., De Saeger, S.: Aid is out there: Looking for help from tweets during a large scale disaster. In: ACL, vol. 1, pp. 1619–1629 (2013)
122. Chakraborty, B., Banerjee, S.: Modeling the evolution of post disaster social awareness from social web sites. In: 2013 IEEE International Conference on Cybernetics (CYBCONF), pp. 51–56. IEEE (2013)
123. Riloff, E., Wiebe, J.: Learning extraction patterns for subjective expressions. In: Proceedings of the 2003 Conference on Empirical Methods in Natural Language Processing, pp. 105–112. Association for Computational Linguistics (2003)

124. Wiebe, J., Riloff, E.: Creating subjective and objective sentence classifiers from unannotated texts. In: Gelbukh, A. (ed.) Computational Linguistics and Intelligent Text Processing, pp. 486–497. Springer, Berlin (2005)
125. Esuli, A., Sebastiani, F.: Sentiwordnet: a publicly available lexical resource for opinion mining. In: Proceedings of LREC, vol. 6, pp. 417–422 (2006)
126. Nielsen, F.Å.: A new anew: evaluation of a word list for sentiment analysis in microblogs (2011). arXiv:1103.2903
127. Granger, C.W.J.: Investigating causal relations by econometric models and cross-spectral methods. Econometrica: J. Econometric Soc. 424–438 (1969)
128. Brooks, M., Kuksenok, K., Torkildson, M.K., Perry, D., Robinson, J.J., Scott, T.J., Anicello, O., Zukowski, A., Harris, P., Aragon, C.R.: Statistical affect detection in collaborative chat. In: Proceedings of the 2013 Conference on Computer Supported Cooperative Work, pp. 317–328. ACM (2013)
129. Hernandez-Farias, I., Buscaldi, D., Priego-Sánchez, B.: Iradabe: adapting english lexicons to the Italian sentiment polarity classification task. In: First Italian Conference on Computational Linguistics (CLiC-it 2014) and the Fourth International Workshop EVALITA2014, pp. 75–81 (2014)
130. Baccianella, S., Esuli, A., Sebastiani, F.: Sentiwordnet 3.0: an enhanced lexical resource for sentiment analysis and opinion mining. In: LREC, vol. 10, pp. 2200–2204 (2010)
131. Socher, R., Perelygin, A., Wu, J.Y., Chuang, J., Manning, C.D., Ng, A.Y., Potts, C.: Recursive deep models for semantic compositionality over a sentiment treebank. In: Proceedings of the Conference on Empirical Methods in Natural Language Processing (EMNLP), vol. 1631, pp. 1642 (2013)
132. Thelwall, M., Buckley, K., Paltoglou, G.: Sentiment strength detection for the social web. J. Am. Soc. Inf. Sci. Technol. 63(1), 163–173 (2012)
133. Dagan, I., Engelson, S.P.: Committee-based sampling for training probabilistic classifiers. In: Proceedings of the Twelfth International Conference on Machine Learning, pp. 150–157 (1995)
134. Topsy Labs: www.topsylabs.com (2012). Accessed 20 Nov 2012
135. Pennebaker, J.W., Francis, M.E., Booth, R.J.: Linguistic inquiry and word count: Liwc 2001. Mahway: Lawrence Erlbaum Associates, p. 71 (2001)
136. Best, D.M., Bruce, J.R., Dowson, S.T., Love, O.J., McGrath, L.R.: Web-based visual analytics for social media. Technical report, Pacific Northwest National Laboratory (PNNL), Richland, WA (US) (2012)

Big Data Sentiment Analysis for Brand Monitoring in Social Media Streams by Cloud Computing

Francesco Benedetto and Antonio Tedeschi

Abstract The rapid growth of the World Wide Web and social media allows users playing an active role in the contents' creation process. Users can evaluate the brands' reputation and quality exploiting the information provided by new marketing channels, such as social media, social networks, and electronic commerce (or e-commerce). Consequently, enterprises need to spot and analyze these digital data in order to improve their reputation among the consumers. The aim of this chapter is to highlight the common approaches of sentiment analysis in social media streams and the related issues with the cloud computing, providing the readers with a deep understanding of the state of the art solutions.

Keywords Big data analyses · Brand monitoring · Cloud-based processing · Computational intelligence · Sentiment analysis · Social media stream

1 Introduction

With the rapid development of Computer Science and Internet technology, users have changed both their communication channels and the way to gather information through the Web. In the era of Web 2.0, users start playing an active role in the creation of digital contents, by using blog and social media [1]. Exploiting the services provided by these new generations of platforms, such as Facebook and Twitter, users first connect each other and then share every kind of information. The information shared between users contains not only updates about their private life, but also updates about their opinions on products, brands (e.g., consumer reviews),

F. Benedetto (✉)
Department of Economics, University of Rome "Roma TRE",
via Vito Volterra, 62, 00146 Rome, Italy
e-mail: francesco.benedetto@uniroma3.it

A. Tedeschi
Department of Engineering, University of Rome "Roma TRE",
via Vito Volterra, 62, 00146 Rome, Italy
e-mail: antonio.tedeschi@uniroma3.it

© Springer International Publishing Switzerland 2016
W. Pedrycz and S.-M. Chen (eds.), *Sentiment Analysis and Ontology Engineering*,
Studies in Computational Intelligence 639, DOI 10.1007/978-3-319-30319-2_14

and/or news [2]. Users exploit these social platforms in order to gather information about any kind of topic, and they influence each other's opinions. Social media now represent one of the main way to access information around the globe. Therefore, sentiment analysis (also known as opinion mining) has become a popular research topic for both Academics and Companies. It refers to the use of natural language processing, text analysis and computational linguistics to identify and extract subjective information in source materials. Sentiment analysis aims at determining the attitude of a speaker or a writer with respect to some topic or the overall contextual polarity of a document. The attitude may be his or her judgment or evaluation, affective state (the emotional state of the author when writing), or the intended emotional communication (the emotional effect the author wishes to have on the reader). Hence, the knowledge obtained from capturing the sentiments and the emotional states of textual information has opened new ways to understand users' needs and feelings on certain topics. Unfortunately, user interactions on social media platforms and mobile applications is usually a source of big data [3]. In fact, big social data is usually high volume, high velocity, and high variety information assets that require new forms of processing to enable enhanced sentiment analysis, trend discovery and marketing prediction. The distillation of knowledge from such a large amount of unstructured information, however, is an extremely difficult task, as the contents of today's Web are perfectly suitable for human consumption, but remain hardly accessible to machines. Big social data analysis grows out of this need and combines disciplines such as social network analysis, multimedia management, social media analytics, trend discovery, and opinion mining, all under the big heading of Computation Intelligence (CI). In particular, CI is defined (by the Montreal Chapter of the IEEE Computational Intelligence Society) as "*a branch of the study of artificial intelligence. Computational intelligence research aims to use learning, adaptive, or evolutionary computation to create programs that are, in some sense, intelligent*". More generally, the IEEE Computational Intelligence Society defines its subjects of interest as neural networks, fuzzy systems and evolutionary computation, including swarm intelligence. This chapter tries to explore how Computation intelligence can enable a more efficient passage from (unstructured) social information to (structured) machine-processable data, in the field of big social data analysis (that is inherently interdisciplinary and spans areas such as machine learning, graph mining, information retrieval, knowledge-based systems, linguistics, common-sense reasoning, natural language processing, and big data computing).

Managing and analyzing big data require an amount of computing resources that are too expensive to buy, especially for Small and Medium Enterprises (SMEs). Such a need led to the definition of the Cloud Computing technology. It relies on the sharing of computing resources without forcing enterprises and users to buy the infrastructure network. Enterprises can rent the computing resources from the cloud provider, according to their needs [4]. Through the cloud computing technology and its service models, the social media enterprises can easily manage users' data, increasing the quality of their services and reducing the data and infrastructure networks' management costs, exploiting the pay-per-use model [5]. Recently, SMEs have started to recognize the benefit derived from the use of both cloud computing technology and social media data analyses. The user-generated contents are becoming a useful resource to understand what people think about a certain brand. A brand is commonly associated with products and services offered by an enterprise for a certain time interval, shaping deeply anchored and clear images in the mind of the end-consumers [6]. Hence, through the analysis of the contents produced by social users, enterprises can: (i) manage their brand; (ii) create ad-hoc marketing campaigns and advertisements analyzing users' feelings [7]. The main benefit derived by these analyses allows enterprises to improve their brands' reputation (brand monitoring) and, consequentially, increase their electronic commerce (or e-commerce) satisfying the consumers' needs. Thanks to these new marketing channels, social media users have also become consumers that can gather information about brands' reputation and quality instantaneously analyzing the data provided by social media, blogs and e-commerce systems. Actually, brands are not only the products and services portfolio of an enterprise but they also represent an emotional differentiation in the form of orientation and the generation of trust. Due to the lack of personal ties, the wealth of information available on the Internet can lead to a depersonalization of relationships and follow the decrease of consumers' loyalty [6]. In addition, negative examples perpetuated in the press can increase the perceived risks of online purchases, especially in the case of less well-known brands. As a matter of fact, with the proliferation of social media and online news channel as principal provider of news and opinions, the traditional market monitoring, based on human-efforts, has become virtually impossible. In fact, it is impossible for individuals to manually spot and analyze all information of particular importance for global large-scale corporations. This need led to the definition of automated market intelligence and tools that allow the monitoring of a brand's reputation in a mechanized fashion. Hence, it has become possible for enterprises to automatically identify the main and upcoming topics of interest and to monitor the reputation of its own brands as well as its competitors [11]. These reputation platforms are based on the idea that consumers' emotions are valid indicators of their satisfaction about the brand and can be used as the basis for customers' second buying. Through these new investigation methods, online businesses should pay more attention to reinforcing the refinement of the brand value and experience design, thus improving customer value [12].

The main contribution of this chapter is to highlight the common approaches of sentiment analysis in social media streams and the related issues with the cloud computing, providing the readers with a deep understanding of the state of the art solutions. We will first discuss about the motivations behind the sentiment analysis approach, describing both the operating scenarios and the related benefits. Then, we will illustrate the main advantages provided by cloud computing in order to manage big data, as provided by social media streams. The third paragraph will describe in details the main categories of computational intelligence approaches used in the literature to compute the polarity related to a sentence or a post on a social media. In particular, we will analyze the knowledge-based approaches, as well as the machine learning techniques. These categories provide Natural Language Processing (NLP), and statistics to extract, identify, and characterize the sentiment content of a text unit. These social analytics tools have become essential in many different application areas, spanning from business, Internet, social networking, politics, and economy. Hence, a wide literature review will be presented in order to give to the reader all the sources of information useful for any in-depth study. In the fourth paragraph, the conventional steps for text pre-processing will be detailed, as well as the methods to compute users' popularity (and how popular users can influence the opinion of other users). Then, with application to brand monitoring campaigns we will introduce some case studies and evaluate the tweets' polarity, also discussing *popularity versus polarity scores* (i.e., how the polarity of textual information modifies by taking into account the popularity of users). The chapter will end with the authors' view about the future developments of opinion mining and sentiment analysis approaches to recognize, interpret and process opinions and sentiments in natural language.

2 Basic Frameworks About Sentiment Analysis

2.1 Motivations and Application Scenarios

Understanding '*what other people think*' has always been a key concept in decision-making processes [13]. Our perception of reality and our choices are generally influenced by other people's opinions and before making a decision we usually ask a friend or an expert for a piece of advice. A similar behavior, but on a larger scale, is adopted by any company that has to deal with consumers and their opinions. Not too many years ago, companies had to conduct surveys or create ad-hoc focus groups in order to gather information and customers' opinions about a given product or service. Since 2000, however, the development of social media and e-commerce web-sites has rapidly changed such a scenario. The advent of consumer-generated media and the proliferation of online news channels on the Web have made the traditional market monitoring model based on human-efforts virtually impossible. Nowadays, companies and consumers can easily and rapidly gather opinions on virtually anything, by simply using the web. In this new scenario, sentiment analysis and opinion mining have become two important research area of natural processing language and have proven to be valid assets in analyzing and understanding what other people think about a given topic, service or product [14].

Recognizing the necessity and the benefits deriving from such process, several companies have promptly developed their own sentiment analysis monitoring tool in order to keep record of their reputation on social media. The demand for this kind of tools has led to the definition of a large number of start-ups and enterprises focused on the creation of new social monitoring services designed to help companies in their monitoring activities. The aim of these tools is therefore not only to evaluate companies' reputation but also to provide a number of value added services, such as brand monitoring and facilities to social events and political elections [14]. For instance, the authors in [15] and [16] investigate the opportunity to involve social media data in order to predict real-world outcomes of future events, which is a crucial task in the present-day business environment. Predicting the success or failure of a product sale, for example, would allow companies to adjust their production planning accordingly [17]. Reliable projections about future events would prove useful in other fields as well. In fact, the authors in [18] and [19] investigate whether or not it would be viable to exploit social media to predict the outcome of political elections. Both these works gather opinions taken from a pool of users on Twitter and develop an ad-hoc classifier for politically-oriented tweets in order to make an in-depth study about the interaction of emotion and sharing. A different field of application was investigated in [20] and [21], defining two different approaches to evaluate consumers' opinions in the domain of tourism. Duan et al. [20] propose an interesting sentiment analysis approach to decompose users' reviews into five dimensions for the evaluation of hotel services. On the other hand, Marrese-Taylor et al. [21] presents a tool, namely OpinionZoom, to help users managing and understanding product reviews in the tourist field. It is interesting to note that both approaches require to pre-process the collected and stored data before detecting the related polarity.

Brand monitoring activities on social media and web platforms have become more and more valuable also for SMEs that want to improve their e-commerce. The authors [22] analyze the state of the art of sentiment analysis tools and systems created by companies and research groups. A number of commercial tools, such as Brandwatch [23], Synthesio [24] and Radian6 [25], are designed for the analysis of web-business marketing. They allow companies to specify a set of keywords to look for in social media to collect valuable data. Keywords are usually based on the name of one of the company's brands/products or on the name of their competitors. Another interesting application of sentiment analysis for products and services evaluation is provided in [26] and [27]. Here, the authors monitor users' sentiments related to applications' reviews on online stores like Google Play Store. In particular, Guzman et al. [27] exploit the NLP to pre-process the text and identify fine-grained features in the application's reviews. Once the features are extracted, the algorithm exploits them to compute users' sentiments.

2.2 Big Social Data Analysis

The analysis of social media streams allows companies and users to gather a virtually unlimited amount of information about a certain topic or product. The collected sets of data can be very large and different in nature, meaning for example that there can be structured and unstructured data with different sizes (from terabytes to zettabytes) [28]. These kinds of data are grouped under the term big data. The term does not only refer to data themselves but also to all the processes, tools and frameworks that can be exploited to create and manage these big data sets. Due to their size and amount, the typical database management systems (DBMS) are not able to collect, manage and analyze such kind of data, as defined by the McKinsey Global Institute [29]. However, no specific size in order of terabytes is defined to assess whether or not a dataset is a big data, since it is assumed that the size of the dataset will increase in time. It is interesting to note that the definition of big data varies, depending on the industry and on what kind of tools and frameworks are available.

In order to define the concept of Big Data, academics and companies consider the concept of big data as divided in four features, namely the four V's of big data [30–32]:

- *Volume*
  It is the largest amount of data that should be stored and processed. The size of datasets has exponentially increased in the last five years due to the growth of social media platforms. Additionally, in order to manage these kinds of datasets, several computing resources as well as scalability in storage are required.
- *Velocity*
  It refers to the frequency of incoming data to analyze. Incoming data must be rapidly processed before becoming old and useless, which means that companies cannot use the well-known batch process to analyze big data. This is why a new paradigm about data managing and processing should be defined: real-time streaming data.
- *Variety*
  The variety is the main feature of the big data era. With the definition of new social media platforms and applications, new kinds of data are designed and can be exploited to acquire new insights. The main data providers are obviously the well-known social media platforms, such as Facebook and Twitter, generating more than 30 billion pieces of content every month and 400 million tweets per day, respectively.
- *Veracity*
  It refers to the trustworthiness of the data available to marketers. Veracity could influence the confidence of the marketers on their own dataset. Hence, the accuracy of the analysis depends on the veracity of the source data which could be poor and could fuel the marketers' uncertainty.

The analyses about big data allow academics and companies to make better and faster decisions, by exploiting structured and unstructured data that were not previously accessible and/or usable. The designing of a big data solution has several advantages:

- it helps finding, visualizing and understanding possible issues and improves decision making processes;
- it improves data warehousing capability in order to manage more data and enhance operational efficiency;
- it allows companies to analyze untapped data sources together with their already existing data. This task is known as big social data analysis.

Big social data analysis stems from the need to quickly process the collected data in real-time and refers to several disciplines, such as social network analysis and analytics, trend discovery, sentiment analysis and opinion mining [33]. In order to process the real-time data streams and assess what users think of a given topic/product at a given time, it is necessary to define new procedures and algorithms for the real-time processing. In the following paragraphs, we briefly introduce and discuss the most interesting solutions at the state-of-art.

The authors highlight in [34] the aim of real-time analytics, which is to answer the query "*what is happening now*" with the slightest delay. To answer such a question, the authors propose a distributed system, namely *Sentinel*. *Sentinel* is entirely developed in Java, and exploits the Apache Storm framework. This is an open source system for the distribution of real-time computation [35]. Through Storm, Sentinel can reliably process unbounded streams of data. In order to process the acquired data in real-time from Twitter, the authors define Sentinel's learning and decision process on the decision tree-learning algorithm. The algorithm is based on the Very Fast Decision Tree (VFDT) that allows to classify data in distributed scenarios.

The Social Data Analytics tool (SODATO) is presented in [36]. Its aim is to gather social data from H&M's official Facebook page and to run the sentiment analysis of posts and users' comments. To do so, the authors present and discuss an integrated modeling approach for the analysis of social media data and then define a sentiment analysis approach based on the Fuzzy set theory. Once the gathered data are prepared through the Fuzzy model and labelled manually, SODATO computes the polarity of the collected data exploiting the Google Prediction API [37].

Mihanović et al. [38] describe a system for the analysis of sentiments related to online reviews of products and tweets gathered from Twitter. Such system is divided in different layers: the first layer provides a crawler based on Apache Nutch [39] and Twitter API to crawl reviews from different sites and tweets, respectively; the second layer provides storage functions based on HBase tables [40] managed via Apache Hadoop server [41]. In details, while Hadoop is an open source software project for the development of large-scale data intensive applications, HBase is a non-relational, distributed database written in Java which efficiently holds unstructured datasets. Since they allow to manage and store considerable amounts of unstructured data in real time, they prove to be valuable in the proposed solution. Finally, the third layer uses the data collected into HBase tables and loads them into KNIME [42] to complete the sentiment analysis task.

It is interesting to note that several recent solutions like [43–45] adopt non-relational database and frameworks, such as HBase or MongoDB and Apache Hadoop or MapReduce, respectively, in order to manage the big data and real-time processing. However, each one of this solutions requires several computing resources that are too expensive to buy and manage for a company. Such drawback has led to the definition of Cloud Computing technology. It is a technology that allows to share computing resources without forcing users to buy the infrastructure network (e.g. servers) and software [46]. Hence, in the Cloud Computing paradigm, end users can buy the computing resources from the cloud provider according to their needs. Through the Cloud Computing technology and its service models, social network companies can store and manage users' data, increasing the quality of their services. One of the most distinctive trait of the cloud computing is the opportunity for companies and academics to choose the service model that better serves their needs. In details, the cloud computing technology provides three different service models [47]:

- Software as a Service (SaaS) allows users to access a specific business application through their own web browser. Through this model, users cannot manage nor control the cloud infrastructure and platform where the application runs, but they can manage the personal data stored in the cloud application and exploit the provided services.
- Platform as a Service (PaaS) allows companies to release their own solution without having to design and manage the infrastructures. The cloud provider sells to companies or developers a cloud platform where the information behind the infrastructure type are not provided. The main benefit of PaaS model is that companies can autonomously choose the frameworks and the programming languages for the development of their application.
- Infrastructure as a Service (IaaS) provides full access to the infrastructure service supplied by the cloud provider. This allows companies to design and choose the desired infrastructure components, such as the Operating System (OS) and the size of servers, to develop their business application. Hence, IaaS provides full control of the entire application.

These service models have led to the definition of new cloud providers which differ from one another in the kind of services and technologies they offer to users [5]. Examples of cloud providers are Amazon Elastic Compute Cloud (EC2) [48], Google App Engine [49] or Microsoft Windows Azure [50]. The cloud computing technology has been immediately exploited for the creation of new and powerful sentiment analysis tools, able to manage large datasets. For instance, works such as [51] and [52] define a cloud based tool that exploits the computing resources and services in order to be faster in the data collecting and processing operations.

However, together with the definition of new kinds of data provided by social media, the necessity to combine the cloud computing resources and the big social data analysis methodologies arises in order to design promising sentiment analysis tools. As a matter of fact, these tools need several computing resources to manage and analyze the large amount of unstructured data provided by social media. Cloud computing can provide scalable resources satisfying this requirement. The relationship between cloud computing and big data is underlined by Hashem et al. in [53]. The authors provide a comprehensive analysis of the correlation between big data and cloud computing through an exhaustive definition of big data characteristics, big data storage systems and exploited technologies. This analysis allows to underline big data requirements to adopt the capabilities provided by a distributed storage technology based on cloud computing. In such scenario, cloud computing has become a service model for big data analysis. Through this combination, large amount of collected data can be stored in no-relational, distributed and fault-tolerant databases, which can increase the cooperation between scalable tools and algorithms. In fact, the stored data are processed through programming languages and tools, thus providing distributed and parallel methods and algorithms in a cluster.

Cloud-based tools for big data analysis have become very popular in recent years, as predicted in [54]. Through the cloud computing, researchers and developers can exploit cloud services to create new big data analytics solutions. In order to simplify the new requirements, it is possible to redefine the cloud service models for data analytics as follows [54]:

- Data analytics IaaS defines a set of virtualized computing resources, such as virtual machines (VMs) and data storage, for designing and developing new frameworks or complex systems for the analysis of big data;
- Data analytics PaaS allows data mining developers and data scientists to exploit predefined frameworks in order to create new applications without managing the infrastructure system. This is the main advantage of this service since it allows data mining users not to design and develop the computing resources of the frameworks defined by the data analytics cloud provider;
- Data analytics SaaS is composed by a set of data mining algorithms (or tools), which can be exploited by an end user via web browser.

Through these new services, data mining developers, data scientists and researchers can easily design and develop their own data mining algorithms in order to extract information from unstructured data. An interesting example of data analytics SaaS is the data analytics application developed by Marozzo et al. [55]. The authors exploited the cloud services and features provided by Microsoft Windows Azure in order to create a framework, namely the Data Mining Cloud Framework, to process a large amount of data via a user-friendly website. Through this web service, end-users can create a workflow of the tasks of interest for their own data mining application selecting the data sources, the data mining algorithms, filters and data splitters. In fact, several works exploit the benefits derived from these two fields. Liu et al. [56] design an ad-hoc Naïve Bayes Classifier (NBC) in Hadoop to analyze the sentiment of movie reviews. In particular, the authors want to test and evaluate the scalability of NBC in the presence of large datasets, designing a big data analyzing system based on Virtual Hadoop cluster. The obtained results show that the proposed fine-grained NBC can easily scale up in the presence of large dataset. The COSMOS platform is defined in [43]. COSMOS is able to analyze both the sentiment and the tension on Twitter data using the cloud environment proposed by OpenNebula [58]. Through the services provided by this cloud platform, the authors design the infrastructure of their tool so it can be distributed and then, in order to manage the data, they define an algorithm based on MapReduce using Hadoop. Differently from COSMOS, the authors in [59] investigate on the scalability of the machine learning approach, by focusing on the Naïve Bayes classifier. Developing their own classifier instead of using a standard library, they obtain a fine-grained control of the analysis procedure. Additionally, they also design a well-defined system to manage big data using MapReduce framework with Hadoop.

The impact of these sentiment analysis tools is very significant. The use of this kind of frameworks and technologies allows to rapidly scale content analysis of opinion documents, significantly reducing the time to acquire and process data.

2.3 Opinion and Tasks Definition of Sentiment Analysis

Previously, we have discussed the motivation and the application scenarios of the sentiment analysis and how it is combined with cloud computing and big social data analysis. Now, we provide the reader with a formal definition of opinion.

As defined in [14] and [60] an opinion is a quintuple $(e_i, a_{ij}, s_{ijkl}, h_k, t_l)$ where

- e_i is the name of an entity;
- a_{ij} is an aspect of e_i;
- s_{ijkl} is the sentiment on aspect a_{ij} of entity e_i;
- h_k is the opinion holder;
- t_l is the time when the opinion is expressed by h_k;

Generally, researchers assume that a sentiment s_{ijkl} can be positive, negative, or neutral. In literature, however, there are several research works, like [22] for example, which identify different levels of intensity expressed through words (e.g. strong or weak positive) or numbers (e.g. from 1 to 5 stars, as commonly used in e-commerce and online stores). The five components listed above allow us to analyze opinions about an entity, considering not only the sentiment component but also the opinion holders and the time in which such opinions are expressed. Therefore, one of the distinctive traits of this definition is that it allows to transform unstructured texts into structured data and, if necessary, to store them easily in a relational database. Despite such advantages, the definition is not exhaustive since it does not cover all the possible facets of the semantic meaning of opinion and could consequently lead to a loss of information. Exploiting the definition of an opinion as a quintuple, Liu defines in [14] and [61] six possible tasks of the sentiment analysis of users' opinions. Such tasks are briefly summarized as follows. Given a set of opinion documents D (e.g. online reviews, tweets and social media posts) the sentiment analysis should: (i) extract all the entities e_i provided in D and divide them in categories; (ii) extract all the aspects a_{ij} related to e_i and cluster them; (iii) detect and extract the opinion holders h_k categorizing them: (iv) extract the time in which the opinion is expressed; (v) detect the sentiment s_{ijkl} of the aspect a_{ij} expressed by the opinion holder h_k; (vi) provide all the opinions quintuples defined in D.

2.4 Issues for a Correct Polarity Classification

Sentiment analysis is a promising field that could turn out to be a powerful asset for companies in the need of improving their reputation. To make this possible, researchers defined sentiment words [14] as valuable indicators of sentiment. Sentiment words such as "*bad*", and/or "*awesome*" express a negative and/or positive sentiment, respectively, and can be involved in a sentiment analysis approach in order to detect the sentiment related to the action of an opinion. However, they cannot be directly applied to the collected opinion documents due to several issues related to the lexicon domain. In particular, differently from a review written for example on an e-commerce site, posts on social media often contain typos, slang, sarcasm and allegories difficult to detect automatically. Here, we propose a brief analysis of the sentiment issues:

- Named Entity Recognition is one of the main subtask of the information extraction and analysis. The addressed problem is how to identify the person/thing/place /brand mentioned in a given text. For instance, *Apple* is both the name of the well-known American multinational technology company but it is also the name of singer *Fiona Apple*. Being able to tell the difference between the two is an extremely important task for a brand analysis tool since it allows to filter non-related results;
- Identify neutral sentences poses another issue in the sentiment analysis. For instance, the question "*is this apple good?*" does not express any sentiment, but the presence of sentiment word "good" in the sentence gives it a positive polarity;
- Detecting the sarcasm and humor in a sentence is an extremely difficult task, e.g. saying "*that's just what I need! Great!*" can be a way to express that something unpleasant has happened;

Abbreviations, lack of capitals, poor spelling, emoticons, and poor grammar are only some of the problems that may arise in the analysis of the opinion document. This kind of issues can find a solution in the definition of a pre-processing algorithm aimed at the normalization of the text.

3 Sentiment Analysis Approaches

In order to evaluate the sentiment related to a sentence or a post, several sentiment analysis approaches, which compute the sentiment behind the users' opinions and sentences, are defined in the literature. Several techniques are proposed in literature [62, 63], which can be classified in two main approaches: (i) machine learning (super-vised and unsupervised) methods; (ii) knowledge-based techniques. The supervised learning is based on a finite set of classes which allow classifying documents and training the available data for each class. Conversely, the unsupervised approach is based on the sentiment orientation of specific sentences within the source document. In addition, the unsupervised algorithms are able to measure how far a word is inclined toward positive and negative polarities [64]. Then, the knowledge-based approach, namely also lexicon or text based approach, is a well-known and diffused strand of literature for detecting the sentiment related to an opinion document exploiting an ontological dictionary of sentiment words.

3.1 Machine Learning Methods

Supervised learning is fairly common in classification problems because the goal is often to get the computer to learn a classification system that we have created. Conversely, unsupervised learning seems much harder: the goal is in fact to have the computer learn how to do something that we don't tell it how to do. Hence, in the following we refer only to the case of supervised learning-based sentiment analysis approaches. In particular, supervised learning is a machine learning approach that successfully manages to train a system in order to solve a specific task automatically. All supervised learning algorithms require the creation of a set of data to use for training the system. The training set should be manually created to define the inputs of the system, the desired output and a learned function, f, which correlates each input with its correct output. Hence, the idea behind supervised learning is that after training the system, said system is able to define a new function which approximates f. If the approximation of f is acceptable, the system should be able to provide results similar to the one provided in the training set. The main issue of this approach is related to the design of the training set. If the training set is composed of few data, the system will probably fail to provide a correct answer, while conversely the system could be too slow due to its overtraining. A supervised learning technique has to perform the following steps in order to solve a given problem (e.g. the analysis of sentiments): (i) determine the type of training examples labelling them manually and then create them; (ii) decide the input representation of the learned function which depends on how the input is represented; (iii) design the structure of the learned functions and algorithm; (iv) train the learning algorithm, and (v) evaluate the accuracy of the learned function. The most common approaches provided by supervised learning and used in literature for sentiment analysis task are Naïve Bayes and Support Vector Machine (SVM) [65, 66].

- Naïve Bayes (NB) is a statistical text classifier based on the Bayes rule. The NB classifier analyzes the content of the opinion document and estimates the probability of a document being positive or negative. In order to do so, it is possible to define a conditional probability that a document d is in class c, $P(d|c)$ [67]:

$$P(d|c) = P(c) \prod_{1 \leq k \leq n_d} P(t_k|c) \qquad (1)$$

  where n_d is the number of unique words in the document d, the is the $P(t_k|c)$ conditional probability of a term t_k occurring in a document of class c and indicates how suitable the term is t_k for class c. $P(c)$ is the probability of a document occurring in class c. The aim of the NB classifier is therefore to detect the best class for a document. In sentiment analysis, it is possible to define two classes: positive or negative. Finally, NB classifier has three different variants, Bernoulli's NB, Gaussian's NB and Multinomial NB, which differ among them for the statistical distribution of words' occurrence in a document. For instance, the multinomial NB assumes that the distribution of words' occurrence conforms to the multinomial distribution [68].

- The SVM is a vector space model classifier where the opinion documents are transformed into feature vectors before the classification task [69]. To do so, the SVM classifier converts a document into multidimensional vectors. Here, the addressed problem is to classify each opinion document represented by a vector into one of the two classes defined for sentiment analysis, positive and negative. Hence, the aim of SVM is to find a decision boundary between two classes that is maximally far from any document in the training data. The distance between the decision and the closest documents assesses the margin of the classifier, and its maximization allows to reduce the uncertainty in the classification.

These approaches are exploited in a large number of research papers. The work [70] is one of the first document that evaluates the performance of SVM and NB classifiers for the classification of movie reviews into positive and negative classes. Troussas et al. [71] exploit the NB classifier for the analysis of the sentiment expressed in the Facebook social media. An interesting probabilistic approach for the classification of the sentiment in Twitter is provided in [72]. Here, the authors investigate on the adoption of the Latent Dirichlet Allocation (LDA) in order to design a new approach for sentiment analysis. Through the use of LDA on a set of opinion document, a Mixed Graph of Terms can be automatically extracted and involved for the classification. The aim of Akaichi et al. [73] is to retrieve useful information about the Tunisian users of Facebook during the "Arabic Spring". The proposed algorithm is based on both the SVM and Naïve Bayes approach with the aim to compare them. It is interesting to note that the authors used WEKA machine learning toolkit to perform their experiments and evaluations. Recently, several works have made a step forward in the application of machine learning approach to sentiment analysis, applying new methods such as Artificial Neural Network. For instance, a sentiment specific word embedding (SSWE) learning method for sentiment classification is provided in [74]. The key idea is to encode the sentiment information into the continuous representation of words. In order to reach this aim, the authors develop three neural networks extending the word embedding learning algorithm. In addition, they provide the SSWE method as a feature of a supervised learning framework for the sentiment classification of tweets. Finally, an interesting discussion about the possible approaches of machine learning techniques for sentiment analysis in provided in [75] and [76].

3.2 Knowledge-Based Methods

The knowledge-based, also known as lexicon-based, approach is commonly used in sentiment analysis techniques. The key idea is to exploit lexical resources, such as ontological dictionaries to assess the sentiment of opinion documents. Such dictionaries are different from common ontological dictionaries because, for each word,

Table 1 Example of
thresholds for three polarity
levels

Score	Polarity
Score > 0	Positive
0	Neutral
Score < 0	Negative

they define a score to represent the semantic orientation of the words, simplifying
the computing of the sentiment score as follows:

$$setiment_{score} = \sum_{i=0}^{n} sentiment(word_i)$$

The possible algorithm then sums up the sentiment scores of the terms in the
review considering negations and intensifiers. Once computed the sentiment score, it
can be compared with a pre-defined threshold in order to assess whether the polarity
of the sentiment is positive or negative. For instance, if the algorithm uses three different classes of polarity (Positive, Neutral and Negative) we could set the thresholds
as depicted in Table 1. However, this kind of approach may lead to bad performances
in case of noise, such as typos or slangs, since no match would be found in the dictionaries. In order to solve this issue, a different approach relies on a well-designed
pre-processing algorithm for opinion documents, as it will be described in the following of this chapter. A distinctive trait of the lexicon-based approach is the opportunity
to adopt existing ontological dictionaries or to create new ontological dictionaries
providing different degrees of freedom in the definition of the sentiment analysis
approach. In addition, the possibility to create a new dictionary of sentiment words
allows to extend the analysis to documents written in many different languages.

In literature, several solutions are exploited for the definition and creation of
new lexical resources for knowledge-based methods. Examples are the well-known
SentiWordNet [28] and WordNet-Affect [77]. SentiWordNet is an English lexicon
resource based on WordNet [45]. Through SentiWordNet, verbs, adverbs, nouns
and adjectives are grouped into a set of cognitive synonyms, namely synsets, each
one expressing a distinct concept and having an assigned score. The synset score
is in the range [0–1] according to positivity, negativity and objectivity. Therefore,
different meanings of the same term may have different sentiment scores. In this
case, it is necessary to couple SentiWordNet with a Word Sense Disambiguation
(WSD) algorithm to identify the most promising meaning. WordNet-Affect [78] is
an English linguistic resource and an extension of WordNet. The idea at the core of
WordNet-Affect is to assign, to a number of sysnsets, one or more affective-labels
(A-Label). For instance, the A-Label emotion is mapped with the noun "*anger*" and
the verb "*fear*". The mapping is performed on the ground of a domain-independent
hierarchy of affective labels automatically built relying on WordNet relationships.
Unfortunately, these lexical resources are used in the analysis of documents written
in English. One of the main challenges of the knowledge-based approach is to design

new corpora for the analysis of texts written in other languages. Strapparava et al. [79] propose a new method to dynamically group a word listed in a given corpus into a lexical resource of sentiment words. This approach allows to create new dictionaries independently from the language of the corpus. An interesting example is [80], in which the authors describe how to design and develop lexical resources currently missing from the Italian language, providing a solution for the detection of sarcasm and irony in the opinion documents.

3.3 Unsupervised Methods

With the proliferation of the big data analysis, supervised learning methods have become unusual in favor of unsupervised learning methods. In the big data context, one of the main issues of supervised learning methods is the time required to create well-defined and labeled training datasets, which are strongly affected by the quantity and the quality of selected data [81]. As a matter of fact, if the training dataset is insufficient or incorrect (e.g. it may be biased), the selected supervised method may fail. To overcome this critical issue, several unsupervised learning approaches are proposed in literature. In this section, we provide a brief definition of these unsupervised learning methods, by describing the main approaches proposed in literature.

The unsupervised learning is based on the need to find hidden structures in unlabeled data. The idea is to find patterns in the unstructured data in order to obtain some knowledge for decision making or classification aims. Differently from supervised methods, the unsupervised methods need large amount of unstructured data to be trained properly without a previous annotation of the data itself [81]. The main unsupervised approaches are the Latent Dirichlet Allocation (LDA) and probabilistic Latent Semantic Analysis (pLSA). These approaches allow to extract latent topics in text documents, where a topic is a feature and it is a distribution over terms.

The LDA approach [82] is defined as a three-level hierarchical Bayesian model and the idea at its core is to represent a document as a mixture over latent topics composed of distributions of words with certain probabilities. The LD assumes that a document is written as the following generative process, for each document in a corpus:

- Choose the number of document words (N) according to a Poisson distribution;
- Choose the mixture of topics for the considered document according to a Dirichlet distribution. For instance, if the document is composed of two topics, e.g. sentiment analysis and big data, it is possible to choose the document consists of 1/3 big data and 2/3 sentiment analysis;
- For each word w_n in the document:
  - Choose the topic z_n according to the multinomial distribution;
  - Choose a word w_n from $p(w_n|z_n, \beta)$, a multinomial probability conditioned on the topic z_n.

Considering this model, the LDA tries to find a set of topics that have generated the corpus. Similarly, the pLSA is the statistical view of the LSA approach for the identification and analysis of different contexts of word usage [83, 84]. The LSA is a mathematical method in Natural Language Processing (NLP) for capturing the meaning among words and it is based on the key idea to reduce the dimension of the word-document co-occurrence matrix and create a latent semantic space. Through LSA, it is possible to classify documents together even if they do not share common words. As seen for the LDA, the pLSA is considered as a generative model for documents consisting of the following steps:

- Select a document d_n with a probability $P(d)$;
- For each word in w_i in d_n: first, it is possible to select the topic z_i from $P(z|d_n)$ which represents the conditional probability of the topic z given the document d_n. Then, a word w_i is selected with a probability $P(w|z_i)$, which is the conditional probability of the word w given the topic z_i.

Exploiting these two main methods, several works define their own approaches to extract latent topics in text documents and, hence, assess the document's polarity through an improvement of the sentiment analysis system [85–89]. For instance, the authors in [87] and the ones in [88] define an extension of the LDA approach, Aspect Detection Model (MDA)-LDA and the hybrid Hierarchical Dirichlet Process (HDP)-LDA. The MDA-LDA is a generative topic model based on the Markov chain for detecting aspects of reviews in a sentiment analysis system. Similarly, the hybrid HDP-LDA approach determines the aspects of the sentences, distinguishing among opinioned and factual words in order to detect the polarity of the sentence. Another interesting unsupervised method is the Joint Sentiment/Topic (JST) approach [89]. The approach is based on the LDA, which is extended in order to construct a better probabilistic model than pLSA. On this statistic premise, the authors add a sentiment analysis layer in order to detect the polarity of analyzed sentences. An alternative method to the JST is the Combined Sentiment-Topic (CST) model proposed in [90]. The model aims to classify the polarity of documents for different domains. As the JST, the CST model is based on the LDA approach and the words are created following two steps: (i) selection of a random topic from a topic distribution; (ii) generation of the word. Additionally, the CST is composed of four layers where sentiment labels, topics and documents are joined together. In order to extract document polarity, the authors involve in their procedure a pre-processing procedure and two lexicons (MPQA and appraisal lexicon). However, these methods work well when extracting information from long texts, such as reviews and blog posts. In social media, though, posts are short and unstructured. In order to get over this issue, some new methods propose to exploit typical emotional signals (e.g. emoticons, which are the graphic representation of the users' mood) in order to assess the polarity of a social media post. For instance, Hu et al. [91] provide a formal definition of the problem of unsupervised sentiment analysis with emotional signals proposing a way to model those signals. The authors create a new framework to identify and exploit emotional signals into an unsupervised method. The key idea is to adopt the orthogonal non-negative matrix tri-factorization model (ONMTF) which is based on the definition of two emotional signals' categories, emotional indicators and emotional correlations. Terrana et al. follow this idea to incorporate emotional signals in their sentiment analysis framework based on Twitter [92]. In order to establish the polarity of the analyzed tweets, the proposed method does not involve any polarity lexicons and manually tagged datasets using only the polarity expressed by emoticons in the training datasets. In addition, the framework will classify as "positive" the tweets containing positive emoticons and as "negative" the tweets composed of negative emoticons.

From this brief discussion, it is possible to assess that the unsupervised methods allow gaining a knowledge about the analyzed data without the adoption of labeled training sets. Despite this advantage, the unsupervised methods require a large amount of data for an accurate training of the system, and their generative model can produce incoherent topics due to the absence of labeled data [81].

3.4 Semi-supervised Methods

In recent years, semi-supervised learning has become an interesting alternative to supervised and unsupervised learning. Works, such as [81], consider the semi-supervised learning as a new machine learning approach to opinion mining. The methods belonging to this category of learning can be trained using both labeled and unlabeled data, creating a better classifier [93–95]. As a matter of fact, semi-supervised methods overcome the limits of both supervised and unsupervised approaches and they can provide high classification performance thanks to the large volume of unlabeled datasets [96]. Several approaches are defined in literature and they are based on different key ideas. In this section, we briefly introduce these approaches and the related works.

The semi-supervised methods can be divided into two main classes as suggested by Yang in [96]. The first class bootstraps the labels of a class using techniques based on self-training, soft self-training based on expectation maximization, co-training and multi-view learning. The second class, instead, is the structural learning method, which aims at creating a good functional structure for the training using unlabeled data. This class provides the graph-based methods, such as Gaussian Random Fields.

The self-training technique allows training the classifier with a small amount of labeled data. Then, the created classifier is adopted to classify the unlabeled data, which are integrated in the training set. Finally, the classifier is re-trained with the new training set. This approach is named self-training because the classifier uses its own prediction to train itself. For instance, He et al. [97] provide a framework which trains the classifier with information extracted from a sentiment lexicon. The classifier is used to detect the sentiment labels of the analyzed unlabeled documents, and the ones with high confidence are used as pseudo-labeled instances for the acquisition of the domain-specific features. Exploiting the pseudo-labeled instance, the authors show how to estimate the word-class distribution of the self-learned features and how to use them to train a second classifier. Finally, they test the performance of the algorithm on a dataset of movie reviews and a multi-domain sentiment data.

Hong et al. [98] recognize the weakness of the self-training method. As a matter of fact, if the classifier makes a classification error, such error can dramatically affect the accuracy of the model due to the second learning phase. In order to overcome this weakness, the authors propose a competitive self-training technique. The key idea is to create three different generative models on the output of the first generated model in order to choose the best one with highest F-measure value. An extension of self-training is the soft self-training which can be based on the generative model or on the Expectation Maximization (EM) method. An interesting semi-supervised approach, which exploits the combination of EM and Naïve Bayes classifier, is designed in [99]. The proposed procedure is composed of the following steps: (i) few available labeled data are used to train the classifier; (ii) the trained classifier is used to assign probabilistically-weighted class labels to each unlabeled document computing the expectation of the missing class labels; (iii) a second classifier is trained using all the documents. In these steps and in the maximum likelihood formulation, the EM performs the hill-climbing in the data space in order to find the parameter to maximize the likelihood of the considered data. Then, the authors combine Naïve Bayes and EM in order to obtain a classifier based on a mixture of multinomial. In co-training technique, the features of examples are split into two sets providing different complementary information about the considered instance. In addition, the two sets are conditionally independent. Then, the two feature sets are used to train two classifiers, respectively. The classifiers are used to classify the unlabeled data in order to predict labels. Then, the most confident unlabeled examples and the predicted labels of each classifier are added to the training set of the other classifier. Finally, the classifiers are re-trained with the new training sets. An example of co-training application is provided by Yang et al. in [100]. The authors design and develop a Lexicon-based and Corpus-based, Co-Training (LCCT) model for semi-supervised sentiment classification. The key idea of LCCT is to combine the lexicon-based and the corpus-based learning in a unified co-training framework. This aspect allows the LCCT model to get the best of the two considered learning, improving the overall performance. In order to reach this aim, a novel semi-supervised sentiment-aware LDA approach is developed for the creation of the lexicon-based classifier, while a stacked denoising auto-encoder is employed for the creation of the corpus-based classifier. A dual-view co-training approach based on dual-view bag-of-word (BOW) representation for semi-supervised sentiment classification is designed by [101]. A dataset of reviews is taken into account in order to create the dual-view bag-of-word representation. For each review, Xia et al. model the text of the original review in order to obtain the antonym through a pair of bags-of-words with opposite views. Adopting the dual-view bag-of-word representation, the dual-view co-training approach is designed. In fact, both the training and, bootstrapping processes are performed by involving two opposites of one review. In addition, the proposed method satisfies the two requirements of co-training method and automatically learns the association among antonyms. A recent co-training approach is proposed in [102]. The authors illustrate the adaptive multi-view selection (AMVS) method in order to improve the accuracy of the labeling process of unlabeled samples. The method is based on two distributions to create multiple discriminative feature views. Adopting these two distributions, several selected features are involved in the training process of classifiers. Differently form the co-training method, the multi-view learning model does not require the same assumptions because it is based on the definition of multiple hypotheses with different inductive biases, such as SVMs or Naïve Bayes. The multiple hypotheses are trained from the same labeled dataset in order to make predictions on the unlabeled instance of the dataset. The semi-supervised nature of multi-view learning allows to employ it in cross-lingual sentiment classification [103, 104] in conjunction with approaches like genetic algorithm. The second class of the semi-supervised method is based on the creation of a graph where each node is represented as labeled and unlabeled examples in the training set, while each edge is the similarity of the two connected nodes. Additionally, the edges may be weighted. This semi-supervised class does not require any parameter and assumes label smoothness over the graph. Several graph methods are defined in literature and may be nonparametric, discriminative and transductive [95]. Zhu describes the main graph methods (e.g. Label Propagation, Modified adsorption [105]). The analysed methods are similar to each other and differ in the choice of the loss function and regularization. A graph-based semi-supervised method is the main semi-supervised method employed in literature for sentiment classification tasks. A graph-based method for sentiment categorization is proposed in [106]. Involving a set of documents (e.g. objects reviews) and related ratings, the authors propose a method to infer numerical ratings for unlabeled documents using the sentiment expressed by the reviews' texts. Lu [107] designs the Semi-supervised Sentiment Analysis using Social relations and Text similarities (SSA-ST) model to improve the microblog sentiment analysis (MSA). Addressing all the issues of MSA, Lu builds a microblog relation graph, which connects the limited labeled data and the large amount of unlabeled data using text similarities and social relation at the same time. This approach defines a finer way to describe relations among microblogs

4 Social Media Analysis for Brand Monitoring: A Case Study

The analyzed sentiment analysis algorithm is based on the use of the knowledge-based approach adopting SentiWordNet, as English lexical resources. Through SentiWordNet, verb, adverbs, nouns and adjectives are grouped into a set of cognitive synonyms, namely synsets, each one expressing a distinct concept and having an assigned score. In the following, the algorithm considered seven polarity levels where each range of score values is matched to a certain sentiment (see Table 2). For instance, the range $[-0.25; 0)$ is the *Weak Negative* level.

Table 2 Seven polarity levels

Score	Polarity
n > 0.75	Strong Positive
$0.25 < n \leq 0.75$	Positive
$0 < n \leq 0.25$	Weak Positive
0	Neutral
$-0.25 \leq n < 0$	Weak Negative
$-0.75 \leq n < -0.25$	Negative
n < -0.75	Strong Negative

However, it is now compulsory to pre-process the tweets' text to improve the accuracy of the algorithm, by minimizing the noise in the tweet (e.g. by removing the presence of slangs and emoticons).

4.1 Text Pre-processing

As described in Sect. 2.4, one of the main problem of sentiment analysis is that it is not possible to directly compute the polarity of an opinion document due to several issues (i.e. the presence of slangs and emoticons which are not in any ontological dictionary or supervised and unsupervised training sets). Several works at the state of art (see [22] and references therein) provide a possible pre-processing algorithm in order to improve the sentiment analysis task. Here, we propose a summary of the pre-processing steps applying this task to the pre-processing of a generic tweet. Following the suggestions and the approaches defined in [108–111], a pre-processing algorithm should be composed by 10 steps, as shown in Fig. 1.

In order to explain step by step how the pre-processing algorithm works, we can consider the following anonymized tweet:

RT @AUTHOR_MENTION Omg, james i h8 you wth you talkin about >:(,cuz it's so coooool!!1! I can imagine him, taking it with meeeee,would be so awesome adding his messages to fav. .. #BRAND_lover btw the problem is my phone is COM-PETITOR_NAME... not BRAND_NAME :(http://t.co/KlwYA4TY

It is easy to asses that the tweet contains several words, such as slangs and typos, which are not simple to identify for an ontological dictionary without any pre-processing step. The consequence is a bad classification of the tweet because several words are not classified in the considered lexical resources or they do not have a polarity score. For instance, the author mention or the brand's name do not have any polarity score which means that they are useless for the classification. Hence, the following pre-processing steps are required.

Step 1: following the strategies proposed in [112, 113], it is necessary to remove from the text the typical elements of a tweet due to their irrelevance to compute the

Fig. 1 Flow diagram of a possible pre-processing algorithm. The proposed algorithm consist of 10 step and it allows to improve the accuracy of the polarity classification task

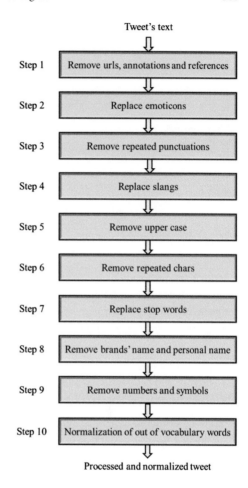

Tweet's text

Step 1 — Remove urls, annotations and references

Step 2 — Replace emoticons

Step 3 — Remove repeated punctuations

Step 4 — Replace slangs

Step 5 — Remove upper case

Step 6 — Remove repeated chars

Step 7 — Replace stop words

Step 8 — Remove brands' name and personal name

Step 9 — Remove numbers and symbols

Step 10 — Normalization of out of vocabulary words

Processed and normalized tweet

associated polarity. URL, author's references (e.g. @username), hashtags and RT characters are identified through pattern matching techniques and removed from the text. The tweet then looks like:

Omg, james i h8 you wth you talking about >:(,cuz it's so coooool and briliant!!1! I can imagine him, following me,would be so aaawesome adding my messages to fav...btw the problem is my phone is COMPETITOR_NAME...not BRAND_NAME :(

Step 2: it consists to replace emoticons with their meaning. In order to achieve this aim, a Western-Style Emoticon Dictionary with 152 mapped emoticons, exploiting the suggestion provided by [114–116], should be exploited. In this dictionary, each emoticon should be mapped with its meaning, as shown for three emoticons in Table 3.

Applying this step, the text of the tweet becomes:

*Omg, james I h8 you wth you talking about **angry**, cuz it's so coooool and briliant!!1! I can imagine him, following me,would be so aaawesome adding my messages to*

Table 3 Example of emoticons dictionary

Emoticon	Meaning
:')	Happiness
D:<	Horror
D:	Disgust

Table 4 Example of slang dictionary

Slang	Meaning
LOL	Laughing out loud
TWITTERPHORIA	Euphoria
IMHO	In my humble opinion

fav...btw the problem is my phone is COMPETITOR_NAME...not BRAND_NAME **_sad_**

Step 3: it requires replacing repeated punctuations with blank space improving the tokenization process without loss of any potential token. For instance, if we remove the punctuation from *"wrong,confidence"*, we obtain *"wrongconfidence"* as token which does not have any matching in the dictionary. The resulted text from this step is:

Omg james I h8 you wth you talking about angry cuz it's so coooool and briliant 1 I can imagine him following me would be so aaawesome adding my messages to fav btw the problem is my phone is COMPETITOR_NAME not BRAND_NAME sad

Step 4: in order to define the slang replacement step, a slang dictionary should be created using the most common slangs and acronyms adopted by social media users [117, 118]. One can also add typical slangs used in Twitter and defined in [119]. In addition, if there is no match, the word in format *"Number + Chars"* are normalized as depicted in [120]. For instance the number 2 and 4 are replaced with *to* and *for*. The Table 4 shows an example of this slang dictionary.

The tweet changes like this:

Oh my god *james I* **_hate_** *you* **_what_** **_the_** **_hell_** *you talking about angry* **_because_** *it's so coooool and briliant 1 I can imagine him following me would be so aaawesome adding my messages to* **_favourites_** **_by_** **_the_** **_way_** *the problem is my phone is COMPETITOR_NAME not BRAND_NAME sad*

Steps 5–6: through these two steps, it is possible to put in lower case all the words and replace the repeated characters with only two occurrences [121]. For instance, *'coooool'* becomes *'cool'*. In addition, if the first character of the word is repeated, it should be replaced with only one occurrence (*'aaawesome'* becomes *'awesome'*). The tweet is hence changed in:

oh my god james I hate you what the hell you talking about angry because it's so **_cool_** *and briliant 1* **_i_** *can imagine him following me would be so* **_awesome_** *adding my* *messages to favourites by the way the problem is my phone is COMPETITOR_NAME not BRAND_NAME sad*

Step 7: we remove the stop words such as: exclamations, personal and possessive pronouns, articles, conjunctions [122]. This step allows to reduce the time spent to analyze words that do not have any polarity score and do not help to compute the sentiment of the opinion document. However, if a knowledge-based approach is adopted, the negations should not be removed from the text because they can subtract points to the final polarity score [123]. Through this step, the tweet becomes:

god james hate hell talking about angry cool briliant 1 can imagine following be awesome adding messages favourites by way teh problem phone COMPETITOR_NAME not BRAND_NAME sad

Step 8–9: these two steps require to remove the names of people and brands designing two different dictionaries as described in [124, 125], respectively:
god hate hell talking about angry cool briliant can imagine following be awesome adding messages favourites by way the problem phone android not sad

Step 10: finally, the last step is the normalization of the Out Of Vocabulary (OOV) words, in order to correct typos. This step is made through two phases: (i) following the approaches described in [126], it is possible to reduce the number of repeated characters by searching for successive reductions in cascade; (ii) now, a search based on the morpho-phonetic approach of the confusion set [127] is performed. For this second phase, both the Levenshtein distance algorithm [128] and an ad-hoc dictionary of common typos provided by [67] and [129] can be adopted. The result is the following normalized tweet:

god hate hell talking about angry cool brilliant can imagine following be awesome adding messages favourites by way the problem phone android not sad

After the application of the preprocessing algorithm, it is possible to compute the polarity level associated to the normalized text of the processed tweet. In order to do that, a second phase is necessary: the Part-Of-Speech (POS) tagging.

4.2 Popularity Versus Polarity

After the application of the preprocessing algorithm, it is possible to compute the polarity level associated to the normalized text of the processed tweet. In order to do that, a second phase is necessary: the Part-Of-Speech (POS) tagging which assigns parts of speech to each word [130]. Through the additional step of the new text preprocessing and the POS tagging, we improve both the precision and the performance of our SA algorithm reducing the number of false positives. Then, we compute the polarity as the sum of each word's polarity after the POS tagging, as follows.

$$message\ score = \sum_{i=0}^{n} sentiment(POSTag(word_i)) \qquad (2)$$

A key factor to keep in mind is the authors' popularity which helps the researcher to understand how the author status and the sentiment associated to his words can influence the opinion of his followers' networks about a certain topic, brand or product. A key factor to keep in mind is the authors' popularity which helps the researcher to understand how the author status and the sentiment associated to his words can influence the opinion of his followers' networks about a certain topic, brand or product. This behavior is explained by the homophily principle in social media [131, 132]. The principle is based on the assumption that contacts among similar people may occur at higher rate than among dissimilar people. Hence, in the social media context, if there is a mutual personal relationship between two users, then the principle allows to underline that the two users are connected and they may hold similar opinions. Recent approaches establish the influence of a user who writes a sentence expressing his opinion. By computing and analyzing the influence of a user, it is possible to determine who the popular users are and how they influence other users with their words. To better describe influence's roles, Chen et al. [133] design five categories of influencers (e.g. fan, celebrity, expert, etc.) and what role they play in social networks. Hence, it is simple to assess that understanding the relationship among users of social media and a user's influence have become a popular research topic which led to the definition of both simple and sophisticated approaches. The main approaches at the state-of-art are designed for Twitter. The basic techniques used to compute the user's popularity exploit both the user's features, such as followers and following count, and the features of the user's tweet, like retweet count. A common and simple approach adopted by several companies, such as twitaholic [134], is to compute the author's popularity exploiting only his followers' count. Authors. Leavit et al. [135] propose to measure the influence of an author by exploiting information like retweets and mentions and by considering the ratio of attention. In order to obtain the rank value, which represents the author's popularity, we suggest considering the followers and following count, the mentions and retweets of each author and his tweets. Finally, the author's influence is obtained by multiplying the rank value with polarity score, obtaining a new polarity level which may be a possible representation of how the author's followers perceive his/her words. For instance, a popular author can write a weak negative tweet, which is perceived by others as a strong negative tweet due to the author's rank. This approach is involved in the analyzed sentiment analysis framework and allows it to identify the influential authors, namely influencer, and understand what they think about a brand or product. The above-mentioned approaches are the basic techniques to detect the influencer in a social media. In literature, it is possible to find techniques that are more sophisticated. For instance, a modified version of PageRank algorithm was proposed in [136, 137]. These works exploit the key idea of the well-known PageRank algorithm, applying it to social media users. Mtibaa et al. [136] describe their modified version of PageRank, namely PeopleRank, which aims to rank the nodes in a social graph using a tunable weighted social information. The authors describe also two variants of PeopleRank which are the Centralized and Distributed PeopleRank. TwitterRank [137] allows to measure the influence of users by taking into account both the link structure and the topical similarity between users. An extension of the modified PageRank is provided in [138]. The authors describe their optimized approach to compute the minimum number of new friends required by a user to have the highest rank value. Deng et al. [139] observe that several approaches do not take into account the hidden and implicit dissimilarity, opposite opinions, and foe relationship between users. In addition, they also observe that these methods can handle only a subgraph related to a single topic. In order to overcome these issues, they design a meta path-based measure to infer pseudo-friendship and dissimilarities among users by using a semi-supervised refining model, encoding similarity and dissimilarity from both user-level and post-level relations. The technique is based on the idea that similar users hold similar opinions, while foes have conflicting opinion with respect to a certain topic or entity, as defined by homophily principle. Tan et al. [140] propose another user-level sentiment analysis. Their idea is to improve sentiment analysis task by exploiting the information about users' relationship. Involving the most famous microblogging network, namely Twitter, they incorporate user information as follows. A model based on the follower or followee network is involved to detect both the dependencies between the opinions of users and their tweets, and the relation between users' opinions and the opinion of their followers. Moreover, they also consider a variant of the prosed model based on the network created using the mention (e.g. "@username"). Through this mention-network-based model, the authors can analyze the dependencies between a user's opinion the users who mention him/her. Similarly, Hu et al. [141] want to exploit social relations for the improvement of MSA by designing a Sociological Approach to handle Noisy and short Texts (SANT) for sentiment classification. The authors propose a model to describe social relations and explain how to use it for supervised sentiment classification. In order to improve the above-mentioned approaches, it is possible to involve the extraction of features such as the centrality. The centrality is the structural characteristic of a user in a network and it allows to compute a score representing the popularity of a user [142]. In particular, centrality can be categorized into two classes: (i) closeness centrality determine how close are two nodes in a weighted graph [143]; (ii) betweenness centrality helps to understand the influence of a node by measuring the amount of traffic flowing through that node to others in the social graph [144].

The aim of the proposed discussion was to underline how the analysis of social influence led to the definition of a new and challenging research topic in the field of computational intelligence. However, the considered studies do not provide a comprehensive analysis about the effectiveness of the features for the estimation of user popularity. In order to fill this void, Mei et al. [144] identify the key features to measure a user's influence by employing the Entropy method, Rank Correlation Analysis and Twitter as social media. The study points out that not only mentions and retweet actions, but also parameters like the number of public lists, the rate of new tweets and the followers to friends ratio, are fairly effective indicators to measure of the user's influence. In addition, the authors noted that the three most important features which define users' influence on Twitter are popularity, engagement and authority.

4.3 Brand Monitoring Campaign: How to Give Meaning to Tweets

In this section, we provide the analysis of some brand monitoring campaigns of sentiment analysis in Twitter. In particular, we show the results about a certain brand (hereinafter named as "Brand") taken in the Technology area, as one of the best-selling mobile phones, during the recent marketing campaign of its new smartphone series. Note that the searches have been anonymized to meet the requirements on disclosure and data responsibility. The adopted sentiment analysis tool allows the collection of tweets through the use of a web crawler, which collects the tweets related to Brand using specific hashtags and the Twitter API. Merging in one dataset the data obtained from different researches, the tool detects the polarity of the analyzed tweets and provides a set of charts and tables to help researchers in the analyses. The number of the analyzed tweets and three different charts are shown in Figs. 2, 3 and 4.

Starting from the two pie-charts, one can immediately understand the overall trend of each polarity level in the range of dates. The Sentiments pie-chart shows that we have 38.9 % of Weak Positive tweets in favor of the analyzed brand, and only 19.1 % of Weak Negative tweets. The Influence pie-chart corroborates these results. Then, it also shows how the author popularity modifies the sentiment perceived by other authors (e.g. his/her followers) increasing the number of both the Positive and Strong Positive tweets. The consequence of such an analysis is twofold: first, we can now understand how much a popular author can influence his/her own network. Then, we can also state that the sentiment perceived by popular authors' followers is stronger than the sentiment actually expressed in the original tweets. In other words, the followers of a popular author are strongly influenced by the author's opinions and words.

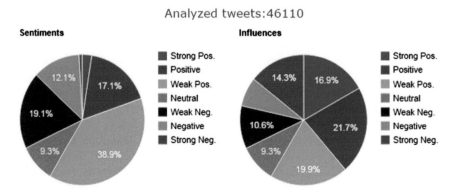

Fig. 2 Sentiment (*left*) and Influences (*right*) pie charts with seven polarity levels showing how the tweets' sentiments are distributed and how they change if the users' popularity is taken into account

Fig. 3 *Line-chart* analyzing the daily trend of sentiments. It evidences how the sentiments of the analyzed tweets change in time and what it is the prevalent polarity

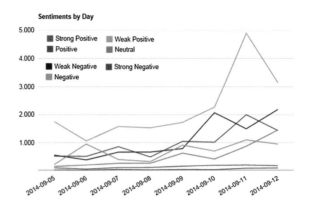

The second type of chart, a line-chart, is shown in Fig. 3. This chart displays the daily trend of each sentiment level, underlining how the sentiments of the analyzed tweets change in time and which is the prevalent polarity in the considered dates. Through this chart, researchers analyze the daily trend of both the polarity level and the tweet count. In the considered case study, we can see that the new marketing campaign of Brand is well considered by Twitter authors. Only during the 10th of September, the Weak Negative line increases enough to reach the Weak Positive one. In the following days, the number of Weak Positive tweets has rapidly increased confirming the positive trend, reaching almost 5000 (weak) positive tweets.

When the authors' popularities are taken into account, it is possible to observe a different and more complex scenario than before. As shown in Fig. 4, now the comparisons between tweets (modified by considering the author's influence) underline an overall prevalence of positive opinions about the selected Brand. Again, it is possible to see a difference between the two line-charts caused by the popularity factor of the considered authors. Analyzing the trend of influences, a researcher may single out the exact moment in which popular authors (in these cases experts, fans and haters) post a comment related to a Brand and the way they influence their networks.

Fig. 4 *Line-chart* analyzing the daily trend of influences. It evidences how the influences of the analyzed tweets change in time and what it is the prevalent polarity

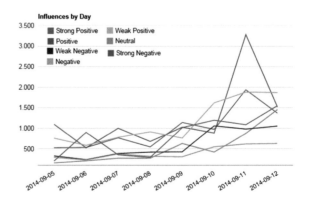

Fig. 5 The column chart
shows a daily comparison
between sentiment values
and influences as well

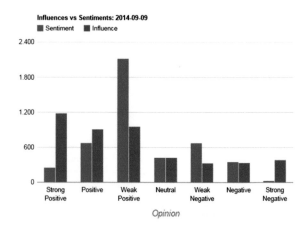

The last but not the least chart is the column chart. This chart daily displays, for each polarity level, a comparison between the sentiment and the influence in order to analyze the consequence of considering the authors' popularity. As shown in Fig. 5, the researcher can now assess how the authors' popularity influence the original sentiments. In this case, the Weak Positive level modifies towards Positive and Strong Positive levels. This chart also helps the researcher to analyze the days of interest studying the variation of tweets count for each polarity level. In addition, once the researcher identifies the day in which the highest number of tweets was registered, he/she can analyze in details how authors' popularity acts and how polarities are distributed over that day.

The last issue that our work addresses is represented by the analyses of the author's popularity (one of the main aspects of social media). We measure the author's popularity through the rank value computing the main features of an author and his tweets, as depicted in Sect. 4.2. The rank value allows us to determine the author's popularity and, hence, his/her influence on the followers' network, by establishing how his words are perceived. The considered tool lists the main influencers that expressed an opinion about the Brand, providing also their feature and some information. Finally, in order to analyze the behavior of the rank and test the validity of the influence approach, we have created a corpus of more than 300.000 tweets discussing several different brands and products. Moreover, the corpus is composed of more than 200.000 authors who have tweeted at least one time. In the rest of this section, we describe the obtained results which are anonymized for privacy reasons.

We have evaluated the rank's range in our corpus in order to define the range of the popularity levels associated to authors. Figure 6 shows this range which is in the interval [0;10] with a high number of occurrences for low values of rank around the value 3. Such data allows us to underline the variability of the rank values. We have computed 225.113 values of rank with at least one occurrence on more than 300.000 tweets. Moreover, this rank distribution has a mean value of 1,27,673 and a variance

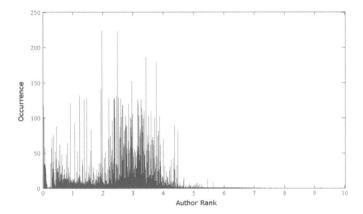

Fig. 6 Number of occurrences associated to each rank values

of 6,90369. We have then monitored the variation of rank based on the number
of retweets. Here, we show the main case of study of our dataset. Analyzing our
corpus, we have found an author (hereinafter *author*) with a high number of different
ranks for the same brand search. This means that *author* has tweeted several times
obtaining different retweet count for his tweets. Figure 7 shows the rank trend of
author and both the minimum rank value (when the author's tweet is not shared by
his followers), which is equal to 2,71, and the maximum rank value of 5,62 obtained
for 71227 retweets.

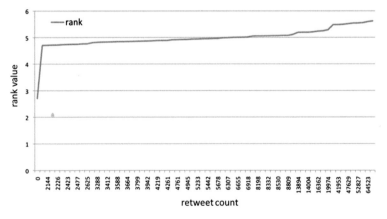

Fig. 7 The trend of rank values based on retweets count for the considered author. In this analysis,
the author has 657 followers and 1652 following

5 Conclusion

This chapter has highlighted the conventional approaches of sentiment analysis in social media streams for brand monitoring, the implications and relations with big data and the cloud computing-based solutions. We have discussed the motivations behind the sentiment analysis approach, describing both the operating scenarios and related benefits. Then, we have depicted in details the main categories of computational intelligence approaches to compute the polarity related to a sentence or a post on a social media. We have proposed a wide literature review in order to give the reader all the sources of information useful for any in-depth study. Finally, with application to a brand monitoring campaign, we have introduced some case studies and evaluated the tweets' polarity, also discussing the case of *popularity versus polarity scores* (i.e., how the polarity of textual information modifies by taking into account the popularity of users).

In particular, we have used Twitter as the target social network, since it is actually a pervasive social news media, due to its features and characteristics. The obtained results have evidenced that our tool can be a valid asset in providing strategic and innovative potentialities to all the small and medium enterprises involved in e-commerce, helping them to be more efficient and effective in satisfying the customers' needs and expectations. Enterprises can enhance their competitiveness satisfying the users' needs and expectations exploiting our tool, detecting the sentiment of a tweet and how it influences people, when the author's popularity is taken into account. The obtained results, showing that branding strategies play an even more important role in online environments, evidence the effectiveness of our approach for application to e-commerce. In addition, we believe that the retweets count in Twitter is a good index to measure the popularity and the influence of an author. Finally, through the statistical analyses on authors' popularity, we have underlined the usefulness of rank value which modifies the real sentiment of a tweet towards the maximum or minimum level of polarity, reducing the data for the middle polarity levels.

References

1. O'Reilly, T.: What is web 2.0—design patterns and business models for the next generation of software. Resource document. http://oreilly.com/web2/archive/what-is-web-20.html (2005). Accessed 15 Sept 2014
2. Kietzmann, J.H., Hermkens, K., McCarthy, I.P., Silvestre, B.S.: Social media? Get serious! Understanding the functional building blocks of social media. Bus. Horizons **54**(3), 241–251 (2011)
3. Noordhuis, P., Heijkoop, M., Lazovik, A.: Mining Twitter in the cloud: a case study. In: IEEE 3rd International Conference on Cloud Computing, pp. 107–114 (2010)
4. Armbrust, M., Fox, A., Griffith, R., Joseph, A.D., Katz, R.H., Konwinski, A., Lee, G., Patterson, D. A., Rabkin, A., Stoica, I., Zaharia, M.: Above the clouds: a berkeley view of cloud computing. Department of Electrical Engineering and Computer Sciences, University of California, Berkeley, Technical Report UCB/EECS, 28 (2009)

5. Shuai, Z., Shufen, Z., Xuebin, C., Xiuzhen, H.: Cloud computing research and development trend. In: 2nd International Conference on Future Networks, pp. 93–97 (2010)
6. Strauss, R.E., Schoder, D., Gebauer, J.: The relevance of brands for electronic commerce results from an empirical study of consumers in Europe. In: 34th Annual Hawaii International Conference on System Sciences, pp. 1–9 (2001)
7. Becker, K., Nobre, H., Kanabar, V.: Monitoring and protecting company and brand reputation on social networks: when sites are not enough. Glob. Bus. Econ. Rev. Inderscience Enterprises Ltd. **15**(2/3), 293–308 (2013)
8. Terzi, N.: The impact of e-commerce on international trade and employment. Procedia—Soc. Behav. Sci. **24**, 745–753 (2011)
9. Wang, L.: Post-crisis era of SMEs management innovation in E-commerce. In: 6th International Conference on Information Management, Innovation Management and Industrial Engineering (ICIII), pp. 471–474 (2013)
10. Xiaofeng, L., Dong, L., Yuanxin, T.: Application of web data mining and data warehouse in e-commerce. In: 2nd International Conference on Software Technology and Engineering (ICSTE), pp. 376–379 (2010)
11. Ziegler, C.N., Skubacz, M.: Towards automated reputation and brand monitoring on the web. In: IEEE International Conference on Web Intelligence, pp. 1066–1072 (2006)
12. Li, L.: Study on the interactive relationship between customer's emotional response and the brand trust – In the view of online shopping. In: IEEE International Conference on Service Operations and Logistics, and Informatics (SOLI), pp. 245–248 (2013)
13. Pang, B., Lee, L.: Opinion mining and sentiment analysis. In: Foundations and Trends in Information Retrieval, pp. 1–135 (2008)
14. Liu, B.: Sentiment analysis and opinion mining. Synth. Lect. Human Lang. Technol. **5**(1), 1–167 (2012)
15. Bothos, E., Apostolou, D., Mentzas, G.: Using social media to predict future events with agent-based markets. IEEE Intell. Syst. **99**, 1, 1, 0
16. Asur, S., Huberman, B.A.: Predicting the future with social media. In: 2010 IEEE/WIC/ACM International Conference on Web Intelligence and Intelligent Agent Technology (WI-IAT), vol. 1, pp. 492–499, 31 Aug–3 Sept 2010
17. Lassen, N.B., Madsen, R., Vatrapu, R.: Predicting iPhone Sales from iPhone Tweets. In: 2014 IEEE 18th International Enterprise Distributed Object Computing Conference (EDOC), pp. 81–90, 1–5 Sept 2014
18. Stieglitz, S., Linh, D.-X.: Political communication and influence through microblogging-an empirical analysis of sentiment in Twitter messages and retweet behavior. In: 2012 45th Hawaii International Conference on System Science (HICSS), pp. 3500–3509, 4–7 Jan 2012
19. Hoang, T.-A., Cohen, W.W., Lim, E.-P., Pierce, D., Redlawsk, D.P.: Politics, sharing and emotion in microblogs. In: 2013 IEEE/ACM International Conference on Advances in Social Networks Analysis and Mining (ASONAM), pp. 282–289, 25–28 Aug 2013
20. Duan, W., Cao, Q., Yu, Y., Levy, S.: Mining online user-generated content: using sentiment analysis technique to study hotel service quality. In: 2013 46th Hawaii International Conference on System Sciences (HICSS), pp. 3119–3128, 7–10 Jan 2013
21. Marrese-Taylor, E., Velasquez, J.D., Bravo-Marquez, F.: Opinion zoom: a modular tool to explore tourism opinions on the web. In: 2013 IEEE/WIC/ACM International Joint Conferences on Web Intelligence (WI) and Intelligent Agent Technologies (IAT), vol. 3, pp. 261–264, 17–20 Nov 2013
22. Tedeschi, A., Benedetto, F.: A cloud-based tool for brand monitoring in social networks. In: Proceedings of the IEEE International Conference on Future Internet of Things and Cloud (FiCloud-2014), pp. 541–546 (2014)
23. Social Media Monitoring Tools—Brandwatch, Brandwatch. Resource document. http://www.brandwatch.com/. Accessed 15 Sept 2014
24. Synthesio. Resource document. http://synthesio.com/corporate/en. Accessed 15 Sept 2014
25. Radian6. Resource document. http://www.exacttarget.com/products/social-media-marketing/radian6. Accessed 15 Sept 2014

26. Sangani, C., Sundaram, A.: Sentiment analysis of app store reviews. Methodology **4**(1)
27. Guzman, E., Maalej, W.: How do users like this feature? A fine grained sentiment analysis of app reviews. In: 2014 IEEE 22nd International Requirements Engineering Conference (RE), 25–29 Aug 2014, pp. 153–162
28. Subu Sangameswar: Big Data—An Introduction
29. McKinsey Global Insitute: Big Data techniques and technologies. In: Big Data: The Next Frontier for Innovation, Competition and Productivity, June 2011
30. James Kobielus Measuring the Business Value of Big Data. http://www.ibmbigdatahub.com/blog/measuring-business-value-big-data
31. Mayer-Schönberger, V., Cukier, K.: Big Data: A Revolution that Will Transform How We Live, Work, and Think
32. IBM: Performance and Capacity Implications for Big Data. http://www.redbooks.ibm.com/redpapers/pdfs/redp5070.pdf
33. Cambria, E., Rajagopal, D., Olsher, D., Das, D.: Big social data analysis, Chapter 13. In: Big Data Computing
34. Rahnama, A.H.A.: Distributed real-time sentiment analysis for big data social streams. In: 2014 International Conference on Control, Decision and Information Technologies (CoDIT), pp. 789–794, 3–5 Nov 2014
35. Apache Storm. https://storm.apache.org/
36. Mukkamala, R.R., Hussain, A., Vatrapu, R.: Fuzzy-set based sentiment analysis of big social data. In: 2014 IEEE 18th International Enterprise Distributed Object Computing Conference (EDOC), pp. 71–80, 1–5 Sept 2014
37. Google prediction API. https://cloud.google.com/prediction/docs/sentiment_analysis
38. Minanovic, A., Gabelica, H., Krstic, Z.: Big data and sentiment analysis using KNIME: Online reviews vs. social media. In: 2014 37th International Convention on Information and Communication Technology, Electronics and Microelectronics (MIPRO), pp. 1464–1468, 26–30 May 2014
39. Apache Nutch. http://nutch.apache.org/
40. HBase. http://hbase.apache.org/
41. Apache Hadoop. https://hadoop.apache.org/
42. KNIME. https://www.knime.org/
43. Conejero, J., Burnap, P., Rana, O., Morgan, J.: Scaling archived social media data analysis using a Hadoop cloud. In: 2013 IEEE Sixth International Conference on Cloud Computing (CLOUD), pp. 685–692, 28 June–3 July 2013
44. Rajurkar, G.D., Goudar, R.M.: A speedy data uploading approach for Twitter trend and sentiment analysis using HADOOP. In: 2015 International Conference on Computing Communication Control and Automation (ICCUBEA), pp. 580–584, 26–27 Feb 2015
45. Jiang, L., Ge, B., Xiao, W., Gao, M.: BBS opinion leader mining based on an improved PageRank algorithm using MapReduce. In: Chinese Automation Congress (CAC), 2013, pp. 392–396, 7–8 Nov 2013
46. Zhang, S., Zhang, S., Chen, X., Huo, X.: Cloud computing research and development trend. In: Second International Conference on Future Networks, 2010. ICFN'10, pp. 93–97, 22–24 Jan 2010
47. Das, N.S., Usmani, M., Jain, S.: Implementation and performance evaluation of sentiment analysis web application in cloud computing using IBM Blue mix. In: 2015 International Conference on Computing, Communication and Automation (ICCCA), pp. 668–673, 15–16 May 2015
48. Amazon Elastic Compute Cloud (EC2). http://aws.amazon.com/it/ec2/
49. Google App Engine https://cloud.google.com/appengine/docs
50. Microsoft Windows Azure. http://azure.microsoft.com/
51. Krishna, P.V., Misra, S., Joshi, D., Obaidat, M.S.: Learning automata based sentiment analysis for recommender system on cloud. In: International Conference on Computer, Information and Telecommunication Systems (CITS), pp. 1–5, 7–8 May 2013

52. Das, N.S., Usmani, M., Jain, S.: Implementation and performance evaluation of sentiment analysis web application in cloud computing using IBM Blue mix. In: 2015 International Conference on Computing, Communication and Automation (ICCCA), pp. 668–673, 15–16 May 2015
53. Hashem, I.A.T., Yaqoob, I., Badrul Anuar, N., Mokhtar, S., Gani, A., Khan, S.U.: The rise of "big data" on cloud computing: review and open research issues. Inf. Syst. **47**, 98–115 (2015). ISSN: 0306–4379
54. Talia, D.: Clouds for scalable big data analytics. Computer **46**(5), 98–101 (2013)
55. Marozzo, F., Talia, D., Trunfio, P.: Large-scale data analysis on cloud systems. In: Special theme: Big Data, ERCIM News, Apr 2012
56. Liu, B., Blasch, E., Chen, Y., Shen, D., Chen, G.: Scalable sentiment classification for Big Data analysis using Naïve Bayes Classifier. In: 2013 IEEE International Conference on Big Data, pp. 99–104, 6–9 Oct 2013
57. OpenNebula: The Open Source Solution for Data Center Virtualization. http://opennebula.org/
58. Liu, B., Blasch, E., Chen, Y., Shen, D., Chen, G.: Scalable sentiment classification for Big Data analysis using Naïve Bayes Classifier. In: 2013 IEEE International Conference on Big Data, pp. 99–104, 6–9 Oct 2013
59. Liu, B.: Sentiment analysis and subjectivity. In: Indurkhya, N., Damerau, F.J. (eds.) Handbook of Natural Language Processing, 2nd edn. (2010)
60. Liu, B.: Web Data Mining: Exploring Hyperlinks, Contents, and Usage Data. Springer (2006 and 2011)
61. Feldman, R.: Techniques and applications for sentiment analysis. Commun. ACM **56**(4), 82–89 (2013)
62. Klein, A., Altuntas, O., Häusser, T., Kessler, W.: Extracting investor sentiment from weblog texts: a knowledge-based approach. In: IEEE 13th Conference on Commerce and Enterprise Computing, pp. 1–9 (2011)
63. Singh, V.K., Piryani, R., Uddin, A., Waila, P.: Sentiment analysis of textual reviews; Evaluating machine learning, unsupervised and SentiWordNet approaches. In: 5th International Conference on Knowledge and Smart Technology, pp. 122–127 (2013)
64. Schölkopf, B.., Burges, C., Smola, A.: Making large-Scale SVM learning practical. In Advances in Kernel Methods—Support Vector Learning. MIT Press (1999)
65. Shawe-Taylor, J., Cristianini, N.: Support Vector Machines. Cambridge University Press (2000)
66. Scarpmaker.com. Common misspellings. Resource document. http://scrapmaker.com/view/language/common-misspellings.txt. Accessed 15 Sept 2014
67. Yoshida, S., Kitazono, J., Ozawa, S., Sugawara, T., Haga, T., Nakamura, S.: Sentiment analysis for various SNS media using Naïve Bayes classifier and its application to flaming detection. In: IEEE Symposium on Computational Intelligence in Big Data (CIBD), pp. 1–6, 9–12 Dec 2014
68. Russel, S., Norvig, P.: Artificial Intelligence: A Modern Approach, 3rd edn. Prentice Hall
69. Bo, P., Lee, L., Vaithyanathan, S.: Thumbs up?: sentiment classification using machine learning techniques. In: Proceedings of Conference on Empirical Methods in Natural Language Processing (EMNLP-2002) (2002)
70. Troussas, C., Virvou, M., Junshean Espinosa, K., Llaguno, K., Caro, J.: Sentiment analysis of Facebook statuses using Naive Bayes classifier for language learning. In: Fourth International Conference on Information, Intelligence, Systems and Applications (IISA), pp. 1–6, 10–12 July 2013
71. Colace, F., De Santo, M., Greco, L.: A probabilistic approach to tweets' sentiment classification. In: Humaine Association Conference on Affective Computing and Intelligent Interaction (ACII), pp. 37–42, 2–5 Sept 2013
72. Akaichi, J., Dhouioui, Z., Lopez-Huertas Perez, M.J.: Text mining facebook status updates for sentiment classification. In: 17th International Conference System Theory, Control and Computing (ICSTCC), pp. 640–645, 11–13 Oct 2013

374   F. Benedetto and A. Tedeschi

73. Tang, D., Wei, F., Yang, N., Zhou, M., Liu, T., Qin, B.: Learning sentiment-specific word embedding for twitter sentiment classification. In: Proceedings of the 52nd Annual Meeting of the Association for Computational Linguistics, vol. 1, pp. 1555–1565 (2014)

74. Neethu, M.S., Rajasree, R.: Sentiment analysis in twitter using machine learning techniques. In: 2013 Fourth International Conference on Computing, Communications and Networking Technologies (ICCCNT), pp. 1–5, 4–6 July 2013

75. Zhang, H., Gan, W., Jiang, B.: Machine learning and lexicon based methods for sentiment classification: a survey. In: 2014 11th Web Information System and Application Conference (WISA), pp. 262–265, 12–14 Sept 2014

76. WordNet-Affect. http://wndomains.fbk.eu/wnaffect.html

77. Strapparava, C., Valitutti, A.: WordNet affect: an affective extension of WordNet. LREC **4** (2004)

78. Jamoussi, S., Ameur, H.: Dynamic construction of dictionaries for sentiment classification. In: Third International Conference on Cloud and Green Computing (CGC), pp. 418–425, 30 Sept 2013–2 Oct 2013

79. Bosco, C., Patti, V., Bolioli, A.: Developing Corpora for sentiment analysis: the case of Irony and Senti-TUT. IEEE Intell. Syst. **28**(2), 55, 63, Mar-Apr 2013

80. Madhoushi, Z., Hamdan, A.R., Zainudin, S.: Sentiment analysis techniques in recent works. In: Science and Information Conference (SAI), 28–30 July 2015, pp. 288–291 (2015)

81. Blei, D.M., Ng, A.Y., Jordan, M.I.: Latent dirichlet allocation. J. Mach. Learn. Res. **3**, 993–1022 (2003)

82. Hofmann, T.: Unsupervised learning by probabilistic latent semantic analysis. Mach. Learn. **42**(1–2), 177–196 (2001)

83. Hong, L.: Probabilistic latent semantic analysis. arXiv:1212.3900

84. Titov, I., McDonald, R.: Modeling online reviews with multi-grain topic models. In: Proceedings of the 17th International Conference on World Wide Web, pp. 111–120. ACM, Apr 2008

85. Liu, Y., Huang, X., An, A., Yu, X.: ARSA: a sentiment-aware model for predicting sales performance using blogs. In: Proceedings of the 30th Annual International ACM SIGIR Conference on Research and Development in Information Retrieval, pp. 607–614. ACM (2007)

86. Bagheri, A., Saraee, M., De Jong, F.: ADM-LDA: An aspect detection model based on topic modelling using the structure of review sentences. J. Inf. Sci. **40**(5), 621–636 (2014)

87. Ding, W., Song, X., Guo, L., Xiong, Z., Hu, X.: A novel hybrid HDP-LDA model for sentiment analysis. In: International Joint Conferences on Web Intelligence (WI) and Intelligent Agent Technologies (IAT), 2013 IEEE/WIC/ACM, pp. 329–336, 17–20 Nov 2013

88. Sowmiya, J.S., Chandrakala, S.: Joint sentiment/topic extraction from text. In: International Conference on Advanced Communication Control and Computing Technologies (ICACCCT), pp. 611–615, 8–10 May 2014

89. Usha, M.S., Indra Devi, M.: Analysis of sentiments using unsupervised learning techniques. In: International Conference on Information Communication and Embedded Systems (ICICES), pp. 241–245, 21–22 Feb 2013

90. Hu, X., Tang, J., Gao, H., Liu, H.: Unsupervised sentiment analysis with emotional signals. In: Proceedings of the 22nd International Conference on World Wide Web, pp. 607–618. International World Wide Web Conferences Steering Committee, May 2013

91. Terrana, D., Augello, A., Pilato, G.: Automatic unsupervised polarity detection on a Twitter data stream. In: IEEE International Conference on Semantic Computing (ICSC), pp. 128–134, 16–18 June 2014

92. Chapelle, O., Schölkopf, B., Zien, A.: Semi-Supervised Learning. The MIT Press (2006)

93. Prakash, V.J., Nithya, D.L.: A Survey on Semi-Supervised Learning Techniques. arXiv:1402.4645 (2014)

94. Zhu,, X.: Semi-supervised learning literature survey (2008)
95. Yang, B.: Semi-supervised Learning for Sentiment Classification

96. He, Y., Zhou, D.: Self-training from labeled features for sentiment analysis. Inf. Process. Manage. **47**(4), 606–616 (2011). ISSN: 0306–4573

97. Hong, S., Lee, J., Lee, J.-H.: Competitive self-training technique for sentiment analysis in mass social media. In: 2014 Joint 7th International Conference on and Advanced Intelligent Systems (ISIS), 15th International Symposium on Soft Computing and Intelligent Systems (SCIS), pp. 9–12, 3–6 Dec 2014

98. Nigam, K., McCallum, A.K., Thrun, S., Mitchell, T.: Text classification from labeled and unlabeled documents using EM. Mach. Learn. **39**(2–3), 103–134 (2000)

99. Yang, M., Tu, W., Lu, Z., Yin, W., Chow, K.P.: LCCT: A Semi-supervised model for sentiment classification. In: The Conference of the North American Chapter of the Association for Computational Linguistics: Human Language Technologies (NAACL-HLT), June 2015. ACL, USA (2015)

100. Xia, R., Wang, C., Dai, X., Li, T.: Co-training for Semi-supervised Sentiment Classification Based on Dual-view Bags-of-words Representation

101. Yang, Z., Liu, Z., Liu, S., Min, L., Meng, W.: Adaptive multi-view selection for semi-supervised emotion recognition of posts in online student community. Neurocomputing **144**, 138–150 (2014). ISSN: 0925–2312

102. Hajmohammadi, M.S., Ibrahim, R., Selamat, A.: Cross-lingual sentiment classification using multiple source languages in multi-view semi-supervised learning. Eng. Appl. Artif. Intell. **36**, 195–203 (2014). ISSN: 0952–1976

103. Lazarova, G., Koychev, I.: Semi-supervised multi-view sentiment analysis. In: Computational Collective Intelligence, pp. 181–190. Springer International Publishing (2015)

104. Yong, R., Nobuhiro, K., Yoshinaga, N., Kitsuregawa, M.: Sentiment classification in under-resourced languages using graph-based semi-supervised learning methods. IEICE Trans. Inf. Syst. **97**(4), 790–797 (2014)

105. Goldberg, A.B., Zhu, X.: Seeing stars when there aren't many stars: graph-based semi-supervised learning for sentiment categorization'. In: Proceedings of the First Workshop on Graph Based Methods for Natural Language Processing (TextGraphs-1). Association for Computational Linguistics, USA, pp. 45–52 (2006)

106. Lu, T.J.: Semi-supervised microblog sentiment analysis using social relation and text similarity. In: 2015 International Conference on Big Data and Smart Computing (BigComp), pp. 194–201, 9–11 Feb 2015

107. Balahur, A.: Sentiment analysis in social media texts. In: 4th Workshop on Computational Approaches to Subjectivity, Sentiment and Social Media Analysis, pp. 120–128 (2013)

108. Habernal, I., Ptácek, T., Steinberger, J.: Sentiment analysis in Czech social media using supervised machine learning. In: 4th Workshop on Computational Approaches to Subjectivity, Sentiment and Social Media Analysis, pp. 65–74 (2013)

109. Kouloumpis, E., Wilson, T., Moore, J.: Twitter sentiment analysis: the good the bad and the omg! In: 5th International Conference on Weblogs and Social Media, pp. 538–541 (2011)

110. Agarwal, A., Xie, B., Vovsha, I., Rambow, O., Passonneau, R.: Sentiment analysis of Twitter data, pp. 30–38. Workshop on Languages in Social, Media (2011)

111. Go, A., Bhayani, R., Huang, L.: Twitter sentiment classification using distant supervision. CS224N Project Report, pp. 1–12 (2009)

112. Bifet, A., Frank, E.: Sentiment knowledge discovery in Twitter streaming data. In: 3th International Conference on Discovery Science, pp. 1–15 (2010)

113. Zhao, J., Dong, L., Wu, J., Xu, K.: Moodlens: an emoticon-based sentiment analysis system for chinese tweets. In: 18th International Conference on Knowledge Discovery and Data Mining, pp. 1528–1531 (2012)

114. Ptaszynski, M., Maciejewski, J., Dybala, P., Rzepka, R., Araki, K.: CAO: a fully automatic emoticon analysis system based on theory of kinesics. IEEE Trans. Affect. Comput. 46–59 (2010)

115. Wikipedia. List of emoticons. Resource document. http://en.wikipedia.org/wiki/List_of_emoticons. Accessed 15 Sept 2014

116. NoSlang.com. Slang Dictionary—Text Slang, Internet Slang, and Abbreviations A guide to everyday acronyms and obscure abbreviations. Resource document. http://www.noslang.com/dictionary/. Accessed 15 Sept 2014
117. Chatslang.com. Social Media slang—Chat slang terms used in social media. Resource document. http://www.chatslang.com/terms/social_media. Accessed 15 Sept 2014
118. Beal, V.: Twitter dictionary: a guide to understanding Twitter Lingo. Resource document. http://www.webopedia.com/quick_ref/Twitter_Dictionary_Guide.asp. Accessed 15 Sept 2014
119. Roberts, K., Roach, M.A., Johnson, J., Guthrie, J., Harabagiu, S.M.: EmpaTweet: annotating and detecting emotions on Twitter. In: Language Resources and Evaluation Conference, pp. 3806–3813 (2012)
120. Balahur, A.: OPTWIMA: Comparing knowledge-rich and knowledge-poor approaches for sentiment analysis in short informal texts. In: The Second Joint Conference on Lexical and Computational Semantics, pp. 460–465 (2013)
121. Hemalatha, I., Varma, G.S., Govardhan, A.: Preprocessing the informal text for efficient sentiment analysis. Int. J. Emerg. Trends Technol. Comput. Sci. 58–61 (2012)
122. Serban, O., Pauchet, A., Rogozan, A., Pecuchet, J.P.: Semantic propagation of contextonyms using SentiWordNet. Resource document. http://asi.insa-rouen.fr/enseignants/apauchet/Files/Publications/WACAI12b.pdf. Accessed 15 Sept 2014
123. Census.gov. Resource document.http://www.census.gov/genealogy/www/data/1990surnames/dist.all.last. Accessed 15 Sept 2014
124. Strategic Name development. A Brand Naming Company. Resource document. http://www.namedevelopment.com/brand-names.html. Accessed 15 Sept 2014
125. Torunoglu, D., Eryigit, G.: A cascaded approach for social media text normalization of Turkish. In: 5th Workshop on Language Analysis for Social Media, pp. 62–70 (2014)
126. Han, B., Baldwin, T.: Lexical normalization of short text messages: makn sens a # Twitter. In: 49th Annual Meeting of the Association for Computational Linguistics: Human Language Technologies, vol. 1, pp. 368–378 (2011)
127. Miller, F.P., Vandome, A.F., McBrewster, J.: Levenshtein Distance. ISBN: 9786130216900
128. Wikipedia. List of common misspellings. Resource document. http://en.wikipedia.org/wiki/Wikipedia:Lists_of_common_misspellings. Accessed 15 Sept 2014
129. Stanford University. Stanford Log-linear Part-Of-Speech Tagger. Resource document. http://nlp.stanford.edu/software/tagger.shtml. Accessed 15 Sept 2014
130. McPherson, M., Smith-Lovin, L., Cook J.M.: Birds of a feather: homophily in social networks. Annu. Rev. Sociol. 415–444 (2001)
131. Lazarsfeld, P.F., Merton, R.K.: Friendship as a social process: a substantive and methodological analysis. In: Berger, M., Abel, T., Page, C.H. (eds.) Freedom and Control in Modern Society, pp. 8–66. Van Nostrand, New York (1954)
132. Chen, C., Gao, D., Li, W., Hou, Y.: Inferring topic-dependent influence roles of Twitter users. In: Proceedings of the 37th International ACM SIGIR Conference on Research & Development in Information Retrieval (SIGIR'14). ACM, New York, NY, USA, pp. 1203–1206 (2014)
133. Twitaholic. Resource document: http://twitaholic.com. Accessed 15 Sept 2014
134. Leavitt, A., Burchard, E., Fisher, D., Gilbert, S.: The influentials: new approaches for analyzing influence on Twitter. Web Ecology project. http://www.webecologyproject.org/wpcontent/uploads/2009/09/influence-report-final.pdf (2009)
135. Mtibaa, A., May, M., Diot, C., Ammar, M.: PeopleRank: social opportunistic forwarding. In: INFOCOM, Proceedings IEEE, pp. 1–5, 14–19 Mar 2010
136. Weng, J., Lim, E.-P., Jiang, J., He: TwitterRank: finding topic-sensitive influential Twitterers. In: Proceedings of the Third ACM International Conference on Web Search and Data Mining (ACM WSDM) (2010)
137. Moreno, F., Gonzalez, A., Valencia, A. Computing the minimum number of new friends required in a social network to get the highest PageRank. Int. J. Complex Syst. Sci. **1**, 37–42 (2012)

138. Deng, H., Han, J., Li, H., Ji, H., Wang, H., Lu, Y.: Exploring and inferring user-user pseudo-friendship for sentiment analysis with heterogeneous networks. Stat. Anal. Data Mining: ASA Data Sci. J. **7**(4), 308–321 (2014)
139. Tan, C., Lee, L., Tang, J., Jiang, L., Zhou, M., Li, P.: User-level sentiment analysis incorporating social networks. In: Proceedings of the 17th ACM SIGKDD International Conference on Knowledge Discovery and Data Mining (KDD'11). ACM, New York, NY, USA, pp. 1397–1405 (2011)
140. Hu, X., Tang, L., Tang, J., Liu. H.: Exploiting social relations for sentiment analysis in microblogging. In: Proceedings of the Sixth ACM International Conference on Web Search and Data Mining (WSDM'13). ACM, New York, NY, USA, pp. 537–546 (2013)
141. Ahsan, M., Singh, T., Kumari, M.: Influential node detection in social network during community detection. In: International Conference on Cognitive Computing and Information Processing (CCIP), pp. 1–6, 3–4 Mar 2015
142. Zhang, J., Ma, X., Liu, W., Bai, Y.: Inferring community members in social networks by closeness centrality examination. In: 2012 Ninth Web Information Systems and Applications Conference (WISA), pp. 131–134, 16–18 Nov 2012
143. Bermingham, A., Conway, M., McInerney, L., O'Hare, N., Smeaton, A.F.: Combining social network analysis and sentiment analysis to explore the potential for online radicalisation. In: Proceedings of the 2009 International Conference on Advances in Social Network Analysis and Miningi, IEEE Computer Society, Washington, DC, USA, pp. 231–236 (2009)
144. Mei, Y., Zhong, Y., Yang, J.: Finding and analyzing principal features for measuring user influence on Twitter. In: 2015 IEEE First International Conference on Big Data Computing Service and Applications (BigDataService), pp. 478–486, 30 Mar 2015–2 Apr 2015

Neuro-Fuzzy Sentiment Analysis
for Customer Review Rating Prediction

Georgina Cosma and Giovanni Acampora

Abstract Consumers often provide on-line reviews on products or services they have purchased, and frequently seek on-line reviews about a product or service before deciding whether to make a purchase. Organisations seek consumer opinions about their products, since this invaluable information allows them to improve future product versions, and to predict sales. The vast amount of on-line customer reviews has attracted research into approaches for intelligently mining these reviews to support decision-making processes. This chapter provides an overview of recent fuzzy-based approaches to sentiment analysis of customer reviews. It also presents a framework which can be utilised for sentiment analysis and review rating prediction tasks. The framework includes methods for preparing the dataset; extracting the best features for prediction via Singular Value Decomposition and a Genetic Algorithm; and constructing a classifier for performing the review rating predictions.

Keywords Neuro-Fuzzy sentiment analysis · Review rating prediction · Rating inference · Dimensionality reduction

1 Introduction

Sentiment Analysis, also known as Opinion Mining, is a research field which aims to discover sentiments, opinions, attitudes, and emotions in textual information. Textual information contains facts and opinions. Facts are objective because they describe expressions about entities, events and their properties. Opinions on the other hand, are usually subjective because they are written statements describing sentiments and feelings towards entities [1]. Opinions amongst people about the same entity may vary, making opinionated textual information broad and subjective. People generally seek others opinions when they want to make a decision and want to hear others

G. Cosma · G. Acampora (✉)
School of Science and Technology, Nottingham Trent University, Nottingham, UK
e-mail: giovanni.acampora@ntu.ac.uk

G. Cosma
e-mail: georgina.cosma@ntu.ac.uk

© Springer International Publishing Switzerland 2016   379
W. Pedrycz and S.-M. Chen (eds.), *Sentiment Analysis and Ontology Engineering*,
Studies in Computational Intelligence 639, DOI 10.1007/978-3-319-30319-2_15

opinions and experiences. The Internet has revolutionised the way that people seek and express their views and opinions. Consumers often use the Internet to provide on-line reviews on products or services provided by organisations; and companies seek consumers opinions about their products, and this invaluable information allows them to improve future product versions and to predict sales. Furthermore, the Internet is considered as an important resource and a major channel of communication to consumers seeking information about a product or service and online consumer reviews are an invaluable resource to consumer decision-making [2, 3]—in general, consumers tend to trust other consumers reviews more than the product reviews posted by sellers. Companies and sellers are more likely to express a subjective opinion about their own products, and may disguise the negative aspects of the products they are promoting, whereas customers honestly evaluate strengths and weaknesses of the product [4].

Companies which use e-commerce to sell products online can adopt a cost-effective way, which helps them to quickly market, and deliver products and ser-vices to customers [5]. However, retaining customers is a challenge to companies and especially to those e-commerce business-to-consumer companies [6]. Word-of-mouth feedback is vital to the reputation of the company and can influence the reputation hence profitability of a company. Essentially, customer satisfaction results after a purchase and it is basically the emotional reaction of a customer to the quality of transaction [7, 8]. Companies use the information obtained from customer reviews to monitor consumer attitudes toward their products/services in real-time mode, and adapt their manufacturing, distribution, and marketing strategies accordingly [9]. For instance, online movie reviews can be used for revenue forecasting, and early volume of online reviews can be used for early revenue forecasting. Furthermore, companies can use online review data to generate estimates of their competitors' sales when sale data are not publicly available [9]. Park et al. [10] examine how product and consumer characteristics influence online consumer reviews on product sales using data from the video game industry, and explain that this information impacts on company online marketing strategies, which should be adjusted accordingly. Under-standing how online reviews affect consumers' purchase decisions is important to e-commerce.

Cui et al. [11] have investigated the effect of online reviews on new product sales for consumer electronics and video games. They have analysed data of 332 new prod-ucts from Amazon.com over nine months, and their results revealed that the volume of reviews has a significant effect on new product sales in the early period and such effect decreases over time. The large volume of reviews that are typically published for a single product makes it very time consuming for consumers and companies to determine the best reviews for understanding the underlying quality of a product [12]. They have conducted an econometric analysis to predict the impact of *reviews on sales* and their *perceived usefulness* based on the degree of subjectivity, infor-mativeness, readability, and linguistic correctness in reviews. Their results revealed that average subjectivity of reviews and higher readability scores (scores generated by applying readability tests on reviews) are associated with an increase in product sales. Increased product sales are also more likely to occur when the reviewer clearly outlines the advantages and limitations of the product. Furthermore, the study also revealed that readability of reviews has a positive impact on review helpfulness and this indicates that consumers consider reviews that mainly confirm the validity of the product description, giving a small number of comments (not giving comments decreases the usefulness of the review) [12]. Other work exists in the literature which considers the quality factors of a review as a predictor for its usefulness. Liu et al. [13, 14] propose a model for predicting the helpfulness of reviews which is based on three factors: the reviewer expertise, writing style, and timeliness. Their findings revealed that these factors influence the performance of helpfulness prediction in the movie domain. One limitation of their work is that each review was analysed independently of all other existing reviews. When predicting the review helpfulness, other reviews should be taken into consideration, and the perceived helpfulness of a review depends on the content of the review but also on how it relates to other reviews of the same product [15].

When a large number of opinions exist about a product, it is very difficult for the human reader to analyse all that information for decision making purposes. For this reason sentiment analysis is needed to automatically analyse the textual information in order to discover and organise the opinions into a usable form. The task of sentiment analysis is a challenging task, due to the subjectivity as well as ambiguity and vagueness naturally found in opinionated text. Ambiguity and vagueness are a problem when analysing opinionated text, such as user reviews, since these are often written using sarcasm (mock or convey or irony), rhetoric or metaphor. Take for example the sentence "A great product, yeah right!"—this may seem like a positive review, but it is indeed a negative review.

There are several approaches in the literature which have been designed to deal with ambiguity and vagueness in natural language text. Among the most well known and widely adopted approach which is highly effective in dealing with vagueness and ambiguity is semantic analysis. Semantic analysis of textual reviews requires the use of mathematical algorithms, which are applied to derive semantic meaning from large collections of textual information. Briefly, a text collection is represented as a matrix of terms-by-documents containing the normalised frequency counts of terms in documents—this is known as the Vector Space Model. A statistical algorithm called Singular Value Decomposition decomposes this term-by-document matrix into separate matrices that capture the similarity between terms and documents across various dimensions in space. The aim is to represent the original relationships between terms in a reduced dimensional space such that noise (i.e. variability in term usage) is removed from the data and therefore uncovering the important relations between terms and documents obscured by noise [16, 17].

One topic of sentiment analysis is concerned with analysing customer reviews and predicting their rating using a rating scale. The task of rating a product or service using a rating scale is known to as rating-inference within the research area of Opinion Mining and Sentiment Analysis. Classifying and predicting review ratings using a multi-point scale is a challenging problem which can only be successfully achieved by computational intelligence algorithms. Computational intelligence approaches to review rating prediction have been adopted mainly for the binary classification of reviews (i.e. positive or negative) and the challenge still remains on the problem of classifying reviews on a multi-point scale [18–20]. The main problem resides in the imprecision and noise found in large customer review datasets. Fuzzy algorithms are well known to deal with imprecise and uncertain data and pose a good solution to the problem of predicting review ratings using a multi-point scale.

2   Basic Concepts and Literature Review

This section provides a literature review about intelligent methods for semantic analysis in the opinion mining area. The emphasis of this section is on fuzzy based approaches.

2.1   Fuzzy Approaches to Mining Customer Reviews

In classical set theory, a set is a container that includes or excludes any given element. For example, consider elements January, April, December, Monday, green, apple. A set containing the months of the year, only includes January, April, and December, and excludes any elements which are not a month for example, Monday, green, apple. Hence with classical set theory an element can solely belong to one set. Fuzzy logic starts with the concept of a fuzzy set. A fuzzy set is a set without a crisp, clearly defined boundary and therefore it can contain elements with partial degrees of membership. Hence, fuzzy sets are different to classical sets, because classical sets can wholly include or wholly exclude any given element, whereas in fuzzy sets an element can belong to multiple sets with varying degree of membership. Fuzzy logic has been successfully applied to tasks which require reasoning with data which contains degrees of uncertainty. In fuzzy logic, the truth of any statement becomes a matter of degree. The major advantage that fuzzy reasoning offers is the ability to reply to yes-no question with a value which falls in between yes or no, so as to resemble human-like reasoning. For example, assume that true is represented by the numerical value of 1 and false the numerical value of 0, fuzzy logic permits in-between values like 0.35 and 0.74. Hence the values represent the degree of membership of an element in set 1 and set 2 respectively. This degree is computed via a membership function. A membership function defines how each point in the input space is mapped to a membership value (or degree of membership) between 0 and 1. Fuzzy logic is a suitable approach to solving sentiment analysis issues, and in particular the rating inference problem which is concerned with the prediction and the classification of opinions on a multi-point scale. Take for example a set of customer reviews, which need to be classified as positive or negative. Surely due to the subjective nature of reviews it may not be possible to classify them as either positive or negative. It would make more sense to classify a review based on its degree of membership in the positive and in the negative set. So for example if only some parts of the review are positive and other parts are negative, then using fuzzy logic the review would receive a membership value within the positive and negative sets.

There have been several attempts to derive hybrid methods by combining fuzzy logic for sentiment analysis related tasks. Li and Tsai [21] propose *Fuzzy Formal Concept Analysis,* a fuzzy conceptualization model for text mining with application in opinion polarity classification. Their proposed classification framework was designed to deal with issues caused by ambiguous terms. Their approach, based on fuzzy formal concept analysis, is able to conceptualise documents into a more abstract form of concepts, and uses these documents to train a classifier to deal with term ambiguity. They have evaluated their proposed model on three opinion polarity datasets. Their evaluation results revealed that applying concept analysis to opinion polarity classification is an effective approach to disambiguation over other state-of-the-art classification methodologies. A comparison of their approach against the J48 decision tree, k-Nearest Neighbors, Support Vector Machine, and Bayesian Network approaches reported good performance in terms of precision. However, Li and Tsai [21] do not discuss in detail the parameters of the methodologies under scrutiny and it is unknown as to whether those approaches would have outperformed their approach when tuned to achieve maximum performance.

Loia and Senatore [22] presented a framework for extracting the emotions and the sentiments expressed in textual data and classifying them in terms of polarity (i.e. positive or negative) and emotions. They applied fuzzy-based modelling of the emotions to enable the identification of linguistic patterns that intensify or reduce such emotions. They have utilised fuzzy logic to model emotions with more flexibility and argue that fuzzy modifiers are an adequate tool to tune these emotions in the text. The interesting aspect of their approach is that it is based on Minsky's conception of emotions [23]. Minsky's conception of emotions consists of four affective dimensions (Pleasantness, Attention, Sensitivity and Aptitude) [23]. The linguistic resources utilised by Loia and Senatore's [22] approach are SentiWordNet and WordNet-Affect (that are mainly extensions of WordNet). Similarly, Dragoni et al. [24] proposed an approach which utilises fuzzy logic for modelling concept polarities, and for addressing the uncertainty associated with concept polarities with respect to different domains. Their approach is also based on the use of a knowledge graph built by combining the WordNet and SenticNet linguistic resources. A graph-propagation algorithm that propagates sentiment information learned from labeled datasets, is applied on the knowledge graph.

Acampora and Cosma [25, 26] proposed a hybrid computational intelligence approach for efficiently predicting customer sentiments in e-commerce reviews. Their neuro-fuzzy framework can analyse customer sentiments in textual reviews and predict their corresponding numerical rating. Their approach addresses the dimension and imprecision found in textual reviews. Experimental results show that their proposed hybrid approach yields better learning performance than other computational intelligence approaches applied to customer review rating prediction.

2.2 Fuzzy Ontology-Based Semantic Analysis

Lau et al. [27] proposed *semi-supervised fuzzy product ontology mining algorithm* and an approach for automating the process of mining of product ontologies. The proposed probabilistic generative model can automatically learn fuzzy product ontologies (i.e. the contexts where consumer sentiments are evaluated) to enhance context-sensitive, aspect-oriented sentiment analysis. Their proposed approach deals with the expensive and time consuming approach of manual labelling of training examples. They evaluated the fuzzy product mining algorithm, embedded in an experimental system (OBPRM), using real-world social media data. The experimental results revealed an improved performance over the baseline ontology learning system and a context-free sentiment analysis system. The OBPRM system outperformed the Association rule, Support Vector Machine, OBPRM-light, and OBPRM approaches. The authors state that the OBPRM system achieves the best performance because of the effectiveness of the Linear Discriminant Analysis-based aspect mining algorithm to learn the aspect-oriented sentiments captured in a product ontology (i.e. the whole hierarchy of aspect-sentiment pairs).

Nie et al.'s [28] method integrates domain sentiment knowledge into the analysis approach to deal with feature-level opinion mining. The method involved constructing a domain ontology called Fuzzy Domain Sentiment Ontology Tree (FDSOT), and then utilising the prior sentiment knowledge of the ontology to improve accuracy in sentiment classification. Particularly, the FDSOT conceptual model represents the semantic relation between features and sentiment terms. The authors have evaluated their approach using a dataset of Chinese product reviews collected and their results demonstrated that their approach is able to automatically identify the domain-dependent polarity for a large subset of sentiment expression and effectively improve the performance of opinion mining.

3 Neuro-Fuzzy Sentiment Analysis Framework

This section summarises the neuro-fuzzy sentiment analysis framework (see Fig. 1) proposed by Acampora and Cosma [25, 26], and explains how the framework can be adopted to perform evaluations with various Fuzzy approaches. The framework [26] is a fully automated approach to sentiment analysis and prediction. It can be applied to classify textual information into clusters; and it can be utilised to make predictions using textual data. Furthermore, the Fuzzy approaches can be replaced by other computational approaches such as Support Vector machine, Naive Bayes, and the Artificial Neural Network. The framework illustrated in Fig. 1 is composed of two main phases, the learning and the prediction phases, and these are described in the subsections below.

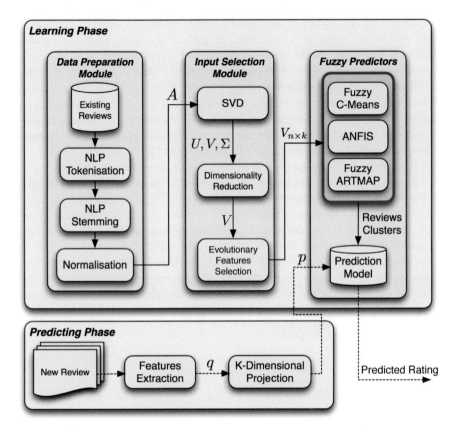

Fig. 1 Neuro-Fuzzy review ratings prediction framework [26]

3.1 Data Preparation Module

Initially, the framework takes a dataset of reviews and applies Natural Language Processing which includes preprocessing the textual reviews and applying tokenisation and stemming [29]. Thereafter, a *Vector Space Model* approach is used to create a term-by-review matrix, which is a representation of the preprocessed reviews, in which each row holds the frequency of a term, and each column represents a textual review.

The second step in preparing the data is to apply normalisation, albeit optional this step is guaranteed to improve results. The term frequencies of the term-by-review matrix are normalised by applying a weighting scheme function. Term frequency normalisation is crucial because it enables the framework to capture information about the importance of each term in describing each textual review. The purpose of term weighting is to increase or decrease the importance of terms using local and global weights in order to improve system performance. There are three types of transformations that are applied to the term-by-review matrix (in terms of granularity, every review is treated as a document) as part of the term-weighting process: local term-weighting, global term-weighting, and document (i.e. review) length-normalisation. *Local weights* determine the value of a term in a particular review, and *global weights* determine the value of a term in the entire review collection. Various local and global weighting schemes exist [30] and these are applied to the term-by-review matrix to give high weights to important terms, i.e. those that occur distinctively across reviews and low weights to terms that appear frequently in the review collection. Long reviews have a larger number of terms and term frequencies than short reviews. When applying *document length normalisation* the term values are adjusted depending on the length of each review in the collection. This improves the similarity values given to reviews since the similarity value is not influenced by their review length. There exist many weighting schemes, and the particular framework adopts the term frequency local weighting; the normal global weighting; and the cosine document length normalisation. Let the value of a term in a review be $l_{i,j} \times g_i \times n_j$, where $l_{i,j}$ is the local weighting for term i in file j, g_i is the global weighting for term i, and n_j is the document-length normalization factor [30]. Let f_{ij} define the number of times (term-frequency) term i appears in review j; the normal weighting is defined by function $1/\sqrt{\sum_j f_{ij}^2}$ and the cosine document length is defined by $\sum_i (g_i l_{ij})^2)^{-1/2}$ [31].

3.2   Input Selection Module

Feature selection is the process of selecting only the relevant features which consequently removes noise from the data. Acampora and Cosma's framework [25] presented in Fig. 1 considers features at term (i.e. term) level, however, this can be replaced by other levels of granularity such as sentence-level, or document-level. There are several approaches to feature selection, the most common ones include: Principal Component Analysis, Singular Value Decomposition, independent component analysis, chi-square testing, correlation analysis, and decision tree induction. The Genetic Algorithm (GA) is among the most effective approaches which can be applied for feature selection either on it's own or in conjunction with other feature selection approaches. A Genetic Algorithm can be applied to search and retrieve the best features for an improved evaluation of reviews or to further reduce the size of the matrix [32]. In general terms, Genetic Algorithms imitate the natural selection process in biological evolution with selection, mating reproduction and mutation. Information about genetic algorithms can be found here [33]. In the proposed framework, the Singular Value Decomposition technique was applied to derive the reviews-by-dimension matrix. It is important when computing Singular Value Decomposition that the number of features to retain must be specified, this process is called dimensionality reduction. The retained features tend to reveal the important relations between terms and reviews [17, 34]. Once the Singular Value Decomposition and dimensionality reduction processes are complete, the framework utilises a Genetic Algorithm [32] for further feature reduction. This technique has proven to be effective in further reducing the size of the data without affecting prediction accuracy [26]. By reducing the size of the data, computation time is also reduced and this increases computational efficiency. The most effective and commonly used method for selecting the number of dimensions is through experimentation. The literature proposes several approaches for determining the optimal dimensionality setting for a dataset. The optimal number of dimensions are those dimensions needed to model the data without loosing important information. Although the literature suggests many techniques for selecting dimensionality, there are no set dimensionality reduction methods or the number of dimensions that should be chosen, because the number of dimensions selected is likely to depend on the dataset in question. Techniques for aiding with selection of dimensionality include Cattell's scree test [35], Kaiser's eigenvalue greater than unity rule [36], the Maximum Likelihood test [37], Horn's parallel analysis [38], Velicer's Minimum Average Partial (MAP) [39] and Bartlett's chi-square test [40]. The literature also contains studies that compare the principal component extraction methods, including [41, 42]. Furthermore, Zwick and Velicer [43, 44] have investigated various principal component extraction methods across various factors including sample size, number of variables, number of components, and component saturation. The experimental results of Zwick and Velicer [43] suggest that Cattell's Scree test and Velicer's MAP test performed better than other tests when tested on a variety of situations. Compared to the other rules for determining the number of components to retain, Cattell's scree test is a graphical test, and is one of the most popular used test for quickly identifying the number of components to retain. Figure 2 shows an example scree test. The optimal number of dimensions reside at the point where the elbowed shape curve starts to stabilise.

Fig. 2 An example scree test [25]

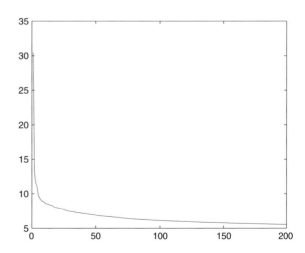

3.3 Fuzzy Predictors

The framework presented in Fig. 1 includes a number of fuzzy predictors, namely the Fuzzy C-Means, Adaptive-Neuro Fuzzy Inference System, and the Fuzzy ARTMAP which could be used for the prediction task. However, the framework could also be used by replacing the fuzzy predictors with non-fuzzy predictors such as the Support Vector Machine, Naive Bayes, Artificial Neural Network or any other computational intelligence algorithm suited for the purpose of review rating prediction.

3.3.1 Fuzzy C-Means

Cluster analysis is a technique for discovering natural groups present in the data. Essentially, cluster analysis groups a given dataset into a set of clusters in such a way that related data is placed in the same cluster. The Fuzzy C-Means (FCM) clustering algorithm was proposed by Bezdek [45]. The Fuzzy C-Means algorithm takes an input a number of clusters and a matrix of data to be clustered. Fuzzy C-Means returns the cluster centres and a membership matrix where each element holds the total membership of a data point (i.e. textual review) belonging to cluster (i.e. a rating class). Fuzzy C-Means updates the cluster centers and the membership grades for each data point (i.e. review), by minimising an objective function.With fuzzy clustering each review belongs to all cluster groups (i.e. review rating groups) with varying degree of membership from 0 to 1. For example, in a reviews data set with clusters 1, 2, 3, 4, a data point can be associated with all clusters with various degree of membership in each cluster. Therefore, instead of defining a definite association of a review with one cluster, fuzzy clustering associates a probability of membership to all clusters.

3.3.2 Adaptive Neuro-Fuzzy Inference System

The Adaptive Neuro-Fuzzy Inference System (ANFIS) was developed by Jang [46]. ANFIS combines Artificial Neural Networks and Fuzzy Logic. Given a set of input data, which in the case of Sentiment Analysis it will be the reviews-by-dimension matrix, ANFIS creates a fuzzy inference system with membership functions generated by adaptive back-propagation learning. In overall ANFIS, consists of five layers [47].

- **Layer 1** each node generates a membership grade of a linguistic variable (in the reviews scenario linguistic variables are the rating classes, i.e. class 1, class 2, class n) using a membership function.
- **Layer 2** the firing strength of each rule is calculated.
- **Layer 3** calculates the ratio of each rule's firing strength to the total of all firing strengths.

- **Layer 4** computes the contribution of each rule toward the overall output, and finally,
- **Layer 5** ANFIS computes the overall output as the summation of the contribution from each rule.

During the learning process, ANFIS adapts the parameters associated with the membership functions and tunes them using a gradient vector which, given a set of parameters, measures the performance of the system by means of how well it models input and output data. When ANFIS is used in conjunction with Fuzzy C-Means, a Fuzzy Inference Structure (FIS) is derived from Fuzzy C-Means clustering which is input into the ANFIS. The FIS parameters are tuned by ANFIS using the input/output training data in order to optimize the prediction model. ANFIS stops the training process once a specified epoch number or the training error goal is reached. The performance of ANFIS is commonly evaluated by computing the root mean square training errors among the predicted and target outputs.

3.3.3 Simplified Fuzzy ARTMAP

The Simplified Fuzzy ARTMAP (SFAM) [48] is a simplification of the fuzzy ARTMAP algorithm which was originally proposed by Carpenter et al. [49]. The Simplified Fuzzy ARTMAP, which is also known as Predictive ART, and can only be applied to solve classification tasks. The advantages of the Simplified Fuzzy ARTMAP over other fuzzy approaches is that it has reduced computational overhead and architectural redundancy (i.e. simplified learning equations, it has a single user selectable parameter and uses a small number of iterations to learn all training patterns) [50]. Basically, the Simplified Fuzzy ARTMAP is a type of a Neural Network classifier. To apply the Simplified Fuzzy ARTMAP for sentiment analysis and prediction of review ratings using the proposed framework, the reviews-by-dimension matrix which was derived after performing Singular Value Decomposition and dimensionality reduction (i.e. input selection module) is passed into the classifier which initially normalizes the input matrix to show the presence or absence of a particular feature in the input pattern. The weights from each of the output category nodes reach down to the input layer. Each input is mapped to a 'winner output' which is basically the category assigned to an input vector (i.e. the predicted rating of a review represented as an input vector).

3.4 Projection of New Reviews for Prediction

A new customer review can be represented as a vector, and projected into the reduced dimensional space, which results from the Input Selection module. Once the query vector is projected into the term-review space it can be compared to all other existing review vectors using a similarity measure such as the *cosine*. Typically, using the cosine measure, the angle between the new review and all other review vectors is computed. This would allow retrieval of reviews which are semantically similar to the new review. Once projected into the dimensional space, the review rating prediction model can be used to predict the review rating of the new review.

4   Evaluation of the Framework

This section describes the results from experiments carried out for testing the performance of the hybrid computational intelligence framework shown in Fig. 1 [26] when using various fuzzy approaches. In addition the impact of applying a Genetic Algorithm, for feature selection, on the review rating performance of classifiers is also investigated. A brief discussion of evaluation measures which are commonly adopted for sentiment analysis evaluations is provided.

4.1   Datasets

An experiment was conducted using two large datasets,[1] namely a software reviews dataset and a wrist watch reviews dataset. The software reviews dataset consists of 95, 084 textual reviews on software products, and the wrist watch reviews dataset consists of 68, 355 textual reviews of wrist watches. Each review is accompanied by a numerical rating provided by the customer who provided the review. The reviews in both datasets were written by various authors. Table 1 shows the characteristics of the datasets. The number of terms is the total number of distinct terms (i.e. distinct words) found across the entire collection of reviews in a particular dataset, and the number of reviews is the total number of reviews found in a dataset.

Table 2 shows the number of reviews in each rating category.

Section 4.2 describes evaluation measures which are commonly adopted for sentiment analysis and prediction tasks. These measures were used to evaluate the performance of the framework presented in Fig. 1 to predict the review ratings on the software reviews and wrist watch reviews datasets.

4.2   Evaluation Measures

This section presents a small scale evaluation for assessing the performance of the framework for predicting the rating classes of reviews, as presented in Acampora and Cosma [26]. Performance can be measured using the *Accuracy* which evaluates

---

[1] Available at http://snap.stanford.edu/data/web-Amazon.html.

Table 1 Stanford dataset characteristics

Stanford review datasets		
Dataset	No. of terms	No. of reviews
Wrist Watches	16,547	68,355
Software	11,210	95,084
Total	27,757	163,439

Table 2 Stanford dataset ratings

	Wrist watches		Software	
Rating class	Count	Percent	Count	Percent
1	5376	7.86	25798	27.13
2	3650	5.34	8076	8.49
3	5398	7.90	8894	9.35
4	14725	21.54	16360	17.21
5	39206	57.36	35956	37.81

the performance of the system as the fraction of its classifications that are correct—it considers the number of true positives and true negatives over all classified cases.

The highest the value of the accuracy is, the better the performance of the system. *Misclassification rate* is also a reliable measure because it gives the fraction of reviews which have been misclassified. Misclassification rate is computed as 1-Accuracy, and the lowest the value the better the performance of the classification system. The functions for accuracy and misclassification rate are shown in function (1) and (2) respectively. Note that for evaluations of non binary classification, the following evaluation measures are applied to each class rating and the average values of performance across all classes is computed.

$$Accuracy = \frac{|TP| + |TN|}{|P| + |N|} \tag{1}$$

$$Misclassification\ rate = \frac{|FP| + |FN|}{|P| + |N|} \tag{2}$$

where for each class of reviews,

- $|TP|$ is the total number of true positive reviews retrieved by the classifier. These are the positive reviews that were correctly labeled by the classifier.
- $|TN|$ is the total the number of true negative reviews retrieved by the classifier. These are the negative reviews that were correctly labeled by the classifier.
- $|FP|$ is the total number of false positive reviews retrieved by the classifier. These are the negative reviews that were incorrectly labeled by the classifier as positive.

- $|FN|$ is the total number of false negative reviews retrieved by the classifier. These are the positive reviews that were incorrectly labeled by the classifier as negative.
- $|P|$ is the total number of positive reviews that exist in the dataset, where $|P| = |TP| + |FN|$.
- $|N|$ is the total number of negative reviews that exist in the dataset, where $|N| = |FP| + |TN|$.

4.3 Experiment Results

Experiments were conducted using the datasets described in Sect. 4.1. The performance of the framework, when utilising different fuzzy approaches, was evaluated using the evaluation measures described in Sect. 4.2. Next the review rating prediction performance of the framework is discussed.

Wrist Watch Reviews dataset: As shown in Table 3, using the *Wrist Watch reviews* dataset, the performance of the FCM and FCM + ANFIS predictors was the same (accuracy = 0.89 for each predictor) regardless whether the Genetic Algorithm was adopted for further feature reduction. However, ANFIS optimises the performance of FCM, where possible, and it is guaranteed to improve the results of FCM, so it is worth adopting ANIFS with FCM. Regarding the Input selection module, and whether the Genetic Algorithm should be adopted, the results show that the accuracy of the SFAM algorithm improved by 0.03 after the Genetic Algorithm was applied. Applying the Genetic Algorithm also meant removing five features which were not needed for prediction. The fact that the input matrix was further reduced had an impact on the speed of training the fuzzy classifier for prediction.

Software Reviews dataset: Table 4 shows the results of the experiments when using the *Software reviews* dataset. The FCM + ANFIS predictor outperformed all

Table 3 Wrist watch reviews dataset results

	Wrist watches reviews dataset					
	Without GA (k = 40)			With GA (k = 35)		
	FCM	FCM + ANFIS	SFAM	FCM	FCM + ANFIS	SFAMP
Accuracy	0.89	0.89	0.83	0.89	0.89	0.86
Misclassification	0.11	0.11	0.17	0.11	0.11	0.14

Table 4 Software reviews dataset results

	Software reviews dataset					
	Without GA (k = 40)			With GA (k = 35)		
	FCM	FCM + ANFIS	SFAM	FCM	FCM + ANFIS	SFAM
Accuracy	0.75	0.79	0.77	0.75	0.78	0.77
Misclassification	0.25	0.21	0.23	0.25	0.22	0.23

Table 5 Average performance across the wrist watches and software reviews datasets

| | Average performance when using different fuzzy predictors | | | | | |
| | Without GA (k = 40) | | | With GA (k = 35) | | |
	FCM	FCM + ANFIS	SFAM	FCM	FCM + ANFIS	SFAM
Accuracy	0.82	0.84	0.80	0.82	0.84	0.81
Misclassification	0.18	0.16	0.20	0.18	0.16	0.19

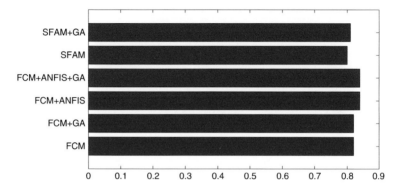

Fig. 3 Performance of datasets measured by accuracy

other predictors when the Genetic Algorithm (GA) was not adopted (accuracy = 0.79) and when it was adopted (accuracy = 0.78) for further feature reduction. Applying the Genetic Algorithm for further feature reduction did not have any impact on the accuracy of the FCM or the SFAM predictors, with accuracy remaining at 0.75 and 0.77 respectively, before and after the Genetic Algorithm was applied. The performance of the FCM + ANFIS very slightly decreased, by less than 0.01 (rounded up as 0.01) when GA was applied.

Overall performance: Table 5 and Fig. 3 show the average results of the experiments across the two datasets. In overall across both datasets, best performance was achieved by $FCM + ANFIS$ (accuracy = 0.84) and $FCM + ANFIS + GA$ (accuracy = 0.84), closely followed by FCM (accuracy = 0.82) and $FCM + GA$ (accuracy = 0.82), $SFAM + GA$ (accuracy = 0.81) and SFAM (accuracy = 0.80). Table 6 shows the difference in accuracy when GA is and is not applied. This difference was computed by a subtraction, $diff = acc\_non\_GA - acc\_with\_GA$. The Genetic Algorithm slightly improved the performance of FCM + ANFIS (by 0.01) and slightly decreased the performance of the SFAM (accuracy increased by 0.01). This finding clearly shows the benefits of including GA for further feature extraction. On large datasets, reducing the number of dimensions reduces computational complexity and the time needed to train a predictor, since it will be trained in a smaller matrix. The fact the performance of each predictor was not affected when applying the Genetic Algorithm is a positive result, which clearly demonstrates the improved

Table 6 Difference in accuracy when GA was and was not applied. Next to each result, a symbol indicates the type of difference: = means that there is no difference, > means that not applying GA returned higher accuracy, < mean that applying GA returned higher accuracy

Stanford datasets

	FCM versus FCM + GA	FCM + ANFIS versus FCM + ANFIS + GA	SFAM versus SFAM + GA
Wrist watch reviews dataset	0.00 (=)	0.00 (=)	−0.03 (<)
Software reviews dataset	0.00 (=)	0.00 (=)	0.00 (=)
Average	0.00 (=)	0.00 (=)	−0.01 (<)

efficiency of the prediction model when the Genetic Algorithm is adopted for the task of further reducing the dimensionality of data.

5 Conclusion

Consumers often provide on-line reviews on products or services they have purchased, and frequently seek on-line reviews about a product or service before deciding whether to make a purchase. Organisations seek consumers opinions about their products, since this invaluable information allows them to improve future product versions, and to predict sales. The vast amount of on-line customer reviews has attracted research into approaches for intelligently mining these reviews to support decision-making processes. This chapter provides an overview of recent ontology and non-ontology fuzzy-based approaches to sentiment analysis of customer reviews. Acampora and Cosma's [26] neuro-fuzzy framework which is suitable for sentiment analysis and review rating prediction is presented and evaluated. The framework includes methods for preparing the dataset, extracting the best features for prediction via Singular Value Decomposition and dimensionality reduction, and reducing further the number of features via the application of a Genetic Algorithm. The framework can be used with various fuzzy-based predictors such as Fuzzy C-Means, a neuro-fuzzy approach combining Fuzzy C-Means and the Adaptive Neuro-Fuzzy Inference System, and the Simplified Fuzzy ARTMAP. The performance of the framework was evaluated using the above mentioned fuzzy predictors applied to two large datasets containing customer reviews. The framework demonstrates a technique of using Singular Value Decomposition technique for reducing the dimensionality and hence noise from a dataset, and then using the Genetic Algorithm for reducing the number of dimensions even further. The results revealed that the Genetic algorithm is effective in further reducing the number of dimensions, after Singular Value Decomposition and dimensionality reduction, without affecting the prediction performance of each algorithm. On large datasets, reducing the number of dimensions reduces computational complexity and the time needed to train a predictor.

References

1. Liu, B.: Sentiment Analysis and Subjectivity, 2nd edn. Taylor and Francis Group (2010)
2. Sparks, B.A., Browning, V.: The impact of online reviews on hotel booking intentions and perception of trust. Tourism Manage. **32**(6), 1310–1323 (2011)
3. Papathanassis, A., Knolle, F.: Exploring the adoption and processing of online holiday reviews: a grounded theory approach. Tourism Manage. **32**(2), 215–224 (2011)
4. Park, D.-H., Lee, J., Han, I.: The effect of on-line consumer reviews on consumer purchasing intention: the moderating role of involvement. Int. J. Electron. Commer. **11**(4), 125–148 (2007)
5. Feicheng, M., Tao, L., Guoliang, Z.: A conceptual model of e-commerce sale service for manufacturing industry. In: International Conference on E-Business and Information System Security, 2009. EBISS '09, pp. 1–5 (2009)
6. Wu, L.-Y., Chen, K.-Y., Chen, P.-Y., Cheng, S.-L.: Perceived value, transaction cost, and repurchase-intention in online shopping: a relational exchange perspective. J. Bus. Res. **67**(1), 2768–2776 (2014)
7. Reynolds, K.L., Harris, L.C.: Dysfunctional customer behavior severity: an empirical examination. J. Retail. **85**(3), 321–335 (2009)
8. Evanschitzky, H., Iyer, G.R., Hesse, J., Ahlert, D.: E-satisfaction: are-examination. J. Retail. **80**(3), 239–247 (2004)
9. Dellarocas, C., Zhang, X.M., Awad, N.F.: Exploring the value of online product ratings in revenue forecasting: the case of motion pictures. J. Interact. Mark. **21**(4), 23–45 (2007)
10. Park, D.-H., Lee, J., Han, I.: The effect of on-line consumer reviews on consumer purchasing intention: The moderating role of involvement. International Journal of Electronic Commerce **11**(4), 125–148 (2007)
11. Cui, G., Lui, H.-K., Guo, X.: The effect of online consumer reviews on new product sales. Int. J. Electron. Commer. **17**(1), 39–58 (2012)
12. Ghose, A., Ipeirotis, P.G.: Estimating the helpfulness and economic impact of product reviews: mining text and reviewer characteristics. IEEE Trans. Knowl. Data Eng. **23**(10), 1498–1512 (2011)
13. Liu, J., Cao, Y., Lin, C.-Y., Huang, Y., Zhou, M.: Low-quality product review detection in opinion summarization. In: Proceedings of the Joint Conference on Empirical Methods in Natural Language Processing and Computational Natural Language Learning (EMNLP-CoNLL), pp. 334–342 (2007). Poster paper
14. Liu, Y., Huang, X., An, A., Yu, X.: Helpmeter: a nonlinear model for predicting the helpfulness of online reviews. In: IEEE/WIC/ACM International Conference on Web Intelligence and Intelligent Agent Technology, 2008. WI-IAT '08, vol. 1, pp. 793–796 (2008)
15. Danescu-Niculescu-Mizil, C., Kossinets, G., Kleinberg, J., Lee, L.: How opinions are received by online communities: a case study on amazon.com helpfulness votes. Proceedings of the 18th International Conference on World Wide Web. WWW '09, pp. 141–150. ACM, NY (2009)
16. Dumais, S., Furnas, G., Landauer, T., Deerwester, S., Harshman, R.: Using latent semantic analysis to improve access to textual information. In: CHI '88: Proceedings of the SIGCHI Conference on Human Factors in Computing Systems, pp. 281–285. ACM, NY (1988)
17. Berry, M.W., Dumais, S.T., O'Brien, G.W.: Using linear algebra for intelligent information retrieval. Technical Report, UT-CS-94-270 (1994)
18. Tang, H., Tan, S., Cheng, X.: A survey on sentiment detection of reviews. Expert Syst. Appl. **36**(7), 10760–10773 (2009)
19. Pang, B., Lee, L.: Opinion mining and sentiment analysis. Foundation Trends in Information Retrieval **2**(1–2), 1–135 (2008)

20. Prabowo, R., Thelwall, M.: Sentiment analysis: a combined approach. J. Informetrics **3**(2), 143–157 (2009)
21. Li, S.-T., Tsai, F.-C.: A fuzzy conceptualization model for text mining with application in opinion polarity classification. Knowl. Based Syst. **39**, 23–33 (2013)
22. Loia, V., Senatore, S.: A fuzzy-oriented sentic analysis to capture the human emotion in web-based content. Knowl. Based Syst. **58**, 75–85 (2014)
23. Cambria, E., Hussain, A., Havasi, C., Eckl, C.: Sentic computing: exploitation of common sense for the development of emotion-sensitive systems **5967**, 148–156 (2010)
24. Dragoni, M., Tettamanzi, A.G.B., da Costa Pereira, C.: Propagating and aggregating fuzzy polarities for concept-level sentiment analysis. Cogn. Comput. 7(2), 186–197 (2015)
25. Acampora, G., Cosma, G.: A hybrid computational intelligence approach for efficiently evaluating customer sentiments in e-commerce reviews. In: 2014 IEEE Symposium on Intellelligent Agents (IA), pp. 73–80 (2014)
26. Acampora, G., Cosma, G.: A comparison of fuzzy approaches to e-commerce review rating prediction. In: IFSA-EUSFLAT-2015, pp. 73–80 (2015)
27. Lau, R.Y.: Li, C., Liao, S.S.: social analytics: learning fuzzy product ontologies for aspect-oriented sentiment analysis. Decis. Support Syst. 65, pp. 80–94 (2014)
28. Nie, X., Liu, L., Wang, H., Song, W., Lu, j.: The opinion mining based on fuzzy domain sentiment ontology tree for product reviews. J. Softw. 8(11)
29. Baeza-Yates, R., Ribeiro-Neto, B.: Modern Information Retrieval. ACM Press/Addison-Wesley (1999)
30. Berry, M., Browne, M.: Understanding Search Engines: Mathematical Modeling and Text Retrieval (Software, Environments, Tools), 2nd edn. In: Society for Industrial and Applied Mathematics. Philadelphia (2005)
31. Singhal, A., Salton, G., Mitra, M., Buckley, C.: Document length normalization. Technical Report. Cornell University, Ithaca (1995)
32. Ludwig, O., Nunes, U.: Novel maximum-margin training algorithms for supervised neural networks. Trans. Neural Netw. **21**(6), 972–984 (2010)
33. Mitchell, M.: An Introduction to Genetic Algorithms. MIT Press, Cambridge (1998)
34. Berry, M.: Large-scale sparse singular value computations. Int. J. Supercomput. Appl. **6**(1), 13–49 (1992)
35. Cattell, R.B.: The scree test for the number of factors. Multivar. Behav. Res. **1**, 245–276 (1966)
36. Kaiser, H.F.: The application of electronic computers to factor analysis. Educ. Psychol. Measur. **20**, 141–151 (1960)
37. Joreskog, K.G.: Some contributions to maximum likelihood in factor analysis. Psychometrika **32**, 433–482 (1967)
38. Horn, J.L.: A rationale and test for the number of factors in factor analysis. Psychometrika 30(179–185)
39. Velicer, W.: Determining the number of components from the matrix of partial correlations. Psychometrika **41**, 321–327 (1976)
40. Bartlett, M.S.: Tests of significance in factor analysis. Br. J. Psychol. **3**, 77–85 (1950)
41. Hubbard, R., Allen, S.J.: An empirical comparison of alternative methods for principal component extraction. J. Bus. Res. **2**(15), 173–190 (1987)
42. Ferr, L.: Selection of components in principal component analysis: a comparison of methods. Comput. Stat. Data Anal. **19**(6), 669–682 (1995)
43. Zwick, W.R., Velicer, W.F.: Factors influencing four rules for determining the number of components to retain. Multivar. Behav. Res. **17**(2), 253–269 (1982)
44. Zwick, W.R., Velicer, W.F.: Factors influencing five rules for determining the number of components to retain. Psychologica Bulleti **99**, 432–442 (1986)
45. Bezdek, J.C.: Pattern Recognition with Fuzzy Objective Function Algorithms. Kluwer Academic Publishers, Norwell (1981)
46. Jang, J.S.R.: Anfis: adaptive-network-based fuzzy inference system. IEEE Trans. Syst. Man Cybern. **23**(3), 665–685 (2002)

47. Jang, J.S.R.: Input selection for anfis learning. In: Proceedings of the IEEE International Conference on Fuzzy Systems, pp. 1493–1499 (1996)
48. Kasuba, T.: Simplified fuzzy artmap. AI Expert **8**(11), 18–25 (1993)
49. Carpenter, G., Grossberg, S., Rosen, D.: Fuzzy art: fast stable learning and categorization of analog patterns by an adaptive resonance system. Neural Netw. **4**(6), 759–771 (1991)
50. Marungsri, B., Boonpoke, S.: Applications of simplified fuzzy artmap to partial discharge classification and pattern recognition. WTOS **10**(3), 69–80 (2011)

OntoLSA—An Integrated Text Mining System for Ontology Learning and Sentiment Analysis

Ahmad Kamal, Muhammad Abulaish and Jahiruddin

Abstract Since the inception of the Web 2.0, World Wide Web is widely being used as a platform by customers and manufactures to share experiences and opinions regarding products, services, marketing campaigns, social events, etc. As a result, there is enormous growth in user-generated contents (e.g. customer reviews), providing an opportunity for data analysts to computationally evaluate users' sentiments and emotions for developing real-life applications for business intelligence, product recommendation, enhanced customer services, and target marketing. Since users' feedbaks (aka reviews) are very useful for products development and marketing, large business houses and corporates are taking interest in opinion mining and sentiment analysis systems. In this chapter, we propose the design of an Ontology Learning and Sentiment Analysis (*OntoLSA*) system for ontology learning and sentiment analysis using rule-based and machine learning approaches. The rule-based approach aims to identify candidate concepts, which are analyzed using a customized HITS algorithm to compile a list of feasible concepts. Feasible concepts and their relationships (both structural and non-structural) are used to generate a domain ontology, which is later on used for opinion mining and sentiment analysis. The proposed system is also integrated with a visualization module to facilitate users to navigate through review documents at different levels of granularity using a graphical user interface.

Keywords Computational intelligence · Opinion mining · Sentiment analysis · Ontology learning · Visualization

A. Kamal
Department of Mathematics, Jamia Millia Islamia, New Delhi 25, India
e-mail: ah.kamal786@gmail.com

M. Abulaish (✉) · Jahiruddin
Department of Computer Science, Jamia Millia Islamia, New Delhi 25, India
e-mail: abulaish@ieee.org

Jahiruddin
e-mail: jahir.jmi@gmail.com

© Springer International Publishing Switzerland 2016
W. Pedrycz and S.-M. Chen (eds.), *Sentiment Analysis and Ontology Engineering*,
Studies in Computational Intelligence 639, DOI 10.1007/978-3-319-30319-2_16

1   Introduction

The emergence of Web 2.0 has caused rapid proliferation of e-commerce and social media contents. Web is widely used as a platform by users and manufactures to share experiences and opinions regarding social events, political movements, products, services, and marketing campaigns. Enormous availability of user-generated contents (e.g., customer reviews) on the Web have attracted researchers to design and develop linguistically motivated computational paradigms to retrieve most cogent and desired information for developing real-life applications, including business intelligence, recommendation system, target marketing, and Web surveillance. However, exponential growth of unstructured and semi-structured user-generated contents on the Web and their uncontrolled generation consisting of various natural language nuances posses a big challenge for research community to fully automate the process of information component extraction. Thus, distillation of knowledge from such sources is a technically challenging task and requires research at the intersection of various disciplines, including natural language processing, ontology engineering, information retrieval and extraction, and computational intelligence.

1.1   Opinion Mining and Sentiment Analysis

Over the past decade, a new research discipline of opinion mining and sentiment analysis as a special case of web content mining has emerged, which mainly focuses to analyze review documents. Sentiment analysis is the computational evaluation of users' opinions, subjectivity, appraisals, emotions, and feedbacks expressed in review documents [28]. Since opinions are very informative in developing marketing and product development plans, large business houses and corporates are taking interest in opinion mining systems for developing business intelligence applications and recommendation systems. Such systems can process users' opinions and sentiments to predict better recommendations for target marketing of products and services. Usage of opinion mining and sentiment analysis has encouraged governments and security agencies in enhancing their abilities to analyze web usage patterns of users on the Web. Intelligence system can be able to track sentiments of Web users on the basis of their usage patterns, friends, and communities they form and participate. Such analysis is very useful in observing and discarding bashing or flames used in social communication and helpful in developing spam detection system [7].

Various research efforts attempted to mine opinions from customer reviews at different levels of granularity, including word-, sentence-, and document-level sentiment analyses. However, development of a fully automated opinion mining and sentiment analysis system is still elusive. A study conducted in [16] reveals that a product feature can appear as an explicit or implicit mention in review documents. Explicit features precisely and clearly appear in one or more review sentences, whereas an implicit feature does not explicitly appear. For example, consider the first and second review sentences of Table 1, in which product features are italicized and opinion words are marked as bold. In first sentence, the features *video quality* and *software* are explicitly mentioned along with their respective opinion bearing words *great*, *amazing* and *friendly*. In contrast, second sentence does not mention any product feature explicitly, rather an opinion word *affordable* is appearing which can be used to infer the implicit feature *price*. Another challenge remains in the fact that a large number of feature words are referenced by anaphoric pronouns present in succeeding sentences of a review document. For example, anaphora word *it* in the fourth sentence of Table 1 is actually referring to the product feature *battery* mentioned in the third sentence.

The challenges of opinion mining and sentiment analysis have been viewed from different perspectives, including feature and opinion extraction to sentiment classification. A study conducted in [33] on movie reviews reveals that sentiment classification problem is different from traditional topic-based classification, as topics are frequently inferable by keywords, whereas sentiment can be expressed in a more subtle or delicate manner. For example, the opinion sentence "how could anyone sit through this movie?," does not contain any single word that reflects negative opinion. This study also reveals that the task of identifying the seed set of keywords representing positive or negative sentiment is less trivial. Moreover, opinions appear with different strength and levels, including some very strong and weak opinions [43]. Therefore, degree of expressiveness of opinion is required for describing the expression level of an opinion [2]. For example, consider the fifth and sixth sentences of Table 1 that contain an opinion bearing word *good*, but the associated modifiers *very* and *almost* are fuzzy qualifiers expressing different levels of customer satisfaction regarding the *battery life* feature of the product.

In general, sentiment analysis task is context-sensitive and domain-dependent despite of the fact that general views of positive and negative opinions remain consistent across different domains [32]. For example, same opinionated word *low* expresses different sentiments (positive or negative) in the feature-opinion pairs ⟨price, low⟩ and ⟨quality, low⟩, respectively. Moreover, presence of negations in review sentences makes the task of automatic sentiment classification a more challenging task. For example, in the review sentence "the picture quality of this camera is not really good", the opinionated word *good* requires inference in negative sense for proper polarity determination with respect to the *picture quality* feature of the product.

S. no.	Sentence
1	The *video quality* is **great**. The *software* is **amazing** and **friendly**
2	So much packed in a small case and very **affordable**!
3	*Battery* goes off so quickly
4	Many times it becomes very **hot** while talking
5	The *battery life* is very **good**
6	The *battery life* is almost **good**

Table 1 Sample opinion bearing sentences

1.2 Role of Ontology in Sentiment Analysis

Ontology provides a knowledge management structure to represent concepts (e.g. product features) and their relationships in an inter-related and comprehendible manner. Such knowledge management structure is useful for both users and software agents. Ontology is helpful in providing structural framework for information expression and enhancing the quality of information extraction. However, ontology development and maintenance is an expensive task and require significant participation of domain experts. Overall task related to relevant knowledge extraction and structuring is both non-trivial for a particular domain. Besides, an effective solution to this problem might be to develop a system that automatically extracts knowledge from source documents for ontology creation and enrichment [1]. Since product features are generally nouns [16] and opinions are adjectives [15], rule-based approaches based on Parts-Of-Speech (POS) information along with some statistical techniques for feasibility analysis can be used to identify concepts (representing features and opinions) and their relationships for ontology learning from text documents.

Though the development of feature-based opinion mining and sentiment analysis systems are gaining momentum, fewer research efforts have been applied in exploiting the benefits of structuring product features for sentiment analysis. It is observed that some of the features exhibit relationship with each others in product domains. According to Cadilhac et al. [6] and Wei et al. [42], the knowledge of hierarchical relationships among product features is not utilized properly in opinion mining and sentiment analysis research. Cadilhac et al. [6] emphasized that knowledge management structure to represent concepts, relationships, and lexical information can be exploited to extract implicit features. Moreover, hierarchical analysis with ontological information can be exploited in propagating feature specific sentiment information to infer sentiments at higher levels of abstraction, specially in the cases where explicit mentions of sentiment information with associated features are missing. In addition, product ontology expedites comprehensive review summarization and facilities in visualizing customer reviews at different levels of granularity.

In this chapter, we propose the design of an Ontology Learning and Sentiment Analysis (*OntoLSA*) system for ontology learning and sentiment analysis using rule-based and machine learning approaches. The proposed system is equipped with a crawler to retrieve relevant reviews from merchant sites and applies various pre-processing techniques for their cleaning, chunking, and parsing using a statistical parser. A rule-based system using POS information and dependency relationships between the words is developed to identify candidate concepts, which are analyzed using a customized HITS algorithm for feasibility analysis [21]. Since ontology is not merely a data store rather a knowledge management tool, domain expert supervision on the list of feasible concepts is performed for validation. Feasible concepts and their relationships (both structural and non-structural) are used to generate an ontology, which later on used to extract opinionated words associated with the feasible concepts. We propose word-level sentiment classification method with the aid of statistical and machine learning techniques to determine users' sentiments. One of the crucial requirements when developing an opinion mining and sentiment analysis system is the ability to browse through the review documents and to visualize various concepts and their relationships along with sentiment information present within the collection in a summarized form. Thus, the proposed system is also integrated with a visualization module to facilitate users to navigate documents collection at different levels of granularity using a graphical user interface.

The rest of the chapter is structured as follows. Section 2 presents a review of the related works on ontology-based opinion mining and sentiment analysis. Section 3 presents the architectural and functional details of the proposed Ontology Learning and Sentiment Analysis (*OntoLSA*) system. Section 4 presents the design of a graphical user interface for review summarization and visualization. The experimental setup and evaluation results are presented in Sect. 5. Finally, Sect. 6 concludes the chapter with future directions of work.

2   Related Works

Opinion mining and sentiment analysis is the computational evaluation and representation of users' sentiments and emotions expressed in customer reviews for developing practical real-life applications. Over the past decade, various research attempts have been made to mine product- or service-specific users' sentiments at document level to classify reviews as positive, negative, or neutral [33, 41]. Document-level sentiment analysis approaches treat each review document as a basic information unit [27]. Such approaches neither acquire insight knowledge regarding product features and their relationships nor exhibit users' feature-specific sentiments. As observed by many researchers, a positive review might contain features that are liked as well as disliked by an opinion holder, but the overall sentiment on the review remains positive. Similarly, a negative review does not mean that the opinion holder dislikes everything regarding the product. Therefore, research work based on feature-based opinion mining is proposed in [16], wherein authors applied a three-step process for feature and opinion extraction. In the first step, product features commented by the end users are identified. In the second step, all opinion sentences containing extracted features are identified and marked as either positive or negative. Finally, review documents are compiled on the basis of individual feature to classify positive and negative opinion sentences. In [34], the authors proposed the design of OPINE system based on an unsupervised pattern mining approach which extracts product features using feature assessor and web PMI statistics. Various extraction rules are applied to associate product features with potential opinion phrases. Relaxation labeling approach is used to classify the sentiment information of opinion phrases. In [20, 25, 45], the authors applied a finer-grained analysis on customer reviews to associate product features and sentiment bearing words on the basis of their explicit co-occurrence at sentence level. In [45], a supervised multi-knowledge based approach is proposed for movie reviews analysis, which applies grammatical rules to identify feature and opinion pairs. In contrast, Hu and Liu in [16] associated each feature word with its nearest opinion bearing word to form feature and opinion word pairs. As a result many ambiguous and noisy pairs were extracted due to complexity of the sentences in movie reviews. A study in [20] explored the link between opinion target (product feature) and opinion expression in review sentences. Using data sets containing car and camera reviews in which features and opinions are manually annotated, a Support Vector Machine (SVM) based classifier is trained for identifying related features and opinions. Considering the algorithm proposed in [4] as a baseline, authors claimed to achieve F-score of 0.70, outperforming the baseline which yields F-score of 0.45 only. A study in [25] emphasized that a complete opinion is always expressed in one sentence along with its relevant feature. Therefore, to avoid false association between features and opinions, it is better to identify them at sentence level. Further, it is highlighted that feature and opinion word pairs retain intra-sentence contextual information for reflecting relevant opinions. Similarly, inter-sentence contextual information is also considered to imitate the relationship among opinions on the same topic or feature. Thus, both intra-sentence and inter-sentence contextual information are exploited for development of effective opinion retrieval method. In [37], a semi-supervised technique based on double propagation approach is proposed to extract opinion words and product features using a small seed of opinion lexicon, and thereafter the newly extracted opinion words and product features are exploited further for feature and opinion words extraction. Syntactic relationships are exploited using dependency parser to associate opinion words with appropriate features.

Applications and usage of ontology-based text information processing for knowledge management structure in a machine-interpretable format have attracted a number of researchers in the field of opinion mining and sentiment analysis. In [8], the authors presented the development of a system (OMINE) for identifying product features using premodeled knowledge. Manually created ontology for car domain (car functions, properties, and components) is used as a set of candidate features. The ontology is dynamically enriched by searching for phrases which match any of the manually defined lexical patterns. In [44], an ontology supported polarity mining technique is proposed for movie reviews in which a hybrid approach for ontology development is adopted. Authors stated that their ontology based approach is not only helpful in improving the efficiency of sentiment analysis task over a standard baseline, but also helps in understanding and management of movie reviews. In [23], a fuzzy ontology based context-sensitive opinion mining system is presented in which a variant of the Kullback-Leibler divergence statistical learning technique is used to predict sentiment polarities. In [40], the authors proposed an ontology-based combined approach for sentiment classification on movie reviews. Ontology development is performed using Web Ontology Language (OWL), and sentiment classification is performed using Support Vector Machine (SVM). In [42], a supervised Hierarchical Learning (HL) process is adopted to propose a manually constructed Sentiment Ontology Tree (SOT) to structure product attributes and associated sentiments from digital camera reviews. Knowledge derived from HL-SOT is utilized to handle sentiment analysis in a hierarchical classification process. Mukherjee and Joshi exploited ontological information in aggregating feature-specific sentiments to derive the overall polarity of review documents [30]. Ontology construction is accomplished using

ConceptNet 5,[1] a very large machine usable semantic network of common sense knowledge. Kontopoulos et al. proposed an ontology based approach for fine-grained sentiment analysis of twitter posts with respect to the subjects discussed in tweets [22]. A sentiment grade for each aspect is obtained, which facilitates in detailed analysis of post opinions with respect to a specific topic. A semi-automatic, data-driven ontology editor OntoGen is used for ontology learning [12] and OPenDover[2] web service is employed for sentiment analysis. However, authors emphasized that use of third party sentiment analysis tool may be considered as one of the limitations of their proposed work as working approach of such tool cannot be verified; the source code and methodology are not publically available.

3 Proposed *OntoLSA* System

In this section, we present the architectural and functional details of our proposed *OntoLSA* system. The major functionalities of the proposed system are—*review document crawling, document preprocessing, concept mining and ontology learning, sentiment analysis,* and *review summarization and visualization.* The functional details of these modules are presented in the following sections.

3.1 Review Document Crawling

For a target review site, the crawler retrieves review documents and stores them locally for further processing. The data samples used in our experimental work consist of review documents on different electronic products crawled from http://www.amazon.com using crawler4j API.[3]

3.2 Document Preprocessing

Crawled review documents are filtered to remove markup language tags and unwanted texts, such as meta-data information containing details regarding source of a review, author name, description, posting date, and star ratings. Filtered review documents are tokenized into record-size chunks (sentences), boundaries of which are decided heuristically on the basis of the presence of special characters. Depending on the application, a record-size chunk may contain a sentence, a paragraph, or a complete document. Thereafter, the sentences are parsed using Stanford parser,[4] which assigns

---

[1] http://conceptnet5.media.mit.edu.

[2] http://opendovr.nl.

[3] http://code.google.com/p/crawler4j.

[4] http://nlp.stanford.edu/software/lex-parser.shtml.

Table 2 Sample review sentences

S. no.	Sentence
S1	During calls the speaker volume is very good
S2	The sound sometimes comes out very clear
S3	Nokia N95 has a pretty screen

Table 3 Sample sentences with POS tags

S. no.	Sentence with POS tags
S1	During/IN calls/NNS the/DT speaker/NN volume/NN is/VBZ very/RB good/JJ./
S2	The/DT sound/NN sometimes/RB comes/VBZ out/IN very/RB clear/JJ./
S3	Nokia/NNP N95/NNP has/VBZ a/DT pretty/JJ screen/NN./

POS tags to every word in a sentence. The POS tag reflects the syntactic category of the words and plays vital role in identification of candidate constituent for ontology learning from texts. The POS tags are also useful to identify the grammatical structure of the sentences, like *noun*, *verb*, *adverb*, and *adjective* phrases and their inter-relationships.

Stanford parser is also used to convert each sentence into a set of dependency relations between the pair of words. The dependency relations between a pair of words w_1 and w_2 is represented as *relationType*(w_1, w_2), in which w_1 is called head or governor and w_2 is called dependent or modifier. The relationship *relationType* between w_1 and w_2 can be of two types—(i) direct relationship, or (ii) indirect relationship [36]. In a direct relationship, one word depends on the other or both of them depend on a third word directly, whereas in an indirect relationship, one word depends on the other through other words or both of them depend on a third word indirectly. A list of sample review sentences along with the POS tags and dependency relationships generated by the Stanford parser are shown in Tables 2, 3, and 4, respectively.

3.3 Concept Mining and Ontology Learning

The functionality of this module is to facilitate the linguistic and semantic analyses of texts to identify candidate constituents for knowledgebase. The candidate concepts are analyzed using a customized HITS algorithm to compile a list of feasible concepts. Feasible concepts and their relationships (both structural and non-structural) are used to generate an ontology, which later on used for opinion mining and sentiment analysis purpose.

Table 4 Sample sentences with dependency relationships

S. no.	Dependency relationships
S1	during(good-8, calls-2)det(volume-5, the-3)nn(volume-5, speaker-4)nsubj(good-8, volume-5)aux(good-8, is-6)advmod(good-8, very-7)
S2	det(sound-2, The-1)nsubj(comes-5, sound-2)det(times-4,some-3)nsubj(comes-5, times-4)dep(clear-8, out-6)advmod(clear-8, very-7)acomp(comes-5, clear-8)
S3	nn(N95-2, Nokia-1)nsubj(has-3, N95-2)det(screen-6, a-4)amod(screen-6, pretty-5) dobj(has-3, screen-6)

3.3.1 Concept Extraction

The concepts or entities normally emerge as a noun phrase in customer reviews. Thus, every n-gram (1- and 2-g) of a review sentence is accessed on the basis of POS information and dependency relationships between the words. For example, in the sample sentence S1, the bigram *speaker volume* is a concept in cell phone domain and can be identified using *nn* tag, which is a noun compound modifier used by the Stanford parser. During candidate concept extraction, various noises are noticed. For elimination of noise during extraction, cleaning steps are performed by removing the stop-words and apostrophes, numerals, special characters, and symbols associated with a word. Cleaning operation proceed further by discarding all noun phrases representing person, organization, or location, as the probability for a noun phrase representing a named entity to be considered as a valid concept in product domain is very low. For named entity annotation, we have used the NER module of Gthis, we have

Apart from candidate concept extraction, opinion mining and sentiment analysis research requires extraction of associated opinion information for sentiment analysis. The dependency relationship between words is very helpful for this purpose. Considering the sample sentence S1, the word *volume* (part of a candidate concept) is related to an adjective *good* with *nsubj* relation (a dependency relationship type used by the Stanford parser). Thus, *good* can be identified as opinion. Further, using *advmod* relation, the adverb *very* can be identified as a modifier to represent the degree of expressiveness of the opinion word *good*. In sentence S2, the noun word *sound* is a nominal subject of the verb *comes*, and the adjective word *clear* is adjectival complement of it. Therefore, *clear* can be extracted as opinion for the concept *sound*. Further discussion on such dependency relationships between words and their usage in sentiment analysis can be found in our previous works [18, 19]. We have defined various rules to tackle different types of sentence structures to identify information components constituting candidate concepts, modifiers, and opinions.

3.3.2 Feasibility Analysis

It is observed during concept extraction phase that various noun, verb, and adjective words extracted from review documents are not relevant features or opinions. Though the ontology concepts normally emerge as a noun phrase in customer reviews, but sometimes verbs are considered as nouns due to parsing error. The basic reason for occurrence of the noises is the presence of ordinary nouns, verbs, and adjectives that are not actual features and opinions, but extracted due to parsing errors and their association with each other. Another issue is that very often several customers comment on same product feature, and in many cases their opinions contradict with each other. To handle these issues, a reliability score, $0 \le r \le 1$, is calculated for each concept-opinion pair with respect to the review documents collection. A higher reliability score reflects a tight integrity between the elements of a pair. We follow the opinion retrieval model based on the Hyperlink-Induced Topic Search (HITS) algorithm [21] used by Li et al. [24]. The extracted concept-opinion pairs are represented as an undirected bipartite graph and treated by the HITS algorithm to generate reliability scores for concept-opinion pairs. For applying iterative HITS algorithm, extracted concept-opinion pairs and review documents are modeled as hubs and authorities, respectively. A hub object (a concept-opinion pair) has links to many authorities (review documents) because the same concept-opinion pair may occur in many review documents. Similarly, an authority object (a review document) contains many concept-opinion pairs, and as a result many hubs (i.e., concept-opinion pairs) are linked to it. Formally, a bipartite graph is represented as a triplet of the form $G = \langle V_p, V_d, E_{dp} \rangle$, where $V_p = p_{ij}$ is the set of concept-opinion pairs that have co-occurrence at sentence level. $V_d = d_k$ is the set of review documents containing concept-opinion pairs, and $E_{dp} = \left\{ e_{i,j}^k | p_{ij \in V_p, \, d_k \in V_d} \right\}$ refers to the correlation between documents and feature-opinion pairs. Each edge $e_{i,j}^k$ is associated with a weight $W_{i,j}^k \in [0, 1]$ denoting the strength or integrity of the relationship between the pair p_{ij} and document d_k. The weight of a pair p_{ij} across all sentences of the document d_k is calculated using Eqs. 1, 2, 3, and 4, where $|d_k|$ is the number of sentences in document d_k and $0 \le \alpha \le 1$ is used as a trade-off parameter.

The feature score is calculated using term frequency (tf) and inverse sentence frequency (isf) in each sentence of the document, $tf(f_i, s_l)$ is the number of times f_i occurs in sentence s_l. N is the total number of sentences in the document, and $sf(f_i)$ is the number of sentences where the feature f_i appears [24, 26, 31]. Similarly, $tf(o_j, s_l)$ is the number of times opinion o_j appears in a sentence s_l and asl is the average number of sentences in the document d_k [3].

$$W_{i,j}^k = \frac{1}{|d_k|} \sum_{P_{ij \in s_l \in d_k}} [\alpha \times fScore(f_i, s_l) + (1 - \alpha) \times oScore(o_j, s_l)] \quad (1)$$

$$fScore(f_i, s_l) = tf(f_i, s_l) \times isf(f_i) \quad (2)$$

$$isf(f_i) = \log\left(\frac{N+1}{0.5 \times sf(f_i)}\right) \tag{3}$$

$$oScore(o_j, s_l) = \frac{tf(o_j, s_l)}{tf(o_j, s_l) + 0.5 + \left\{1.5 \times \frac{len(sl)}{asl}\right\}} \tag{4}$$

The authority score $AS^{(t+1)}(d_k)$ of document d_k and hub score $HS^{(t+1)}(p_{ij})$ of p_{ij} in $(t+1)$th iteration are computed as a function of the hub scores and authority scores obtained in tth iteration, using Eqs. 5 and 6, respectively.

$$AS_{(d_k)}^{(t+1)} = \sum_{P_{ij} \in V_p} w_{ij}^k \times HS_{(p_{ij})}^{(t)} \tag{5}$$

$$HS_{(p_{ij})}^{(t+1)} = \sum_{d_k \in V_d} w_{ij}^k \times AS_{(d_k)}^{(t)} \tag{6}$$

In order to apply HITS algorithm, initial weight for each feature-opinion pair and review document is set to 1, and the process is repeated till the convergence point is achieved. In line with [26], the convergence is reached when score computed for two successive iterations for any review document or feature-opinion pair falls below a given threshold. In our experiment, the threshold value is set to 0.0001. After convergence, the generated hub scores of (f_i, o_j) pairs represent the soundness of the integration of the respective features and opinions. Thereafter, the reliability score r_{ij} for a pair (f_i, o_j) is calculated by normalizing the hub score using *min-max* normalization given in Eq. 7, where $HS(p_{ij})$ denotes the hub score of p_{ij}, and *NewMin* and *NewMax* are set to 0 and 1, respectively. The reliability score of a feature-opinion pair represents the reliability of the relatedness of the feature and opinion.

$$r_{ij} = \frac{HS(p_{ij}) - min_{xy}\{HS(p_{xy})\}}{max_{xy}\{HS(p_{xy})\} - min_{xy}\{HS(p_{xy})\}} \times (NewMax - NewMin) + NewMin \tag{7}$$

3.3.3 Relationship Extraction

After identification of feasible concepts, their validation is done by the domain expert to make sure that only valid concepts are added in the ontology—as ontology is not a general data store, rather a knowledge management tool. A relationship is defined as a specific association between two ontology concepts and assumed to be binary in nature. It can be structural or generic. The structural relations (IS-A, HAS-PART, etc.) also known as conceptual semantic and lexical relations are obtained using WordNet [29], which is a large lexical database containing English words. In wordNet, nouns, verbs, adjectives and adverbs are grouped into synsets, each representing a distinct concept. Synsets are interconnected through conceptual-semantic and lexical relations.

In order to identify generic relationships between ontological concepts, a two-steps process is devised. In the first step, parts-of-speech information and dependency relationship between terms are exploited to identify candidate relations that associate ontology concepts; whereas in second step, a feasibility analysis is applied to retain only valid relations between the concepts. For example, consider the sample sentence S3 of Table 4, wherein "N95" (a constituent of the bigram "Nokia N95") is the nominal subject of the verb "has" and "screen" is its direct object. Thus "screen" and "Nokia N95" can be identified as concepts related through "has" relationship.

3.4   Sentiment Analysis

Sentiment analysis can be treated as a ternary classification problem in which the opinion words are mapped into one of the *positive*, *negative*, or *neutral* class, with respect to a particular concept or feature. In this section, we present a two-step process for word-level sentiment classification. In first step, popular statistical text classification methods are used to generate score for target opinion words, whereas in second step, generated scores are used to engineer an enriched set of features to develop a word-level sentiment classification system using supervised machine learning techniques. Thereafter, sentiments at higher levels of abstraction are determined using domain ontology.

3.4.1   Statistical Feature Identification

This section presents the proposed statistical approach for feature extraction. As proposed in [35], popular association measures, including Pointwise Mutual Information (PMI) [9], Mutual Information (MI) [11], Chi-square (commonly known as Karl Pearson's Chi-square), and Log Likelihood Ratio (LLR) [10] are used to compute score for each feasible opinion words. For association measures, a set of positive seed-words ($N^{(+)}$) and a set of negative seed-words ($N^{(-)}$) are compiled. Thereafter, for each opinion word, positive opinion score ($Score^{(+)}$), negative opinion score ($Score^{(-)}$), and final opinion score ($OpnScore$) are calculated using Eqs. 8, 9, and 10, respectively; wherein "AssociationFunction" represents PMI, MI, Chi-square, or LLR. Similarly, $SeedPos_j$ and $SeedNeg_j$ represent the jth positive and negative seed words, respectively.

$$Score^{(+)}(w_i) = \sum_{j=1}^{|N^{(+)}|} AssociationFunction(w_i, SeedPos_j) \qquad (8)$$

$$Score^{(-)}(w_i) = \sum_{j=1}^{|N^{(-)}|} AssociationFunction(w_i, SeedNeg_j) \qquad (9)$$

$$OpnScore(w_i) = Score^{(+)}(w_i) - Score^{(-)}(w_i) \qquad (10)$$

The value of $OpnScore(w_i)$ is used in training data set to determine the sentiment polarity of opinion words. If $OpnScore(w_i)$ value is greater than 0, it is an indication that the word w_i has higher association with positive seeds set and its sentiment orientation is considered as positive. Similarly, a negative score refers higher association of the word w_i with the negative seeds set and reflects negative sentiment. An opinion score value as zero represents that w_i is equally associated with both positive and negative seeds set, and its polarity is considered as neutral.

3.4.2 Sentiment Determination Using Machine Learning Techniques

In addition to a rich set of statistical features discussed in the previous section, some linguistic features, including negation, tf-idf, and modifiers are also considered for classification purpose. The proposed sentiment analysis system is implemented as a two-phase process—model learning (aka training phase) and classification (aka testing phase). The training phase uses the feature vectors generated from training data set to learn classification models, whereas the classification phase is used to determine the sentiments of words from test data set. Four different classifiers, including Naive Bayes [13], Decision Tree (J48) [38], Multilayer Perceptron (MLP) [39], and Bagging [5] are considered initially, but finally settled with Bagging and Decision Tree (J48) algorithms implemented as a part of WEKA [14] due to their best performance. Once the sentiment class of individual opinionated words is determined, the semantic orientation at higher levels of abstraction is determined with the help of domain ontology.

3.4.3 Ontology-Based Sentiment Propagation

As discussed earlier, knowledge management structure provided by ontology in a machine-interpretable format is very helpful in propagating feature-specific sentiment information to infer sentiments at higher levels of abstraction, especially in the case where explicit mentions of sentiment information with associated features are missing. Motivated by the work presented in [30, 42], we propose an unordered sentiment propagation ontology tree T with a finite set of one or more nodes, in which there exists a specially designated node called *root*, and remaining nodes are partitioned into $n \geq 1$ disjoint subsets $T_1, T_2, \ldots, T_n$, where each T_i ($i = 1, 2, \ldots, n$) is a tree and $T_1, T_2, \ldots, T_n$ are called subtrees of the root. Each node of the proposed ontology tree is a quadruple of the form $\langle c, m, o, s \rangle$, where c is a *concept*, m is an

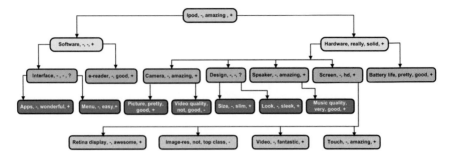

Fig. 1 Sentiment propagation using ontology tree

optional modifier representing degree of expressiveness of opinion word *o*, and *s* is the sentiment information referring *positive*(+), *negative*(−), or *neutral*(=) orientation. Figure 1 shows an exemplar sentiment propagation ontology tree generated using the documents from IPod domain, wherein opinion words and subsequently sentiment information are missing for the concept node *software* and its child node *interface*. Propagating sentiment in a bottom-up manner, the sentiment orientation of the node *interface* can be derived as *positive*, since both of its left and right child nodes at depth one (i.e., *Apps* and *Menu*) are referring positive sentiment. Similarly, sentiment orientation of the concept node *software* can be derived as *positive*.

4 Review Summarization and Visualization

This section presents the design of Opinion Summarization and Visualization (OSV) module to present extracted information components in a graphical form which facilitates users to have a quick glance over the mined concepts, opinions, and sentiments polarity without reading the pile of review documents. As mentioned in one of our previously published papers [17], OSV module is capable to visualize mining results both from single as well as multiple review documents. It also provides graphical tools for end users to explore and visualize summarized information components using bar charts and pie charts generated using Google chart API.[5]

Extracted information components along with opinion summary statistics are presented using Java Script Object Notation (JSON)[6] object, which is a language independent, lightweight text-data interchangeable format. Figure 2 shows the JSON representation of an object describing information component and opinion summary statistics. The object uses string fields for feature, modifier, opinion, and sentiment polarity; a number field for reliability score, and contains an array of objects for opinion score. During execution, OSV retrieves all required information from data-

---

[5]http://code.google.com/apis/ajax/playground/#chart_wrapper.

[6]http://www.json.org/.

Fig. 2 JSON representation
of information components
and opinion scores

```
{
    "concept": "Speaker quality",
    "modifier": "very",
    "opinion": "bad",
    "scoreReliabilityPair": 0.0108,
    "scoreOpinion": [
        {
            "type": "pmi",
            "number": -0.7344
        },
        {
            "type": "mi",
            "number": -109.3725
        },
        {
            "type": "chi",
            "number": -850.0066
        }
    ],
    "orientation": "negative"
}
```

base to form JSON object and uses the same as an input for visualization purpose. Figure 3 (taken from one of our previous research paper [17]) shows the main screen of OSV module, consisting of two rows viz. the upper-row and the lower-row. The upper-row is divided into three panels—upper-left, upper-middle, and upper-right. The upper-left panel contains list of reviews crawled from merchant sites. When a user selects a particular review from the upper-left panel, its description and metadata appear in the upper-middle and upper-right panels, respectively. Metadata of a review consists of information such as source from where the review was crawled, domain, author name, description, date of posting, and star rating. The lower-row of the main screen is also divided into two panels viz. lower-left and lower-right. The lower-left panel uses pie chart for opinion summarization of a particular review selected by the user from upper-left panel. The pie chart makes use of different colour combination mainly blue, red, and green to visualize the number of positive, negative, and neutral opinions, respectively. For a selected review document, the lower-right panel presents the list of extracted results that includes feature, modifier (if any), opinion, sentiment polarity (positive, negative, or neutral) and opinion indicator of the feature-opinion pair. For visibility purpose, colour scheme is used to highlight the information components extracted from review documents. On clicking a feature word appearing in the lower-right panel, the constituents of the corresponding feature-opining pair is highlighted using orange and yellow colors, respectively, and the relevant snippets (containing feature and opinion words) of the review document

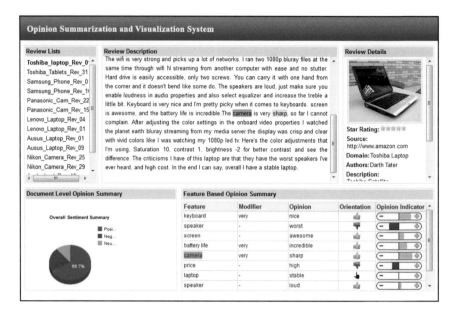

Fig. 3 Opinion summarization and visualization using OSV module

also accentuates. When a user clicks to a highlighted snippet representing product feature in upper-middle panel, a pop-up window appears visualizing the percentage of positive, negative, and neutral opinions using pie-chart.

Figure 4 (taken from one of our previous research paper [17]) shows the percentage of opinions expressed on a product feature from a corpus of customer reviews. As discuss earlier, OSV module facilitates users to navigate through the pile of customer reviews in an efficient way to produce feature-based opinion summary. Thus, pop-up window appearing in the above mentioned step contains a *view more* option, clicking which causes the window to expand in size, visualizing opinion score summary for the respective product feature. Figure 5 shows an expanded pop-up window, where size of each slice in the 3D pie-chart represents the degree of expressiveness of opinion. Opinion scores are calculated using Chi-square value due to its best performance. Higher the score for an opinion, larger the size of the slice in 3D pie-chart.

5 Experimental Setup and Results

In this section, we present the experimental results of the proposed opinion mining and sentiment analysis system. The data set used in our experiment consists of 1200 review documents related to different electronic products crawled from http://www. amazon.com. Information component extraction mechanism is implemented using a rule-based system. Reliability score is calculated for each concept and associated

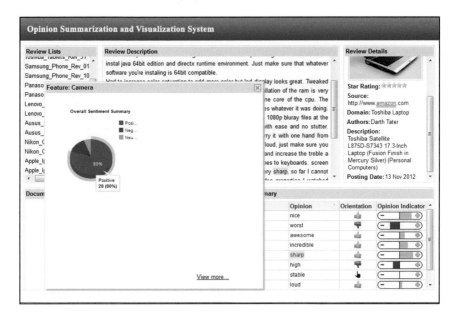

Fig. 4 Feature-based opinion summarization using OSV module

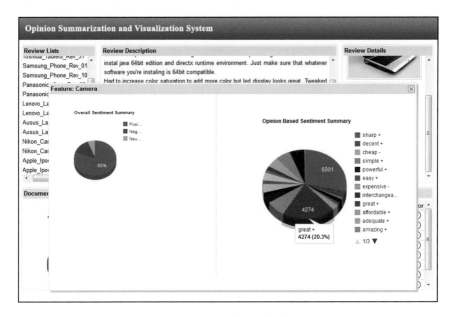

Fig. 5 Opinion-based sentiment summary using OSV module

Table 5 A partial list of candidate information components including concepts, opinions, and modifiers

Product	Concept	Modifier	Opinion
Digital Camera	Lens	Too	Heavy
	Battery	–	Standard
	Picture	Really	Great
	Price	Very	Reasonable
	Access	–	Quicker
IPod	Ipod touch	Not	Great
	Retina display	–	Beautiful
	Sound	–	Great
	Resolution	–	Fine
	Battery life	Very	Good
Laptop	Screen	Too	Glossy
	Keyboard	Very	Nice
	Software	–	Expensive
	Battery life	–	Incredible
	Picture quality	Just	Great
Cell Phone	OS	–	Beautiful
	Touch screen	–	Awesome
	Feel	–	Great
	Picture	Quite	Good
	Battery life	Not	Bad
Tablet	Scrolling	Very	Smooth
	Processor	–	Faster
	Screen	–	Beautiful
	File transfer	–	Painless
	Speaker	Really	Good

opinion using a reliability score generator implemented in Java. A partial list of feasible concepts along with their opinions and modifiers are presented in Table 5. After analysis, we found that occurrence frequency of genuine concepts is very high in review documents, which is due to the tendency that various reviewers refer to same product feature with different opinion words to express distinct sentiments.

An opinion score generator implemented in Java is applied to compute the scores for the feasible opinion words. Table 6 presents statistical feature values for a partial list of opinion words. Thereafter, each opinion word is characterized using the features described in Sect. 3.4.2. It can be observed from Table 6 that majority of the opinion scores obtained using Log Likelihood Ratio (LLR) is found negative. Thus, it is discarded from further analysis. Subsequently, a feature vector generator is implemented in Java to generate feature values for each opinion bearing word, and a ternary classification model is learned to classify polarity of a word as positive, neg-

Table 6 A partial list of opinionated words and their opinion scores obtained using different statistical measures

Opinionated word	Opinion score			
	PMI	MI	Chi-square	LLR
Bad	−0.7344	−109.3725	−850.0066	−10419.1723
Expensive	−0.2984	−51.8493	−556.9257	−11182.3347
Poor	−0.5560	−54.0277	−378.8968	−2483.0364
Slow	−0.6935	−66.1516	−369.4841	−6.5339
Horrible	−0.9389	−34.7363	−240.8866	−8247.2498
Bittersweet	0.0000	0.0000	0.0000	0.0000
Unbelievable	0.0000	0.0000	0.0000	0.0000
Amazing	2.3403	172.7484	1177.2141	−47112.7543
Bright	0.3932	47.6693	245.7392	−21448.2128
Beautiful	1.0260	76.5412	485.5965	−15660.7247
Fantastic	1.4603	98.6912	607.5986	−24541.5964
Wonderful	2.0459	75.8369	419.2278	−10558.8426

ative, or neutral. Evaluation of the experimental results is performed using standard Information Retrieval (IR) metrics *Precision*, *Recall*, and *F-score* which are defined in Eqs. 11, 12, and 13, respectively. In these equations, TP represents *true positives*, FP represents *false positives*, and FN represents *false negative*.

$$Precision = \frac{TP}{TP + FP} \qquad (11)$$

$$Recall = \frac{TP}{TP + FN} \qquad (12)$$

$$F\text{-}Score = \frac{2 \times precision \times recall}{precision + recall} \qquad (13)$$

5.1 Evaluating Concept and Opinion Extraction Process

To the best of our knowledge, no benchmark data is available in which concepts and related opinion information are marked simultaneously for electronic product domain. Therefore, manual evaluation is performed to analyze the overall performance of the proposed system. From a corpus of 1200 review documents, a total of 120 documents (Cell Phone: 30, Laptop: 30, IPod: 30, Digital Camera: 15, and Tablet: 15) are randomly selected for testing purpose. Our rule-based method is applied to extract concepts and related opinion words. Initially, the total count obtained for

Table 7 Performance evaluation of information component extraction process

Product category	TP	FP	FN	Precision	Recall	F-Score
Cell Phone	181	51	120	78.02	60.14	67.92
Laptop	319	37	99	89.61	76.32	82.43
Camera	104	23	68	81.89	60.47	69.57
IPod	129	124	96	50.99	57.34	53.98
Tablet	108	38	67	73.98	61.72	67.29
Macro-Average	841	273	450	75.50	65.15	69.94

TP, FP, and FN are 841, 866, and 450, respectively. We have observed that direct and strong relationship between words causes extraction of nouns, verbs, and adjectives, representing irrelevant concepts and opinion words. As a result, the number of false positives (FP) increases, which has an adverse effect on the precision value. To overcome this problem, post-processing step is applied to remove noisy extraction by performing feasibility analysis discussed in Sect. 3.3.2. After removal of the noisy concepts and associated opinions, the total count of FP reduces to 273. Macro-averaged values are calculated to present a synthetic performance measure by simply averaging the results of different categories of the products. Table 7 summarizes the performance measure values in the form of a misclassification matrix.

Since most of the reviewers use informal approach while commenting, reviews are generally lack in grammatical correctness and pose a number of challenges for natural language parser. The recall value is lower than the precision is an indication of system inability to extract certain concepts and associated opinions correctly.

5.2 Evaluating Sentiment Classification Process

In our experiment, we have considered 1000 and 200 opinion bearing words for training and testing purpose, respectively. A feature vector generator is implemented in Java to generate numeric values of the features for each opinion bearing word. Table 8 shows the information gain ranking of attributes on the basis of WEKA's attribute evaluator, in which the Chi-square feature seems to be most discriminative followed by Mutual Information (MI).

Table 8 A ranked list of sentiment classification features based on information gain values

Attribute name	Information gain
Chi-square	0.3609
MI	0.3542
PMI	0.1680
Negation	0.0657
TF-IDF	0.0236
Modifier	0.0103

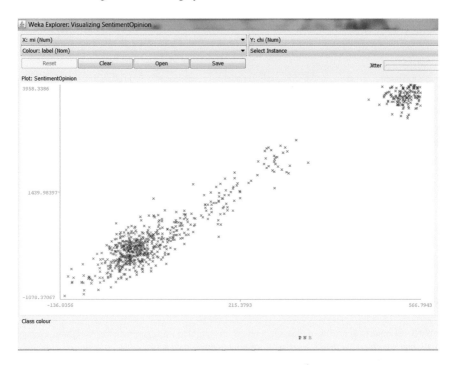

Fig. 6 Visualization of sentiment classification results based on χ^2 and MI values

Figure 6 visualizes the effect of classifying opinion bearing words on the basis of Chi-square (χ^2) and MI scores placed on X-axis and Y-axis, respectively.

We have experimented with some prominent classifiers that are best suited for the classification task, but settled with Bagging and Decision Tree (J48) algorithm due to their best performance during training and testing phases. Figure 7 presents Receiver Operating Characteristics (ROC) curves of all four classifiers, visualizing their comparative accuracy in terms of true positive rate and false positive rate during training. The ROC curve is generated using WEKA by plotting false positive rate and true positive rate on X-axis and Y-axis, respectively. Best classification performance of Bagging (curve consisting of + symbol) can be observed easily due to its appearance at the extreme left and higher in the ROC space.

A ternary classification model is learned to determine the class of an opinion word as *positive*, *negative*, or *neutral*. To judge the overall performance of the classifiers used in our experiment, Weighted Average Precision ($P_{\omega a}$), Weighted Average Recall ($R_{\omega a}$), and Weighted Average F-score ($F_{\omega a}$) values are calculated using Eqs. 14, 15, and 16, respectively. In these equations, P_{pos}, R_{pos}, and F_{pos} represent the Precision, Recall, and F-score, respectively of the positive instances; P_{neg}, R_{neg}, and F_{neg} represent the Precision, Recall, and F-score, respectively of the negative instances; and P_{neu}, R_{neu}, and F_{neu} represent the Precision, Recall, and F-score, respectively of the neutral instances. Similarly, ω_{pos}, ω_{neg}, and ω_{neu} represent the weights (ratio of the

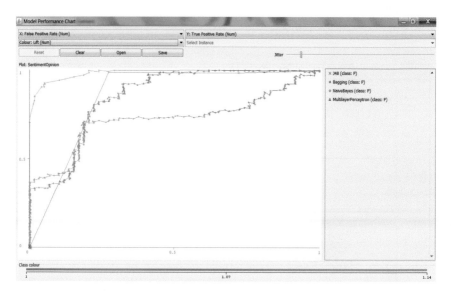

Fig. 7 ROC curves of classifiers for sentiment analysis on training data set

Table 9 Comparison results of classifiers based on weighted average precision, weighted average recall, and weighted average f-score over training and test data sets

Classifier	Weighted average result (on training data set)			Weighted average result (on test data set)		
	Precision	Recall	F-Score	Precision	Recall	F-Score
NB	0.902	0.703	0.779	0.840	0.617	0.683
J48	0.935	0.948	0.939	0.885	0.872	0.855
MLP	0.919	0.940	0.929	0.821	0.857	0.833
Bagging	0.952	0.951	0.942	0.875	0.865	0.848

number of instances of a particular category to the total number of instances) of the positive, negative, and neutral instances, respectively. Table 9 presents the comparison results of the classifiers in terms of these metrics calculated over the training and test data sets.

$$P_{\omega a} = \frac{(P_{pos} \times \omega_{pos}) + (P_{neg} \times \omega_{neg}) + (P_{neu} \times \omega_{neu})}{\omega_{pos} + \omega_{neg} + \omega_{neu}} \quad (14)$$

$$R_{\omega a} = \frac{(R_{pos} \times \omega_{pos}) + (R_{neg} \times \omega_{neg}) + (R_{neu} \times \omega_{neu})}{\omega_{pos} + \omega_{neg} + \omega_{neu}} \quad (15)$$

$$F_{\omega a} = \frac{(F_{pos} \times \omega_{pos}) + (F_{neg} \times \omega_{neg}) + (F_{neu} \times \omega_{neu})}{\omega_{pos} + \omega_{neg} + \omega_{neu}} \quad (16)$$

6 Conclusion and Future Works

Rapid growth in user-generated contents, mainly unstructured or semi-structured textual data, on the Web and their uncontrolled generation containing various natural language nuances provides an opportunity for data analysts to apply computational techniques for determining users' sentiments and emotions with respect to various features of products. Despite various research efforts attempted worldwide, development of a fully automatic opinion mining and sentiment analysis system is still elusive. It has been observed that overall problems associated with the development of opinion mining and sentiment analysis system is non-trivial and requires more research exploration. The main contribution of this work is to propose linguistically motivated computational paradigm for ontology learning and sentiment analysis using rule-based and machine learning approaches. Natural language processing techniques along with statistical analysis are applied to identify feasible product features and opinions, whereas machine learning techniques are applied for sentiment polarity determination. The electronic products ontology developed in this work, helps to structure and manage product features, and improves the performance of sentiment analysis task by propagating sentiment information from lower to higher levels among related concepts. Consideration of other structural information of ontology can be used for sentiment propagation along ontology tree, which is one of our future directions of work. Moreover, handling implicit concepts that are very common with customer reviews is an area in which we wish to direct our future research.

References

1. Abulaish, M.: An ontology enhancement framework to accommodate imprecise concepts and relations. J. Emerg. Technol. Web Intell. **1**(1), 22–36 (2009)
2. Abulaish, M., Jahiruddin, M.N.D., Ahmad, T.: Feature and opinion mining for customer review summarization. In: Proceedings of the 3rd International Conference on Pattern Recognition and Machine Intelligence (PReMI'09), pp. 219–224 (2009)
3. Allan, J., Wade, C., Bolivar, A.: Retrieval and novelty detection at the sentence level. In: Proceedings of the 26th Annual International ACM SIGIR Conference on Research and Development in Information Retrieval, pp. 314–321 (2003)
4. Bloom, K., Garg, N., Argamon, S.: Extracting appraisal expressions. In: Proceedings of Human Language Technologies 2007: the Conference of the North American Chapter of the Association for Computational Linguistics, pp. 308–315 (2007)
5. Breiman, L.: Bagging predictors. Mach. Learn. **24**(2), 123–140 (1996)
6. Cadilhac, A., Benamara, F., Aussenac-Gilles, N.: Ontolexical resources for feature-based opinion mining: a case-study. In: Proceedings of the 6th Workshop on Ontologies and Lexical Resources, pp. 77–86 (2010)
7. Cambria, E., Schuller, B., Xia, Y., Havasi, C.: New avenues in opinion mining and sentiment analysis. IEEE Intell. Syst. **28**(2), 15–21 (2013)
8. Cheng, X., Xu, F.: Fine-grained opinion topic and polarity identification. In: Proceedings of the Poster Session of the Sixth International Conference on Language Resources and Evaluation, pp. 2710–2714 (2008)
9. Church, K.W., Hank, P.: Word association norms, mutual information, and lexicography. Comput. Linguist. **16**(1), 22–29 (1990)

10. Dunning, T.: Accurate methods for the statistics of surprise and coincidence. Comput. Linguist. **19**(1), 61–74 (1993)
11. Fano, R.M.: Transmission of information: a statistical theory of communications. Am. J. Phys. **29**(11), 793–794 (1961)
12. Fortuna, B., Grobelnik, M., Mladenic, D.: Ontogen: semi-automatic ontology editor. In: Proceedings of the 2007 Conference on Human Interface, pp. 309–318 (2007)
13. Friedman, N., Geiger, D., Goldszmidt, M.: Bayesian network classifiers. Mach. Learn. **29**(2–3), 131–163 (1997)
14. Hall, M., Frank, E., Holmes, G., Pfahringer, B., Reutemann, P., Witten, I.H.: The weka data mining software: an update. ACM SIGKDD Explor. Newsl. **11**(1), 10–18 (2009)
15. Hatzivassiloglou, V., McKeown, K.R.: Predicting the semantic orientation of adjectives. In: Proceedings of the 35th Annual Meeting of the Association for Computational Linguistics and Eighth Conference of the European Chapter of the Association for Computational Linguistics, pp. 174–181 (1997)
16. Hu, M., Liu, B.: Mining and summarizing customer reviews. In: Proceedings of the Tenth ACM SIGKDD International Conference on Knowledge Discovery and Data Mining, pp. 168–177 (2004)
17. Kamal, A.: Review mining for feature based opinion summarization and visualization. Int. J. Comput. Appl. **119**(17), 6–13 (2015)
18. Kamal, A., Abulaish, M.: Statistical features identification for sentiment analysis using machine learning techniques. In: Proceedings of the 2013 International Symposium on Computational and Business Intelligence, pp. 178–181 (2013)
19. Kamal, A., Abulaish, M., Anwar, T.: Mining feature-opinion pairs and their reliability scores from web opinion sources. In: Proceedings of the 2nd International Conference on Web Intelligence, Mining and Semantics, pp. 15:1–15:7 (2012)
20. Kessler, J.S., Nicolov, N.: Targeting sentiment expressions through supervised ranking of linguistic configurations. In: Proceedings of the Third International AAAI Conference on Weblogs and Social Media, pp. 90–97 (2009)
21. Kleinberg, J.M.: Authoritative sources in a hyperlinked environment. J. ACM **46**(5), 604–632 (1999)
22. Kontopoulosa, E., Berberidisa, C., Dergiadesa, T., Bassiliadesa, N.: Ontology-based sentiment analysis of twitter posts. Expert Syst. Appl. **40**(10), 4065–4074 (2013)
23. Lau, R.Y., Lai, C.C., Ma, J., Li, Y.: Automatic domain ontology extraction for context-sensitive opinion mining. In: Proceedings of the 13th International Conference on Information Systems, pp. 35–53 (2009)
24. Li, B., Zhou, L., Feng, S., Wong, K.: A unified graph model for sentence-based opinion retrieval. In: Proceedings of the 48th Annual Meeting of the Association for Computational Linguistics, pp. 1367–1375 (2010)
25. Li, B., Zhou, L., Feng, S., Wong, K.-F.: A unified graph model for sentence-based opinion retrieval. In: Proceedings of the 48th Annual Meeting of the Association for Computational Linguistics, pp. 1367–1375 (2010)
26. Li, F., Tang, Y., Huang, M., Zhu, X.: Answering opinion questions with random walks on graphs. In: Proceedings of the Joint Conference of the 47th Annual Meeting of the ACL and the 4th International Joint Conference on Natural Language Processing of the AFNLP, pp. 737–745 (2009)
27. Liu, B.: Sentiment analysis and subjectivity. In: *Handbook of Natural Language Processing*, pp. 627–666. Taylor and Francis Group, Boca (2010)
28. Liu, B.: *Sentiment Analysis and Opinion Mining*. Morgan & Claypool (2012)
29. Miller, G.A.: Wordnet: a lexical database for english. Commun. ACM **38**(11), 39–41 (1995)
30. Mukherjee, S., Joshi, S.: Sentiment aggregation using conceptnet ontology. In: Proceedings of the 6th International Joint Conference on Natural Language Processing, pp. 570–578 (2013)
31. Otterbacher, J., Erkan, G., Radev, D.R.: Using random walks for question-focused sentence retrieval. In: Proceedings of the Conference on Human Language Technology and Empirical Methods in Natural Language Processing, pp. 915–922 (2005)

32. Pang, B., Lee, L.: Opinion mining and sentiment analysis. Found. Trends Inf. Retr. **2**(1–2), 1–135 (2008)
33. Pang, B., Lee, L., Vaithyanathan, S.: Thumbs up? Sentiment classification using machine learning techniques. In: Proceedings of EMNLP, pp. 79–86 (2002)
34. Popescu, A.M., Etzioni, O.: Extracting product features and opinions from reviews. In: Proceedings of the 2005 Conference on Empirical Methods in Natural Language Processing (EMNLP05), pp. 339–346 (2005)
35. Prabowo, R., Thelwall, M.: Sentiment analysis: a combined approach. J. Inf. **3**(2), 143–157 (2009)
36. Qiu, G., Liu, B., Bu, J., Chen, C.: Expanding domain sentiment lexicon through double propagation. In: Proceedings of the 21st International Joint Conference on Artifical Intelligence, pp. 1199–1204 (2009)
37. Qiu, G., Liu, B., Bu, J., Chen, C.: Opinion word expansion and target extraction through double propagation. Comput. Linguist. **37**(1), 9–27 (2011)
38. Quinlan, J.R.: *C4.5: Programs for Machine Learning*. Morgan Kaufmann Publishers Inc., San Francisco, CA, USA (1993)
39. Rumelhart, D.E., Hinton, G.E., Williams, R.J.: *Parallel distributed processing: explorations in the microstructure of cognition, vol. 1*, chapter Learning internal representations by error propagation, pp. 318–362. MIT Press, Cambridge, MA, USA (1986)
40. Shein, K.P.P.: Ontology based combined approach for sentiment classification. In: Proceedings of the 3rd International Conference on Communications and Information Technology, pp. 112–115 (2009)
41. Turney, P.D.: Thumbs up or thumbs down? Semantic orientation applied to unsupervised classification of reviews. In: Proceedings of the 40th Annual Meeting on Association for Computational Linguistics, pp. 417–424 (2002)
42. Wei, W., Gulla, J.A.: Sentiment learning on product reviews via sentiment ontology tree. In: Proceedings of the 48th Annual Meeting of the Association for Computational Linguistics, pp. 404–413 (2010)
43. Wilson, T., Wiebe, J., Hwa, R.: Just how mad are you? Finding strong and weak opinion clauses. In: Proceedings of the 19th National Conference on Artifical Intelligence, pp. 761–767 (2004)
44. Zhou, L., Chaovalit, P.: Ontology-supported polarity mining. J. Am. Soc. Inf. Sci. Technol. **59**(1), 98–110 (2008)
45. Zhuang, L., Jing, F., Zhu, X.-Y.: Movie review mining and summarization. In: Proceedings of the 15th ACM International Conference on Information and Knowledge Management, pp. 43–50 (2006)

Knowledge-Based Tweet Classification for Disease Sentiment Monitoring

Xiang Ji, Soon Ae Chun and James Geller

Abstract Disease monitoring and tracking is of tremendous value, not only for containing the spread of contagious diseases but also for avoiding unnecessary public concerns and even panic. In this chapter, we present a near real-time sentiment analysis service of public health-related tweets. Traditionally, it is impossible for humans to effectively measure the degree of public health concerns due to limited resources and significant time delays. To solve this problem, we have developed a computational intelligence approach for Epidemic Sentiment Monitoring System (ESMOS) to automatically analyze the disease sentiments and gauge the Measure of Concern (MOC) expressed by Twitter users. More specifically, we present a knowledge-based approach that employs a disease ontology to detect the outbreak of diseases and to analyze the linguistic expressions that convey subjective expressions and sentiment polarity of emotions, feelings, opinions, personal attitudes, etc. with a sentiment classifier. The two-step sentiment classification method utilizes the subjective vocabulary corpus (MPQA), sentiment strength corpus (AFINN), as well as emoticons and profanity words that are often used in social media postings. It first automatically classifies the tweets into personal and non-personal classes, eliminating many tweets such as non-personal "retweets" of news articles from further consideration. In the second stage, the personal tweets are classified into Negative and non-Negative sentiments. In addition, we present a model to quantify the public's Measure of Concern (MOC) about a disease, based on sentiment classification results. The trends of the public MOC are visualized on a timeline. Correlation analyses between MOC timeline and disease-related sentiment category timelines show that the peaks of the MOC are

X. Ji · J. Geller
Department of Computer Science, New Jersey Institute of Technology,
323 Martin Luther King Blvd, Newark, NJ, USA
e-mail: xj25@gmail.com

J. Geller
e-mail: james.geller@njit.edu

S.A. Chun (✉)
Information Systems and Informatics, City University of New York,
College of Staten Island, 2800 Victory Blvd,
Staten Island, NY, USA
e-mail: soon.chun@csi.cuny.edu

© Springer International Publishing Switzerland 2016
W. Pedrycz and S.-M. Chen (eds.), *Sentiment Analysis and Ontology Engineering*,
Studies in Computational Intelligence 639, DOI 10.1007/978-3-319-30319-2_17

weakly correlated with the peaks of the News timeline without any appreciable time delay or lead. Our sentiment analysis method and the MOC trend analyses can be generalized to other topical domains, such as mental health monitoring and crisis management. We present the ESMOS prototype for public health-related disease monitoring, for public concern trending and for mapping analyses.

Keywords Computational intelligence · Sentiment analysis · Public health concern monitoring · Social data analytics

1 Introduction

Disease monitoring and tracking is of tremendous value, not only for containing the spread of contagious diseases but also for avoiding unnecessary public concerns and even panic. In this chapter, we focus on studying the Twitter users' concerns about diseases instead of the outbreak of the disease itself, which has been extensively studied [1–5]. Recent outbreaks of Ebola in Africa, measles on the West Coast of the USA and MERS in South Korea have shown how important it is to monitor and understand the public sentiments in addition to tracking the location, trend and potential trajectory of disease outbreaks. Public health concerns can also develop into dangerous panic states, resulting in irrational behaviors, unjustified fear, discrimination against patients, mistrust in governments' containment efforts as well as a negative overall economic impact. For instance, the Korean MERS incidents scared the public and prompted school and hospital closures, as well as decreased businesses activities and cancelled tourist visits. These, in turn, paralyzed the economy severely within a very short period of time and the central bank had to cut interest rates to prevent a further downward spiral (Wikipedia 2015).

Another example illustrating the consequence of public health concerns is that since the Ebola outbreak in 2014, the immigration examination and the medical system's ability to deal with Ebola have been widely mistrusted by the general public. Even the president of the United States addressed that issue [6], due to a series of mistakes when a traveler was issued a visitor visa and was not diagnosed by the local hospital. For example, a tweet complained that, "I know. Our government is a FAIL. People were more upset about Ebola than they are about thugs killing our cops." Another case is the SARS outbreak in China in 2003. Zhu, Wu, Miao and Li [7] reviewed the mental state changes of the general public during the SARS outbreak in China.

As shown by these examples, it is of great value if public health specialists and government decision makers can actively monitor public health concerns. However, the existing public health concern surveillance methods, such as questionnaires and clinical tests, are not able to cover a large number of respondents due to their expenses and furthermore the survey results are often published with significant time delays. A last example is the public's reaction to Japan's nuclear emergency in March 2011 [8]. Text messages about nuclear plumes spread throughout Asia. In China, Vietnam, and the Philippines rumors spread about possible disastrous consequences.

Ginsberg et al. [9] used search engine logs, in which users submitted queries in reference to issues that they were concerned about, to approach the disease monitoring problem. Their thread of research led to the realization that an aggregation of large numbers of queries might show patterns that are useful for the early detection of epidemics. However, comprehensive access to search engine logs is limited to the search engine providers. Twitter, a popular social network site, has more than 500 million users out of which more than 302 million are active users [10]. This shows Twitter's potential to address the limitations of traditional public health surveillance methods, and of search keyword logs. A percentage of Twitter messages are publicly available and researchers are able to retrieve the tweets as well as related information through the Twitter API [11].

We have developed a method to gauge the Measure of Concern (MOC) expressed by Twitter users for public health specialists and government decision makers [12]. More specifically, we developed a two-step sentiment classification approach. Firstly, personal tweets are distinguished from News tweets. News tweets are considered as Non-Personal, as opposed to Personal tweets posted by individual Twitter users. In the second stage, the Personal tweets are further classified into Personal Negative tweets or Personal Non-Negative tweets. The two-step sentiment classification problem addressed in this chapter is different from the traditional Twitter sentiment classification problem, which classified tweets into positive/negative or positive/neutral/negative tweets [4, 13–16] without distinguishing Personal from Non-Personal tweets first. Although News tweets may also express concerns about a certain disease, they tend not to reflect the direct emotional impact of that disease on people.

The sentiment classification method presented in this chapter is able to identify Personal tweets and News (Non-Personal) tweets in the first place. More importantly, using the sentiment classification results, we quantified the Measure of Concern (MOC) using the number of Personal Negative tweets per day. The MOC increases with the relative growth of Personal Negative tweets and with the absolute growth of Personal Negative tweets. Previous research [17, 18] visually noticed that sentiment surges co-occurred with health events on a timeline. Different from the previous work, which is based on visual observation, we correlated the peaks on the MOC timeline (i.e., change over time) and the peaks of the News timeline and also the peaks of the Non-Negative timeline and the peaks of News timeline using the Jaccard Coefficient [19]. Government officials can use MOC to track public health concerns on the timeline to help make timely decisions, disprove rumors, intervene, and prevent unnecessary social crises at the earliest possible stage. More importantly, public health concern monitoring using social network data is faster and cheaper than the traditional method.

The rest of this chapter is organized as follows. In Sect. 2, we discuss an intelligent system for health sentiment monitoring with background and related. In Sect. 3, sentiment classification methods and results are introduced in detail. In Sect. 4, the sentiment timeline trend analysis results are illustrated, interpreted, and discussed. Section 5 contains the chapter summary.

2   Cognitive Approach for Monitoring Public Health Sentiments

To enable public health sentiment monitoring, an intelligent system called the Epidemic Sentiment Monitoring System (ESMOS) is proposed. It monitors social media data (specifically, Twitter data) and "recognizes" and collects tweets about public-health related diseases. ESMOS then analyzes the data to automatically classify them into sentiment categories, and calculates the public degree of concerns for a particular disease outbreak and its spread. ESMOS also provides visual tools such as the intensity map (heat map) of sentiments to understand the geographic distribution and concentration of public concerns, and a health concern trending map to be able to track the public health concerns over time. The trending map will also allow users to compare collected tweets with the disease-associated news events to qualitatively investigate how the news coverage may influence the public concern. Figure 1 shows the component architecture of ESMOS.

The goal of computational intelligence is to build cognitive systems that can act like and interact with humans, to provide insights and intelligence that are needed in complex problem solving and decision-making situations. Cognitive systems

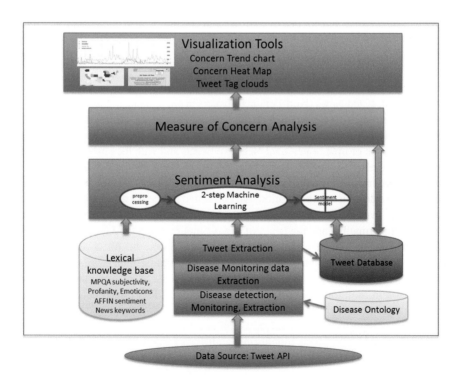

Fig. 1 Epidemic Sentiment Monitoring System (ESMOS) architecture

process [20] structured and unstructured data and learn specific domain knowledge by experience, much in the same way humans do. Cognitive systems use Natural Language Processing and image and speech recognition to understand the world and to interact with humans and other smart systems seamlessly. Unlike many expert systems, cognitive systems not only match and fire pre-defined rules with anticipated possible actions, but also can be trained using AI and Machine Learning methods to sense, integrate, analyze, predict and reason as human brains do for various tasks.

ESMOS utilizes both a linguistic knowledge base and automated machine learning methods to exhibit a degree of computational intelligence. It also utilizes a disease ontology to identify the disease names to collect and monitor tweet data. In addition, the machine learning algorithm used for sentiment classification utilizes automated labeling for the training data set using the linguistic knowledge base, instead of having human-labeled, supervised training data. It distinguishes personal tweets from non-personal tweets such as news articles. In the second step, it automatically labels the positive or negative sentiment tweets to generate a training dataset, avoiding another human labeling step. The training data sets are used to build the sentiment analysis model. The system uses unsupervised learning to learn classifiers using the lexical knowledge.

2.1 Disease Ontology

ESMOS contains 8,043 inherited, developmental and acquired human disease names from the Disease Ontology (DO), an open-source medical ontology, developed by Schriml et al. [21].

A partial structure of the ontology used by ESMOS is shown in Fig. 2. The extracted disease vocabulary is used to supply the keywords to monitor disease tweets for signs of epidemics. The current prototype system monitors a handful of infectious diseases. However, all DO disease terms may be used, given sufficient computational resources. The core component uses the Twitter Streaming API for collecting epidemics-related real-time tweets. The advantage of using the Disease Ontology is that it is linked to medical terminology codes in other ontologies with a set of synonyms. For instance, the Ebola virus is known under different labels. It has the following synonyms and cross-references with three of the most important medical terminology collections: NCI, SNOMED CT and UMLS:

```
id: DOID:4325
name: Hemorrhagic Fever, Ebola
synonym:"Ebola Hemorrhagic Fever" EXACT [NCI2004_11_17:C36171]
synonym: "Ebola virus disease" EXACT [SNOMEDCT_2005_07_31:186746000]
xref: UMLS_CUI:C0282687
```

Fig. 2 Partial structure of
the disease ontology used to
extract disease keywords

The use of other collections of medical terms (e.g. Wikipedia's list of ICD-9 codes 001–139: infectious and parasitic diseases [22]) may further support the purpose. The major issue with the use of formal ontologies of medical terms is that the laymen's disease terms (e.g. Ebola) may have to be matched with the scientific terms used in medical terminologies (e.g. Ebola Hemorrhagic Fever) [23]. This requires fuzzy matching and similarity matching, which are not addressed in this chapter.

2.2 Lexical Knowledge

Understanding human emotions (also called sentiment analysis) is a difficult task for a machine, for which the computational intelligence approach may provide better results. Often, linguistic expressions as well as paralinguistic features in spoken languages (e.g., pitch, loudness, tempo, etc.) reveal the sentiments or emotional states of individuals. Prior research studies have developed sentiment lexicons using a dictionary approach and a corpus approach [24].

The MPQA (Multi Perspective Question Answering) lexicon [25] was constructed through human annotations of a corpus of 10,657 sentences in 535 documents that contain English-based news from 187 different sources in a variety of countries, dated from June 2001 to May 2002. The annotated lexicon represents opinions and other private states, such as beliefs, emotions, sentiments, speculations, etc. The subjective and objective expressions were also annotated with values of intensity and polarity to indicate the degree of subjectivity, and a negative, positive or neutral sentiment. AFINN is another affective lexicon with a list of English words with sentiment valence ratings between minus five (negative) and plus five (positive). These words have been manually labeled and evaluated in 2009–2011 [26]. Most of the positive words were labeled with +2 and most of the negative words with

−2, and strongly obscene words were rated with either −4 or −5. AFINN includes 2477 words with 1598 negative words (65 %), 878 positive words and one neutral word. Another lexical resource is SentiWordNet [27], which associates each synset of WordNet with three numerical scores Obj(s), Pos(s) and Neg(s). The numerical scores describe the polarity for the terms in the synset, i.e., how Objective, Positive, or Negative the terms contained in the synset are.

In this chapter, we applied the MPQA lexicon to distinguish the Personal from Non-Personal tweets (such as News article tweets) since the subjectivity of the lexicon helps to distinguish Personal expressions from Non-Personal ones. We used the valence ratings in the AFINN lexicon to further distinguish the Personal Negative from Personal Non-Negative tweets. Our goal is to develop an automated sentiment analysis system for the social heath tweets generated by the general public. This system can provide public health officials with the capability to monitor the viral effects in social media communication, so that they can take early actions to prevent unnecessary panic regarding public health related diseases.

2.3 Twitter Sentiment Classification

Since the 2000s, with the tremendous amount of user-generated data from various data sources, such as blogs, review sites, News articles, and micro-blogs becoming available, researchers have become interested in mining high-level sentiments from the user data. Pang and Lee [28] reviewed the sentiment analysis work. This thread of research, depending on the analysis target, can be classified into the one of these levels: document-level [29], blog-level [30], sentence-level [31], tweet-level [32–34] with the sub-category non-English tweet level [35], and tweet-entity-level [36]. Since 2009, extensive research has been carried out on the topic of Twitter sentiment classification. [15, 33, 37–41].

Most of this thread of research used Machine Learning-based approaches, such as Naïve Bayes, Multinomial Naïve Bayes, and Support Vector Machine. The Naïve Bayes classifier is a derivative of the Bayes decision rule [42], and it assumes that all features are independent from each other. Good performance of Naïve Bayes (NB) was reported in several sentiment analysis papers [33, 37, 41]. Multinomial Naive Bayes (MNB) is a model that works well on sentiment classification [38, 39, 41]. MNB takes into account the number of occurrences and relative frequency of each word. Support Vector Machine [43] is also a popular ML-based classification method that works well on tweets [33, 40]. In Natural Language Processing, SVM with a polynomial kernel is more popular [44].

There are two drawbacks of the previous sentiment classification work. Firstly, Twitter messages were classified into either positive/negative or positive/negative/ neutral with the assumption that all Twitter messages express ones' opinion. However, this assumption does not hold in many situations, especially when the tweets are about epidemics or more broadly, about crises. In these situations, as we found when we randomly sampled 100 tweets, many tweets (up to 30 %) of the samples, are repetitions of the News without any personal opinion. Since they are not explicitly labeled with re-tweet symbols, it is not easy for a stop-word based pre-processing filter to detect them. We attempt to solve a different problem, which is how to classify tweets into three categories: Personal Negative tweets, Personal Non-Negative tweets, and News tweets (tweets that are non-Personal tweets).

Recently, some researchers identified irrelevant tweets. Brynielsson, Johansson, Jonsson and Westling [33] used manual labeling to classify tweets into "angry," "fear," "positive," or "other" (irrelevant). Salathe and Khandelwal [45] also identified irrelevant tweets together with sentiment classifications. Without considering irrelevant tweets, they calculated the H1N1 vaccine sentiment score from the relative difference of positive and negative messages. In our two-step classification method, we can automatically extract News tweets and perform the sentiment analysis. The results of sentiment classification are used for computing the correlation between sentiments and News trends. In this way, the goals of sentiment classification and measuring the public concern can be achieved in an integrated framework. Secondly, although sophisticated models were developed by the above research, the results of the sentiment classification were not utilized to provide insights. We provide the Measure of Concern timeline trends as a useful sentiment monitoring tool for public health specialists and government decision makers.

2.4 Quantifying and Visualizing Twitter Sentiment Trends on Timeline

Sentiment quantification is a method to process unstructured text and generate a numerical value or a timeline of numerical values to gain insights into the sentiment trends. Zhuang et al. [46] generated a quantification of sentiments about movie elements, such as special effects, plot, dialogue, etc. Their quantification contains a positive score and a negative score towards a specific movie element. For tweet-level sentiment quantification on a timeline, Chew and Eysenbach [47] used a statistical approach to computing the relative proportion of all tweets expressing concerns about H1N1 and visualized the temporal trend of positive/negative sentiments based on their proportion. Similar research was done by O'Connor et al. [48], in which they calculated a daily (positive and negative) sentiment score by recording the number of positive and negative words of one tweet appearing in the subjectivity lexicon of OpinionFinder [31]. Sha et al. [17] found that the sentiment fluctuations on Sina Weibo were associated with either the new policy announcements or government actions.

The drawbacks of the existing Twitter sentiment quantification research are twofold. Firstly, the lexicon-based sentiment extraction models have limited coverage of words. As pointed out by Wiebe and Riloff [49], identifying positive or negative tweets by counting words in a dictionary or lexicon usually has high precision but low recall. In the case of Twitter sentiment analysis, the performance suffers more. The lexicon or dictionary does not contain the slang words that are common in social media. For example, LMAO (Laughing My A** Off), is a positive "word" in Twitter, but it does not match any word in MPQA [25], which is a popular sentiment dictionary. In this study, we consider these profanity or slang words as well as the emoticons. Secondly, the existing sentiment quantification work [17, 48] has shown the correlation between sentiments and real-world events (e.g. news) through observing their co-occurrence on a timeline, but has not provided a comprehensive, quantitative correlation between the sentiment timeline trend and the News timeline trend. To the best of our knowledge, there is no prior work that both quantitatively and qualitatively studies these correlations between Twitter sentiment and the News in Twitter to identify concerns caused by diseases and crises.

In the next two sections, we first describe the machine learning approach to automatically labeling the training datasets to generate the sentiment classifiers, and then the quantitative model for measuring the Measure of Concern and its trend line analysis to see any correlations with news events on the traditional broadcast media.

3   Two-Step Tweet Sentiment Classification

In this section, we present our two-step sentiment classification method. As discussed earlier, our approach to sentiment classification is different from the classic sentiment classification of Tweets. The first step of our method involves training a Twitter sentiment classifier for distinguishing Personal tweets from News (Non-personal) tweets. The second step builds the sentiment classifier using only the personal tweets to identify the Negative versus Non-Negative tweets. The formal definitions of Personal Tweet, News Tweet, Personal Negative Tweet, and Personal Non-Negative Tweet are shown below.

Definition 1 (*Personal Tweet*) A Personal Tweet is a tweet that conveys its author's private states [31, 50]. A private state can be a sentiment, opinion, speculation, emotion, or evaluation, and it cannot be verified by objective observation. In addition, if a tweet talks about a fact observed by the Twitter user, it is also defined as a Personal Tweet. The goal of this definition is to distinguish the tweets written by the Twitter users from scratch from the News tweets that are re-tweeted in the Twittersphere.

Example (Personal Tweet)
"Since when does a commercial aircraft accident become a matter of National Security Interests? #diegogarcia#mh370"

Definition 2 (*News Tweet*) A News Tweet (denoted with NT) is a tweet that is not a Personal Tweet. A News Tweet states an objective fact.

Example (News Tweet):
"#UPDATE Cyanide levels 350x standard limits detected in water close to the site of explosions in China's Tianjin http://u.afp.com/Z5ab"

Definition 3 (*Personal Negative Tweet and Personal Non-Negative Tweet*) A tweet is a Personal Negative Tweet (denoted as PN) if it conveys negative emotions or attitude and it is a Personal Tweet. Otherwise, it is a Personal Non-Negative Tweet (denoted as PNN). Personal Non-Negative Tweets include personal neutral or personal positive tweets. A Personal Tweet is either a PN or a PNN. A Personal Negative Tweet expresses a user's negative sentiment, such as panic, depression, anxiety, etc. Note that this definition is focused on the user's negative emotional state as opposed to expressing the absence of an illness, e.g., getting a negative test result. Two examples are as follows.

Example (Personal Negative Tweet):
"Apparently #Ebola doesn't read our textbooks - it keeps changing the rules as it goes along. Frightening news :("

Example (Personal Non-Negative Tweet):
"To ensure eliminating the #Measles disease in the country, it is important to vaccinate people who are at risk."

As many News tweets are re-tweeted in Twitter, classifying the tweets into Personal and News tweets in the first step can help consider only Personal tweets in a sentiment analysis in the next step (Negative versus Non-Negative classification). An overview of sentiment classification method is shown in Fig. 3. We only consider English tweets, which were automatically detected during the data collection phase in this chapter. As shown in Fig. 3, the sentiment classification problem is approached in two steps. First, for all English tweets we separated Personal from News (Non-Personal) tweets. Second, after the Personal tweets were extracted by the most successful of the Personal/News Machine Learning classifier, these Personal tweets were used as input to another Machine Learning classifier, to identify Personal Negative tweets and Personal Non-Negative tweets. After the Personal Negative tweets, Personal Negative tweets, and News tweets were all identified, they were utilized to compute the Measure of Concern and the quantitative correlation between the sentiment timeline trend and the News timeline trend.

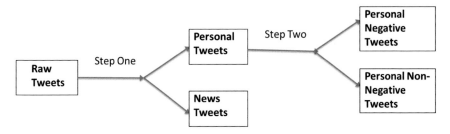

Fig. 3 Overview of the two-step sentiment classification method

3.1 Data Collection

We developed a real-time Twitter data collector with the Twitter API version 1.1 and Twitter4J library [51]. It collects real-time tweets containing the pre-defined public health-related keywords (e.g., listeria). We can describe the overall data collection process as "ETL" (Extract-Transform-Load) pipeline, which is a frequently used term in the area of Data Warehousing. In the first step, the data collector collected tweets in JSON format with Twitter Streaming API. (Extract step). In the second step, the data in JSON format was parsed into relational data, such as tweets, tweet_mentions, tweet_place, tweet_tags, tweet_urls, and users (Transform step). In the last step, the relational data was stored into our MySQL relational database (Load step).

We monitored 12 diseases including infectious diseases: listeria, influenza, swine flu, measles, meningitis, and tuberculosis; four mental health problems: Major depression, generalized anxiety disorder, obsessive-compulsive disorder, and bipolar disorder; one crisis: Air disaster; and one clinical science issue: Melanoma experimental drug. The preprocessing step filters out re-tweets and converts special characters into corresponding unigrams. More specifically, all tweets starting with "RT," were deleted because "RT" indicates that they are re-tweets without comments to avoid duplications. For the tweets that have a non-starting string "RT," the "RT" was removed.

One member of each pair of tweets that contain the same tokens (words) in the same order was deleted. For example, of the below two tweets only one is kept in the database.

(1) "27 test positive for tuberculosis at high school. http://t.co/Ss4QT1EPP2#CNN"

(2) "27 test positive for tuberculosis at high school. http://t.co/M4D6rgzYaI-@CNN."

3.2 Sentiment Classification of Twitter Data

3.2.1 Automatic Tweet Labeling Based on Subjective Clue Corpus

The clue-based classifier parses each tweet into a set of tokens and matches them with a corpus of Personal clues. There is no available corpus of clues for Personal versus News classification, so we used a subjective corpus MPQA [25] instead, on the assumption that if the number of strongly subjective clues and weakly subjective clues in the tweet is beyond a certain threshold (e.g., two strongly subjective clues and one weakly subjective clue), it can be regarded as Personal tweet, otherwise it is a News tweet. The MPQA corpus contains a total of 8,221 words, including 3,250 adjectives, 329 adverbs, 1,146 any-position words, 2,167 nouns, and 1,322 verbs. As for the sentiment polarity, among all 8,221 words, 4,912 are negatives, 570 are neutrals, 2,718 are positives, and 21 can be both negative and positive. In terms of strength of subjectivity, among all words, 5,569 are strongly subjective words, and the other 2,652 are weakly subjective words.

Twitter users tend to express their personal opinions in a more casual way compared with other documents, such as News, online reviews, and article comments. It is expected that the existence of any profanity might lead to the conclusion that the tweet is a Personal tweet. We added a set of 340 selected profanity words [52] to the corpus described in the previous paragraph. US law, enforced by the Federal Communication Commission prohibits the use of a short list of profanity words in TV and radio broadcasts [53]. Thus, any word from this list in a tweet clearly indicates that the tweet is not a News item.

We counted the number of strongly subjective terms and the number of weakly subjective terms, checked for the presence of profanity words in each tweet and experimented with different thresholds. A tweet is labeled as Personal if its count of subjective words surpasses the chosen threshold; otherwise it is labeled as a News tweet. If the threshold is set too low, the precision might not be good enough. On the other hand, if the threshold is set too high, the recall will be decreased. The advantage of a clue-based classifier is that it is able to automatically extract Personal tweets with more precision when the threshold is set to a higher value. Because only the tweets fulfilling the threshold criteria are selected for training the "Personal versus News" classifier, we would like to make sure that the selected tweets are indeed Personal with high precision. Thus the threshold that leads to the highest precision in terms of selecting Personal tweets is the best threshold for this purpose.

The performance of the clue-based approach with different thresholds on human-annotated test datasets is shown in Table 1 [12]. Among all the thresholds, s3w3 (3 strong, 3 weak) achieves the highest precision on all three human annotated datasets. In other words, when the threshold is set so that the minimum number of strongly subjective terms is 3 and the minimum number of weakly subjective terms is 3, the

Table 1 Results of Personal tweets classification with thresholds (Precision/Recall) [12]

Threshold	Dataset		
	Epidemic	Mental health	Clinical science
s1w0	0.61/0.69	0.55/0.74	0.48/0.58
s1w1	0.64/0.48	0.53/0.63	0.51/0.52
s1w2	0.70/0.24	0.53/0.38	0.61/0.40
s1w3	0.75/0.18	0.50/0.20	0.58/0.22
s2w0	0.86/0.37	0.53/0.40	0.75/0.42
s2w1	0.86/0.28	0.53/0.38	0.73/0.38
s2w2	0.91/0.15	0.51/0.24	0.76/0.26
s2w3	0.91/0.15	0.37/0.10	0.80/0.16
s3w0	1.00/0.21	0.79/0.21	0.89/0.16
s3w1	1.00/0.21	0.79/0.21	0.88/0.14
s3w2	1.00/0.15	0.84/0.15	0.86/0.12
s3w3	**1.00/0.15**	**1.00/0.07**	**1.00/0.01**

clue-based classifier is able to classify Personal tweets with the highest precision of 100% but with a low recall (15% for epidemic, 7% for mental health, 1% for clinical science).

3.2.2 Personal versus News Tweet Classification Based on Machine Learning

To overcome the drawback of low recall in the clue-based approach, we combined the high precision of clue-based classification with Machine Learning-based classification in the Personal versus News classification, as shown in Fig. 4. Suppose the collection of Raw Tweets of a unique type (e.g. tuberculosis) is T. After the preprocessing step, which filters out non-English tweets, re-tweets and near-duplicate tweets, the resulting tweet dataset is $T' = \{tw_1, tw_2, tw_3, \ldots, tw_n\}$, which is a subset of T, and is used as the input for the clue-based method for automatically labeling datasets for training a Personal versus News classifier as shown in Fig. 4, where the blue part is the automatic labeling of tweets with lexicons (highlighted in green) and the yellow part is the Machine Learning classifiers for classifying Personal tweets from News tweets. We choose three Machine Learning classifiers, Naïve Bayes, Multinomial Naïve Bayes, and Support Vector Machine, as these classifiers achieved good results for similar tasks as discussed in Sect. 2.

In the lexicon-based step for labeling training datasets, each tw_i of T' is compared with the MPQA dictionary [25]. If tw_i contains at least three strongly subjective clues and at least three weakly subjective clues, tw_i is labeled as a Personal tweet. Similarly, tw_i is compared with a News stop word list [54] and a profanity list [52].

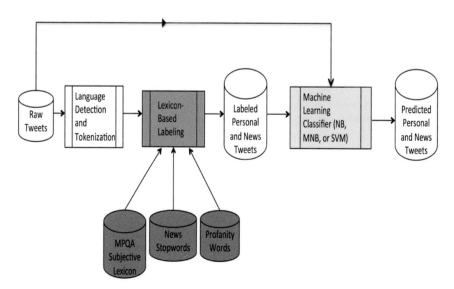

Fig. 4 Personal versus News (Non-Personal) classification

The News stop word list contains 20+ names of highly influential public health News sources and the profanity list has 340 commonly used profanity words. If tw_i contains at least one word from the News stop word list and does not contain any profanity word, tw_i is labeled as a News tweet. For example, the tweet "*UN official: Ebola epidemic could be defeated by end of 2015–41 NBC News* http://bit.ly/1J2rQ1F #world#health" is labeled as a News tweet, because it contains at least one word from the News stop word list and does not contain any profanity word. We mark the set of labeled Personal tweets as T'_p, and the set of labeled News tweets as T'_n, note that $T'_p \cup T'_n \subseteq T'$.

The next step is the Machine Learning-based method. The two classes of T'_p data and T'_n from the clue-based labeling are used as training datasets to train the Machine Learning models. We used three popular models: Naïve Bayes, Multinomial Naïve Bayes, and polynomial-kernel Support Vector Machine. After the Personal versus News classifier is trained, the classifier is used to make predictions on each tw_i in T', which is the preprocessed tweets dataset. The goal of Personal versus News classification is to obtain the Label for each tw_i in the tweet database T', where the Label is either *Personal* or *NT* (News Tweet). Personal could be *PN* or *PNN*.

3.2.3 Negative Versus Non-Negative Sentiment Classifier

Ji, Chun, Wei and Geller [12] discussed automatic labeling of Personal Negative and Personal Non-Negative tweets using a sequential approach. In this method, firstly, a profanity list is used to test if a tweet contains any word from the profanity list. If the tweet contains a profanity word, it is labeled as a Personal Negative tweet. Secondly, for the tweets that do not contain a profanity word, negative and non-negative emoticon lists are used to test whether the tweet contains a negative emoticon or a non-negative emoticon. A partial list of emoticons is shown in Table 2. If the tweet contains a negative emoticon, it is labeled as a Personal Negative tweet, and if the tweet contains a non-negative emoticon, it is labeled as a Personal Non-Negative tweet.

This approach has limitations in terms of coverage and sentiment strength. For coverage, this previous method only considered the existence of profanity and emoticons, but it did not take into account the frequency of them. A single use of a profanity word is relatively common for Twitter users to express their emotions, but multiple

Table 2 Partial list of the emoticons used

Negative	Non-Negative
;(	x-D
:'(	:')
:{	:-))
:$	}{'
:'-(	;-)

Table 3 Whitelist of stop words for building TR-NN

	Negative	Non-Negative
Whitelist	Negative emoticons, profanities, AFINN negative words	Neutral and positive emoticons, AFINN non-negative words

uses of profanity words indicate a strong negative sentiment. In addition, the number of profanity words or emoticons is relatively small, since the profanity list contains "only" 340 words and the emoticon list consists of 33 emoticons. It is quite possible to miss potential Personal Negative or Personal Non-Negative tweets with this approach. For the sentiment strength, this previous method only considered the existence of profanity or emoticon, but did not consider the various sentiment strengths of words, which are good indicators for the tweet sentiment detection.

In this chapter, to address these previous limitations, we have developed a new Personal Negative versus Personal Non-Negative labeling method. This new method uses metrics generated from the AFINN lexicon as shown in Table 3, in addition to emoticons and profanities. AFINN [26] is a publicly available list of 2,477 English words and phrases rated for valence with an integer between -5 (negative) and $+5$ (positive). To label a Personal tweet as Personal Negative or Non-Negative, we aggregated the frequencies of profanity words, emoticons, and AFINN words into two metrics: Negative Score and Non-Negative Score. Subsequently we used a threshold to determine the label of the tweet. The Negative Score and Non-Negative Score are defined as follows.

$$NegativeScore = \frac{PR*5 + NC_3*3 + NC_4*4 + NC_5*5 + NE*3}{PR*5 + NC_3*3 + NC_4*4 + NC_5*5 + NE*3 + NNC_3*3 + NNC_4*4 + NNC_5*5 + NNE*3}$$
$$NonNegativeScore = \frac{NNC_3*3 + NNC_4*4 + NNC_5*5 + NNE*3}{PR*5 + NC_3*3 + NC_4*4 + NC_5*5 + NE*3 + NNC_3*3 + NNC_4*4 + NNC_5*5 + NNE*3}$$

In the above formulas, PR, NC_3, NC_4, NC_5, and NE are the numbers of profanity words, words having -3 valence in AFINN, words having -4 valence, words having -5 valence, and the number of negative emoticons in a tweet. Analogously, NNC_3, NNC_4, NNC_5, NNE are the numbers of words having $+3$ valence in AFINN, $+4$ valence, $+5$ valence, and the number of non-negative emoticons in a tweet. The two metrics express the proportion of negative words and the proportion of non-negative words, respectively.

The threshold is set to 0.95, which means that if the Negative Score of a tweet is greater than or equal to 0.95, it is labeled as a Personal Negative tweet. A Personal Non-Negative tweet is labeled similarly. These two metrics use a larger number of words to label tweets than our previous method (2850 compared with 373). This allows us to generate a larger size of training data and to use sentiment strength to label tweets more accurately. The experimental results that compare the current method and our previous method are shown in Sect. 3.3.2.

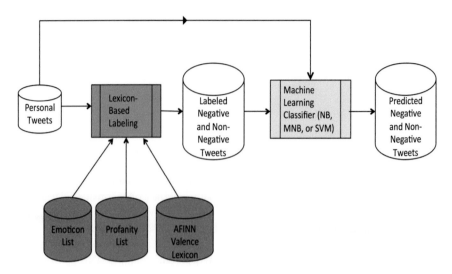

Fig. 5 Negative versus Non-Negative classification

Figure 5 shows the process of Negative versus Non-Negative classification, where the blue part represents the automatic labeling of tweets; the green part represents lexicons; the yellow part represents the Machine Learning classifiers. In the rest of this section, Negative is used to refer to the Personal Negative and Non-Negative is used to refer to the Personal Non-Negative tweets.

Tweets were labeled as *PN* (Personal Negative) or *PNN* (Personal Non-Negative) using the labeling method described above. These two categories (*PN* and *PNN*) of labeled tweets were combined into the training dataset TR-NN for Negative versus Non-Negative classification. Table 4 shows examples of tweets in TR-NN. The set of labeled *PN* tweets is marked as T''_{ne}, and the set of labeled *PNN* tweets is marked as T''_{ne}, and the set of labeled *PNN* tweets is marked as T''_{nn}, and $(T''_{ne} \cup T''_{nn}) \subseteq T'$. Similarly, T''_{ne} and T''_{nn} are used to train the Negative versus Non-Negative classifier, and the classifier is used to make predictions on each tw_i in T'', which is the set of Personal tweets. The goal of Negative versus Non-Negative classification is to obtain the Label for each tw_i in the tweet database T'', where the Label $O(tw_i)$ is either *PN* (Personal Negative) or *PNN* (Personal Non-Negative). (There are no News tweets at this stage.)

After step 1 (Personal tweets classification) and step 2 (sentiment classification), for a unique type of tweets (e.g. tuberculosis), the Raw Tweet dataset T is transformed into a series of Tweet Label datasets TS_i. TS_i is the tweet label dataset for time i, and $TS_i = \{ts_1, ts_2, ts_3, \ldots, ts_n\}$, where the label of ts_i is either *PN* (Personal Negative), or *PNN* (Personal Non-Negative), or *NT* (News Tweet).

Table 4 Examples of Personal Negative and Personal Non-Negative tweets in training dataset TR-NN

Personal Negative	I remember the Ebola panic days .. Damn. If you were ill during that time, everyone would look at you funny legionnaires disease is killing #newyorkcity one block at a time :(
Personal Non-Negative	So excited for @CancerCenter's #GenomicsChat that starts in ONE HOUR! :) Who's joining me? Learn about new precise treatments for #melanoma. Did it! Woke early today to run 10 miles! First time since the melanoma on my knee 2 years ago :) Praise God!

3.3 Experimental Results of the Two-Step Sentiment Classification

The dataset was collected from March 13 2014 to June 29 2014 and was used by our previous work [12]. The statistics of the collected datasets are shown in Table 5. Only English tweets are used in our experiments. Some datasets have a larger portion of non-English tweets, for example, influenza, swine flu, and tuberculosis compared with other datasets.

Table 5 The statistics of the collected dataset [12]

Dataset id	Tweet type	Total number of tweets	Number of non-English tweets	Number of tweets after preprocessing
1	Listeria	13,572	1,979	4,544
2	Influenza	1,509,609	716,901	527,489
3	Swine flu	73,974	35,970	20,430
4	Measles	166,555	8,808	60,016
5	Meningitis	159,393	52,824	42,229
6	Tuberculosis	215,083	147,350	33,030
7	Major depression	2,269,885	121,649	884,304
8	Generalized anxiety disorder	380,094	271,758	71,978
9	Obsessive-compulsive disorder	434,571	168,061	171,211
10	Bipolar disorder	51,520	7,416	20,915
11	Air disaster	15,871	681	5,765
12	Melanoma experimental drug	86,757	9,858	40,261

3.3.1   Evaluation Based on Human Annotated Test Dataset

To compare the Naïve Bayes, Two-Step Multinomial Naïve Bayes, and Two-Step Polynomial-Kernel Support Vector Machine classifiers, we created a test dataset using human annotation. Weka's implementations [55] of these classifiers was used. We extracted three test data subsets by random sampling from all tweets from the three domains epidemic, clinical science, and mental health, collected in the year 2015. Each of these subsets contains 200 tweets. Note that the test tweets are independent from the training tweets that were collected in the year 2014.

Each tweet was annotated by three people out of a set of six contributors. In the evaluation, 1 was assigned to Personal and a 0 was assigned to News. If a tweet was labeled as a Personal tweet, the annotator was asked to label it as Personal Negative or Personal Non-Negative tweet. Fleiss' Kappa [56] was used to measure the inter-rater agreement between the three annotators. Table 6 [12] presents the agreement between human annotators. For each tweet, if at least two out of three annotators agreed on a Label (Personal Negative, Personal Non-Negative, or News), we labeled the tweet with this sentiment.

3.3.2   Sentiment Classification Results

The results of the two-step classification approach are shown in this section. In order to evaluate the usability of two-step classification, Personal versus News classification and Negative versus Non-Negative classification were also evaluated with human annotated datasets. For Personal versus News Classification, we compared our Personal versus News classification method with three baseline methods.

- A random selection method
- The clue-based classification method described above
- A URL-based method, in which a tweet that contains an URL is classified as a News tweet; otherwise a Personal tweet.

The classification accuracies of different methods are presented in Table 7 [12]. The results show that 2S-MNB and 2S-NB outperforms all three baselines in most of the cases. Overall, all methods exhibit a better performance on the epidemic dataset than on the other two datasets. In addition, as we compare the ML-based approaches (2S-MNB, 2S-NB, 2S-SVM), the ML-based approaches outperform the base line clue-based classification approaches in most of the cases.

Table 6 Agreement between human annotators [12]

Domains	Epidemic	Clinical Science	Mental Health
Total number of tweets	200	200	200
At least two annotators agree	192/200	194/200	188/200
Fleiss? kappa coefficient	0.4	0.54	0.33

Table 7 Accuracy of Personal versus News classification on human annotated datasets [12]

Dataset	Random	Clue-Based	URL-Based	2S-MNB	2S-NB	2S-SVM
Epidemic	0.52	0.77	0.82	0.86	0.87	0.71
Mental Health	0.48	0.56	0.68	0.72	0.78	0.59
Clinical Science	0.49	0.82	0.72	0.74	0.71	0.36

Some unigrams are learned by the ML-based methods and are shown to be useful for the classification. To better understand this effect, ablation experiments were carried out with the Personal versus News classification on the human annotated datasets. The classifier 2S-MNB was used, since it took much less time to train than the best classifier 2S-NB on the human-annotated test dataset. More precisely, it was trained with the automatically generated data from the Epidemic, Mental Health, and Clinical Science domains collected in 2014.

The trained classifiers were used to classify the sentiments of human annotated datasets from the year 2015, where unigrams were removed from the test dataset one at a time, in order to study each removed unigram's effect on accuracy. The change of accuracy was recorded each time, and the unigram that leads to the largest decrease in accuracy (when removed) is the most useful one for predictions. Table 8 shows the results of ablation experiments for Personal versus News classification. For example, the unigrams "i", "http", "app", "url" are not in MPQA corpus but are learned by the ML classifier 2S-MNB as the most important unigrams contributing to classification.

The second step in the two-step classification algorithm is to separate Negative tweets from Non-Negative tweets. As discussed in Sect. 3.2.1, the training datasets are automatically labeled, if the tweet has a higher-than-threshold Negative Score or Non-Negative Score. Both scores are calculated by considering the occurrence frequency of words from the profanity list, words from AFINN, and emoticons from the emoticon lists. We compare the performance of Negative versus Non-Negative classification with previous labeling method and with the current labeling method. In this chapter, we call the previous method EPE, as it is an **E**xistence-based method using **P**rofanity list and **E**moticons, and we called the current labeling method FPEA, as it is a **F**requency-based method using **P**rofanity, **E**moticon, and **A**FINN. The classifier is trained by each of the three models, Multinomial Naïve Bayes (MNB), Naïve Bayes (NB), and Support Vector Machine (SVM). The accuracies of Negative versus Non-Negative classification and confusion matrices of the best classifiers

Table 8 Most important unigrams in Personal versus News classification

Dataset	Unigrams with most importance
Epidemic	url, i, case, but
Mental Health	url, disorder, often, bipolar
Clinical Science	melanoma, health, http, co, risk, prevention, app

Table 9 Negative versus Non-Negative Classification Results on Human Annotated Datasets (EPE/FPEA)

Dataset Id	2S-MNB	2S-NB	2S-SVM
Epidemic	0.73/**0.80**	0.59/0.59	0.59/0.66
Mental Health	0.63/0.61	0.65/**0.67**	0.57/0.57
Clinical Science	0.64/0.73	0.73/0.68	0.68/0.68

Table 10 Confusion Matrices of The Best Personal Negative versus Personal Non-Negative Classifier on Human Annotated Datasets (Positive Class is Personal Negative and Negative Class is Personal Non-Negative)

Dataset id	Best classifier	True positive	False negative	False positive	True negative
Epidemic	2S-MNB (FPEA)	19	6	6	28
Mental Health	2S-NB (FPEA)	19	15	15	43
Clinical Science	2S-MNB (FPEA)	4	6	6	28

for human annotated datasets are shown in Tables 9 and 10, respectively. Overall, the **F**requency-based method with **P**rofanity, **E**moticon, and **A**FINN increases the accuracy of the best classifier by 7% in the epidemic dataset, and by 2% in the mental health dataset compared with the previous **E**xistence-based method using **P**rofanity list and **E**moticons. Overall, 2S-MNB (FPEA) achieved the best Negative versus Non-Negative result in terms of accuracy while being faster than 2S-SVM and 2S-NB.

3.3.3 Error Analysis of Sentiment Classification Output

We analyzed the output of sentiment classification. As discussed in Sect. 3.3.1, we manually annotated 600 tweets as Personal Negative, Personal Non-Negative, and News. We used 2S-MNB, which achieved the best accuracy in our experiments described in Sect. 3.3.2, to classify each of the 600 manually annotated tweets as Personal Negative, Personal Non-Negative, or News. Then we analyzed the tweets that were assigned different labels by 2S-MNB and by the human annotators.

For the Personal versus News classification, we found two major types of errors. The first type of error is that the tweet is a Personal tweet, but is classified as a News tweet. By manually checking the content, we found that these tweets are often users' comments on News items (Pointed to by a URL) or users are citing the News. There are 27 out of all 140 errors belonging to this type.

One possible solution to reduce this type of error is that we can calculate what percentage of the tweet text that appears in the web page pointed to by the URL. If the percentage is low, it is probably a Personal tweet since most of the tweet text is the user's contribution. If the percentage is high, approaching 100%, it is more likely a News tweet since tweeters often paste the title of a news article into their messages.

The second type of error is that the tweet is in fact a News item, but is classified as a Personal tweet. In total, 48 out of all 140 errors are of this type. A suggested solution is to check the similarity between the tweet text and the title of the web page content pointed to by the URL. If both are very similar to each other, the tweet is more likely a News item. Those two types of errors together cover 54% (75/140) of the errors in Personal versus News classification. For Negative versus Non-Negative classification, in 50% (30/60) of all errors the tweet is in fact Negative, but is classified as Non-Negative. One possible improvement is to incorporate "Negative phrase identification" to complement the current Machine Learning paradigm. The appearance of negative phrases such as "I did not like XYZ at all" and "I will not do XYZ any more" are possible indicators of Negative tweets.

3.3.4 Twitter Data Bias

As pointed out in the work of Bruns and Stieglitz [57], there are two questions to be addressed in terms of generalizing collected Twitter data.

- Does Twitter data represent Twitter?
- Does Twitter represent society?

According to the documentation of Twitter [11], the Twitter Streaming API returns at most 1% of all the tweets at any given time. Once the number of tweets matching the given API parameters (keywords, geographical boundary, user ID) goes beyond the 1% threshold, Twitter will return a sample of the data to the user. To address this problem, we used domain specific keywords (e.g. h1n1, h5n1) for each tweet type (e.g. listeria) to increase the coverage of collected data [58].

As for the question whether Twitter postings are representative of the society at large, Mislove et al. [59] have found that the Twitter users significantly over-represent the densely populated regions of the US. This might be due to the better availability of high-speed internet in large cities. Twitter users are also overwhelmingly male, and highly biased with respect to race distribution and ethnicity distribution. To reduce the first bias of the collected Twitter data, we defined the Measure of Concern in relative terms. It depends on the fraction of all tweets that have been classified as "Personal Negative" tweets. We assume that as long as the sample of tweets is representative, the Measure of Concern, which is the Personal Negative portion of all tweets, should be similar across different samples sizes, e.g., 1, 10, 100%, etc.

4 Public Health Concern Trend Analysis

We are interested in making the sentiment classification results available for public health monitoring, especially the results of computing the *Measure of Concern,* to monitor public sentiments towards different types of diseases. The definitions of Measure of Concern, Non-Negative Sentiment, News Count, and Peak are shown below as English text. For a more formal treatment, refer to work by Ji et al. [12].

Definition 4a (*Measure of Concern*) *Measure of Concern (MOC)* M_i is the square of the total number of Personal Negative tweets that are posted at time i, divided by the total number of Raw Tweets of a particular type at the same time i. The Measure of Concern increases with the relative growth of Personal Negative tweets and with the absolute growth of Personal Negative tweets.

The reason for including the relative and the absolute growth of personal tweets into one measure is that, for example, a ratio of 9 : 10 Personal Negative tweets: Personal Tweets appears high, but a lower ratio of 7000 : 10000 should contribute more to the Measure of Concern, because a greater number of the "tweeting public" is involved in this social media discourse.

Definition 4b (*Non-Negative Sentiment*) Similarly, the *Non-Negative Sentiment* NN_i is the square of the total number of Personal Non-Negative tweets that are posted at time i, divided by the total number of raw tweets of a particular type at the same time i.

Definition 4c (*News Count*) Finally, the *News Count* NE_i is the total number of News Tweets at the time i. Note that the News Count is not normalized by the total number of raw tweets. The reason is that we are interested in studying the relationship between sentiment trends and News popularity trends. An absolute News Count is able to better represent the popularity of News.

Definition 5 (*Peak*) Given a timeline of numerical values, a value X_i on the timeline is defined as a peak if and only if X_i is the largest value in a given time interval $[i - b, i + a]$. The time intervals $a > 0, b > 0$ can be chosen according to each specific case to limit the number of peaks. Peaks are defined for MOC timelines, Non-Negative timelines, and News Count timelines.

The method for computing the quantitative correlation is shown in Fig. 6. There are three inputs for the correlation process. The News tweets are the outputs of the first step, as shown in Fig. 4; the Personal Negative tweets and the Personal Non-Negative tweets are the outputs in the second step, as shown in Fig. 5. The Jaccard Coefficient is used for computing the correlation.

After the two-step sentiment classification method has been applied to the raw tweets, we can produce three timelines: Measure of Concern timeline, Non-Negative sentiment timeline, and News timeline, respectively. Next, peaks P_1, P_2, and P_3 are generated from these three timelines. The time interval is set to seven days. We are

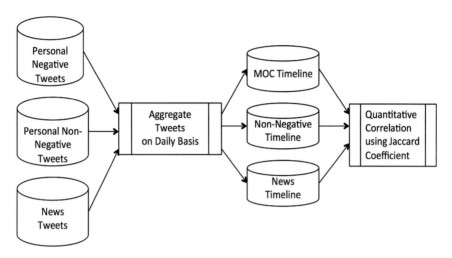

Fig. 6 Correlation between sentiment trends and News trends

interested in the correlation between P_1 and P_2 (peaks of News and peaks of MOC), and the correlation between P_1 and P_3 (peaks of News and peaks of Non-Negative sentiments).

We hypothesized that there might be a time delay between the sentiment peaks and the News timeline peaks. For example, an alarming news report might lead to many Twitter users expressing their negative emotions. On the other hand, social media are nowadays often ahead of official news reports, as shown in Broersma and Graham [60] that the news media now often pick up tweets for their news coverage. In the first case, News peaks would precede tweet sentiment peaks. In the second case, tweet sentiment peaks would precede News peaks. We attempted to quantify these alternative choices using the Jaccard Coefficient for this purpose. Thus we defined two correlations as follows:

$$JC(MOC, NEWS, t) = \frac{|P_{2,c+t} \cap P_{1,c}|}{|P_{2,c+t} \cup P_{1,c}|}$$

$$JC(NN, NEWS, t) = \frac{|P_{3,c+t} \cap P_{1,c}|}{|P_{3,c+t} \cup P_{1,c}|}$$

$P_{2,c+t}$ is meant to assign a time lag or time lead of t days (depending on the sign of t) to the collection of MOC peaks, thus the News peak at date c will be compared with the MOC peak at date $c + t$. Similarly, $P_{3,c+t}$ is meant to assign a time lag or time lead of t days to the collection of Non-Negative peaks, thus the News peak at date c will be compared with the Non-Negative peak at date $c + t$.

By its definition, the Jaccard Coefficient has a value between 0 and 1. The closer the value is to 1, the better the two time series correlate with each other. To illustrate the calculation of Jaccard Coefficient, we use the following example.

Assume a MOC timeline and a News timeline, where the MOC timeline has nine peaks and the News timeline has eight peaks. Three peaks of MOC and another three peaks of News are pair-wise matched. Note that two peaks match with each other if and only if the two peaks happen on exactly the same day. This is an arbitrary definition, which may be replaced by a finer grain size (hours) or possibly a larger grain size (e.g., weekends). The remaining six peaks of MOC and the remaining five peaks of News are not matched with each other.

Then the Jaccard Coefficient between the peaks of MOC and the peaks of News is calculated by the size of the intersection divided by the size of the union. Therefore,

$$JC = 3/(9 + 8 - 3) = 0.21$$

The best Jaccard Coefficient between MOC peaks and News peaks for a given dataset was computed as follows: First we directly computed the JC between MOC peaks and News peaks without any time delay or lead, and we recorded the result. Then we added one, two, or three days of lead to the original MOC, computed the correlation between the revised MOC peaks and the original News peaks respectively, and recorded these three results. Thirdly, we added one, two, or three days of delay to the original MOC, and we recorded three more results. Finally, we chose the highest measure from the above seven results as the best correlation between MOC and News. The peaks of MOC and the peaks of NN (Non-Negative) were correlated with the peaks of News in all datasets with a Jaccard Coefficient of 0.2–0.3.

In order to study how an observable increase in MOC relates to an actual health event (e.g., News count), we quantified the timeline trends of daily MOC and daily News Count for listeria, a potentially lethal foodborne illness, as shown in Fig. 7.

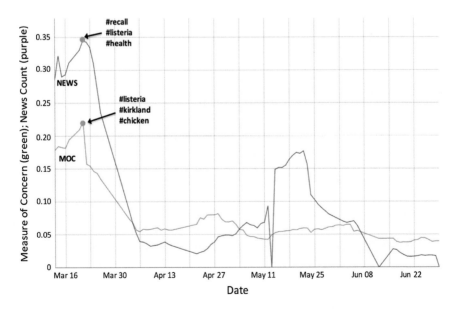

Fig. 7 Comparison between Measure of Concern and News Count timeline trend

The News Count is 0–1 normalized and the top 3 most frequent hash tags for the peak date are shown. The News Count (purple) Peak occurred on March 21st, because on that same day, several food items produced by Parkers Farm were recalled due to a listeria contamination. We note that there was an observable increase of MOC (green) as well, which shows that the general public seemed to express negative emotions according to the news during this circumstance.

Prototype System

We have developed a prototype system of ESMOS (Epidemic Sentiment Monitoring System) to monitor the timeline and topic distribution of public concern, as a part of the Epidemics Outbreak and Spread Detection System [12]. ESMOS displays (1) a concern timeline chart to track the public concern trends on the timeline; (2) a tag cloud for discovering the popular topics within a certain time period with a capability to drill down to the individual tweets; and (3) a public health concern map to show the geographic distribution of particular disease concentrations with different granularity (e.g. state, county or individual location level). Figure 8 shows the different visual analytics tools. The public health specialists can utilize the concern timeline chart, as shown in Fig. 8a, to monitor (e.g. identify concern peaks) and compare public concern timeline trends for various diseases. Then the specialists might be interested

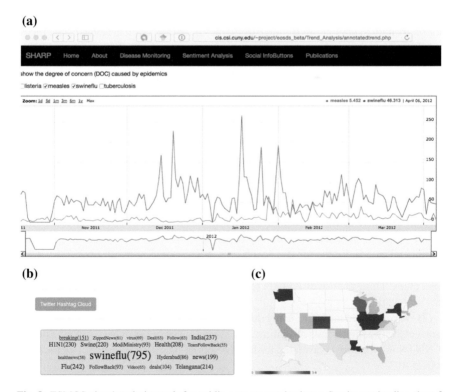

Fig. 8 ESMOS visual analytics tools for public concern monitoring: **a** Sentiment timeline chart, **b** Topics cloud and **c** Public concern distribution map

in what topics people are discussing on social media during the "unusual situations" discovered with the help of the concern timeline chart. To answer this question, they can use the tag cloud, as shown in Fig. 8b to browse the top topics within a certain time period for different diseases, and individual tweets. The public health concern heat map in Fig. 8c shows the state-level public concern levels.

The ESMOS prototype is currently implemented to monitor limited diseases (c.f. Table 5), but our proposed model can use a disease ontology, such as a dedicated epidemic ontology or a UMLS ontology, to monitor any disease of interest.

5 Chapter Summary

In this chapter, we explored the potential of mining social network data, such as tweets, to provide a tool for public health specialists and government decision makers to gauge a Measure of Concern (MOC) expressed by Twitter users under the impact of diseases. To derive the MOC from Twitter, we developed a two-step classification approach to analyze sentiments in disease-related tweets. We first distinguished Personal from News (Non-Personal) tweets. In the second stage, the sentiment analysis was applied only to Personal tweets to distinguish Negative from Non-Negative tweets. In order to evaluate the two-step classification method, we created a test dataset by human annotation for three domains: epidemic, clinical science, and mental health. The Fleiss's Kappa values between annotators were 0.40, 0.54, and 0.33, respectively. These moderate agreements illustrate the complexity of the sentiment classification task, since even humans exhibit relatively low agreement on the labels of tweets.

Our contributions are summarized as follows.

(1) We developed a two-step sentiment classification method by combining clue-based labeling and Machine Learning (ML) methods by first automatically labeling the training datasets, and then building classifiers for Personal tweets and classifiers for tweet sentiments. The two-step classification method shows 10 % and 22 % increase of accuracy over the clue-based method on epidemic and mental health dataset, respectively in Personal versus Non-Personal classification. In Negative versus Non-Negative classification, the frequency-based method FPEA, which uses the AFINN lexicon, increases the accuracy of the best classifier by 7 % for the epidemic dataset and by 2 % for the mental health dataset compared with our previously used method EPE which had a list of profanities and a list of emoticons as the only sentiment clues. Thus, the use of AFINN resulted in a measurable improvement over our previous work.

(2) We quantified the MOC using the results of sentiment classification, and used it to reveal the timeline trends of sentiments of tweets. The peaks of MOC and the peaks of NN (Non-Negative) correlated with the peaks of News with Jaccard Coefficients of 0.25–0.3.

(3) We applied our sentiment classification method and the Measure of Concern to other topical domains, such as mental health monitoring and crisis management. The experimental results support the hypothesis that our approach is generalizable to other domains.

Future work involves the following.

(1) The Measure of Concern (MOC) is currently based on the number of Personal Negative tweets and total number of tweets on the same day. The Measure of Concern was used to define the fraction of tweets that are Personal Negative tweets. We plan to fine-grain this definition to quantify the number of tweets expressing real concern. To achieve this goal, we need to extend the simplistic Negative/Non-Negative categories to a wider range of well recognized emotions, such as "concern", "surprise", "disgust", or "confusion".

We plan to employ an ontology engineering approach to construct an emotion ontology from our collected Twitter messages. The emotion ontology will contain the basic emotions such as anger, confusion, disgust, fear, concern, sadness, etc. along with their representative words or phrases. With the constructed emotion ontology, we will be able to detect the tweets expressing real concern and to more accurately quantify the trend of the Measure of Concern.

(2) To improve the performance of classification, we plan to extend the current feature set to include more features specific to micro-blogs, such as slang terms and intensifiers to capture the unique language in micro-blogs. In Personal versus News classification, we chose to work in the Machine Learning-based paradigm. However, we note that some lightweight knowledge-based approaches could possibly produce competitive results. For example, if the tweet is of the form "TEXT URL" and the TEXT appears on the web page that the URL points to, the tweet is likely a News tweet. The intuition behind this approach is that the title of a news article is often pasted into the tweet body followed by the URL to that news article. We plan to perform a quantitative comparison of these knowledge-based approaches with our ML approach in the future.

(3) The prototype ESMOS implementation needs to be scaled to detect and monitor diseases on a large scale using a full scale Disease Ontology. The sentiment analysis should be performed as the data is captured so that the tracking of public concerns happens in real-time. Public concern about health in general is not limited on infectious diseases but concerns may also be expressed about particular drugs or treatments, and even current and proposed health policies. To promote an understanding of all the contexts related to a disease outbreak, the system needs to consider many different data sources.

(4) Although it is difficult to find the ground truth for sentiment trends, we would like to conduct a systematic experiment on comparing the sentiments derived by our methods with the epidemic cases reported by other available tools, and with authoritative data sources, such as HealthMap and CDC reports. The sentiment trends for topics will also be studied by combining the sentiment analysis algorithms with topic modeling algorithms.

(5) All our work so far used epidemics, mental health and clinical science as domains. Thus, all of our experiments were health-related. However, there are other areas where tweets and the traditional news compete with each other. These areas include politics, e.g. presidential candidate debates, the economy, e.g., a precipitous fall of the Dow Jones index, natural disasters, e.g. typhoons and hurricanes, acts of terrorism and war, e.g., roadside bombs, and spontaneous protests, as they were common during the Arab Spring. All these areas are excellent targets for testing theories about the interplay between news and social media.

Acknowledgments This research has been partially funded by the PSC-CUNY Research Foundation under the awards 42-64266 and 43-65232, and by the Leir Charitable Foundations through the School of Management at NJIT.

References

1. Brownstein, J.S., Freifeld, C.C., Reis, B.Y., Mandl, K.D.: Surveillance sans frontieres: internet-based emerging infectious disease intelligence and the HealthMap project. PLoS Med. **5**, e151 (2008)
2. Collier, N., Doan, S.: Syndromic classification of Twitter messages. Electron. Healthc. **91**, 186–195 (2012)
3. Signorini, A., Segre, A.M., Polgreen, P.M.: The use of Twitter to track levels of disease activity and public concern in the U.S. during the influenza A H1N1 pandemic. PloS One **6**, e19467 (2011)
4. Aramaki, E., Maskawa, S., Morita, M.: Twitter catches the flu: detecting influenza epidemics using Twitter. In: Proceedings of the Conference on Empirical Methods in Natural Language Processing, pp. 1568–1576 (2011)
5. Lampos, V., Cristianini, N.: Tracking the flu pandemic by monitoring the Social Web. In: Proceedings of 2nd International Workshop on Cognitive Information Processing, pp. 411–416 (2010)
6. Reuters News. http://www.reuters.com/article/2014/10/18/us-health-ebola-usa-idUSKCN0I61BO20141018
7. Zhu, X., Wu, S., Miao, D., Li, Y.: Changes in emotion of the Chinese public in regard to the SARS period. Soc. Behav. Personal. **36**, 447–454 (2008)
8. Guardian News. http://www.guardian.co.uk/world/2011/mar/17/chinese-panic-buy-salt-japan
9. Ginsberg, J., Mohebbi, M.H., Patel, R.S., Brammer, L., Smolinski, M.S., Brilliant, L.: Detecting influenza epidemics using search engine query data. Nature **457**, 1012–1014 (2009)
10. Twitter. http://www.twitter.com
11. Twitter Documentation. https://dev.twitter.com/docs
12. Ji, X., Chun, S.A., Wei, Z., Geller, J.: Twitter sentiment classification for measuring public health concerns. Soc. Netw. Anal. Min. **5**, 1–25 (2015)
13. Zhang, L., Ghosh, R., Dekhil, M., Hsu, M., Liu, B.: Combining lexicon-based and learning-based methods for Twitter sentiment analysis (2011)
14. Liu, B., Zhang, L.: A survey of opinion mining and sentiment analysis. Mining Text Data, pp. 415–463 (2012)
15. Mohammad, S.M., Kiritchenko, S., Zhu, X.: NRC-Canada: Building the state-of-the-art in sentiment analysis of tweets (2013). arXiv preprint arXiv:1308.6242
16. Saif, H., Fernandez, M., He, Y., Alani, H.: Evaluation datasets for twitter sentiment analysis. In: Proceedings of 1st Workshop on Emotion and Sentiment in Social and Expressive Media (2013)

17. Sha, Y., Yan, J., Cai, G.: Detecting public sentiment over PM2.5 pollution hazards through analysis of Chinese microblog. In: The 11th International Conference on Information Systems for Crisis Response and Management, pp. 722–726 (2014)
18. Ji, X., Chun, S.A., Geller, J.: Monitoring public health concerns using Twitter sentiment classifications. In: Proceedings of IEEE International Conference on Healthcare Informatics, pp. 335–344 (2013)
19. Liben Nowell, D., Kleinberg, J.: The link prediction problem for social networks. J. Am. Soc. Inf. Sci. Technol. **58**, 1019–1031 (2007)
20. Hollnagel, E., Woods, D.D.: Cognitive systems engineering: new wine in new bottles. Int. J. Man Mach. Stud. **18**, 583–600 (1983)
21. Schriml, L.M., Arze, C., Nadendla, S., Chang, Y.W.W., Mazaitis, M., Felix, V., Feng, G., Kibbe, W.A.: Disease Ontology: a backbone for disease semantic integration. Nucleic Acids Res. **40**(D1), D940–D946 (2012)
22. List of ICD-9 Codes. https://en.wikipedia.org/wiki/List_of_ICD-9_codes_001%E2%80% 93139:_infectious_and_parasitic_diseases
23. Chun, S., Geller, J.: Evaluating ontologies based on the naturalness of their preferred terms. In: Proceedings of the 41st Annual International Conference on System Sciences, pp. 238–238 (2008)
24. Liu, B.: Sentiment analysis and opinion mining. Synth. Lect. Hum. Lang. Technol. **5**, 1–167 (2012)
25. Riloff, E., Wiebe, J.: Learning extraction patterns for subjective expressions. In: Proceedings of the 2003 Conference on Empirical Methods in Natural Language Processing, pp. 105–112 (2003)
26. Hansen, L.K., Arvidsson, A., Nielsen, F.Å., Colleoni, E., Etter, M.: Good friends, bad news-affect and virality in twitter. Future information technology, pp. 34–43. Springer, Berlin (2011)
27. Esuli, A., Sebastiani, F.: Sentiwordnet: a publicly available lexical resource for opinion mining. In: Proceedings of LREC, pp. 417–422 (2006)
28. Pang, B., Lee, L.: Opinion mining and sentiment analysis. Found. Trends Inf. Retr. **2**, 1–135 (2008)
29. Pang, B., Lee, L., Vaithyanathan, S.: Thumbs up?: sentiment classification using machine learning techniques. In: Proceedings of the ACL-02 Conference on Empirical Methods in Natural Language Processing, vol. 10, pp. 79–86 (2003)
30. Mishne, G.: Experiments with mood classification in blog posts. In: Proceedings of ACM SIGIR 2005 Workshop on Stylistic Analysis of Text for Information Access (2005)
31. Wilson, T., Wiebe, J., Hoffmann, P.: Recognizing contextual polarity in phrase-level sentiment analysis. In: Proceedings of Human Language Technologies Conference on Empirical Methods in Natural Language Processing, pp. 347–354 (2005)
32. Johansson, F., Brynielsson, J., Quijano, M.N.: Estimating citizen alertness in crises using social media monitoring and analysis. In: Proceedings of European Intelligence and Security Informatics Conference, pp. 189–196 (2012)
33. Brynielsson, J., Johansson, F., Jonsson, C., Westling, A.: Emotion classification of social media posts for estimating people's reactions to communicated alert messages during crises. Secur. Inf. **3**, 1–11 (2014)
34. Saif, H., Fernández, M., Alani, H.: Automatic stopword generation using contextual semantics for sentiment analysis of Twitter. In: Proceedings of 13th International Semantic Web Conference (2014)
35. Refaee, E., Rieser, V.: An Arabic twitter corpus for subjectivity and sentiment analysis. In: Proceedings of the Ninth International Conference on Language Resources and Evaluation, pp. 2268–2273 (2014)
36. Saif, H., He, Y., Alani, H.: Semantic sentiment analysis of Twitter. The Semantic Web-ISWC, pp. 508–524 (2012)
37. Barbosa, L., Feng, J.: Robust sentiment detection on Twitter from biased and noisy data. In: Proceedings of the 23rd International Conference on Computational Linguistics, pp. 36–44 (2010)

38. Bifet, A., Frank, E.: Sentiment knowledge discovery in twitter streaming data. In: Proceedings of the 13th International Conference on Discovery Science, pp. 1–15. Springer (2010)
39. Pak, A., Paroubek, P.: Twitter as a corpus for sentiment analysis and opinion mining. In: Proceedings of the Seventh Conference on International Language Resources and Evaluation, pp. 1320–1326 (2010)
40. Jiang, L., Yu, M., Zhou, M., Liu, X., Zhao, T.: Target-dependent twitter sentiment classification. In: Proceedings of the 49th Annual Meeting of the Association for Computational Linguistics: Human Language Technologies, vol. 1, pp. 151–160 (2011)
41. Zhou, Z., Zhang, X., Sanderson, M.: Sentiment analysis on twitter through topic-based lexicon expansion. Databases Theory and Applications, pp. 98–109. Springer International Publishing, Switzerland (2014)
42. Fukunaga, K.: Introduction to Statistical Pattern Recognition, 2nd edn. Academic Press, New York (1990)
43. Cortes, C., Vapnik, V.: Support-vector networks. Mach. Learn. **20**, 273–297 (1995)
44. Chang, C.-C., Lin, C.-J.: LIBSVM: a library for support vector machines. ACM Trans. Interact. Intell. Syst. **2**, 1–27 (2011)
45. Salathe, M., Khandelwal, S.: Assessing vaccination sentiments with online social media: implications for infectious disease dynamics and control. PLoS Comput. Biol. **7**, e1002199 (2011)
46. Zhuang, L., Jing, F., Zhu, X.-Y.: Movie review mining and summarization. In: Proceedings of the 15th ACM International Conference on Information and Knowledge Management, pp. 43–50 (2006)
47. Chew, C., Eysenbach, G.: Pandemics in the age of Twitter: content analysis of Tweets during the 2009 H1N1 outbreak. PloS One **5**, e14118 (2010)
48. O'Connor, B., Balasubramanyan, R., Routledge, B.R., Smith, N.A.: From tweets to polls: Linking text sentiment to public opinion time series. In: Proceedings of International Conference on Weblogs and Social Media, pp. 122–129 (2010)
49. Wiebe, J., Riloff, E.: Creating subjective and objective sentence classifiers from unannotated texts. In: Proceedings of the 6th international conference on Computational Linguistics and Intelligent Text Processing, pp. 486–497 (2005)
50. Wilson, T., Wiebe, J.: Annotating opinions in the world press. In: Proceedings of 4th SIGdial Meeting on Discourse and Dialogue, pp. 13–22 (2003)
51. Twitter 4J. http://twitter4j.org/en/
52. Profanity List. http://web.njit.edu/xj25/eosds_beta/files/profanity_list.txt
53. FCC Guide. http://www.fcc.gov/guides/obscenity-indecency-and-profanity
54. News Stopwords. http://web.njit.edu/xj25/eosds_beta/files/news_stopwords.txt
55. Hall, M., Frank, E., Holmes, G., Pfahringer, B., Reutemann, P., Witten, I.H.: The WEKA data mining software: an update. SIGKDD Explor. Newsl. **11**, 10–18 (2009)
56. Fleiss, J.L.: Measuring nominal scale agreement among many raters. Psychol. Bull. **76**, 378–382 (1971)
57. Bruns, A., Stieglitz, S.: Twitter data: what do they represent? Inf. Technol. **56**, 240–245 (2014)
58. Morstatter, F., Pfeffer, J., Liu, H., Carley, K.M.: Is the sample good enough? comparing data from twitter's streaming api with twitter's firehose (2013). arXiv preprint arXiv:1306.5204
59. Mislove, A., Lehmann, S., Ahn, Y.-Y., Onnela, J.-P., Rosenquist, J.N.: Understanding the demographics of Twitter users. In: Proceedings of the 5th International AAAI Conference on Weblogs and Social Media, pp. 554–557 (2011)
60. Broersma, M., Graham, T.: Twitter as a news source: how Dutch and British newspapers used tweets in their news coverage, 2007–2011. Journal. Pract. **7**, 446–464 (2013)

Index

© Springer International Publishing Switzerland 2016
W. Pedrycz and S.-M. Chen (eds.), *Sentiment Analysis and Ontology Engineering*,
Studies in Computational Intelligence 639, DOI 10.1007/978-3-319-30319-2